Edited by
Peter Ramm,
James Jian-Qiang Lu, and
Maaike M.V. Taklo

Handbook of Wafer Bonding

Related Titles

Klauk, H. (ed.)

Organic Electronics II

More Materials and Applications

2012

ISBN: 978-3-527-32647-1

Franssila, S.

Introduction to Microfabrication

Second edition

2010

ISBN: 978-0-470-74983-8

Saile, V., Wallrabe, U., Tabata, O. (eds.)

LIGA and its Applications

2009

ISBN: 978-3-527-31698-4

Garrou, P., Bower, C., Ramm, P. (eds.)

Handbook of 3D Integration

Technology and Applications of 3D Integrated Circuits

2008

ISBN: 978-3-527-32034-9

Hierold, C. (ed.)

Carbon Nanotube Devices

Properties, Modeling, Integration and Applications

2008

ISBN: 978-3-527-31720-2

Tabata, O., Tsuchiya, T. (eds.)

Reliability of MEMS

Testing of Materials and Devices

2008

ISBN: 978-3-527-31494-2

Klauk, H. (ed.)

Organic Electronics

Materials, Manufacturing and Applications

2006

ISBN: 978-3-527-31264-1

Edited by Peter Ramm, James Jian-Qiang Lu, and Maaike M.V. Taklo

Handbook of Wafer Bonding

WILEY-VCH Verlag GmbH & Co. KGaA

The Editors

Dr. Peter Ramm
Fraunhofer Research Institution for
Modular Solid State Technologies
EMFT
Hansastrasse 27d
80686 Munich
Germany

Prof. Dr. James Jian-Qiang Lu
Rensellaer Polytechnic Institute
110 8th Street
Troy, NY 12180-3590
USA

Dr. Maaike M.V. Taklo
SINTEF ICT
Gaustadalléen 23 C
0314 Oslo
Norway

All books published by **Wiley-VCH** are carefully produced. Nevertheless, authors, editors, and publisher do not warrant the information contained in these books, including this book, to be free of errors. Readers are advised to keep in mind that statements, data, illustrations, procedural details or other items may inadvertently be inaccurate.

Library of Congress Card No.: applied for

British Library Cataloguing-in-Publication Data
A catalogue record for this book is available from the British Library.

Bibliographic information published by the Deutsche Nationalbibliothek
The Deutsche Nationalbibliothek lists this publication in the Deutsche Nationalbibliografie; detailed bibliographic data are available on the Internet at <http://dnb.d-nb.de>.

© 2012 Wiley-VCH Verlag & Co. KGaA, Boschstr. 12, 69469 Weinheim, Germany

All rights reserved (including those of translation into other languages). No part of this book may be reproduced in any form – by photoprinting, microfilm, or any other means – nor transmitted or translated into a machine language without written permission from the publishers. Registered names, trademarks, etc. used in this book, even when not specifically marked as such, are not to be considered unprotected by law.

Cover Design   Adam-Design, Weinheim
Typesetting   Toppan Best-set Premedia Limited, Hong Kong

Printing and Binding   Markono Print Media Pte Ltd, Singapore
Printed in Singapore
Printed on acid-free paper

Print ISBN: 978-3-527-32646-4
ePDF ISBN: 978-3-527-64424-7
oBook ISBN: 978-3-527-64422-3
ePub ISBN: 978-3-527-64423-0
Mobi ISBN: 978-3-527-64425-4

Contents

Preface *XV*
Obituary *XVII*
List of Contributors *XXI*
Introduction *XXV*

Part One   Technologies *1*

A.   Adhesive and Anodic Bonding *3*

1   **Glass Frit Wafer Bonding** *3*
           Roy Knechtel
1.1   Principle of Glass Frit Bonding   *3*
1.2   Glass Frit Materials   *4*
1.3   Screen Printing: Process for Bringing Glass Frit Material onto Wafers   *5*
1.4   Thermal Conditioning: Process for Transforming Printed Paste into Glass for Bonding   *8*
1.5   Wafer Bond Process: Essential Wafer-to-Wafer Mounting by a Glass Frit Interlayer   *11*
1.6   Characterization of Glass Frit Bonds   *14*
1.7   Applications of Glass Frit Wafer Bonding   *15*
1.8   Conclusions   *16*
           References   *17*

2   **Wafer Bonding Using Spin-On Glass as Bonding Material**   *19*
           Viorel Dragoi
2.1   Spin-On Glass Materials   *19*
2.2   Wafer Bonding with SOG Layers   *21*
2.2.1   Experimental   *21*
2.2.2   Wafer Bonding with Silicate SOG Layers   *22*
2.2.3   Wafer Bonding with Planarization SOG   *28*

2.2.4	Applications of Adhesive Wafer Bonding with SOG Layers	29
2.2.5	Conclusion	30
	References	31

3 Polymer Adhesive Wafer Bonding   33
Frank Niklaus and Jian-Qiang Lu

3.1	Introduction	33
3.2	Polymer Adhesives	34
3.2.1	Polymer Adhesion Mechanisms	34
3.2.2	Properties of Polymer Adhesives	36
3.2.3	Polymer Adhesives for Wafer Bonding	38
3.3	Polymer Adhesive Wafer Bonding Technology	42
3.3.1	Polymer Adhesive Wafer Bonding Process	43
3.3.2	Localized Polymer Adhesive Wafer Bonding	50
3.4	Wafer-to-Wafer Alignment in Polymer Adhesive Wafer Bonding	52
3.5	Examples for Polymer Adhesive Wafer Bonding Processes and Programs	54
3.5.1	Bonding with Thermosetting Polymers for Permanent Wafer Bonds (BCB) or for Temporary Wafer Bonds (mr-I 9000)	54
3.5.2	Bonding with Thermoplastic Polymer (HD-3007) for Temporary and Permanent Wafer Bonds	56
3.6	Summary and Conclusions	57
	References	58

4 Anodic Bonding   63
Adriana Cozma Lapadatu and Kari Schjølberg-Henriksen

4.1	Introduction	63
4.2	Mechanism of Anodic Bonding	64
4.2.1	Glass Polarization	64
4.2.2	Achieving Intimate Contact	65
4.2.3	Interface Reactions	66
4.3	Bonding Current	67
4.4	Glasses for Anodic Bonding	68
4.5	Characterization of Bond Quality	69
4.6	Pressure Inside Vacuum-Sealed Cavities	70
4.7	Effect of Anodic Bonding on Flexible Structures	71
4.8	Electrical Degradation of Devices during Anodic Bonding	71
4.8.1	Degradation by Sodium Contamination	72
4.8.2	Degradation by High Electric Fields	73
4.9	Bonding with Thin Films	75
4.10	Conclusions	76
	References	77

B. Direct Wafer Bonding 81

5 Direct Wafer Bonding 81
Manfred Reiche and Ulrich Gösele
5.1   Introduction 81
5.2   Surface Chemistry and Physics 82
5.3   Wafer Bonding Techniques 84
5.3.1 Hydrophilic Wafer Bonding 84
5.3.2 Hydrophobic Wafer Bonding 86
5.3.3 Low-Temperature Wafer Bonding 88
5.3.4 Wafer Bonding in Ultrahigh Vacuum 89
5.4   Properties of Bonded Interfaces 90
5.5   Applications of Wafer Bonding 93
5.5.1 Advanced Substrates for Microelectronics 93
5.5.2 MEMS and Nanoelectromechanical Systems 95
5.6   Conclusions 95
      References 96

6 Plasma-Activated Bonding 101
Maik Wiemer, Dirk Wuensch, Joerg Braeuer, and Thomas Gessner
6.1   Introduction 101
6.2   Theory 102
6.2.1 (Silicon) Direct Bonding 102
6.2.2 Mechanisms of Plasma on Silicon Surfaces 103
6.2.3 Physical Definition of a Plasma 104
6.3   Classification of PAB 104
6.3.1 Low-Pressure PAB 105
6.3.2 Atmospheric-Pressure PAB 106
6.4   Procedure of PAB 107
6.4.1 Process Flow 107
6.4.2 Characterization Techniques 108
6.4.3 Experiments and Results 110
6.5   Applications for PAB 111
6.5.1 Pressure Sensor 112
6.5.2 Optical Microsystem 112
6.5.3 Microfluidics Packaging 113
6.5.4 Backside-Illuminated CMOS Image Sensor 113
6.5.5 CMOS Compatibility of Low-Pressure PAB 114
6.6   Conclusion 115
      References 115

C. Metal Bonding 119

7 Au/Sn Solder 119
Hermann Oppermann and Matthias Hutter
7.1   Introduction 119

7.2	Au/Sn Solder Alloy *120*
7.3	Reflow Soldering *127*
7.4	Thermode Soldering *130*
7.5	Aspects of Three-Dimensional Integration and Wafer-Level Assembly *132*
7.6	Summary and Conclusions *135*
	References *136*

8 Eutectic Au–In Bonding *139*
Mitsumasa Koyanagi and Makoto Motoyoshi

8.1	Introduction *139*
8.2	Organic/Metal Hybrid Bonding *140*
8.3	Organic/In–Au Hybrid Bonding *142*
8.3.1	In–Au Phase Diagram and Bonding Principle *142*
8.3.2	Formation of In–Au Microbumps by a Planarized Liftoff Method *144*
8.3.3	Eutectic In–Au Bonding and Epoxy Adhesive Injection *146*
8.3.4	Electrical Characteristics of In–Au Microbumps *148*
8.4	Three-Dimensional LSI Test Chips Fabricated by Eutectic In–Au Bonding *149*
8.5	High-Density and Narrow-Pitch Mircobump Technology *152*
8.6	Conclusion *157*
	Acknowledgment *157*
	References *157*

9 Thermocompression Cu–Cu Bonding of Blanket and Patterned Wafers *161*
Kuan-Neng Chen and Chuan Seng Tan

9.1	Introduction *161*
9.2	Classification of the Cu Bonding Technique *162*
9.2.1	Thermocompression Cu Bonding *162*
9.2.2	Surface-Activated Cu Bonding *162*
9.3	Fundamental Properties of Cu Bonding *163*
9.3.1	Morphology and Oxide Examination of Cu Bonded Layer *163*
9.3.2	Microstructure Evolution during Cu Bonding *164*
9.3.3	Orientation Evolution during Cu Bonding *165*
9.4	Development of Cu Bonding *166*
9.4.1	Fabrication and Surface Preparation of Cu Bond Pads *166*
9.4.2	Parameters of Cu Bonding *167*
9.4.3	Structural Design *168*
9.5	Characterization of Cu Bonding Quality *169*
9.5.1	Mechanical Tests *169*
9.5.2	Image Analysis *170*
9.5.3	Electrical Characterization *171*
9.5.4	Thermal Reliability *171*

9.6	Alignment Accuracy of Cu–Cu Bonding	171
9.7	Reliable Cu Bonding and Multilayer Stacking	172
9.8	Nonblanket Cu–Cu Bonding	174
9.9	Low-Temperature (<300 °C) Cu–Cu Bonding	176
9.10	Applications of Cu Wafer Bonding	178
9.11	Summary	178
	References	179

10 Wafer-Level Solid–Liquid Interdiffusion Bonding 181
Nils Hoivik and Knut Aasmundtveit

10.1	Background	181
10.1.1	Solid–Liquid Interdiffusion Bonding Process	181
10.1.2	SLID Bonding Compared with Soldering	182
10.1.3	Material Systems for SLID Bonding	183
10.2	Cu–Sn SLID Bonding	189
10.2.1	Cu–Sn Material Properties and Required Metal Thicknesses	190
10.2.2	Bonding Processes	191
10.2.3	Pretreatment Requirements for SLID Bonding	195
10.2.4	Fluxless Bonding	196
10.3	Au–Sn SLID Bonding	199
10.3.1	Au–Sn Material Properties and Required Metal Thicknesses	199
10.3.2	Bonding Processes	199
10.4	Application of SLID Bonding	201
10.4.1	Cu–Sn Bonding	201
10.4.2	Au–Sn Bonding	204
10.5	Integrity of SLID Bonding	207
10.5.1	Electrical Reliability and Electromigration Testing	207
10.5.2	Mechanical Strength of SLID Bonds	207
10.6	Summary	210
	References	212

D. Hybrid Metal/Dielectric Bonding 215

11 Hybrid Metal/Polymer Wafer Bonding Platform 215
Jian-Qiang Lu, J. Jay McMahon, and Ronald J. Gutmann

11.1	Introduction	215
11.2	Three-Dimensional Platform Using Hybrid Cu/BCB Bonding	217
11.3	Baseline Bonding Process for Hybrid Cu/BCB Bonding Platform	220
11.4	Evaluation of Cu/BCB Hybrid Bonding Processing Issues	222
11.4.1	CMP and Bonding of Partially Cured BCB	222
11.4.2	Cu/BCB CMP Surface Profile	223

11.4.3	Hybrid Cu/BCB Bonding Interfaces	224
11.4.4	Topography Accommodation Capability of Partially Cured BCB	227
11.4.5	Electrical Characterization of Hybrid Cu/BCB Bonding	231
11.5	Summary and Conclusions	232
	Acknowledgments	233
	References	233

12 Cu/SiO$_2$ Hybrid Bonding   237
Léa Di Cioccio

12.1	Introduction	237
12.2	Blanket Cu/SiO$_2$ Direct Bonding Principle	239
12.2.1	Chemical Mechanical Polishing Parameters	239
12.2.2	Bonding Quality and Alignment	243
12.3	Blanket Copper Direct Bonding Principle	245
12.4	Electrical Characterization	251
12.5	Die-to-Wafer Bonding	255
12.5.1	Daisy Chain Structures	256
12.6	Conclusion	257
	Acknowledgment	257
	References	258

13 Metal/Silicon Oxide Hybrid Bonding   261
Paul Enquist

13.1	Introduction	261
13.2	Metal/Non-adhesive Hybrid Bonding–Metal DBI®	261
13.3	Metal/Silicon Oxide DBI®	262
13.3.1	Metal/Silicon Oxide DBI® Surface Fabrication	263
13.3.2	Metal/Silicon Oxide DBI® Surface Patterning	264
13.3.3	Metal/Silicon Oxide DBI® Surface Topography	264
13.3.4	Metal/Silicon Oxide DBI® Surface Roughness	264
13.3.5	Metal/Silicon Oxide DBI® Surface Activation and Termination	265
13.3.6	Metal/Silicon Oxide DBI® Alignment and Hybrid Surface Contact	265
13.3.7	Metal Parameters Relevant to DBI® Surface Fabrication and Electrical Interconnection	268
13.3.8	DBI® Metal/Silicon Oxide State of the Art	270
13.4	Metal/Silicon Nitride DBI®	271
13.5	Metal/Silicon Oxide DBI® Hybrid Bonding Applications	273
13.5.1	Pixelated 3D ICs	273
13.5.2	Three-Dimensional Heterogeneous Integration	275
13.5.3	CMOS (Ultra) Low-k 3D Integration	276
13.6	Summary	276
	References	277

Part Two Applications *279*

14 Microelectromechanical Systems *281*
Maaike M.V. Taklo
14.1 Introduction *281*
14.2 Wafer Bonding for Encapsulation of MEMS *282*
14.2.1 Protection during Wafer Dicing *282*
14.2.2 Routing of Electrical Signal Lines *282*
14.3 Wafer Bonding to Build Advanced MEMS Structures *284*
14.3.1 Stacking of Several Wafers *284*
14.3.2 Post-processing of Bonded Wafers *285*
14.4 Examples of MEMS and Their Requirements for the Bonding Process *286*
14.5 Integration of Some Common Wafer Bonding Processes *287*
14.5.1 Fusion Bonding of Patterned Wafers *287*
14.5.2 Anodic Bonding of Patterned Wafers *290*
14.5.3 Eutectic Bonding of Patterned Wafers: AuSn *293*
14.6 Summary *297*
References *297*

15 Three-Dimensional Integration *301*
Philip Garrou, James Jian-Qiang Lu, and Peter Ramm
15.1 Definitions *301*
15.2 Application of Wafer Bonding for 3D Integration Technology *303*
15.3 Motivations for Moving to 3D Integration *305*
15.4 Applications of 3D Integration Technology *307*
15.4.1 Three-Dimensional Applications by Evolution Not Revolution *307*
15.4.2 Microbump Bonding/No TSV *308*
15.4.3 TSV Formation/No Stacking *310*
15.4.4 Memory *312*
15.4.5 Memory on Logic *321*
15.4.6 Repartitioning Logic *322*
15.4.7 Foundry and OSAT Activity *323*
15.4.8 Other 3D Applications *323*
15.5 Conclusions *325*
References *325*

16 Temporary Bonding for Enabling Three-Dimensional Integration and Packaging *329*
Rama Puligadda
16.1 Introduction *329*
16.2 Temporary Bonding Technology Options *330*
16.2.1 Key Requirements *331*
16.2.2 Foremost Temporary Wafer Bonding Technologies *332*

16.3	Boundary Conditions for Successful Processing	*337*
16.3.1	Uniform and Void-Free Bonding	*337*
16.3.2	Protection of Wafer Edges during Thinning and Subsequent Processing	*337*
16.4	Three-Dimensional Integration Processes Demonstrated with Thermomechanical Debonding Approach	*338*
16.4.1	Via-Last Process on CMOS Image Sensor Device Wafers	*338*
16.4.2	Via-Last Process with Aspect Ratio of 2:1	*341*
16.4.3	Via-Last Process with 50 μm Depth Using High-Temperature TEOS Process	*341*
16.4.4	Die-to-Wafer Stacking Using Interconnect Via Solid–Liquid Interdiffusion Process	*342*
16.5	Concluding Remarks	*343*
	Acknowledgments	*344*
	References	*344*

17 Temporary Adhesive Bonding with Reconfiguration of Known Good Dies for Three-Dimensional Integrated Systems *347*
Armin Klumpp and Peter Ramm

17.1	Die Assembly with SLID Bonding	*347*
17.2	Reconfiguration	*348*
17.3	Wafer-to-Wafer Assembly by SLID Bonding	*349*
17.4	Reconfiguration with Ultrathin Chips	*351*
17.5	Conclusion	*352*
	Acknowledgments	*353*
	References	*354*

18 Thin Wafer Support System for above 250 °C Processing and Cold De-bonding *355*
Werner Pamler and Franz Richter

18.1	Introduction	*355*
18.2	Process Flow	*356*
18.2.1	Release Layer Processing	*357*
18.2.2	Carrier Wafer Processing	*357*
18.2.3	Bonding Process	*357*
18.2.4	Thinning	*359*
18.2.5	De-bonding Process	*360*
18.2.6	Equipment	*361*
18.3	Properties	*361*
18.3.1	Device Wafer Thickness	*361*
18.3.2	Thickness Uniformity	*361*
18.3.3	Stability	*362*
18.4	Applications	*362*
18.4.1	Bonding of Bumped Wafers	*363*

18.4.2	Packaging of Ultrathin Dies 363
18.4.3	TSV Processing 364
18.4.4	Re-using the Carrier 364
18.5	Conclusions 364
	Acknowledgments 365
	References 365

19 Temporary Bonding: Electrostatic 367
Christof Landesberger, Armin Klumpp, and Karlheinz Bock

19.1	Basic Principles: Electrostatic Forces between Parallel Plates 367
19.1.1	Electric Fields and Electrostatic Forces in a Plate Capacitor 368
19.1.2	Electrostatic Attraction in a Bipolar Configuration 369
19.1.3	Johnsen–Rahbek Effect 370
19.2	Technological Concept for Manufacture of Mobile Electrostatic Carriers 371
19.2.1	Selection of Substrate Material 371
19.2.2	Selection of Thin-Film Dielectric Layers 372
19.2.3	Electrode Patterns: Materials and Geometry 374
19.2.4	Examples of Mobile Electrostatic Carriers 375
19.3	Characterization of Electrostatic Carriers 376
19.3.1	Electrical and Thermal Properties, Leakage Currents 376
19.3.2	Possible Influence of Electrostatic Fields on CMOS Devices 378
19.4	Electrostatic Carriers for Processing of Thin and Flexible Substrates 379
19.4.1	Handling and Transfer of Thin Semiconductor Wafers 379
19.4.2	Wafer Thinning and Backside Metallization 380
19.4.3	Electrostatic Carriers in Plasma Processing 380
19.4.4	Electrostatic Carriers Enable Bumping of Thin Wafers 380
19.4.5	Electrostatic Carriers in Wet-Chemical Environments 381
19.4.6	Electrostatic Handling of Single Dies 381
19.4.7	Processing of Foils and Insulating Substrates 381
19.5	Summary and Outlook 382
	References 383

Index 385

Preface

One may ask if we need another book on wafer bonding. The answer is a clear yes. The research and development on wafer bonding has truly sped up in the last few years, motivated by the extended use of wafer bonding in new technology areas with a variety of materials. It is very desirable to summarize the recent advances in wafer bonding fundamentals, materials, technologies, and applications in a handbook format, rather than just focusing on scientific fundamentals and/or applications.

So far there have been several books and review articles on wafer bonding, such as

- Tong, Q.-Y. and Gösele, U. (1999) *Semiconductor Wafer Bonding: Science and Technology*, John Wiley & Sons, Inc.;
- Alexe, M. and Gösele, U. (eds) (2004) *Wafer Bonding: Applications and Technology*, Springer;
- Plößl, A. and Kräuter, G. (1999) Wafer direct bonding: tailoring adhesion between brittle materials. *Materials Science and Engineering*, **R25**, 1–88.

We do need an update. The change is mainly due to the fast pace of research and development in three-dimensional (3D) integration, temporary bonding, and microelectromechanical systems (MEMS) with new functional layers.

Formerly, wafer bonding was applied for manufacturing silicon-on-insulator wafers, for fabrication of sensors and actuators, and for various fluidic systems. Today, manufacturers of IC wafers have also learnt the terminologies related to wafer bonding. As Moore's law seems to come to an end, or at least to meet some resistance, memory and logic devices are being stacked in the third dimension to increase the density of transistors and improve performance and functionality. IC manufacturers work on larger wafers and produce wafers in huge quantities, so they have truly challenged lately the vendors of wafer bonding tools. Their interest in wafer stacking has resulted in increased alignment precision, tools for larger wafers, an increased focus on new materials, lower cost and higher throughput, etc.

Based on the tremendous progress in wafer bonding in recent years, we invited world experts to contribute chapters to this wafer bonding handbook, covering a variety of technologies and applications. The wafer bonding technologies are presented in Part One. We have grouped them into (i) adhesive and anodic bonding, (ii) direct wafer bonding, (iii) metal bonding, and (iv) hybrid metal/dielectric bonding. Several other possible ways of sorting the technologies are possible, but the sorting approach taken here distinguishes the materials, the approaches, and their possible applications. In Part Two, some key wafer bonding applications are summarized, that is, 3D integration, MEMS, and temporary bonding, to give readers a flavor for where the wafer bonding technologies are significantly applied.

This handbook focuses on wafer-level bonding technologies including chip-to-wafer bonding. However, some of the technologies can also apply to chip-to-chip bonding, probably with some modifications.

Peter Ramm
James Jian-Qiang Lu
Maaike M.V. Taklo

Obituary

In Honor of Ulrich Gösele (1949–2009)

The editors would like to honor Professor Ulrich Gösele for his great contributions to wafer bonding, and are proud to have his chapter on "Direct Wafer Bonding" – his last authored article – in this book.

Ulrich M. Gösele
25 January 1949–8 November 2009
The photo was taken in July 2009 (source: MPI Halle)

Professor Ulrich Gösele passed away on 8 November 2009. His death was unexpected and is a great loss for his family and his many friends and colleagues all over the world.

His research interests covered different areas and were of impact for the science and technology of wafer bonding, diffusion and defects in semiconductors, semiconductor nanoparticles and nanowires, complex oxide films on semiconductors, silicon photonics, photonic crystals, and self-organized nanoscale structures.

Ulrich M. Gösele was born on 25 January 1949, in southern Germany in the city of Stuttgart. He studied physics at the University of Stuttgart and at the Technical University of Berlin and obtained his diploma in 1973. His PhD work was carried out at the University of Stuttgart and at the Max Planck Institute for Metals Research. In 1975 he completed his PhD thesis and was afterwards a scientific staff member of the Max Planck Institute for Metals Research until 1984. During this time he was also a visiting scientist at the Atomic Energy Board, Pretoria (South Africa) in 1976–77 and at the IBM Watson Research Center, Yorktown Heights (NY, USA) in 1980–81. He finished his Habilitation in 1983 at the University of Stuttgart. From 1984 to 1985 he was with Siemens Corporation, Munich, before he accepted a professorship of materials science in 1985 at Duke University, Durham (NC, USA). In 1991 he was a visiting scientist at the NTT LSI Laboratories, Atsugi (Japan). From 1993 he was a Director and Scientific Member at the Max Planck Institute of Microstructure Physics, Halle (Germany). He was also an Adjunct Professor at the Martin-Luther University, Halle-Wittenberg (from 1994) and at Duke University (from 1998).

He started his career as a theoretician, working on topics like diffusion-controlled reaction kinetics, radiation damage in metals, and transfer of electron excitation energy in liquids and solids. In 1975 he became interested in point defects and diffusion in silicon. His first paper in this area was published together with H. Föll and B.O. Kolbesen on agglomerates of intrinsic point defects, the so-called swirl defects [1]. Especially, his time at the IBM Watson Research Center and the intensive cooperation with T.Y. Tan resulted in numerous publications about point defects and diffusion in silicon.

His theoretical education and an evolving deep understanding and appreciation of experimental work may be a reason that he started research in the field of wafer bonding in the late 1980s at Duke University. Especially, the support by Dr. Takao Abe from Shin-Etsu Handotai Co., Isobe (Japan) enabled such experiments. A first result was the construction and application of a micro-cleanroom setup in 1988 allowing the bonding of wafers in a particle-free ambient under environmental conditions [2]. The principle of the micro-cleanroom setup was the basis of one of his most important patents [3] and was transferred to Karl Süss GmbH, a manufacturer of semiconductor equipment in the city of Garching close to Munich (Germany) resulting in one of the first commercially available wafer bonding tools in the early 1990s.

The increasing number of new students in his group allowed him to study different aspects of wafer bonding. First experiments on the wafer bonding of silicon to glass or to sapphire were carried out. Furthermore, aspects of wafer thinning processes by applying etch stop layers (carbon ion-implantation, etc.) were investigated. One of the most remarkable studies at this time was the analysis of defects formed in the interface of bonded wafer pairs [4]. All these investigations resulted in a first model of (hydrophilic) silicon wafer bonding by him together with Stengl and Tan [5]. In the early 1990s his research was focused on the preparation of silicon-on-insulator wafers by wafer bonding techniques. Besides numerous publications, another important patent concerns hydrogen-induced layer transfer [6].

Gösele continued his research activities after joining the newly founded Max Planck Institute of Microstructure Physics in 1993. Based on the support given by the Max Planck Society additional activities were undertaken in the research of wafer bonding. One example was the installation of an ultrahigh-vacuum tool allowing the wafer bonding under ultrahigh-vacuum conditions. Combined with computer simulations, molecular dynamic models of interface processes during wafer bonding were developed. Furthermore, the installation of various pieces of equipment such as cleanroom facilities resulted in numerous other research activities in the field of wafer bonding. These activities included, for instance, wafer bonding via designed monolayers, the bonding of different III–V compounds, and the development of methods for low-temperature wafer bonding.

His enormous range of research activities resulted in the publication of more than 700 articles in refereed journals, and the granting of numerous patents. He was one of the organizers of the first Symposium on Semiconductor Wafer Bonding: Science, Technology, and Applications held during the Autumn Meeting of the Electrochemical Society in October 1991 in Phoenix, AZ, USA. In addition to C.E. Hunt, H. Baumgart, S.S. Iyer, and T. Abe, he was an organizer of the 3rd Symposium on Semiconductor Wafer Bonding held during the Spring Meeting of the Electrochemical Society in 1995 in Reno, NV, USA. He was coauthor and coeditor, respectively, of the famous monographs *Semiconductor Wafer Bonding: Science and Technology* [7] and *Wafer Bonding: Applications and Technology* [8].

Ulrich Gösele's work and personality were appreciated all over the world and recognized by many honors and awards. For instance, he obtained the Electronics Division Award of the Electrochemical Society (1999), he was on the Board of Directors of the Materials Research Society (USA), and was a Fellow of the American Physical Society and a Fellow of the Institute of Physics (UK).

References

1 Föll, H., Gösele, U., and Kolbesen, B.O. (1977) The formation of swirl defects in silicon by agglomeration of self-interstitials. *J. Cryst. Growth*, **40**, 90.

2 Stengl, R., Ahn, K.-Y., and Gösele, U. (1988) Bubble-free silicon wafer bonding in a non-cleanroom environment. *Jpn. J. Appl. Phys.*, **27**, L2364.

3 Gösele, U. and Stengl, R. (1988) Method for bubble-free bonding of silicon wafers. US Patent 4,883,215, filed 19 December 1988, issued 28 November 1989.

4 Mitani, K., Lehmann, V., Stengl, R., Feijoo, D., Gösele, U., and Massoud, H. (1991) Causes and prevention of temperature-dependent bubbles in silicon wafer bonding. *Jpn. J. Appl. Phys.*, **30**, 615.

5 Stengl, R., Tan, T., and Gösele, U. (1989) A model for the silicon wafer bonding process. *Jpn. J. Appl. Phys.*, **28**, 1735.

6 Gösele, U. and Tong, Q.-Y. (1997) Method for the transfer of thin layers of monocrystalline material to a desirable substrate. US Patent 5,877,070, filed 31 May 1997, issued 2 March 1999.

7 Tong, Q.-Y. and Gösele, U. (1999) *Semiconductor Wafer Bonding: Science and Technology*, John Wiley & Sons, Inc., New York.

8 Alexe, M. and Gösele, U. (eds) (2004) *Wafer Bonding: Applications and Technology*, Springer Verlag, Heidelberg.

List of Contributors

Knut Aasmundtveit
Vestfold University College
Department of Micro and Nano Systems Technology
Faculty of Science and Engineering
PO Box 2243
3103 Tønsberg
Norway

Karlheinz Bock
Fraunhofer Research Institution for Modular Solid State Technologies EMFT
Hansastrasse 27d
80686 Munich
Germany

Joerg Braeuer
Fraunhofer ENAS
Department of System Packaging
09126 Chemnitz
Germany

Kuan-Neng Chen
National Chiao Tung University
Department of Electronics Engineering
Hsinchu 300
Taiwan

Léa Di Cioccio
CEA-Leti, MINATEC
Département Intégration Hétérogène Silicium
17 rue des Martyrs
38054 Grenoble Cedex 9
France

Viorel Dragoi
EV Group
E. Thallner Straße 1
4782 St Florian
Austria

Paul Enquist
Ziptronix, Inc.
800 Perimeter Park
Morrisville, NC 27560
USA

Philip Garrou
Microelectronic Consultants of NC
3021 Cornwallis Road
Research Triangle Park
NC 27709-2889
USA

Thomas Gessner
Fraunhofer ENAS
Department of System Packaging
09126 Chemnitz
Germany
and
TU Chemnitz
Center for Microtechnologies
09126 Chemnitz
Germany

Ulrich Gösele
Max Planck Institute of
Microstructure Physics
Weinberg 2
06120 Halle
Germany

Ronald J. Gutmann
Rensselaer Polytechnic Institute
Department of Electrical, Computer,
and Systems Engineering
CII-6015, 110 8th Street
Troy, NY 12180
USA

Nils Hoivik
Vestfold University College
Department of Micro and Nano
Systems Technology
Faculty of Science and Engineering
PO Box 2243
3103 Tønsberg
Norway

Matthias Hutter
Fraunhofer IZM
Gustav-Meyer-Allee 25
13355 Berlin
Germany

Armin Klumpp
Fraunhofer Research Institution for
Modular Solid State Technologies
EMFT
Hansastrasse 27d
80686 Munich
Germany

Roy Knechtel
X-FAB Semiconductor Foundries AG
Haarbergstraße 67
99097 Erfurt
Germany

Mitsumasa Koyanagi
Tohoku University
Graduate School of Engineering
6-6-04 Aramaki Aza, Aoba Ku
Sendai 808578
Japan

Christof Landesberger
Fraunhofer Research Institution for
Modular Solid State Technologies
EMFT
Hansastrasse 27d
80686 Munich
Germany

Adriana Cozma Lapadatu
SensoNor Technologies AS
Knudsrødveien 7
3192 Horten
Norway

James Jian-Qiang Lu
Rensselaer Polytechnic Institute
110 8th Street
Troy, NY 12180-3590
USA

List of Contributors

Jian-Qiang Lu
Rensselaer Polytechnic Institute
Department of Electrical, Computer, and Systems Engineering
CII-6015, 110 8th Street
Troy, NY 12180
USA

J. Jay McMahon
Rensselaer Polytechnic Institute
Department of Electrical, Computer, and Systems Engineering
CII-6015, 110 8th Street
Troy, NY 12180
USA

Makoto Motoyoshi
ZyCube Co. Ltd
4259-3 Nagatsuta-cho, Midori-ku
Yokohama 226-8510
Japan

Frank Niklaus
KTH – Royal Institute of Technology
Microsystem Technology (MST)
School of Electrical Engineering
100 44 Stockholm
Sweden

Hermann Oppermann
Fraunhofer IZM
Gustav-Meyer-Allee 25
13355 Berlin
Germany

Werner Pamler
Thin Materials AG
Hansastraße 27d
80686 Munich
Germany

Rama Puligadda
Brewer Science Inc.
Brewer Drive
Rolla, MO 65401
USA

Peter Ramm
Fraunhofer Research Institution for Modular Solid State Technologies
EMFT
Hansastrasse 27d
80686 Munich
Germany

Manfred Reiche
Max Planck Institute of Microstructure Physics
Weinberg 2
06120 Halle
Germany

Franz Richter
Thin Materials AG
Hansastraße 27d
80686 Munich
Germany

Kari Schjølberg-Henriksen
SINTEF
Department of Microsystems and Nanotechnology
0314 Oslo
Norway

Maaike M.V. Taklo
SINTEF ICT
Instrumentation
PO Box 124 Blindern
0314 Oslo
Norway

Chuan Seng Tan
Nanyang Technological University
50 Nanyang Avenue
Singapore 639798
Singapore

Maik Wiemer
Fraunhofer ENAS
Department of System Packaging
09126 Chemnitz
Germany

Dirk Wuensch
TU Chemnitz
Center for Microtechnologies
09126 Chemnitz
Germany

Introduction

Peter Ramm, James Jian-Qiang Lu, and Maaike M.V. Taklo

As editors of this book, we consider that there is a strong need to update the state of the art on wafer bonding [1–3]. The major reason is that corresponding development has truly sped up in recent years, motivated by the extended use of wafer bonding in new technology areas, such as three-dimensional (3D) integration, temporary bonding, and microelectromechanical systems (MEMS) with new functional layers. With novel areas of application, new objectives for wafer bonding technologies arise. For IC fabrication the traditionally much finer pitch used – compared to MEMS applications – has demanded sophisticated wafer bonding solutions. Metals, such as copper that has been used widely in the IC world, have also received a lot more attention for wafer bonding after wafer bonding became of interest for the IC community. For true 3D vertical integration where the IC wafers contain through-silicon vias (TSVs), electrical interconnects are needed between the wafers at their interfaces and can be realized by metal bonding. The interface interconnects and the TSVs can be formed either before or after bonding. This handbook covers a variety of wafer bonding technologies and applications and is structured according to these two aspects.

Technologies

In Part One, the technologies are sorted into four categories:

A) adhesive and anodic bonding;
B) direct wafer bonding;
C) metal bonding;
D) hybrid metal/dielectric bonding.

Other ways of sorting the technologies are possible, but the sorting taken here distinguishes the materials, approaches, and applications. We have attempted to include all bonding technologies that have received a reasonable level of attention from both the research and commercial communities. We are fully aware that some variants have been left out. These are not forgotten or excluded, but some level of standardization is believed to be beneficial for most of the communities.

This being said, it can be a hard decision whether to further optimize existing bonding technologies or to move to a new bonding regime. To move from one bonding technology to another can take years, something companies have painfully learned. Take MEMS as an example: it seems that the MEMS manufacturers are gradually moving from glass-based methods to metal-based methods. A product can benefit from a reduced bond frame width in the case of metal bonding, but the inclusion of metal as a bond frame material may have consequences for the complete process flow of the product. For yet other products in the same manufacturer's portfolio, the former glass-based solution may still be a better solution. The result is that a manufacturer has to deal with two bonding technologies rather than only one; this normally increases the manufacturing cost since it always takes more engineering effort to keep two processes up and running. Some choices have to be made: whether to always keep changing or upgrading the bonding technology for a given product, or to have one robust and flexible bonding technology and to modify all products accordingly. In this handbook the reader will find an overview of the most commonly used bonding solutions, and the arguments for these choices, as of today.

In Part One (A) we consider adhesive, anodic, glass frit, and spin-on glass (SOG) bonding. A glass material is common for anodic, glass frit, and SOG bonding. Commonly, the bonding material is spun or otherwise coated for adhesive, glass frit, and SOG bonding. Part One (A) focuses on these materials.

Adhesive bonding has in some cases been referred to as "simply" gluing wafers. The associations with glue seem something less advanced than other bonding methods; however, this is not true, and the reader will get a clear understanding from Chapter 3. Even though a finally optimized adhesive bonding process may appear straightforward as the surface topography can be compensated to a large extent and the bonding material appears homogeneous, one should not be fooled; a successful adhesive bonding is typically a result of combined advanced material development and careful bonding process optimization. Applications of adhesive bonding range from bonding of delicate logic wafers to bonding of low-cost wafers with large line widths for fluidic systems. In addition, temporary bonding is to a large extent based on the knowledge of adhesive bonding. However, some new materials have been developed for temporary bonding, in particular as the need for debonding was introduced. Some alternative process modification techniques were presented by SOITEC/LETI some years ago for debonding (the roughness could be tuned with a precision of nanometers), but wax seems to dominate this field at the present time.

The history of anodic bonding is long – normally considered to start with the article of Wallis and Pomerantz back in 1969. The robustness of the process is probably what has kept the technology in use. It is a popular bonding technique especially for MEMS manufacturing, and the technology is described in this book in Chapter 4, the authors of which have decades of experience using anodic bonding for industrial applications. This bonding technology is demanding with regard to surface topography, but to a level acceptable for most MEMS manufacturers. Smart designs of horizontal conductors have allowed cavities to be hermetically sealed using anodic bonding despite the demands of a low surface roughness.

The challenge of sealing cavities with crossing conductor lines has come up repeatedly. Glass frit bonding, here described in Chapter 1, has offered a way around this since more surface roughness is tolerable. Since the final bond is glass in both cases, the two technologies have several common issues with regard to reliability (leak rates, fracture mechanics, etc.) and can be considered as competing technologies. However, today we see several glass wafer manufacturers offering wafers with silicon or metal vertical vias. When the conductors are no longer crossing the bond line, there is a possibility that anodic bonding will actually gain some new interest. Spin-on glass is not as widely used as the other techniques mentioned, but there are some advantages with this technique, which justify a separate chapter (Chapter 2).

In Part One (B) we have gathered direct wafer bonding (DWB) methods. We have limited this to techniques where the wafers, either bare or oxidized, are contacted directly. We consider fusion bonding and plasma-activated bonding (PAB) to be subgroups of DWB, and the topics are thoroughly described in this book in Chapters 5 and 6, respectively. In both cases, the wafers are pre-bonded at room temperature and post-bond annealing is performed in a furnace or hot chamber depending on required temperatures. The pretreatment of the wafers differs in the two cases, but typically the surfaces are rendered either extremely hydrophilic or hydrophobic by wet or dry processing. The post-bond treatment temperature ranges from low temperature (or no heating) for PAB to heating above 1000 °C for fusion bonding. Obviously, the temperature treatment sets the limit for how a process sequence can be organized, but also the extreme demands of DWB with regard to surface roughness set strong limitations on how wafers can be treated before bonding. These issues are discussed in both the technology and the application chapters dealing with DWB.

In Part One (C) we have grouped four chapters on metal bonding methods. There is a wide range of metals that have been demonstrated to be applicable for wafer-level bonding within the semiconductor industry. It is certainly a benefit when metals that are already part of the CMOS industry can be used for bonding, but this has not been an absolute limitation for the choice of metals. When bonding is performed as one of the last process steps, compatibility issues with existing materials are less of a concern – as long as a satisfactory diffusion barrier is incorporated. But for bonding as an intermediate process step, the choice of metal is crucial. The application of copper as a bonding material has lately become popular for compatibility issues among other reasons. Copper is a demanding material with regard to wafer bonding; a high bonding pressure and/or extreme surface control may be required. However, these obstacles have been overcome as described, for example, in Chapter 9. Other metals or metal systems that are of interest are combinations of tin, copper, gold, and indium. Important knowledge about the complexity of the corresponding phase diagrams of these systems has, to a large extent, come from the knowledge of researchers working in the field of electronic packaging. Chapters 7, 8, and 10 describe how this knowledge has been applied to "tame" the devils in the detailed diagrams.

The idea of hybrid bonding, as described in Part One (D), is basically to take the best of two worlds and combine them into a wafer bonding method, resulting in mechanical and electrical connection between wafers at one step. The challenge necessarily is to find process parameters that fit the two combined bonding methods at once. As more than one solution to this challenge has been presented, several chapters are dedicated to this topic. Chapters 12 and 13 both describe how metals like copper or nickel can be combined with inorganic dielectrics like oxides; copper combined with a polymer is presented in Chapter 11. A core part of these discussions is how to tune the polishing processes to two materials at the same time to achieve the required surface smoothness for high-quality bonding.

Applications

In Part Two, recent applications of wafer bonding are described.
In general, wafer bonding technologies are applied over a wide range, such as:

- silicon-on-insulator wafers or variations for semiconductor applications;
- MEMS for nonhermetic, hermetic, and extremely hermetic attachment of parts, and combined possibilities;
- fluidic systems for a variety of applications in thermal management or biomedical devices;
- handling of thin wafers/chips for further processing in semiconductor processes and packaging;
- 3D ICs for stacking memories or memories/microprocessors for computation and communication systems;
- 3D heterogeneous integration and packaging to integrate a variety of components (such as CMOS image sensors and readout circuits) into one compact system.

This handbook is not intended to give a comprehensive summary of wafer bonding applications because many applications are described in other books and more applications are still emerging. Besides the applications discussed in the chapters in Part One, chapters in Part Two summarize the key wafer bonding applications developed recently, that is, 3D integration, MEMS, and temporary bonding, to give readers a taste of where the wafer bonding technologies are applied significantly.

Definitions

Many terminologies are defined in individual chapters in this handbook. Here we focus only on a few terminologies, which we thought should be further clarified.
Several chapters in this handbook deal with wafer bonding by use of metal systems. Besides direct metal bonding (Chapters 6 and 9), there are two bonding

methods using liquid phases: eutectic bonding and solid–liquid interdiffusion (SLID) bonding (Chapters 7 and 10). Both use a combination of low- and high-melting metals, and unfortunately are not always correctly distinguished in the literature. Eutectic soldering is a very well-known technology that has been used for many years in microelectronic packaging, for example in so-called flip-chip bonding. The soldering process of heating to the melting temperature and cooling is, in principle, reversible: by heating again above the eutectic temperature the metal system re-melts. In contrast, SLID bonding is irreversible, utilizing the diffusion of solid material of a high-melting metal into the liquid phase of a low-temperature metal at constant temperature and resulting in a thermomechanically stable intermetallic compound. It may be slightly confusing that in the literature different terms are used for this dedicated soldering principle, as for example "transient liquid-phase bonding." As the process was described first in 1966 by Bernstein and Bartholomew [4], the editors prefer to use the original term SLID defined there.

Three-dimensional integration is generally defined as the fabrication of stacked and vertically interconnected device layers. Numerous corresponding technology concepts have been introduced since the 1980s (e.g., Siemens and Fraunhofer Institute Munich) and several companies and research organizations have developed full 3D process demonstrations. However, only a few of them have already come to production. One of the reasons is the concern about introduction of sophisticated advanced processes into production lines. Reliable and robust wafer bonding technologies are considered to play a key role in the fabrication of 3D integrated products.

Chapter 15 reports on diverse applications of 3D integration with the need for specific bonding technologies. Categories for a great variety of 3D technologies are defined there. Of particular interest are TSV technologies. A TSV is defined as an electrical interconnection between two sides of a silicon substrate. Besides the TSV process, wafer thinning, thin wafer handling, backside processing, and 3D stacking are the basic technology modules. Wafer bonding technologies are needed here for both thin wafer handling and 3D stacking:

- permanent bonding for 3D stacking (Chapters 2–14);
- temporary bonding for thin wafer handling (Chapters 16, 18, and 19).

The choice of the wafer bonding technologies has to be made by considering full process conditions such as temperature limitations and the sequence of the technology modules, which may vary for different applications. The resulting large variety of process flows is categorized according to ITRS [5] and explicitly corresponds to the order of TSV processing, wafer thinning, and wafer bonding. The order of the TSV process with respect to device fabrication is one of the main criteria, and today commonly characterizes TSV fabrication:

- *prior* silicon front-end-of-line (FEOL) – "via first";
- *post* silicon back-end-of-line (BEOL) – "via last";
- *post* FEOL but *prior* BEOL – "via middle."

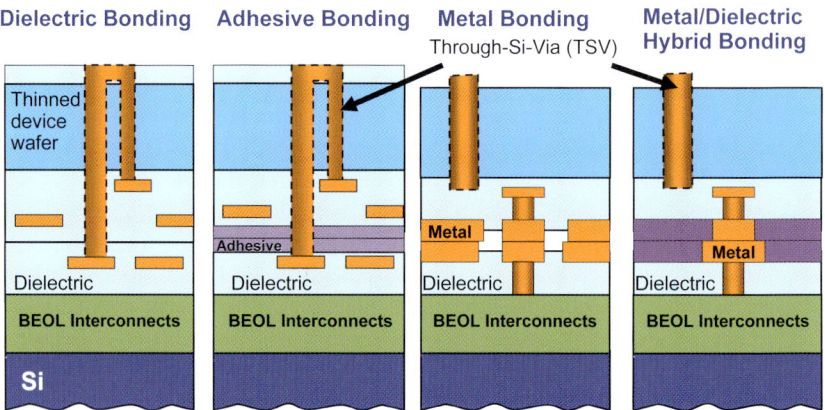

Figure 1 Schematic of four typical wafer bonding schemes for 3D integration with TSVs (not to scale). The thinned device wafer usually has FEOL devices and BEOL interconnects (not shown).

The choice of the corresponding TSV technology for fabrication of a 3D integrated product has significant consequences for the wafer bonding technology to be applied, as shown in Figure 1. Here, the two bonding approaches on the left-hand side are "via last" 3D integration technologies, where TSVs are formed after wafer bonding and thinning. The two bonding approaches on the right-hand side can form electrical interconnections during the bonding process, while TSVs can be made at any stage of the processing flow, depending on the applications and processing constraints. Detailed materials and bonding conditions are described in the corresponding chapters.

In Chapter 15 the large variety of 3D integration technologies are defined and categorized. The most common wafer bonding technologies for application in 3D integration are there specified and exemplarily described regarding the interaction between bonding processes and 3D processing flows.

Conclusions

This handbook summarizes a variety of wafer bonding technologies and applications. In particular, the fundamental knowledge needed for wafer bonding is highlighted in individual chapters. As polymers are applied more widely in electronics and for sensors to reduce cost, knowledge of chemistry is required in addition to the physics and electronics knowledge that has dominated in some research areas. Eutectic bonding demands a thorough understanding of metallurgy in order to get a good understanding of how various phases are formed during heating and cooling. For adhesive bonding it is rather an understanding of how the molecules crosslink and how the materials soften with heat that is important. For both bonding solutions, wetting of surfaces is critical, so this is another topic that should be well understood. For hybrid metal/dielectric bonding, knowledge of chemistry, physics, and surface planarization and treatment is critical. When it comes to characterization, it is often fracture mechanics that is the subject. It is important to know how the bond strength of interfaces should be characterized in order to ensure a certain reliability of the bonding interface.

For future applications, more knowledge may be needed for a variety of topics, such as larger or thinner wafers and fine-pitch interwafer interconnections (such as TSVs), and hence better alignment, better wafer thinning control, as well as thermomechanical constraints. We expect that more detailed research and development will be conducted by companies for specific products and applications.

References

1 Tong, Q.-Y. and Gösele, U. (1999) *Semiconductor Wafer Bonding: Science and Technology*, John Wiley & Sons, Inc., New York.
2 Plößl, A. and Kräuter, G. (1999) Wafer direct bonding: tailoring adhesion between brittle materials. *Mater. Sci. Eng.*, **R25**, 1–88.
3 Alexe, M. and Gösele, U. (eds) (2004) *Wafer Bonding: Applications and Technology*, Springer, Berlin.
4 Bernstein, L. and Bartholomew, H. (1966) Applications of solid–liquid interdiffusion (SLID) bonding in integrated-circuit fabrication. *Trans. Metall. Soc. AIME*, **236**, 405–412.
5 Semiconductor Industry Association (2009) *The International Technology Roadmap for Semiconductors*, SEMATECH, Austin, TX (2010 update).

**Part One
Technologies**

A. Adhesive and Anodic Bonding

1
Glass Frit Wafer Bonding
Roy Knechtel

Glass frit wafer bonding is widely used in industrial microsystems applications where fully processed wafers have to be bonded. This end-of-process-line bonding must fulfill some very specific requirements, such as: process temperature limited to 450 °C to prevent any temperature-related damage to wafers, no aggressive cleaning to avoid metal corrosion, high process yield since wafer processing to this stage is expensive, bonding of wafers with certain surface roughness or even surface steps resulting from metal lines electrically running at the bonding interface to enable electrical connections into the cavity sealed by the wafer bonding, as well as a mechanically strong, hermetically sealed, reliable bond. All of these requirements are fulfilled by the glass frit wafer bonding process, which additionally can be very universally applied since it can be used to bond almost all surfaces common in microelectronics and microsystem technologies.

1.1
Principle of Glass Frit Bonding

The basic principle of glass frit bonding is the use of glass as a special intermediate bonding layer. This glass must have a low-temperature melting point. In the bonding process, the glass between the wafers to be bonded is heated, so that its viscosity continually decreases until the so-called wetting temperature is reached, at which point the glass is soft enough and liquid enough to flow and wet the wafer surface. In this flowing process, the glass comes into contact with the surface to be bonded at the atomic level, and flows into surface roughnesses and around surface steps. As a result, perfect sealing behavior is achieved by the liquid glass; hence, this bonding process is also called seal glass bonding. On cooling the glass in the wafer stacks, it re-solidifies, to finally form a mechanically strong, hermetically sealed bond. Due to its excellent bonding and flowing behavior, glass frit bonding is very often used to cap microelectromechanical systems (MEMS) at the wafer level by simultaneous sealing under vacuum or low pressure, which is required for resonant devices such as gyroscopes. Figure 1.1 shows an example of a glass-frit-bonded gyroscope. The flowed glass frit is highlighted by the materials

Handbook of Wafer Bonding, First Edition. Edited by Peter Ramm, James Jian-Qiang Lu, Maaike M.V. Taklo.
© 2012 Wiley-VCH Verlag GmbH & Co. KGaA. Published 2012 by Wiley-VCH Verlag GmbH & Co. KGaA.

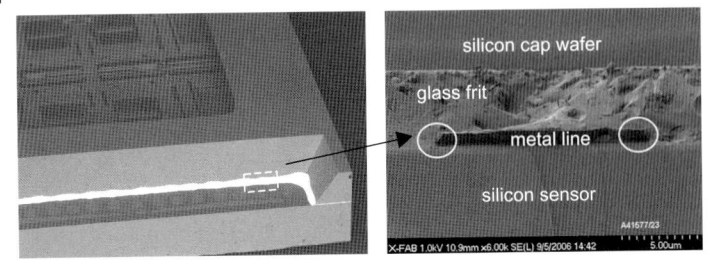

Figure 1.1 Example of glass frit bonding as a process for MEMS capping: flowing and sealing behavior at signal metal lines into a sealed cavity.

contrast of the scanning electron microscope (SEM), so that its sealing behavior is well illustrated.

1.2
Glass Frit Materials

Essential for glass frit bonding is the glass material which ultimately forms the bond between wafers. Low-melting-point glasses which can reflow in the temperature range of 400 to 450 °C are very special glass materials with a rather unstable glass matrix. Such materials are very difficult or nearly impossible to deposit by classical layer deposition techniques such as sputtering, spin-on deposition, or chemical vapor deposition (CVD). Additionally, a layer thickness of 5 µm or greater is required to allow a good reflow of the glass during bonding, to ensure excellent sealing by the glass flowing into surface roughnesses and around surface structures. Since glasses can be transformed into pastes, screen printing is the best method for bringing the glass frit onto one of the wafers to be bonded. To prepare such pastes for screen printing, a low-melting-point glass, typically a lead zinc silicate glass or lead borate glass, is ground to particles with a size less than 15 µm. This glass powder is mixed with an organic binder, to form a paste. In order to achieve optimum mixing, the mixing is done with a milling tool. Additionally, the paste contains solvents for tuning the viscosity to obtain the best screen printing results (300–600 P (30–60 Pa s)). Because the thermal expansion of the low-melting-point glasses is typically significantly higher than the glass surfaces to be bonded, filler particles are added to tune the thermal expansion of the final glass frit. These filler particles have a much higher melting temperature than the active glass component, so they are built into the refused glass frit and reduce the thermal expansion. For example, by adding barium silicate glass ceramic fillers to lead zinc silicate glass frits, the thermal expansion coefficient can be reduced from 10×10^{-6} to $8 \times 10^{-6}\,\text{K}^{-1}$ which finally gives a stress reduction in the glass-frit-bonded wafer stack. Glass frit pastes are offered by various vendors of glass or electronic materials, such as Ferro Corporation, Schott Glass, and DIMAT, typically as ready-to-use pastes. For high-temperature glass frits, there is a wide range of different materials

- Lead Zinc Silicate Glass as a low melting point active glass frit component
- Barium silicate glass fillers to lower the thermal expansion coefficient of the glass frit
- Organic binder with solvent

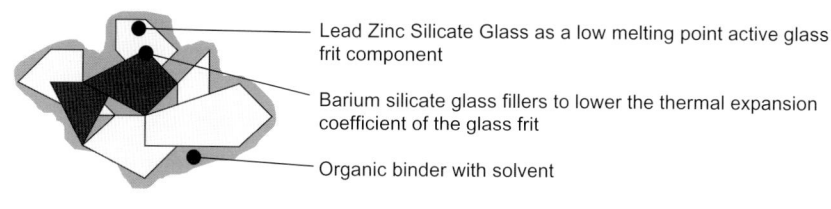

Figure 1.2 Main composition of Ferro FX-11-036 glass frit paste.

available, but for the low-temperature range (<450 °C) that is required to bond processed wafers, only a few materials are available. The most commonly used paste is Ferro FX-11-036 [1], which is also upon which the following exposition is based. The reason for the popularity of this material is that it does not tend to crystallize. This facilitates the thermal processing of the glass frit, since, otherwise, crystallization would change the glass dramatically by increasing the reflow temperatures, which would make final bonding impossible. The composition of Ferro FX-11-036 is shown schematically in Figure 1.2.

In addition to the ready-to-use commercial glass frit pastes, there is also the option to mix special glass frit pastes based on glass material from special melts or special glass powders. However, the mixing of a screen-printable glass paste is very difficult since the binder has to have a rather high viscosity.

From the environmental protection point of view, the lead in glasses used to reach very low melting points is critical. In the RoHS directive there is still an exception for lead in glasses of electronic components, and the total amount of lead in glass-frit-bonded chips is too small to be in conflict with automotive limitations. Nevertheless, the lead in glass frits should be replaced in the medium term. For example, at Ferro, lead-free glass frits for wafer bonding are in development and should be considered for any new product or process developments.

1.3
Screen Printing: Process for Bringing Glass Frit Material onto Wafers

For glass frit bonding, glass layers that are several micrometers thick are required to allow a reflow of the material in the wafer bonding process, which ultimately allows the planarizing and sealing of wafer topographies. Such thick layers are relatively easy to deposit using screen printing technology, which offers the additional benefit that the material is structured directly with the deposition. Using this method, glass frits can also be printed on structured wafers, for example on cap wafers with through-holes. During screen printing, the structures to print are realized as openings in a screen film layer which is supported by a mesh structure stretched onto a supporting frame. The film structure is realized by photolithographic methods. The thickness of the screen wires and the screen defines the thickness of the print. The mesh width is related to the paste to be printed, and should be three times the size of the particles. The minimal structure width should

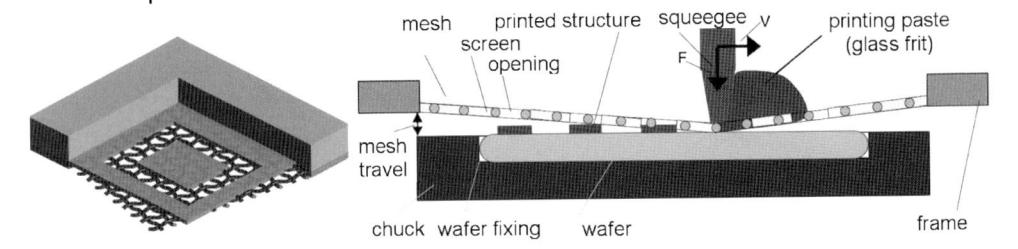

Figure 1.3 Principle of screen printing for glass frit bonding.

include three meshes, to allow a safe screen printing process. For typical glass frit materials, with particle sizes up to 15 μm, the mesh width is 45 μm at a wire thickness of about 35 μm and a film thickness of about 10 μm. Polymer meshes are mainly used. Structures of width in the range 150–300 μm at a wet thickness of about 45 μm can be realized with glass frit pastes. Details of screen materials construction and selection are available at www.koenen.de. For screen printing, the wafer to be printed on is placed underneath the mesh. The wafer is aligned to the mesh by an automatic vision system using special marks on both the wafer and the mesh. In the first step, a squeegee floods the mesh with the glass paste, to fill the mesh openings with the paste. In the second step, the wafer is printed by pressing the mesh to the wafer with the squeegee, and moving it over the mesh area. Figure 1.3 illustrates the screen printing process.

The high-precision screen printing that is required for glass frit bonding (the bond frames should be as small as possible and exactly positioned, since the bond area increases the chip size and, as a result, the cost) is a very sophisticated process. Structure size, mesh travel, squeegee force and speed, as well as paste properties such as viscosity and particle size have a strong influence on the structure definition and placement accuracy. To define general design rules is very difficult, but in practice the following tolerance chain should be considered:

- mesh opening as designed;
- increasing of structure at printing (underflow of paste to mesh);
- misalignment at screen printing and misalignment by screen stretching.

Based on this, an area of up to 400 μm could be required for printing minimal designed frames of 150 μm width on unstructured silicon wafers. For this, the tolerances of the bonding process (glass out-flow and misalignment) must be additionally considered, which finally leads to an area requirement of up to 500 μm in width for microsystem wafers which should be capped by glass frit bonding. Of course, this is much too high an area consumption, so that a self-aligned screen printing process is recommended. Here, structures etched into the wafer to be printed on – mostly such wafers have to be structured anyway, so that this mask layer can also be used for this – define the bond frames. Since screen printing is based on wetting of the surface with the paste, the printing is limited by the structures, even if the mesh opening is wide. With this approach (Figure 1.4), very

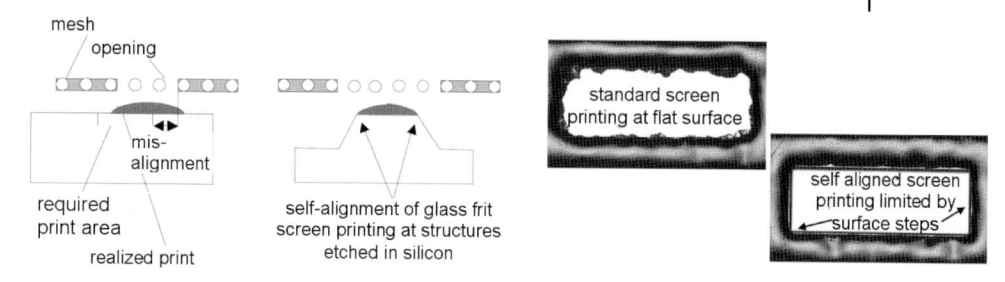

1.3 Screen Printing: Process for Bringing Glass Frit Material onto Wafers

Figure 1.4 Principle of self-aligned screen printing.

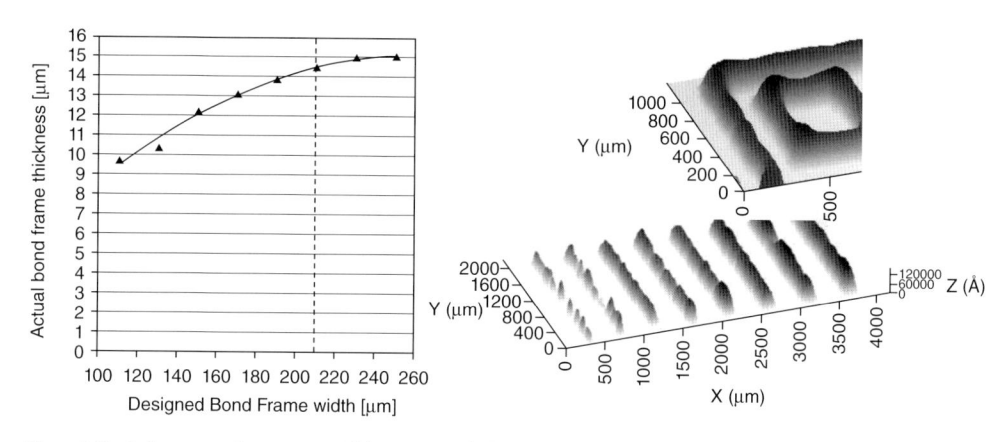

Figure 1.5 Influences of structure width on printed glass frit thickness.

well-defined structures with minimal width and without screen alignment influences can be realized, and the total area consumption on the system wafer is reduced to about 250–300 µm in width.

Finally, the printed glass frit frame must be considered as a three-dimensional object. The thickness of the glass is mainly defined by the mesh construction (wire and film thickness), but is also influenced by many other factors. Therefore, the final thickness after bonding typically varies in the range 5–15 µm, and should not be used to define functional gaps in microdevices. For the bonding, the actual glass thickness is not very critical, as long it is above 5 µm, to ensure the sealing of surface profiles. More critical is the variation of the glass thickness within the structures which should create a hermetic bond. Even if the glass flows during the bonding process, the planarizing behavior is limited. Since the main influence of the printed structure thickness is the width (Figure 1.5), particularly for small bond frames, it must be ensured in the design that all structures on one wafer have one and the same width. Even corners of rectangular structures show increased thickness, but these are within the limits of the glass flow, and no special

1.4
Thermal Conditioning: Process for Transforming Printed Paste into Glass for Bonding

When bond frames are printed with glass frit paste onto one of the wafers to be bonded, the material is not already usable for bonding. The material is a compound of glass and filler particles held in a binder. If this configuration were be bonded at the process temperature at which the glass compound is melted and wets surfaces, a bond would occur, but it would be neither strong nor with hermetic sealing, because the organics from the binder and solvents would cause large bubbles in the glass, as shown in Figure 1.6.

Therefore, a thermal preprocess to burn out the primer and to transform the paste glass into real pre-melted glass is essential for a high-quality bond. This thermal process consists of three steps. First, and directly after screen printing, the glass paste has to be dried and stabilized at 120 °C. In this step, most of the solvents disappear and the polymer binder is internally crosslinked and stabilized; with this step the glass paste becomes stable enough for storage, easy handling, and shipment, if required. Second, the binder is burned out at 360 °C. At this temperature the glass is not really melted, but its viscosity is significantly reduced, so that the glass particles take on a round, droplet-like shape and form an initial adhesion to each other and to the base of the wafer on which they are situated. This microstructure is very important for two reasons: (i) it is an open sintered-like structure, which allows complete burning out of the binder and (ii) it has a certain mechanical stability to maintain the shape of the printed structure. Third, after

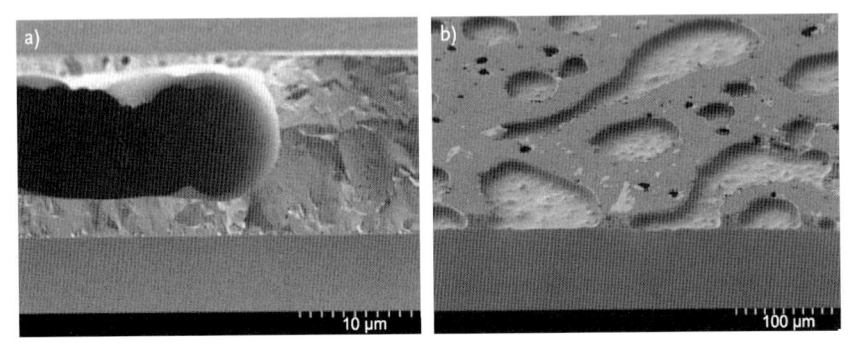

Figure 1.6 Bubble generation in glass frit band if binder is not completely burned out: (a) cross section; (b) after debonding.

the paste is completely burned out, which takes about 10 min, the temperature is increased to 450 °C; the glass particles are completely melted at this temperature and fuse into solid glass. If filler particles for thermal expansion coefficient management are in the paste, these are enclosed by the glass but not fused into it, which is not required for the targeted effect. Table 1.1 illustrates the thermal conditioning process; the actual temperatures must be cross-checked with the glass frit paste specification, but temperature tolerances of ±10 K are not critical.

In thermal conditioning, the glass paste is not only transformed into a bulk glass, but also the structure is densified. Thus, the structure thickness after screen printing is reduced by a factor of about three, but the structure shape and width are not changed. Even if the viscosity of the glass in the pre-melt is low enough to wet the surface, it is not low enough to allow the glass to flow. As a result, the lateral structure is finally defined during the screen printing. Due to the surface tension, the free surface of the glass frit is bowed. Figure 1.7 shows the densification during thermal conditioning.

This multiple-temperature glass-conditioning process can either be done at one profiled temperature step or as three separate processes. It is recommended to keep the wafers horizontal during the thermal processing in order to prevent gravitation-related flowing of the glass structures. Some glass frit materials have a tendency to crystallize. Glass crystallization must be prevented during the conditioning through the use of precise temperatures and defined rates of temperature change, because the glass properties are dramatically altered by crystallization. In particular, after crystallization, the processing temperatures are increased, so that the glass cannot be re-melted in the bonding process, the result being that finally the glass frit is no longer bondable. Crystallization, or deglazing, occurs if on cooling the glass reaches a viscosity range in which the amount of internal glass nucleation and crystallization speed are maximal. Figure 1.8 shows a glass crystallization diagram [2] and an example of a crystallized glass frit. For glass frit materials deposited as glass paste, nucleation is increased due to organic residues coming from the binder, so the probability for crystallization is increased. The crystallization of the glass can be prevented by the glass chemistry: if the number of nucleation points and the crystallization speed do not have their maxima in the same viscosity range, crystallization practically does not occur. This explains why some glasses have a tendency to crystallization and others do not. If crystallization is observed during processing of glass frit materials, the cooling rate of the thermal processing should be varied experimentally. Normally faster cooling should prevent crystallization, but provides a risk of thermal cracks in the glass. On the other hand, glass crystallization can be actively used in the later bond process: if the glass crystallizes on cooling after bonding, a thermally much more stable bond is created. However, the process window for such bonding appears quite narrow.

After this multiple-step thermal treatment, the wafers with the glass frit are finally ready to bond.

10 *1 Glass Frit Wafer Bonding*

Table 1.1 Overview of glass frit conditioning process.

Thermal conditioning step	Process temperature (°C)	SEM images	Schematic images	Remarks
Drying and binder polymerizing	120	Flown binder at printed structure edge; Structure of glass and filler particles	Glass frit paste; Printed paste; Paste with polymerized binder	– Solvent drying – Binder polymerizing – No change in glass or binder
Binder burn-out	360	Binder removed by burning out; Glass particles start melting; Filler particles not melted	Binder burn-out	– Sintered-like structure – Glass melting starts, but with high viscosity – Fixing glass particles to substrate and to each other – No change in filler particles – Binder burn-out
Glass pre-melting	450	Glass particles in the melted glass; Edge of screen printed structure; Pre-melted glass surface; Filler particles in the glass structure	Start of pre-melting; Completed pre-melting	– Complete pre-melting of the glass – Starting from wafer as heat carrier – Void-free glass – Filler particles sealed but not fused into the glass

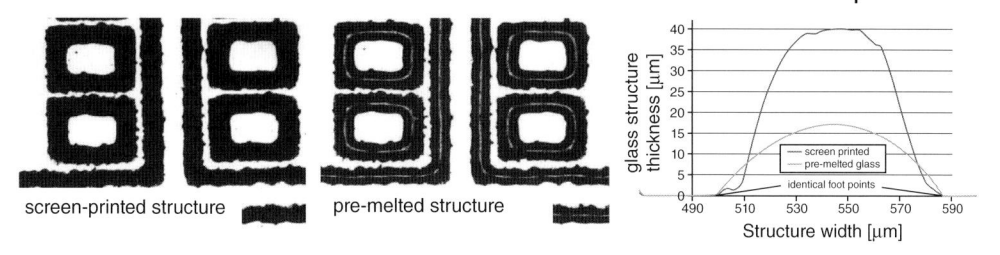

Figure 1.7 Changes in screen-printed glass frit structures during thermal conditioning.

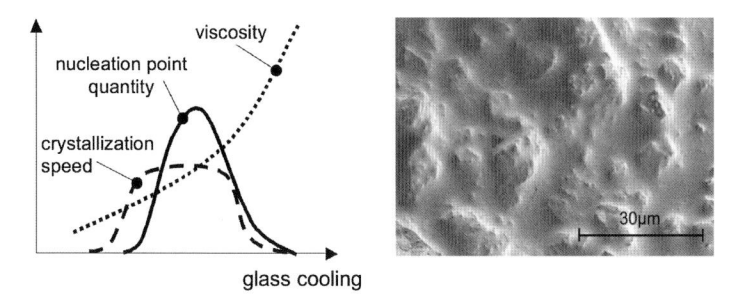

Figure 1.8 Critical glass frit crystallization: relation to glass processing and SEM image of crystallized glass frit.

1.5
Wafer Bond Process: Essential Wafer-to-Wafer Mounting by a Glass Frit Interlayer

The actual bonding process is comparable to other wafer bonding technologies. Since glass frit bonding is typically done at structured surfaces, both of the wafers to be bonded must be aligned before they are loaded into a standard wafer-bonding chamber. In the chamber, the wafer pair is heated to the wetting temperature of the glass frit, which is normally in the range 440–450 °C. At that temperature, the glass flows and wets the surfaces to be bonded. The bonding process, which is, from the physics, driven by temperature, has to be supported by mechanical pressure on the wafers, in order to compensate for wafer bow influences and screen printing inhomogeneities. In the bonding process there is another geometrical transformation of the glass, where it really flows from the bond frame as shown in Figure 1.9. When the glass has flowed and wetted the surfaces to be bonded, the wafer stack can be cooled. The glass frit solidifies and forms a mechanically strong bond.

The parameters of temperature and pressure for this bonding process must be very carefully adjusted. If the temperature and pressure are too low, there is no or just local bonding. If one of these parameters is too high, the glass may flow too much and reach functional structures, where it may block or destroy them. This tuning process has to be done for each bonding machine, and when its

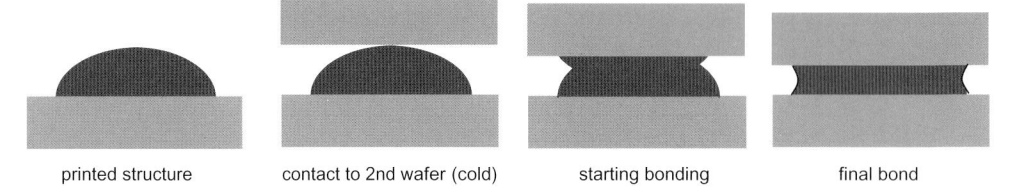

Figure 1.9 Schematic of progress of glass frit bonding (cross section).

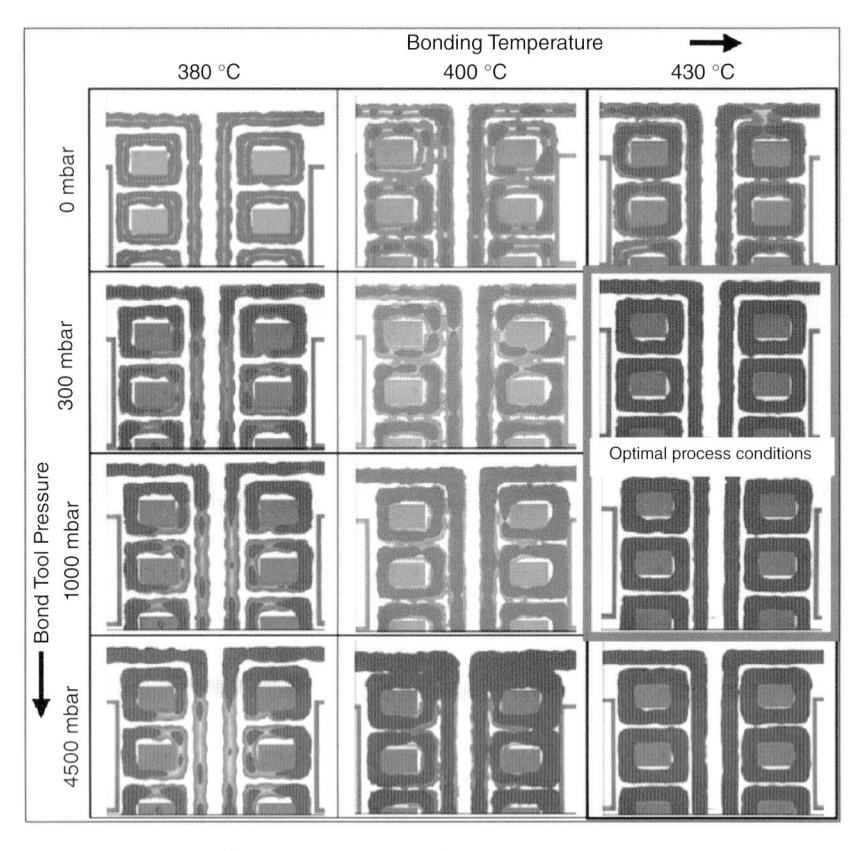

printed structure   contact to 2nd wafer (cold)   starting bonding   final bond

Figure 1.10 Parameter field evaluation of glass frit bonding process.

configuration is changed. An essential requirement here is that the chuck and tool plate are very parallel to ensure homogeneous bonding and glass flow. Figure 1.10 illustrates this kind of parameter optimizing process. Here, the silicon wafer with a glass frit is bonded to a glass wafer, which allows very easy inspection of the glass frit flow.

The flowing glass allows hermetic bonding of rough or structured surfaces, even if they also have surface steps such as from metal layers. However, to ensure hermetic bonding, the glass must be hot enough to flow well.

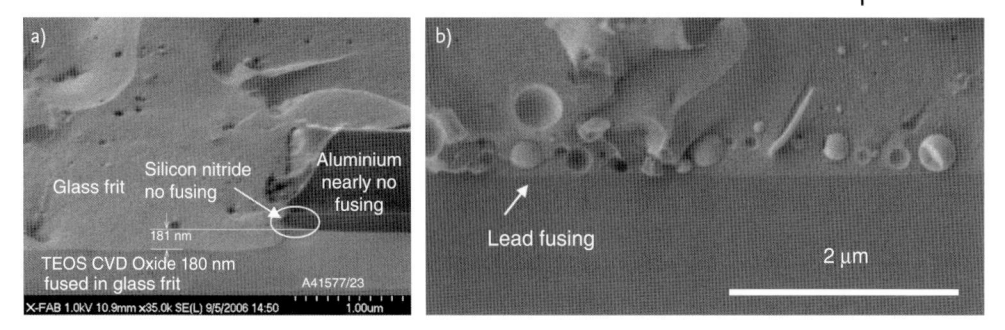

Figure 1.11 Fusing effect of surface layers: (a) CVD oxide; (b) silicon.

The physics of glass frit bonding is related to the glass chemistry [2, 3]. When the glass wets the surface to be bonded at high temperature, thin layers of the material to be bonded are fused into the glass, so that finally after cooling, a continuous material interaction occurs which forms a strong bond. This layer fusing can be investigated using SEM cross sections (Figure 1.11). For CVD oxides (e.g., tetraethylorthosilicate), layer fusing can reach several tens of nanometers and can be evaluated if there is any masking material (e.g., metal lines in the bonding interface). On bonding silicon this kind of fusing cannot be observed, but lead precipitates (ball shaped) indicate that a redox reaction has taken place for which some silicon has to be fused into the glass frit. These lead balls are not critical for bonding strength [4], but could cause leakage currents if unpassivated metal lines are sealed with the glass frit.

It is very critical in glass frit bonding to maintain the wafer-to-wafer alignment accuracy during the bonding process. As mentioned earlier, the wafers are slightly pressed together when the glass is soft. Any sideways force can cause a shift of the wafers very easily. To prevent this, the following measures are recommended:

- The glass bonding should be done with closed fixture clamps to fix the wafers until bonded.

- The bonder chuck and tool plate should be as parallel as possible to prevent wafer-to-wafer sliding.

- Chuck and tool plate should be made from silicon or silicon carbide to reduce thermal mismatch with the wafers to be bonded for preventing lateral forces which can cause wafer-to-wafer shifting and thus poor post-bond alignment.

If all of these points are considered, wafer alignment accuracies of better than 5 µm can be achieved as state of the art for glass frit bonding, as confirmed in Figure 1.12.

If bonding under vacuum or defined gas pressures is required, it is not necessary to use spacers. The glass frit shows a surface topography before bonding that allows the passage of gasses when the wafers are in contact at the cold stage.

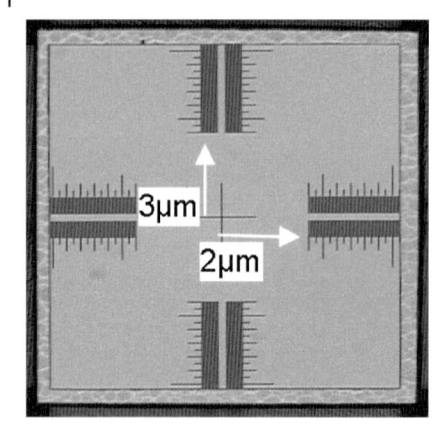

Figure 1.12 Alignment accuracy after glass frit bonding.

1.6
Characterization of Glass Frit Bonds

The bonding interface must be strong enough to withstand subsequent processing steps such as dicing and packaging. Furthermore, the bonding has to ensure good protection of the active structure during its lifetime. It has been shown that almost all relevant layers for microsystem processes can be bonded with high yields using glass frit bonding. Stud pull tests have shown that the bonding interface is strong enough to fulfill these requirements. Figure 1.13 illustrates the results of pull tests on diced chips. The fracture load varies a little, but is high enough for most applications and comparable to the bonding strength that can be achieved with anodic bonding.

Hermetic sealing is often required to ensure the correct functioning and sufficient reliability of capped micromachined components. Therefore, hermeticity investigations were performed on glass-frit-bonded devices. Because of the lack of MEMS hermeticity testing standards, resonant structures were chosen, whose frequency behavior depends on the gas pressure in the space surrounding the structures. The structures we used show only resonant behavior when moving in vacuum (Figure 1.14) and therefore provide a very sensitive pressure indication. Testing of several resonant structures has shown that a vacuum of about 1–5 mbar can be sealed using glass frit bonding. The minimum sealing pressure is related to glass frit outgassing of remaining organics. After storage of several weeks, no shift of the sealed cavity pressure was observed.

Besides these general wafer bonding requirements and, further, bonding strength and yield, hermetic bonding, and universal usability, glass fit bonding shows some additional advantages:

- bonding of CMOS wafers is possible without adversely influencing them;
- direct sealing of unpassivated metal lines is possible;

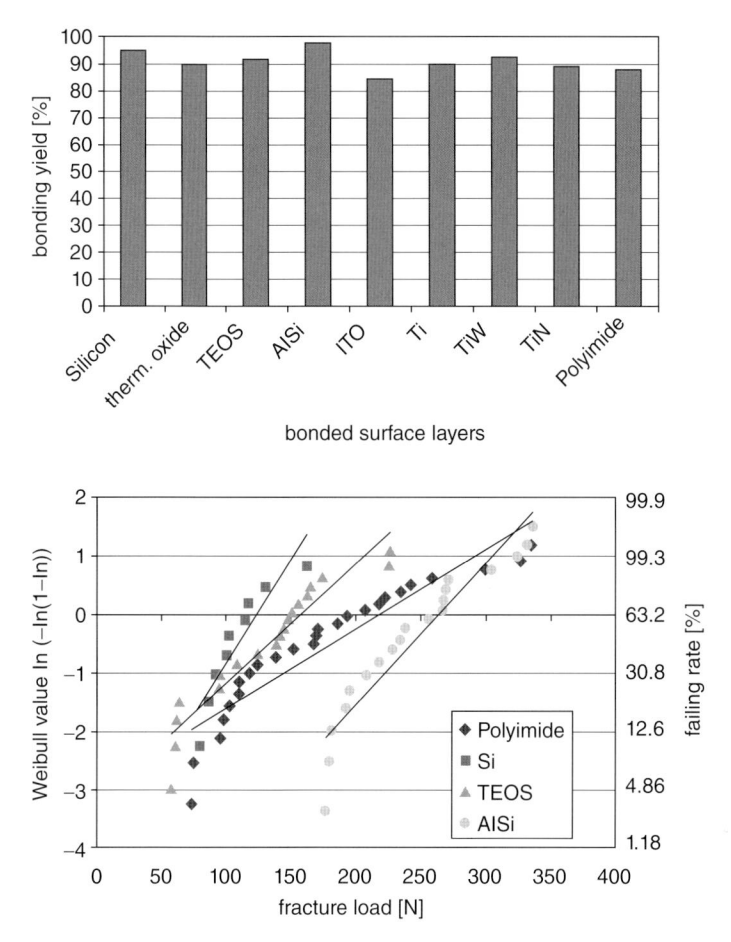

Figure 1.13 Results of bonding of various surface layers. (TEOS, tetraethylorthosilicate; ITO, indium tin oxide.)

- very good reliability;
- very low mechanical stress on bonded mechanical components;
- very safe, reproducible process;
- bonding of structures with through holes is possible;
- no surface activation is required.

1.7
Applications of Glass Frit Wafer Bonding

Many different applications for glass frit bonding are possible, because of the universal characteristics of this bonding technology. The main application for glass frit bonding is the encapsulation of surface micromachined sensors such as

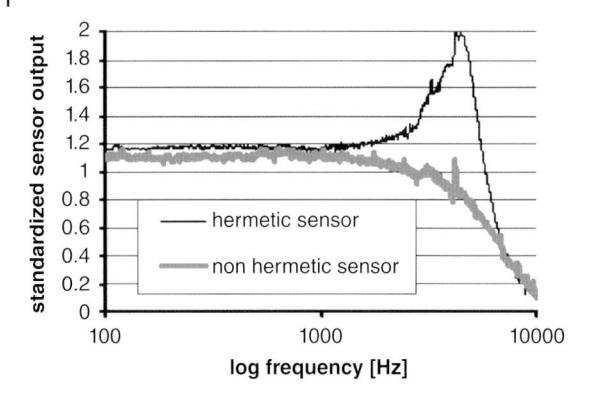

Figure 1.14 Hermeticity evaluation using resonant structures.

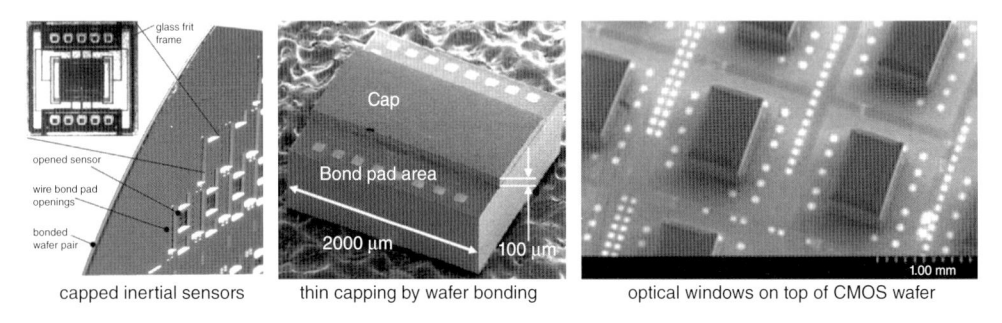

Figure 1.15 Examples of wafer bonding applications.

gyroscopes (Figure 1.15) and acceleration sensors. With these sensors, the main advantages of glass frit bonding, such as hermetic sealing, metallic lead-throughs, narrow bonding frames, and high bonding yield, play an important role. Furthermore, fully processed cap wafers with cavities above the sensor structures and holes for wire bond pad access can be bonded, so that the cap bonding is the last wafer process step. This sensor technology using glass frits has been introduced into production and is available as a MEMS foundry process. Glass frit bonding can also be used successfully for several additional applications, such as bulk micromachined sensors, the sealing of absolute pressure sensor cavities, the mounting of optical windows, and the capping of thermally active devices.

1.8
Conclusions

Glass frit bonding technology provides a wide range of possibilities for the bonding of wafers at process temperatures below 450 °C. The screen printing of the bonding glass as a paste allows *in situ* deposition and structuring on processed cap wafers.

The structured bonding layer protects moveable structures from parasitic bonding. Following this, organic materials are eliminated from the paste by temperature cycling. The bonding is a thermo-compressive process. The bond is formed at 430 °C with a slight pressure applied. Almost all surface layers commonly used in silicon micromachining can be bonded using glass frits. The main advantages of glass frit bonding are hermetic sealing, high process yield, low mechanical stress at the bonding interface, possibility of metallic lead-throughs, high bonding strength, and good reliability. For various applications, the technological potential and the universal use of glass frit bonding have been shown. Critical for some applications may be the high process costs related to the glass frit material and the complex processing.

References

1 Ferro (2002) MEMS and sensor materials, 11-036 sealing glass specification, Rev. 1200, Ferro Electronic Materials, Santa Barbara, CA.

2 Zinke, A. (1996) Technologie der Glasverschmelzungen, Technisch-physikalische Monographien Band 12, Akademische Verlagsgesellschaft Geest und Portig K-G, Leipzig.

3 HVG (1995) *Fügen von Glas: HVG Fortbildungskurs*, Verlag der Deutschen Glastechnischen Gesellschaft, Frankfurt am Main.

4 Boettge, B., Dresbach, C., Graff, A., and Petzold, M. (2008) Mechanical characterization and microstructure diagnostics of glass frit bonded interfaces. *Electrochem. Soc. Trans.*, **16** (8), 441–448.

2
Wafer Bonding Using Spin-On Glass as Bonding Material
Viorel Dragoi

Among the most used wafer bonding processes, adhesive wafer bonding attracts a high level of interest due to its specific features:

- Compensation of surface defects: if wafer bonding is used to join patterned wafers there is a risk of generating surface defects during wafer preparation (e.g., scratches, local high roughness), which may have an impact on bonding results.

- Compensation of particle contamination: small amounts of particles remaining on the surface can be incorporated in the bonding layer if the particle diameter is smaller than the layer thickness.

- Low processing temperature compared to fusion bonding (<300 °C).

- Relatively simple process flow (standard sol–gel process) compared with other thin-film deposition methods requiring vacuum deposition (e.g., evaporation, sputtering).

Due to the large variety of polymer materials designed for various microelectronics processes, the selection pool for bonding materials is very broad. However, there is only a small part of this large amount of materials which fulfills the requirements for wafer bonding.

This chapter reviews some basic aspects related to the use of spin-on glass (SOG) materials as bonding layers in adhesive wafer bonding. An SOG-based adhesive wafer bonding process is presented [1].

2.1
Spin-On Glass Materials

Spin-on glasses represent a category of materials used mainly as interlevel dielectrics in microelectronic applications or simply for surface planarization. Typically the material is applied to a substrate using a sol–gel process flow consisting of a deposition step (spin coating or spray coating) followed by a baking step for

Handbook of Wafer Bonding, First Edition. Edited by Peter Ramm, James Jian-Qiang Lu, Maaike M.V. Taklo.
© 2012 Wiley-VCH Verlag GmbH & Co. KGaA. Published 2012 by Wiley-VCH Verlag GmbH & Co. KGaA.

removal of solvents and finally a thermal curing step. After cure, SOG films should show good uniformity, crack resistance, low stress, high thermal stability, and good adhesion to the substrate. SOGs can fill narrow spaces in pre-metal and metal levels simultaneously planarizing the surfaces.

The term "surface planarization" may have different meanings depending on the application area: microelectronics or wafer bonding.

In microelectronics the starting semiconductor wafers for device fabrication are ideally flat or planar prior to undergoing various fabrication steps. This series of material growth, deposition, and patterning steps decreases the flatness or planarity of the wafer surface.

The recent trends towards device fabrication that involve an increasing number of metal layers aggravate the problem of wafer nonplanarity. The shift to narrower and narrower metal lines has also prompted the emergence of thicker metal lines in order to meet current device requirements.

At least two major problems arise from the decrease in surface flatness: (i) full step coverage of fine lines so that no breakage in line continuity occurs becomes more difficult and (ii) fine-featured pattern imaging (e.g., in photolithography steps) is more difficult, if not impossible.

There are two categories of planarization techniques: local planarization and global planarization. Local planarization refers to techniques that increase planarity over short distances. Global planarization consists of techniques that decrease long-range variations in wafer surface topology, especially those that occur over the entire image field of the stepper.

Planarization techniques include: (i) oxidation; (ii) chemical etching; (iii) deposition of oxide or nitride layers; (iv) deposition of low-melting-point glass layers; (v) use of polyimide films; (vi) use of resins and low-viscosity liquid epoxies; (vii) use of SOG materials; (viii) sacrificial etch-back; and (ix) as a stand-alone step or associated with layer deposition techniques, mechanical–chemical polishing of wafers.

The use of SOG materials for surface planarization in microelectronics is not new. The early SOG films were typically silicates. Today's mainstream technologies using multiple interconnect levels employ mainly siloxane SOG films. Siloxanes became mainstream in SOG technology due to their excellent crack resistance and ability to fill gaps as small as 0.1 µm while ensuring complete local planarization. Phosphosilicates can be used as well in order to further improve crack resistance.

Some SOG materials can also be used for new applications as adhesive layers for wafer bonding. One of the main challenges of using such materials for wafer bonding is the ability to obtain a perfectly flat surface of the SOG layer across the entire area that is desired to be bonded.

If in microelectronics manufacturing the term "planar surface" refers to areas as large as a stepper field area which would show a small waviness, for wafer bonding it would refer to a perfectly flat surface across the entire wafer area.

2.2
Wafer Bonding with SOG Layers

An example of wafer bonding process development using a silicate SOG material as bonding layer is now presented. The target for this process was for it to be applicable for bonding Si and GaAs substrates. The challenge in this case was the thermal mismatch illustrated by the significant difference in thermal expansion coefficient (TEC$_{GaAs}$ ≈ 2TEC$_{Si}$). It was experimentally determined that a process temperature of 200 °C would be acceptable for the final application if the bond strength achieved would be enough such that a GaAs wafer would survive mechanical thinning (grinding).

2.2.1
Experimental

In order to check the concept of wafer bonding via SOG intermediate layers, the process was first applied to an Si/Si system. The experiments were carried out on p-type, (100)-oriented Si wafers, 100 and 150 mm in diameter. The Si wafers were first cleaned using standard RCA 1 and RCA 2 procedures, followed by a deionized water rinse and drying in nitrogen flow.

Two different processes were used for SOG deposition on substrates: spin coating and spray coating. The most commonly used process is spin coating. In this approach, a liquid precursor is dispensed on the wafer (static – the wafer is not rotated during dispense; or dynamic – the wafer is slowly rotated during dispense). The wafer is then rotated at low speed for uniform spread of the material on the entire wafer surface. Finally the wafer is rotated at high speed to remove excess material and for layer drying. The film thickness is highly dependent on the rotation speed used in the drying step.

Spray coating is an alternative to spin coating and can be especially useful for high-topography wafers, noncircular substrates, or substrates with etched trough via holes. The film thickness achieved in the spray-coating process is dependent on the amount of dispensed material as well as on the programmed motion of the dispense arm (stroke time and arm speed) and spinner chuck (rotation speed). While the wafer rotates slowly at 30–100 rpm, the atomizing nozzle starts spraying right outside the wafer area and moves over the wafer center to the other side. To get a uniform layer thickness over the whole wafer, the speed of the nozzle increases towards the wafer center and decreases from there on outwards according to a controlled speed profile.

SOG films were deposited by spin coating using the following steps:

- dispense of the liquid SOG precursor on the Si surface;
- obtaining the gel film by spinning at low rotation speeds;
- spin-drying of the solvent for solvent evaporation;
- pyrolysis of the gel film onto a hot plate in air at temperatures between 150 and 180 °C.

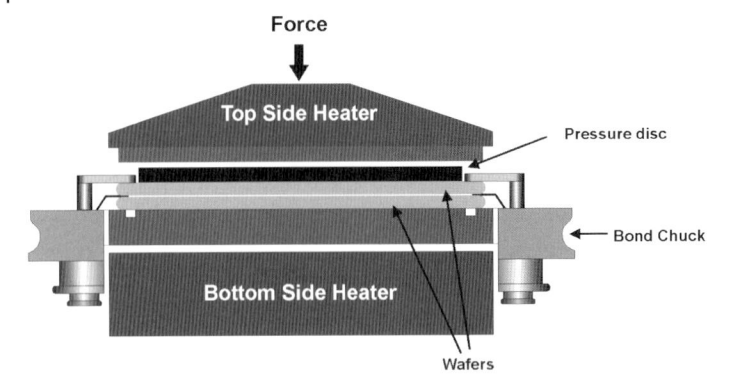

Figure 2.1 Bond chamber setup for adhesive bonding.

For spray coating, the SOG material liquid precursor was diluted with ethanol and the process was optimized for different layer thicknesses.

The thickness of the SOG layers was measured with an Alpha-Step 500 profiler from KLA Tencor.

The wafer bonding step was performed in an EVG®520 semi-automated bonding system. The experimental setup used is shown schematically in Figure 2.1. The two wafers were loaded clamped together onto the bond chuck, with spacers keeping them separated. Mechanical alignment of wafer edges was performed using a chuck with alignment pins. The bond chamber was first evacuated down to a few tens of millibars in order to avoid air being trapped between the wafers when moving them into contact. The wafers were then brought into contact. After contacting the wafers the chamber was purged with nitrogen and the two heaters started to heat the wafer stack. A force of up to 7 kN could be applied using a piston mounted together with the top heater.

SOG-coated Si wafers were bonded at room temperature with clean blank Si wafers using the bonding procedure described above. After bonding, the pairs were annealed at temperatures starting from 200 °C in order to optimize the process.

Bonded interface quality was investigated using infrared (IR) transmission with an EVG®20 infrared inspection station and scanning acoustic microscopy (SAM) with a Sonoscan D9000 system. The surface energy was evaluated using the crack-opening method [2] and by tensile testing (Si/GaAs bonds).

2.2.2
Wafer Bonding with Silicate SOG Layers

Spin- and spray-coating methods were evaluated for a silicate SOG material. The film thickness for spin coating varied in the range 280–125 nm for a spin-drying speed range 2000–5000 rpm (500 rpm increments).

Spray-coating tests showed relatively good uniformity for thicker SOG layers (~600 nm) but bond tests showed for spray-coated layers that the uniformity was

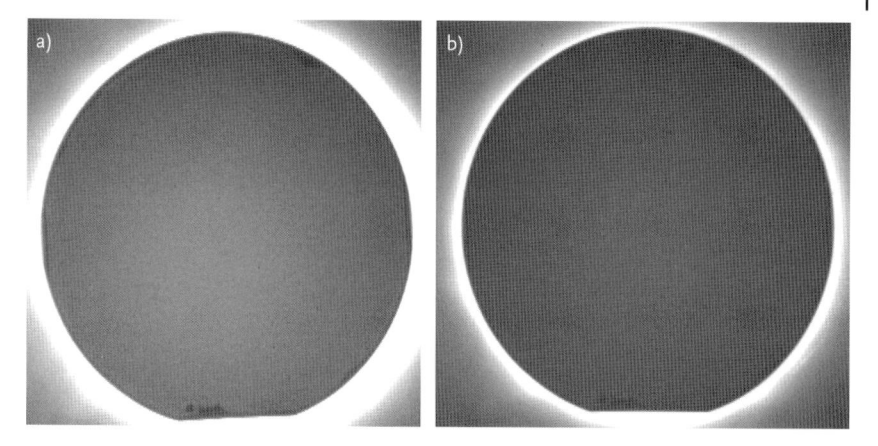

Figure 2.2 IR transmission images of a bonded wafer pair (100 mm in diameter): (a) room temperature bonded; (b) annealed at 200 °C for 10 h.

not enough for good-quality bonding. In order to improve the quality of the spray-coated layers additional work would be required in studying various precursor dilutions and deposition conditions. As there was no specific need for spray coating in this application (it was investigated here only for comparison with standard spin coating), this method was not further pursued due to the good results obtained for spin coating.

In order to avoid the occurrence of interface defects, the SOG deposition process was optimized for obtaining defect-free layers. Figure 2.2 shows IR images of a bonded wafer pair immediately after room temperature bonding and after annealing at 200 °C for 10 h. The only defects observed are located around the laser marks of the two wafers (close to the wafer flat), which in this case are in the bonded interface. It can be observed that for one wafer there is an unbonded area around the marks (SOG-coated wafer – "shadow" defect in the SOG layer) while for the other the marked numbers (blank Si) can be clearly observed.

It can be observed that no additional defects are generated, and those existing after room temperature pre-bonding do not increase in size, showing there is no outgassing at the bonded interface. This test reveals the importance of the SOG layer quality for successful bonding.

The bond strength measured after room temperature bonding was found to be in the range 0.4–0.45 J m^{-2}, about 3–4 times higher than for normal hydrophilic Si/Si bonding which is about 0.1–0.15 J m^{-2} [3]. This very high surface energy can be the result of the chemistry at the SOG–Si interface, which is very different from standard Si–Si bonding. IR spectroscopy measurements (Figure 2.3) revealed the existence of CH_3 radicals in the SOG films baked at temperatures below 200 °C. The residual organic radicals remaining from the solvent can easily hydrolyze with the water molecules adsorbed at the GaAs surface and create strong covalent bonds. Even after a baking procedure at 150 °C for 5 min the SOG surface also contains silanol groups, which may generate adhesion at room temperature due

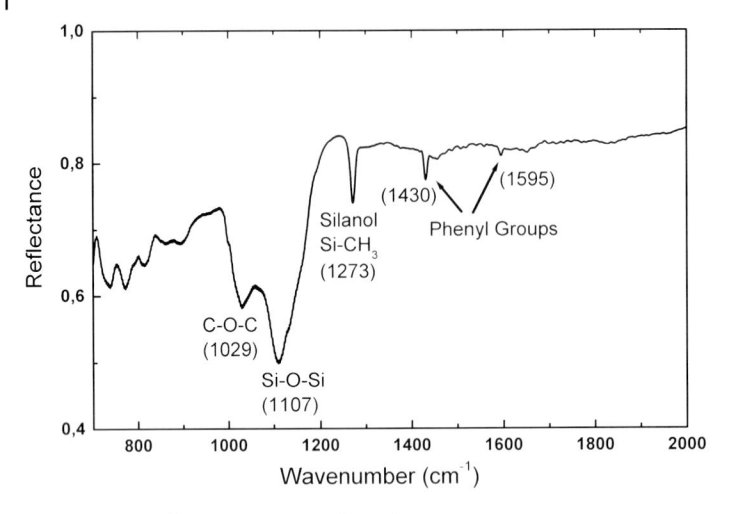

Figure 2.3 IR reflection spectrum of SOG layer.

to their ability to form chemical bonds with molecules from the other surface directly or via water molecules.

Thermal annealing optimization was performed by measuring the dependence of the surface energy on the annealing time at 200 °C and by trying different temperatures. To allow the diffusion of the gaseous products generated at the interface, the heating/cooling rate was less than 1 °C min^{-1}. A low temperature ramping is also required for the final application (Si/GaAs bonding) in order to avoid thermal shocks which may result in wafer breakage.

Surface energy increased to 1.8 J m^{-2} for annealing at 200 °C for 6 h, and after 18 h the surface energy reached 2.3 J m^{-2}, a value close to the fracture energy of Si. An increase of the annealing temperature to 300 and 400 °C produced no increase of the surface energy. Further experiments showed that the maximum surface energy value is obtained after a thermal annealing of 10 h at 200 °C.

The influence of the SOG thickness on the surface energy was studied by measuring the energy at room temperature and after 200 °C/10 h annealing for samples having SOG layers with various thicknesses. A slight increase of the surface energy is observed with increasing SOG thickness for the room temperature bonded pairs (Figure 2.4a). After completion of thermal annealing, no influence of the SOG layer thickness on the surface energy could be observed (Figure 2.4b); all measured values were found to be within the measurement error range.

In order to allow for a qualitative estimation of the bonding strength, bonded pairs were submitted to a mechanical thinning procedure consisting of grinding followed by chemical–mechanical polishing (CMP). One of the Si bonded wafers was thinned down to about 30 μm by grinding and then polished down to 5 μm by CMP. The harsh mechanical thinning process does not affect the quality of the bonded interface.

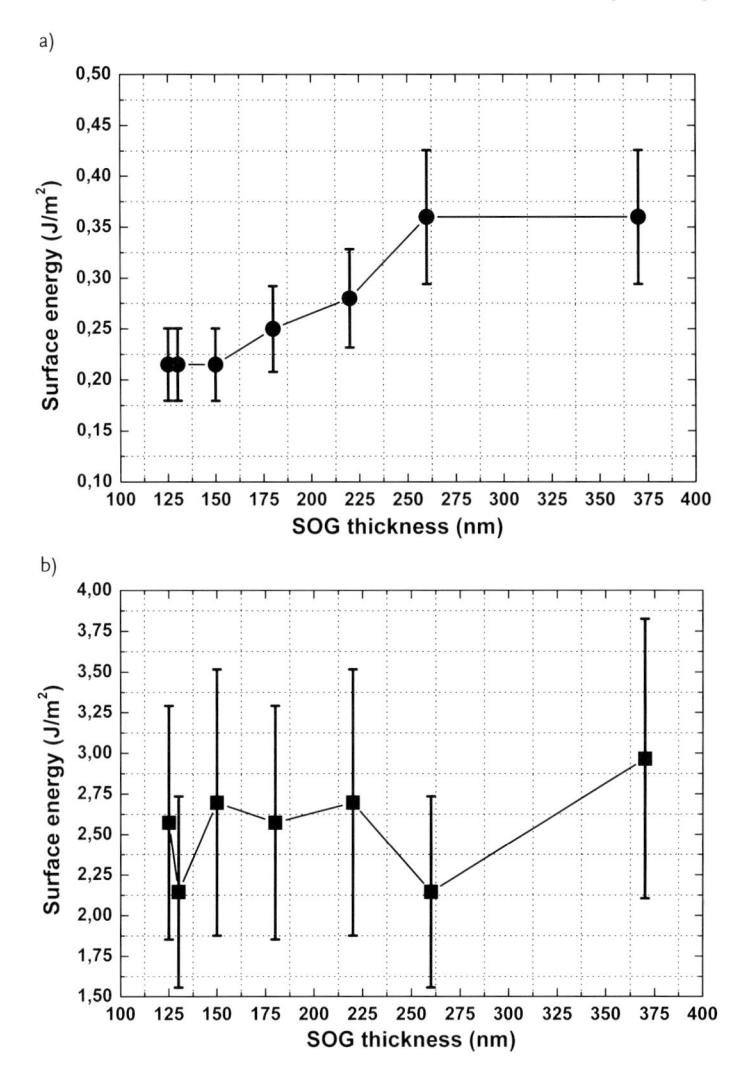

Figure 2.4 Surface energy versus SOG thickness: (a) room temperature bonded wafers; (b) after annealing at 200 °C for 10 h.

The process described above was further applied for the fabrication of GaAs/Si heterostructures. The main issue in this case is the thermal mismatch of the two materials. Also, GaAs requires processing temperatures below 450 °C in ambient air atmosphere or vacuum, or As-rich atmosphere at higher temperatures. The use of a temperature of only 200 °C eliminates the complications of using special atmospheres based on toxic gases requiring special handling.

Silicon (p-type, (100)-oriented) wafers were coated with SOG layers and then bonded with semi-insulating (100)-oriented GaAs wafers. The surface energy for

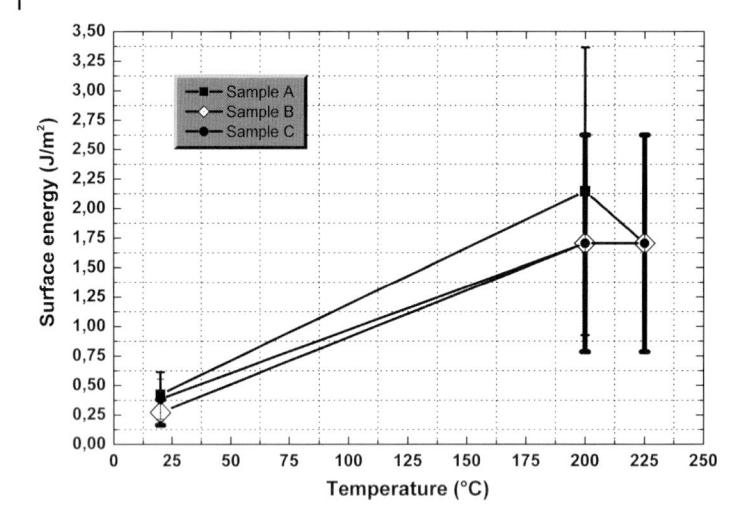

Figure 2.5 Surface energies measured for three different GaAs/Si bonded pairs.

room temperature bonded samples was about $0.4\,\mathrm{J\,m^{-2}}$, also high if compared to Si/Si hydrophilic bonding.

The bonded pairs were then annealed at 200 and 225 °C for 10 h (increasing the temperature above 225 °C will result in a high stress which leads to debonding and GaAs shattering). Figure 2.5 presents the surface energy for three identically processed GaAs/Si bonded pairs. After annealing at 200 °C, the surface energy reached values of about $2\,\mathrm{J\,m^{-2}}$. Annealing at 225 °C or increasing the annealing time did not result in any significant increase of the surface energy.

Due to the high thermal mismatch of the two materials, the behavior of the bonded pairs during annealing was investigated by *in situ* measurement of the bow. A large bow developed during heating, reaching 500 μm at 200 °C (with "zero" considered to be the room temperature bow value). On cooling the bonded pair from 200 °C to room temperature, the bow values decreased back to zero following exactly the same path as during heating, showing no hysteresis (Figure 2.6). On repeating the heating/cooling cycle for the same bonded pair, the bow values were found to be the same. No degradation of the bonded interface was observed during this heating/cooling cycling.

This behavior is induced by the SOG layer, which can compensate the high thermally induced mechanical stress. To determine the maximum temperature at which the GaAs/Si heterostructure can be heated, the bow was measured for the heating of the annealed bonded wafers pairs. At temperatures higher than 280 °C the wafers debonded and/or broke. Similar to the Si/Si bonding via SOG intermediate layers, some GaAs/Si bonded pairs were submitted to a mechanical thinning process. The GaAs wafers were thinned by grinding down to 40 μm and then by CMP to 10 μm. This mechanical thinning process did not affect the interface quality, and after thinning the GaAs/Si bonded pairs were heated at 450 °C without any debonding or defect generation. The contribution of the thin wafer to the

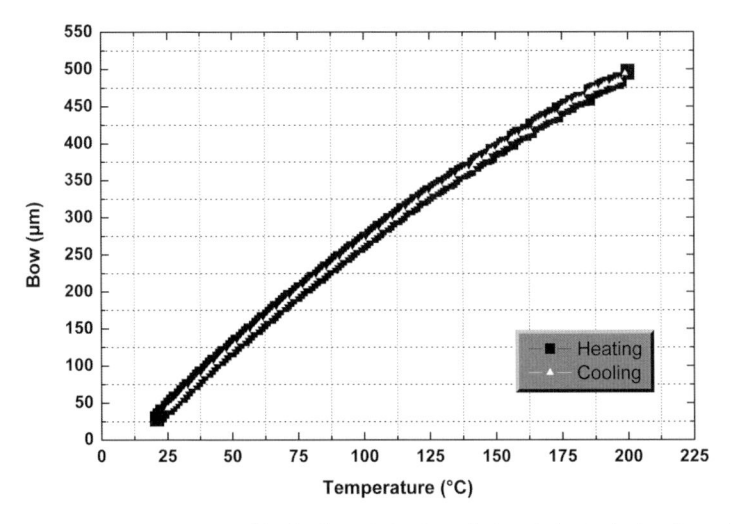

Figure 2.6 Bow measured *in situ* for a GaAs/Si wafer bonded pair during thermal annealing.

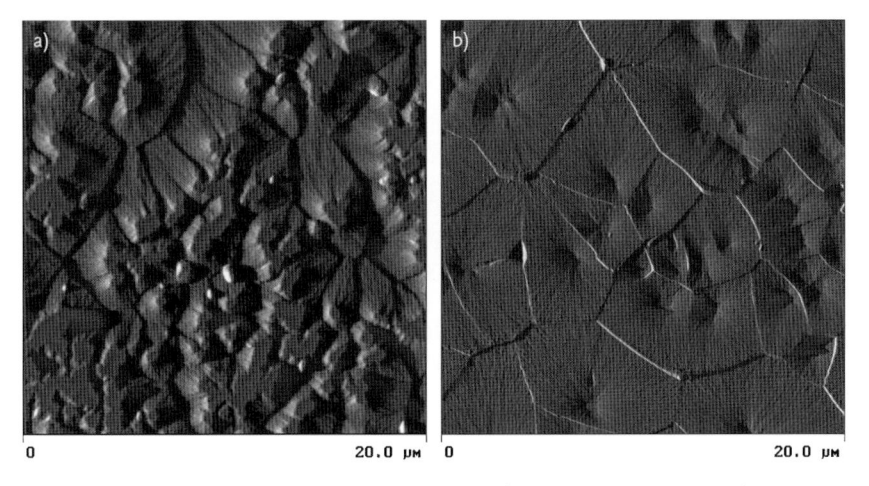

Figure 2.7 AFM images of the two surfaces resulting after separation during tensile tests: (a) SOG on Si; (b) SOG on GaAs.

induced stress became less important and the bonded pair could stand higher temperatures than the initial full-thickness wafers.

Tensile testing was used for bond strength quantification. For this test, $6 \times 6\,mm^2$ specimens were diced from bonded wafer pairs, which were then loaded in a pulling machine (pull force perpendicular to the bonded interface. After the two wafer pieces were separated it was determined visually that separation occurred due to a crack in the SOG layer and not due to SOG delamination. Further atomic force microscopy (AFM) investigation of the two surfaces resulting after breakage confirmed SOG fracture as the root cause of the separation (Figure 2.7). This result

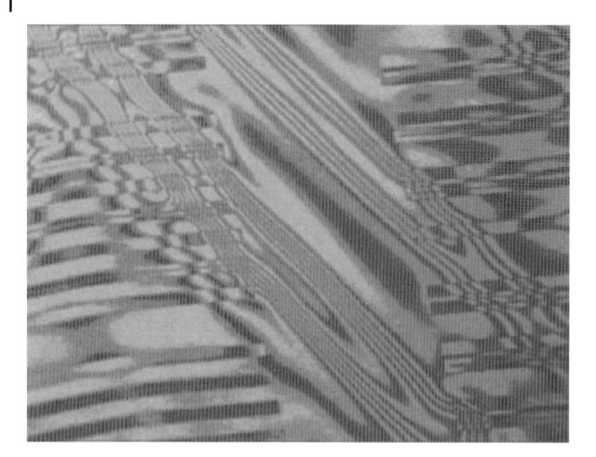

Figure 2.8 Optical photograph of the surface of a planarization SOG layer spin coated on a blank GaAs wafer.

shows the bonding strength is higher than the bulk fracture strength of the SOG material.

2.2.3
Wafer Bonding with Planarization SOG

Despite its flowing properties induced in order to enhance the filling ability of SOG material (crucial for its main function), an integrated circuit (IC) planarization SOG may not be usable for wafer bonding even when deposited on the flat surface of a prime wafer. Figure 2.8 shows a photograph of a planarization SOG spin-coated on a prime-grade GaAs wafer and baked at 180 °C. The alternating lines visible in Figure 2.8 are generated by the surface waviness of the SOG layer. Even when coated on a nonpatterned surface, the SOG surface was found not to be flat. In such a case the full area bonding of the two surfaces is not possible. AFM studies were performed on SOG surfaces similar to that in Figure 2.8. The thickness variation of the SOG layer was found to be up to 80 nm for an SOG thickness of ~500 nm. The average microroughness between waves was around 0.5 nm. Similar results were obtained for bonding Si wafers.

Figure 2.9 shows a GaAs/InP (nonpatterned) wafer pair bonded with this type of SOG as adhesive layer. The two wafers are only partially bonded, in some dot-like small areas (dark gray contrast in Figure 2.9), probably only on top of the wave topography of the SOG layer.

A typical procedure for planarization is to perform a CMP planarization step on surfaces with topography too high to be tolerated for wafer bonding. In the particular case of the planarization SOG used, this method was not considered suitable due to the fact the SOG is processed below the curing temperature (<400 °C) and CMP was found not to be effective due to material softness. SOG processing

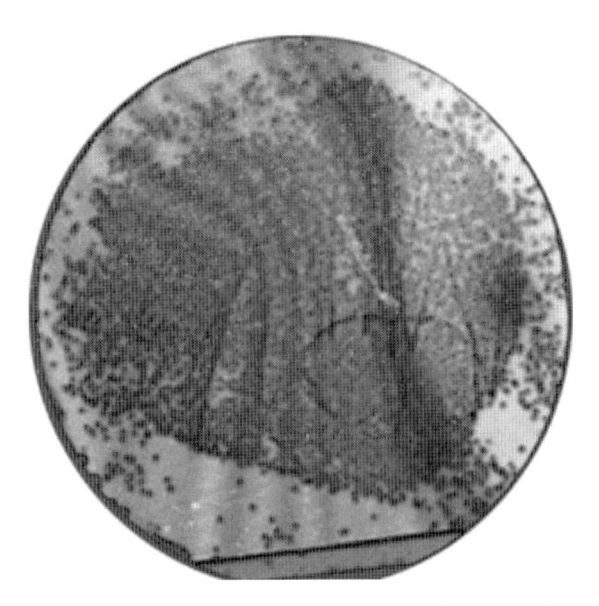

Figure 2.9 SAM image of a GaAs/InP pair bonded with a planarization SOG layer.

above 450 °C is not desirable due to the main target of the process development being to set up a low-temperature process. Additionally the thermal curing process of planarization SOG is based on a well-defined thermal profile which would add significant time to the process flow.

The behavior described here is common to various types of materials produced by different vendors. This makes the selection of SOG materials suitable for wafer bonding very difficult, as there are many SOG materials available but only a few types give good bonding results.

2.2.4
Applications of Adhesive Wafer Bonding with SOG Layers

The process development described above was used for an application requiring the transfer of an epitaxial structure grown on a GaAs substrate to a CMOS wafer with the final goal of on-chip optical interconnect fabrication [4]. The bonding process for this application had to be compliant with two requirements:

- low-temperature process imposed by the use of CMOS wafer (<400 °C);
- thermal mismatch of the two substrates (Si for CMOS and GaAs).

The process described in the previous sections was found to be compatible with this application.

In the first step, a CMOS wafer was planarized for direct bonding. Planarization consisted of an oxide chemical vapor deposition followed by CMP. After the oxide surface was ready for bonding, a thin SOG layer (~250 nm) was deposited as a bonding layer. The bonding process was performed as for blank Si and GaAs wafers and thermal annealing was performed at 200 °C.

After the wafer bonding step, the GaAs substrate was removed by etching and the epitaxial structure transferred on the CMOS wafer was patterned for further processing.

In this type of application the SOG material is used for its low-temperature processing and for its ability to provide stress compensation between thermally mismatched substrates.

Besides low-temperature processing, the use of SOG bonding layers is also attractive for nonstandard applications in which the simplicity of SOG processing may be good from both technical and cost perspectives.

Examples of various uses of SOG materials have been reported in the literature: a "compliant layer" between Si and Si_3N_4 [5], a low-temperature silicon-on-insulator manufacturing process with SOG replacing the thermal silicon oxide [6], as well as an intermediate layer allowing anodic bonding of two Si substrates [7].

Due to the increasing amount of wafer bonding-based applications, SOG manufacturers have started to produce materials optimized for wafer bonding. However, not all of these materials are field-proven for wafer bonding applications, and process development is required when starting a new application.

2.2.5
Conclusion

Adhesive wafer bonding based on SOG layers was investigated. The main benefits expected from the use of such bonding layers are low-temperature processing (<400 °C), "clean" process (CMOS-compatible), elastic behavior to allow stress compensation, and compatibility with a wide range of substrate materials.

Tests were performed first using Si wafers and then applied to various compound semiconductors wafers.

Two general types of SOG materials were evaluated: a standard planarization SOG used in IC manufacturing and a silicate SOG material with thermal, optical, and mechanical properties of the cured material similar to those of thermally grown SiO_2.

The silicate SOG showed good-quality surfaces when spin- or spray-coated on a flat surface and baked in the temperature range 150–300 °C. During process optimization, an optimum baking temperature of 180 °C immediately after substrate coating was determined and a bonding temperature of 200 °C.

Bond tests using Si/Si pairs with a thin SOG layer coated on one of the two wafers were performed for process condition optimization. Process results were evaluated in terms of defects (IR transmission and SAM) and bond strength (crack-opening method and tensile tests). The surface quality of the bonding layers was high enough to allow spontaneous bonding under ambient conditions and defect-free interfaces after thermal annealing. The bond strength obtained after thermal annealing was very high, in the range of the bulk fracture strength of the substrates. Tensile tests revealed no delamination but a fracture into the bonding layer, showing SOG fracture strength was less than substrates fracture strength.

A final process was applied to bond Si to GaAs wafers in order to check bonding layer behavior for substrate combination with a thermal expansion coefficient ratio of about 2. Silicate SOG exhibited an elastic behavior allowing the two substrates to bend during thermal annealing (bonded pair bow reached about 500 µm) and recover their flat shape once cooled to room temperature. The contribution of wafer thickness to bonded wafer behavior during heating was proven by thinning the GaAs wafer from 625 µm down to 10 µm after annealing at 200 °C. In this case the bonded pair could survive thermal cycling from room temperature to 450 °C, significantly higher compared to "thick" wafer behavior, when GaAs breakage was typically observed at temperatures below 300 °C.

The process was successfully applied for bonding CMOS wafers to compound semiconductor wafers at a maximum temperature of 200 °C [3].

The planarization SOG evaluated showed behavior significantly different from that of the silicate material. Even when coated onto a blank surface this material showed certain waviness of the surface after a baking step. This also showed the major difference between the planarization concept in IC manufacturing, where waviness is acceptable, and the wafer bonding requirement of a perfectly flat surface.

CMP was not effective in this case due to the softness of the SOG material which was only baked for solvent removal (~180 °C) and not fully cured at temperatures greater than 400 °C. Due to the aim of a low-temperature process, fully curing the SOG layer was not an option in this case.

As a result of the surface waviness the bond between the two substrates occurred only partially, proving the material is not suitable for the bonding process.

The results presented in this chapter were obtained using a very small selection of SOG materials from the large variety available. However, the main characteristics of the materials presented here apply to a large number of different SOG materials from various vendors, as the chemistries used are similar in principle.

The choice of a bonding layer material suitable for a specific application is a relatively difficult and requires, in most cases, experimental evaluation and process optimization considering the boundary conditions of each application.

References

1 Dragoi, V., Alexe, M., Reiche, M., Radu, I., Thallner, E., Schaefer, C., and Lindner, P. (2001) Si/GaAs heterostructures fabricated by direct wafer bonding. *MRS Proc. Ser.*, **681E**, I.5.3.1.

2 Maszara, W.P., Goetz, G., Caviglia, A., and McKitterick, J.B. (1988) Bonding of silicon wafers for silicon-on-insulator. *J. Appl. Phys.*, **64**, 4943.

3 Tong, Q.-Y. and Gösele, U. (1999) *Semiconductor Wafer Bonding: Science and Technology*, John Wiley & Sons, Inc., New York.

4 Georgakilas, A., Deligeorgis, G., Aperathitis, E., Cengher, D., Hatzopoulos, Z., Alexe, M., Dragoi, V., Gösele, U., Kiriakos-Blitzaros, E.D., Minoglou, K., and Halkias, G. (2002) Wafer-scale integration

of GaAs optoelectronic devices with standard Si integrated circuits using a low temperature bonding procedure. *Appl. Phys. Lett.*, **81** (27), 5099.

5 Yamada, A., Kawasaki, T., and Kawashima, M. (1987) Bonding silicon wafer to silicon nitride with spin-on glass as adhesive. *Electron. Lett.*, **23** (7), 314.

6 Yamada, A., Kawasaki, T., and Kawashima, M. (1987) SOI by wafer bonding with spin-on glass as adhesive. *Electron. Lett.*, **23** (1), 33.

7 Quenzer, H.J., Dell, C., and Wagner, B. (1996) Silicon-silicon anodic-bonding with intermediate glass layers using spin-on glasses, in *Proceedings of the 9th Annual International Workshop on Micro Electro Mechanical Systems*, IEEE, New York, pp. 272–276.

3
Polymer Adhesive Wafer Bonding

Frank Niklaus and Jian-Qiang Lu

3.1
Introduction

In polymer adhesive bonding, an intermediate polymer layer is used to create a bond between two surfaces to hold them together. In most commonly used polymer adhesive wafer bonding processes, a well-defined and defect-free polymer layer is applied to one or both of the wafer surfaces to be bonded. After joining the wafer surfaces, pressure is applied to force the wafer surfaces into intimate contact. The polymer is then converted from a liquid or viscoelastic state into a solid state, typically done by heating the polymer [1, 2].

The key advantages of polymer adhesive wafer bonding are the relatively low bonding temperatures (between room temperature and 450 °C, depending on the polymer material), the insensitivity to the topography of the wafer surfaces, the compatibility with standard CMOS wafers, and the ability to join practically any type of wafer materials. Special wafer surface treatments such as planarization and excessive cleaning are not required. Structures and particles at the wafer surfaces can be tolerated and accommodated to some extent by the polymer layer. While polymer adhesive wafer bonding is a comparatively simple, robust, and low-cost process, concerns need to be considered such as limited temperature stability and limited data about long-term stability of many polymer adhesives in demanding environments. Moreover, polymer adhesive wafer bonding does typically not provide bonds that are hermetically sealed towards gases and moisture [1–3].

Although successfully used to join various similar and dissimilar materials in many industries including the airplane, aerospace, and automobile manufacturing industries, polymer adhesive bonding has only recently been widely applied for bonding of semiconductor wafers. The development of reliable and high-yield wafer bonding processes has made polymer adhesive wafer bonding a generic and in some cases enabling bonding technique for a variety of new applications, specifically for applications that require low bonding temperatures. Polymer adhesive wafer bonding is widely used for bonding wafers to temporary handle wafers, enabling the handling and support of thin wafers or devices during and after the grinding or etching processes [4] and during manufacturing processes

Handbook of Wafer Bonding, First Edition. Edited by Peter Ramm, James Jian-Qiang Lu, Maaike M.V. Taklo.
© 2012 Wiley-VCH Verlag GmbH & Co. KGaA. Published 2012 by Wiley-VCH Verlag GmbH & Co. KGaA.

for through-substrate vias [5]. Other applications of polymer adhesive wafer bonding include the fabrication of three-dimensional integrated circuits (ICs) [6], heterogeneous integration of microelectromechanical systems (MEMS) with ICs [7–10], heterogeneous integration of photonics (e.g., III–V materials) with silicon-based waveguide or IC wafers [11], heterogeneous integration of PZT layers [12], and manufacturing of thin-film solar cells [13] and laser systems [14]. Polymer adhesive wafer bonding is also used for packaging of CMOS imaging sensors [15, 16], fabrication of microcavities in packaging applications [17–19], manufacturing of radiofrequency MEMS devices [20], manufacturing of liquid crystal on silicon components [21], and fabrication of bio-MEMS and micro total analysis systems [22]. Details of many of these applications are presented in other chapters of this book. The present chapter reviews the state of the art of wafer bonding technologies using synthetic organic polymer layers as intermediate adhesive. An elaborate list of references concerning polymer adhesive wafer bonding is given in [2].

3.2
Polymer Adhesives

3.2.1
Polymer Adhesion Mechanisms

In polymer adhesive wafer bonding, a polymer adhesive is placed between the pair of wafers to be bonded, bearing the forces involved to hold the wafer surfaces together. Like most bonding techniques, polymer adhesive wafer bonding is based on the fact that atoms and molecules adhere to each other when they are brought into sufficiently close contact. The cohesion of atoms or molecules within polymers and the adhesion of atoms or molecules between polymers and wafer materials are ensured by one or more different basic intermolecular bond and interaction types: (i) covalent bonds, (ii) ionic bonds, (iii) dipole–dipole interactions (including hydrogen bonds), and (iv) van der Waals interactions. These are all chemical bonds that are based on electromagnetic forces.

To form covalent or van der Waals bonds, the atoms of two opposing surfaces must be less than 0.3–0.5 nm apart [2]. The resulting bonds have varying energies that depend on the surface materials and the distance between the atoms of the surfaces, but none of the bonds extend further than 0.5 nm [1, 23]. Macroscopically flat surfaces, such as the surfaces of polished silicon (Si) wafers, have a root mean square roughness of 0.3–0.8 nm. Nevertheless, the profile depth (peak to valley) of these surfaces is several nanometers, which typically prevents bonding over larger surface areas. Figure 3.1a shows a schematic of the contact interface of two solid surfaces that are macroscopically flat, but exhibit surface roughness on a microscopic level. In order to bring two surfaces in sufficiently close contact to achieve bonding, at least one of the surfaces must deform to fit the other. This deformation may be accomplished by plastic or elastic deformation, by diffusion of a solid material, or by wetting of a surface with a liquid material. In bonding with inter-

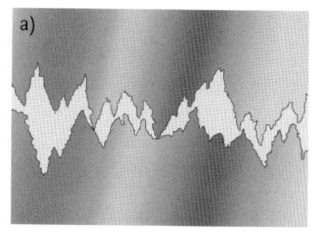

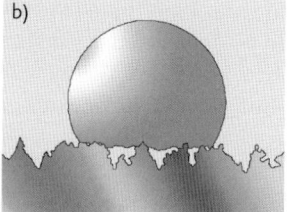

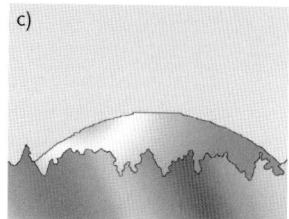

Figure 3.1 (a) Contact interface of two macroscopically flat solid surfaces. (b) Boundary layer of a solid surface and a liquid that does not wet the surface. (c) Boundary layer of a solid surface and a liquid that wets the surface [2].

mediate polymer adhesives, it is naturally the polymer that deforms and adapts to the topography of the wafer surface.

Several theories for polymer adhesive bonding have been proposed, including (i) adsorption theory, (ii) chemical bonding, (iii) diffusion theory, (iv) electrostatic attraction, (v) mechanical interlocking, and (vi) weak boundary layer theory. The adsorption theory relates adhesion to the interatomic and intermolecular attractive forces and has found substantial experimental support. A detailed discussion of the proposed alternative theories for adhesion can be found in [1, 24]. In the adsorption theory, the wetting of a surface by the adhesive is a key factor in determining the strength of the adhesive bond. For atoms and molecules to adhere, they must be brought to a distance of less than 0.5 nm. When an intermediate polymer adhesive is used to join two solid-state surfaces, the polymer adhesive deforms to fit the surfaces to be bonded. Polymer adhesives are typically in a liquid or semi-liquid phase during part of the bonding process and wet the surfaces to be bonded by flowing into the troughs of the surface profile. The liquid polymer adhesive must then harden into a material that is capable of bearing the forces involved to hold the surfaces together. Wetting of the surfaces by the liquid or semi-liquid polymer adhesive is critical in bonding. Figure 3.1b shows a schematic of a liquid that does not wet the surface and Figure 3.1c shows a liquid that does wet the surface. For wetting to occur, the solid surface must have a greater surface energy than the liquid. The surface energy is a result of unbalanced cohesive forces at the material surface. A higher cohesive force between the atoms or molecules of a material correlates with a higher surface energy. A liquid can wet a solid material only if the liquid has a lower surface energy than the solid. A detailed discussion of surface energy measurements and surface wettability can be found in [24].

The degree of wetting of a surface with a liquid polymer adhesive can be reduced by surface contaminants (such as weakly adsorbed organic molecules) or condensed moisture, and influenced by the microscopic surface profile or dust particles. Clean and contaminant-free surfaces can be achieved with cleaning procedures using solvents, oxidants, strong acids, or bases. Surface pretreatment with adhesion promoters can significantly improve the wettability of a surface. Adhesion promoters typically consist of a very thin coating of a few monolayers of a material that bonds well to the surface on one side and that enhances the bonding of the polymer adhesive to the surface on the other side. Specific adhesion promoters are often recommended by the material suppliers for certain combinations of surface material and polymer adhesive. The more completely the polymer adhesive flows into and fills the troughs of a surface profile, the better the resulting bond quality and the long-term stability of the bond. Polymer adhesives that have low viscosity and low shrinkage during hardening generally achieve better filling of the troughs of a surface profile which decreases the amount of unfilled space at the bond interface. Small molecules such as water or gas molecules can creep or diffuse in the unfilled space at the boundary layer between the adhesive and the surface and can decrease the bond energy or affect the materials at the boundary layer.

3.2.2
Properties of Polymer Adhesives

Polymers are large molecules (macromolecules) consisting of large numbers of linked small molecules (monomers). The joining process of the monomers is called polymerization. The molecular chains, typically 0.2–1 nm wide and up to several hundreds of nanometers long [25, 26], and their internal structure determine the specific properties of a polymer. Polymers can be placed into the four broad material classes: (i) thermoplastic polymerss, (ii) thermosetting polymers, (iii) elastomers, and (iv) hybrid polymers. Thermoplastic polymers solidify by cooling and can be re-melted. Thermosetting polymers undergo crosslinking between the polymer chains to form a three-dimensional network and, unlike thermoplastics, cannot be re-melted or reshaped. However, before they are crosslinked they are thermoplastics, semi-liquids, or liquids and they do flow for a short time when heated the first time to achieve crosslinking between the polymer chains. The distinguishing characteristics of elastomeric materials are their ability to sustain large deformations (5 to 10 times the unstretched dimensions) with relatively low stresses and their ability to spontaneously recover their original shape without rupturing. Hybrid polymers are alloys and blends of polymers from the three previous classes, which form new materials whose properties and characteristics can be quite different from those of the individual components. In principle, polymers from all four material classes can be used as adhesive materials [25, 26].

A polymer adhesive must exist in a liquid, semi-liquid, or viscoelastic phase during the bonding process to achieve sufficiently close contact with the surfaces to be bonded. The polymer adhesive must then transform into a solid material to bear the forces involved and achieve a lasting bond. Three basic ways for polymer adhesives to harden and transform from a liquid phase into a solid phase are [1]:

- Polymers that are dissolved in water or in solvents harden when the water or solvents are evaporated. Polymer adhesives based on this principle are called physical drying polymer adhesives.

- Thermoplastic polymers melt when heated to their melting temperature and solidify upon cooling below their melting temperature Polymer adhesives based on this principle are called hot-melts.

- Polymer precursors cure (polymerize or crosslink) by chemical reactions that form larger molecules or crosslinked molecular chains, with the polymer precursor either in a liquid phase (e.g., resins) prior to curing or transforming from a solid into a liquid phase for some time during the curing process. Depending on the specific polymer, the curing process can be triggered or amplified by various mechanisms, such as:

    - mixing of two or more components (e.g., two-component epoxies),

    - heating (e.g., many thermosetting adhesives and epoxies),

    - illumination with light (e.g., ultraviolet (UV) light-curable adhesives),

    - presence of moisture (e.g., some polyurethanes and cyanoacrylates),

    - absence of oxygen (e.g., anaerobic adhesives).

For many polymer adhesives, the above mentioned hardening and curing principles are combined in various ways. For example, solvent-based thermosetting polymers (e.g., B-stage epoxies) both dry and cure. The solvents in the thermosetting polymers are often employed to realize polymers that have very low viscosities, with the solvents being evaporated before or during the curing (polymerization or crosslinking) process. Another example of combining the hardening and curing principles is in two-component polymer adhesives for which the start of the curing process is triggered with UV light illumination. The curing process continues to proceed even after the UV light illumination is removed. Very often, the polymerization process of UV-curable polymers can be supported and intensified by additional heat treatment. Tacky, pressure-sensitive adhesives, such as used on tapes, are highly viscous polymers that deform and flow very slowly into surface troughs to bond to a surface. These types of polymer adhesives remain highly viscous, do not harden, and provide comparably low bond strengths.

Most polymers can be used as adhesives and a large number of polymer materials are commercially available that have widely varying material properties and chemistries [25, 26]. Polymers typically have excellent cohesive properties and adhere well to a large variety of substrate materials. In general, polymers are hard and brittle at room temperature, but soften when heated. The transformation of a polymer from a hard (glassy) state to a rubber-like state is called the glass transition; the temperature at which this occurs is the glass transition temperature (T_g).

All polymers creep if influenced by a load, which is called the viscoelastic effect. The amount of creep is dependent on the ambient temperature, the time during which the load is present, and the polymer type [1, 25, 26].

Polymers are subject to similar environmental concerns as other materials such as glass and metals. They may, or may not, be affected by chemicals, temperature, radiation (UV and gamma radiation), stress, and biological deterioration, and thus their properties can change over time [1].

Polymers are typically several orders of magnitude more permeable to gases and moisture than glass or metals [3]. Water molecules with dimensions of slightly more than 0.1 nm diffuse in the free space between the molecular chains of polymers. Thus, polymers cannot be directly used to achieve hermetically sealed bonds and cavities [1, 3].

Thermoplastic polymers have a useful temperature range up to 200–300 °C and are limited at the low-temperature end by their brittleness. Thermoplastic polymers can be elongated and deformed to a large extent when heated and if the temperature is further increased, they are converted to a viscous melt. Typically they have poor creep resistance but good peel strength. Chemical resistance ranges from poor to excellent depending on the polymer [1, 25, 26].

Thermosetting polymers can operate at temperatures up to 300–450 °C, are more rigid than thermoplastics, and generally offer better chemical resistance. Fully crosslinked thermosetting polymers cannot flow but continue to soften until degradation occurs when exposed to increasing temperatures. Typically they have good creep resistance but only fair peel strength [1, 25, 26].

Elastomeric polymers can operate over a broad temperature range up to about 260 °C. They have high peel strength, low overall strength, and high flexibility. Chemical resistance is variable depending on the elastomer [1, 25, 26].

Hybrid polymers can have the properties of all the other material classes but with a more balanced combination. Some high-performance polymers, for example, polybenzimidazoles, can survive temperatures of up to 760 °C for short times without degradation.

3.2.3
Polymer Adhesives for Wafer Bonding

With the varying properties of different polymers, several aspects must be considered when selecting a polymer for a specific wafer bonding application. Selection of previously accepted polymer materials eases incorporation of any new industrial process, especially in the electronics industry. In particular, good availability, minimum material and process incompatibilities, and applicable a priori characterization can be expected. The polymer adhesive, including its solvents and impurities, must be compatible with the wafer surface materials and devices (e.g., CMOS circuitry), as well as with previously deposited films and post-bond processing steps. The physical properties of polymer adhesives, such as thermal stability, mechanical stability, and creep strength, have to be considered. Chemical resistance to acids, bases, or solvents is another important factor. Many processes in electronic and MEMS fabrication technologies involve solvents and etchants to which the polymer adhesive may be exposed. In applications where the polymer adhesive remains as a functional material on the device, chemical stability and aging effects are critical. In applications where the wafer bonding is of a temporary nature, the polymer adhesive at the bond interface should be easy to etch or dissolve. In these cases the long-term stability and aging effects of the polymer are not critical. For many microfluidic and bio-MEMS applications, the polymer must be inert or biocompatible. Table 3.1 lists polymers that have been proposed for polymer adhesive wafer bonding in various application areas [2, 4–22, 27–38]. These include polymers with hardening methods based on the evaporation of solvents (drying), thermal curing, two-component curing, UV light curing, and the combination of evaporation of solvents together with thermal curing or UV light curing. Material suppliers that provide polymers suitable for polymer adhesive wafer bonding include Dow Chemical Company of the USA (e.g., benzocyclobutene (BCB)), Brewer Science Inc. of the USA (e.g., WaferBOND), Micro Resist Technology GmbH of Germany (e.g., mr-I 9000), HD MicroSystems of the USA (e.g., HD-3007), Gersteltec Engineering Solutions of Switzerland (e.g., SU8), MicroChem Corporation of the USA (e.g., SU8), Delo of Germany (e.g., KATIOBOND), 3M of the USA (e.g., LC-3200), Dow Corning Corporation of the USA (e.g., silicones), and others. Many of the suppliers offer adhesion promoters together with their polymer materials to enhance the adhesion between specific substrate materials and the polymer adhesive.

Polymers that use the evaporation of solvents or water for polymer hardening during the bonding process are generally not suitable for adhesive wafer bonding applications. Because semiconductor wafers are typically not porous or permeable to liquids and gases, the volatile substances cannot escape the thin bond line between the wafers and get trapped as voids and deteriorate the bond interface. The same is true for polymer adhesives that outgas or otherwise produce byproducts during the hardening process after the wafers are joined [2, 29]. For example, many polyimide coatings produce water vapor as a byproduct during the curing (imidization) process [2, 29]. Thus, only thermoplastic polyimides that are fully imidized prior to wafer bonding and that can melt again during the bonding process would provide void-free bond interfaces. Drying or outgassing polymers may only be used if at least one of the two bonded substrate materials is permeable to gases or if ventilation channels are incorporated in the bond line [38], allowing the volatile substances from the bond interface to be discharged. For polymer adhesives in which the evaporation of solvents or water does not occur in the final hardening step, the evaporation can be done before the wafers are joined for bonding [10].

Thermal curing of thermosetting polymers [2] or melting of thermoplastic polymers [2] are suitable mechanisms for use in bonding of wafers that consist of identical materials or wafers that consist of materials with similar coefficient of thermal expansion (CTE). When two wafers of dissimilar materials with different CTEs are bonded at an elevated temperature, the bonded wafer stack will bend after cooling to room temperature. The wafer with the higher CTE is expanded more during heating and consequently shrinks more during cooling to room temperature than the wafer with the lower CTE. The resulting stresses in the wafer stack may even cause cracking of the wafers. Bonding at room temperature with

Table 3.1 Polymers that have been proposed for polymer adhesive wafer bonding.

Polymer adhesives	Features	References
Epoxies	• Thermosetting materials	[2, 27]
	• Thermal curing or two-component curing	
	• Strong and chemically stable bond	
UV-curable epoxies (e.g., SU8)	• Thermosetting materials	[2, 16]
	• UV curing (one of the substrates has to be transparent to UV light)	
	• Strong and chemically stable bond	
	• Bonding with patterned films	
Nano-imprint resists	• Thermosetting versions (optional UV curable)	[28]
	• Thermoplastic versions	
	• Optimized for good reflow around surface structures, thus typically suitable for wafer bonding	
Positive photoresists	• thermoplastic materials	[2, 29]
	• hot melt	
	• typically void formation at the bond interface, weak bond	
Negative photoresists	• Thermosetting materials	[2, 29]
	• Thermal curing or UV curing	
	• Typically weak bond, low thermal stability	
	• Bonding with patterned films	
BCB	• Thermosetting materials	[2, 6, 11, 18, 19, 29]
	• Thermal curing	
	• High yield on wafer scale	
	• Very strong, chemically and thermally stable bond	
	• Bonding with patterned films	
PMMA	• Thermoplastic materials	[2, 30]
	• Hot melt	
Polydimethylsiloxane (PDMS)	• Elastomeric materials	[2, 31]
	• Thermal curing	
	• Suitable for plasma-activated bonding	
	• Biocompatible	

Table 3.1 (Continued)

Polymer adhesives	Features	References
Fluoropolymers (e.g., Teflon, Flare)	• Thermoplastic and thermosetting materials • Thermal curing or hot melt • Chemically very stable bond • Bonding with patterned films	[2, 32]
Polyimides (thermosetting)	• Thermosetting materials • Thermal curing • For many polyimides, void formation from the imidization process • Bonding with patterned films • Mainly chip-scale processes	[2, 29]
Polyimides (thermoplastic)	• Thermoplastic materials • Hot melt • High-temperature stability • Temporary bonds	[10]
Polyetheretherketone (PEEK)	• Thermoplastic materials • Hot melt	[2, 33]
Thermosetting copolyesters	• Thermosetting materials • Thermal curing	[2, 34]
Parylene	• Thermoplastic materials • Hot melt	[2, 35]
Liquid crystal polymers	• Thermoplastic materials • Hot melt • Good moisture barrier • Typically not available as a liquid polymer precursor	[2, 36]
Waxes	• Thermoplastic materials • Hot melt • Low thermal stability • Mainly for temporary bonds	[2, 37]

two-component or UV-curable epoxies can prevent thermally induced stresses. However, when using UV-curable polymer adhesives at least one of the substrate materials must be transparent to UV light [2, 16].

If the polymer adhesive is in a solid or gel-like state prior to the curing process and if curing of the polymer is initiated by curing parameters other than time only, wafers with deposited adhesive coatings may be stored between the adhesive deposition process and the bonding process, which can be beneficial in a production environment. Polymer adhesives that have such characteristics are thermoplastic polymers or solvent-based thermosetting polymers (B-stage polymers).

Polymers that work specially well for polymer adhesive wafer bonding applications are, for example, B-stage polymers (e.g., BCB, SU8, and the nano-imprint resist mr-I 9000) [2, 6, 11, 16, 18, 19, 28, 29] and many thermoplastic adhesives (e.g., polymethylmethacrylate (PMMA), copolymers, and waxes) [2, 10, 30, 34, 37]. As an example of the properties of a typical B-stage thermosetting polymer, the properties of BCB are shown in Figure 3.2. Thermosetting polymers such as BCB undergo crosslinking between polymer chains during curing to form a stable polymeric network. They are typically mobile for a short time during the curing process to achieve crosslinking and cannot be re-melted or reshaped after curing. Figure 3.2a shows the polymer viscosity as functions of temperature during curing using three different heating rates. In this example, the minimum viscosity of 1000 P (100 Pa s) occurs at about 170–190 °C. For thermosetting polymers, the viscosity as a function of temperature is changed permanently as soon as the crosslinking level of the polymer is changed. This is in contrast to thermoplastic polymers for which the viscosity as a function of temperature remains unchanged even after repeated temperature cycles. Thermosetting polymers are typically supplied in the form of a liquid polymer precursor which is initially crosslinked to a certain degree (e.g., 20–50%) and optionally dissolved in a solvent. Figure 3.2b shows the percentage of crosslinking of a thermosetting polymer as a function of the curing time and temperature. For example, a pre-curing temperature of 190 °C together with a pre-curing time of 30 min results in an increase of the crosslinking level from the initial 35% to about 43%.

3.3
Polymer Adhesive Wafer Bonding Technology

To achieve bonding results with repeatable high quality, the bonding process and parameters must be precisely controlled. Parameters such as the polymer material, bonding pressure, bonding temperature, chamber pressure, and temperature ramping profile have a significant impact on the resulting bonding quality and defects at the bond interface. The qualitative influence and mechanisms of the relevant bonding parameters are presented in this section. Basic bonding process schemes for wafer bonding with nonpatterned polymer adhesives and for localized bonding are presented.

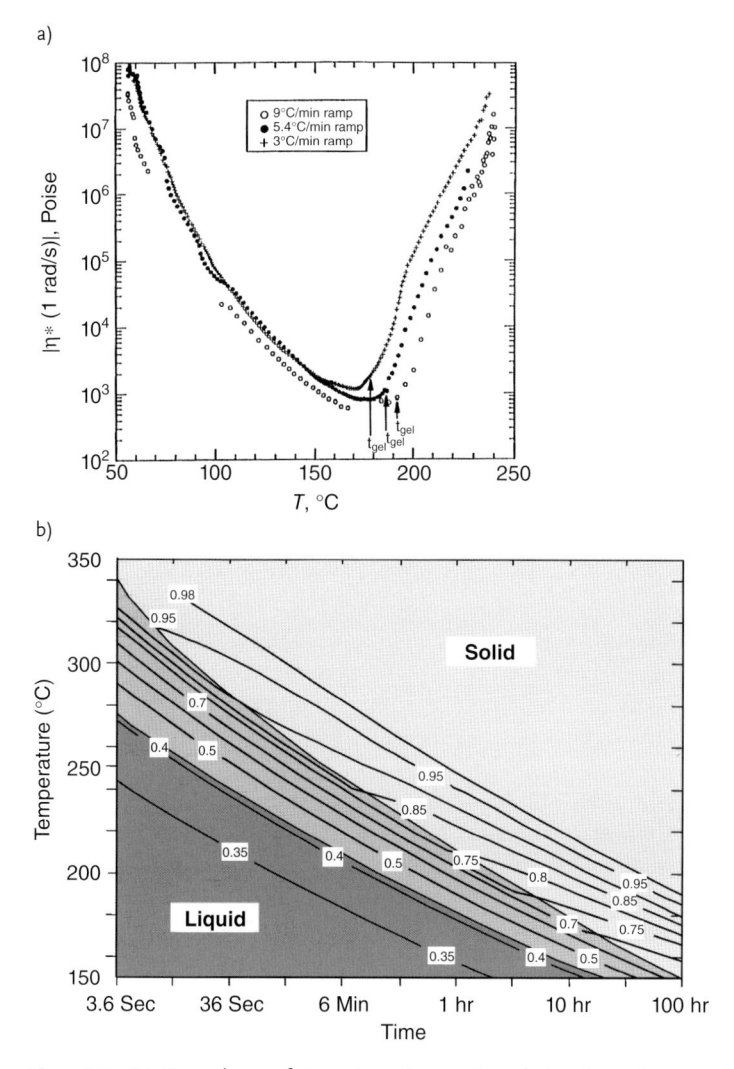

Figure 3.2 (a) Dependence of viscosity on temperature during the curing process of a heat-curable thermosetting polymer. (b) Percentage of crosslinking of a thermosetting polymer as a function of curing time and temperature [6].

3.3.1
Polymer Adhesive Wafer Bonding Process

Polymer adhesive wafer bonding can be performed with standard commercial wafer bonding or lamination equipment. Standard wafer bonding equipment typically consists of a vacuum chamber, a mechanism for joining wafers inside the vacuum chamber, and two bond chucks that can apply a controlled force and heat to the wafer stack. The bond chucks typically are stiff flat plates. Soft plates or sheets (e.g., graphite or silicone) can optionally be placed between one or both bond chucks and the wafer stack. Soft plates or sheets typically adapt better to nonuniformities of the wafer stack, and thus may distribute the pressure more evenly over the wafer stack. Table 3.2 details the basic process steps for polymer adhesive wafer bonding with thermoplastic or thermosetting polymer adhesives using commercial wafer bonding equipment.

Several approaches can be used to apply the optional adhesion promoter and the polymer adhesive. Most applications in microelectronics and MEMS require uniform thicknesses of the intermediate polymer bonding material of 0.1–100 µm. The most common method is spin coating of a liquid polymer precursor on a wafer [39]. Highly uniform coatings with well-defined thicknesses and smooth surfaces can be achieved by spin coating. Spin coating of a liquid polymer precursor also has a planarizing effect on an existing wafer surface topography. A more planar polymer surface reduces the required reflow of the polymer adhesive during bonding to achieve contact between the surfaces and thereby can improve the bond quality. Spray coating, electrodeposition, stamping, screen printing, brushing, and dispensing of liquid polymer precursors are alternative methods to deposit polymer coatings [2, 39]. However, these methods often do not achieve the uniformity and thickness control of spin-coated layers. Yet other polymer deposition methods are chemical vapor deposition and atomic layer deposition processes [2, 40]. Some polymers are available as thin films or sheets [2], which can be laminated to the wafer surfaces.

The bond quality and the amount of void formation at the bond interface are influenced by the polymer properties (e.g., reflow capabilities and viscoelastic behavior during bonding), the size and amount of particles at the wafer surfaces, the wafer surface topography, the thickness of the intermediate polymer adhesive layer, the bonding pressure (force applied with the bond chucks divided by the bond area), the temperature ramping and the bonding temperature, the atmospheric condition in the bond chamber before the bonding of the wafers, and the wafer stiffness. These bonding parameters and their qualitative influences on the resulting wafer bonds are detailed in Table 3.3.

Void-free bond interfaces can be achieved more easily if the intermediate polymer adhesive attains a low-viscosity state during bonding, and thus readily reflows and deforms during bonding. However, in these cases the polymer adhesive tends to flow from areas of high pressure towards areas of lower pressure, which can lead to significant thickness nonuniformities of the polymer adhesive layer after bonding [6, 28]. If wafer surface topographies are small and very uniform intermediate polymer adhesive layer thicknesses after bonding are required, void-free bond interfaces can be achieved by using polymer adhesives that do not significantly reflow during bonding. Examples of this approach are the use of thermoplastic polymer adhesives in combination with bonding temperatures that are below their melting point or the use of thermosetting polymer adhesives that are crosslinked before bonding to an extent that does not allow

Table 3.2 Basic process steps for polymer adhesive wafer bonding [2, 6, 28, 29].

No.	Process step	Purpose of the process step
1	Cleaning and drying of the wafers	Remove particles, contaminations and moisture from the wafer surfaces. Exposure to an ultrasound bath to remove particles from the wafer surfaces is typically sufficient.
2	Treating the wafer surfaces with an adhesion promoter (optional)	Adhesion promoters can enhance the adhesion between the wafer surfaces and the polymer adhesive.
3	Applying the polymer adhesive to the surface of one or both wafers and optionally patterning the polymer adhesive	The most commonly used application method is spin coating. Alternative application methods are described in the text. Polymer patterning is described in Section 3.3.2.
4	Soft-baking or partial crosslinking of the polymer, depending on the type of polymer	Solvents and volatile substances are removed from the polymer coating. Thermosetting adhesives should not be fully crosslinked to remain bondable.
5	Placing the wafers in the bond chamber, establishing a low-pressure atmosphere, and joining the wafers inside the bond chamber	The wafers are joined in a low-pressure atmosphere to prevent voids and gases from being trapped at the bond interface. The low-pressure atmosphere can also be established after the wafers are joined, as long as trapped gasses can be pumped away from the bond interface before the bond is initiated.
6	Applying bonding pressure to the wafer stack with the bond chucks	The wafer and polymer adhesive surfaces are forced into intimate contact over the entire wafer. For thermosetting polymer adhesives, the bonding pressure has to be applied before the polymer is crosslinked. For thermoplastic polymers, the bond pressure may be applied before or after the polymer reflow temperature is reached.
7	Heating the wafer stack with the bond chucks	The reflow or crosslinking of the polymer adhesive is typically initiated by heating the bottom and/or top bond chucks. Depending on the selected polymer adhesive, crosslinking may also happen at or near room temperature.
8	Cooling, bond pressure release, and chamber purge	End bonding process. The sequence of cooling, bond force release, and chamber purge is largely interchangeable. However, for thermoplastic polymers, the bond pressure should not be fully released before cooling, thus ensuring polymer solidification before releasing the bond pressure.

significant polymer reflow during bonding [6, 28]. Figure 3.3 shows two bonded wafer pairs in which the top wafers have been sacrificially removed to expose the intermediate polymer adhesive layer. Both wafer pairs are bonded with the same thermosetting polymer but with different levels of crosslinking of the polymer before bonding. For example, the wafer shown in Figure 3.3a is bonded with a lower polymer viscosity during bonding than the wafer shown in Figure 3.3b. Both bond interfaces are void-free; however, the wafer depicted in Figure 3.3a shows fringes that indicate significant thickness variations of the polymer adhesive layer, which are not present on the wafer shown in Figure 3.3b.

Table 3.3 Influence of various polymer adhesive wafer bonding parameters on the resulting wafer bond [2, 28, 29].

Parameter	Influence on bond defects	Influence
Properties of the polymer adhesive	• The type of polymer adhesive, the involved solvents and the associated bonding temperatures must be compatible with the wafer materials.	Strong
	• For wafer materials not permeable to gases, the polymer adhesive must not outgas or release solvents or byproducts during the bonding process since volatile substances that are trapped at the bond interface form voids. Drying or outgassing polymers may be used if at least one of the substrates is permeable to gases or if ventilation channels are incorporated at the bond interface, allowing the volatile substances to be discharged.	
	• The polymer adhesive must provide sufficient wetting and bonding to the wafer surface materials.	
	• The polymer adhesive must reflow or achieve a viscoelastic state to deform during the bonding process to be able to adapt to the opposing wafer surfaces. The reflow capabilities of thermosetting polymers may be varied by selecting different levels of partial crosslinking prior to bonding. The reflow capabilities of thermoplastic polymers can be varied to some extent by selecting different reflow and bonding temperatures.	
	• Polymer adhesives tend to flow from areas of high pressure towards areas of lower pressure if they are in a very low-viscosity phase. This can lead to substantial thickness nonuniformities of the polymer adhesive layer in between the bonded wafers.	
	• For large CTE mismatches between two wafers, low-temperature bonding processes are preferred to reduce stresses between the bonded wafers.	

Table 3.3 (*Continued*)

Parameter	Influence on bond defects	Influence
Amount and size of particles at the wafer surfaces	• Particle-free surfaces are the key to good bonding results as the presence of large particles at the bond interface may cause extended unbonded areas. To some extent small particles may be embedded in the polymer adhesive.	Strong
Wafer surface topography	• If the wafer surface topography is high compared to the thickness of the polymer layer, unbonded areas can result. • Polymer deposition processes that have a planarizing effect on surface topographies, such as spin coating, can reduce the tendency for unbonded areas.	Strong
Polymer adhesive layer thickness	• Polymer adhesive layers that are thick in comparison to wafer surface topographies can more easily reflow around the surface structures, reducing the tendency for void formation or unbonded areas. • Very thin polymer layers compensate to a lesser extent for small surface nonuniformities and particles at the bond interface. • Stresses due to CTE mismatch between the bonded wafers can be reduced to some extent by the increased viscoelastic deformation capabilities of a thicker polymer layer.	Strong
Bonding pressure (force applied with the bond chucks divided by the bond area)	• The bonding pressure helps to deform the intermediate polymer adhesive and the wafers to adapt to the surface topography and wafer thickness nonuniformities, thus bringing the surfaces in sufficiently close contact to achieve bonding. The bonding pressure typically has a very significant impact on reducing void formation due to surface topographies and can be adjusted in a very flexible way. • Excessive bonding pressures may cause high stress, leading to wafer cracking or damage of structures at the wafer surfaces. • The use of deformable sheet(s) (e.g., graphite or silicone foam sheets) between the wafer stack and the rigid bond chuck(s) can even out the bond pressure, counter void formation at the bond interface, and reduce the risk for wafer cracking.	Strong

(*Continued*)

Table 3.3 (Continued)

Parameter	Influence on bond defects	Influence
Temperature ramping profile and bonding temperature	• The temperature ramping and the final bonding temperature must be tailored to the selected polymer adhesive. • For thermosetting polymers, the final bonding and curing temperature in combination with the curing time must be high enough for the polymer to achieve sufficient crosslinking. Polymer reflow with reduced viscosities takes place at temperatures lower than the polymer crosslinking temperature. Thus, a holding point at the reflow temperature may be introduced during the temperature ramping process to allow sufficient polymer reflow before the polymer is fully crosslinked. • For thermoplastic polymers, the bonding temperature has to be selected so that a low enough viscosity of the polymer is provided to sufficiently reflow and redistribute the polymer at the bond interface. The viscosity of a thermoplastic polymer and its resulting reflow capabilities can be influenced to some extend by the bonding temperature. • The effective polymer redistribution is a function of the reflow temperature and the time that is available for the reflow process. Extremely fast temperature ramping with extremely short hold times may result in a tendency to form voids at the bond interface. • Extremely fast temperature ramping cycles can also cause nonuniform and incomplete heating of the wafer stack, which may cause incomplete adaptation of the polymer adhesive to the wafer surfaces or excessive stresses. • For wafers that have large differences in CTE, very slow heating and cooling cycles may be used to reduce the stresses and the risk for wafer cracking.	Strong/ medium
Atmospheric condition in the bond chamber before the wafer bonding is initiated	• Gas pressures in the bond chamber of below 100 mbar before joining the wafers are typically sufficient to prevent gases from being trapped at the bond interface. • For many polymer adhesives it is also possible to pump out gases trapped between two wafer surfaces after the wafers are joined together but before they are bonded. In these cases the wafer surfaces can be joined before a low-pressure atmosphere is established in the bond chamber.	Medium
Wafer stiffness	• Thin wafers and wafers consisting of materials with a low Young's modulus are more easily deformed by the bonding pressure to compensate for surface nonuniformities at the bond interface.	Medium

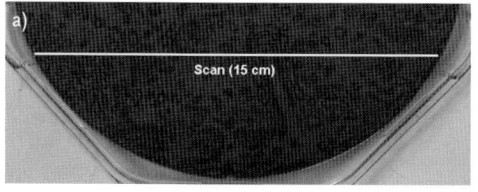

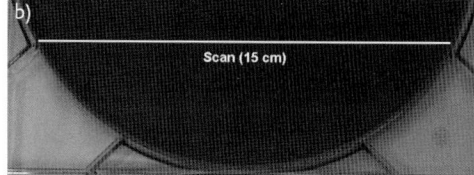

Figure 3.3 Bonded wafer pairs in which the top wafer has been sacrificially removed to expose the intermediate polymer adhesive at the bond interface. The fringes on the wafer in (a) result from thickness variations of the polymer adhesive layer that are caused by polymer redistribution during bonding [6].

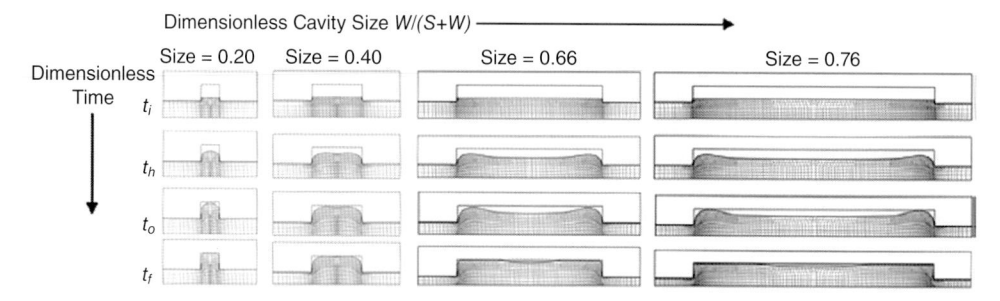

Figure 3.4 Example of a simulation of the step-by-step polymer flow into surface cavities with different aspect ratios during both nano-imprinting and polymer adhesive wafer bonding [42].

In nano-imprint lithography, a substrate with surface structures and cavities (the nano-imprint mold) is imprinted into a thin polymer layer with the requirement that the polymer readily flows around the surface structures and cavities, thus replicating them in the polymer. Detailed simulation models for nano-imprint processes have been developed that describe the polymer flow from the nanometer to millimeter length scale around the structures and cavities of the nano-imprint molds [28, 41, 42]. Figure 3.4 shows simulation results of a step-by-step polymer flow into surface cavities with different aspect ratios during nano-imprinting. These simulation models are directly applicable to polymer adhesive bonding of wafers with surface topographies and they can detail and qualitatively predict the effects of void formation due to surface topographies. In nano-imprinting and in polymer adhesive wafer bonding, the polymer underneath the protruding surface structures must be displaced and transported to nearby cavities or trenches. Void formation occurs when the cavities or trenches cannot be completely filled with the polymer adhesive. The simulation models predict that the dimensions, heights, and aspect ratios of the protruding surface structures and cavities or trenches influence the ability to fill the cavities or trenches during polymer reflow. These qualitative predictions have been confirmed in polymer adhesive wafer bonding experiments where microvoids tend to form more easily in the wide, high-aspect-ratio trenches between large protruding structures than in narrow trenches between the same protruding structures. In addition, these simulation models confirm the influencing bonding parameters listed in Table 3.3, for example, that an increased bond pressure, an increased thickness of the intermediate polymer layer, an increased flow ability of the intermediate polymer, and an increased time for polymer reflow enhance filling of the trenches and cavities, and thus prevents or reduces void formation. An increased time for the polymer reflow can be implemented in the polymer adhesive wafer bonding process by a slow temperature ramping or by introducing a hold time at the polymer reflow temperature as detailed in Table 3.3. However, the most generic and convenient parameter to prevent void formation is an increased bonding pressure.

In addition to the basic bonding scheme discussed above, adhesive wafer bonding with UV-curable thermosetting polymers is also frequently used. An important advantage of UV-curable polymer adhesives is that they can be crosslinked at room temperature by exposure to UV light, which is specifically useful for bonding of wafers that have different CTEs. However, at least one of the two bonded wafers must be transparent to UV light and the polymer crosslinking process cannot be initiated with standard wafer bond equipment. In addition, a number of other techniques have been implemented to avoid heating of the entire wafer stack during bonding, and thus to reduce thermally induced damages or stresses. These techniques include localized heating of the polymer adhesive at the bond interface, for example, by integrated electrical heaters at the bond interface [2], by heat generation at the bond interface using chemical reactions, or by absorption of energetic radiation at the bond interface [35]. Specialized bonding process schemes, for example those based on polymer surface pretreatment with a plasma or with solvents, have also been reported [2, 31].

3.3.2
Localized Polymer Adhesive Wafer Bonding

In localized or selective polymer adhesive wafer bonding only predefined parts of the wafer surfaces are bonded. Localized polymer adhesive wafer bonding can be achieved by applying the polymer adhesive only on areas where bonding is desired [18], by using structured surfaces that contact the corresponding bonding surface only at predefined areas [43], by special surface treatments on areas at the wafer surface where no bond should be established [44], or by locally heating the bond interface with, for example, integrated heaters at the bond interface to create local bonding at the desired areas [2]. Figure 3.5 shows four schematic examples of how localized bonding can be implemented with patterned polymer adhesive layers or patterned wafer surfaces to create cavities or other three-dimensional structures.

To achieve small and well-defined dimensions of the localized bonding areas, the patterning of these areas can be done using photolithographic techniques. One approach is that the intermediate polymer adhesive itself is patterned, as shown in Figures 3.5a and d, for example, by polymer etching with a lithographically defined mask, by using photosensitive polymer adhesives, by lithographic lift-off processes, or by selective polymer deposition processes. Another approach is that

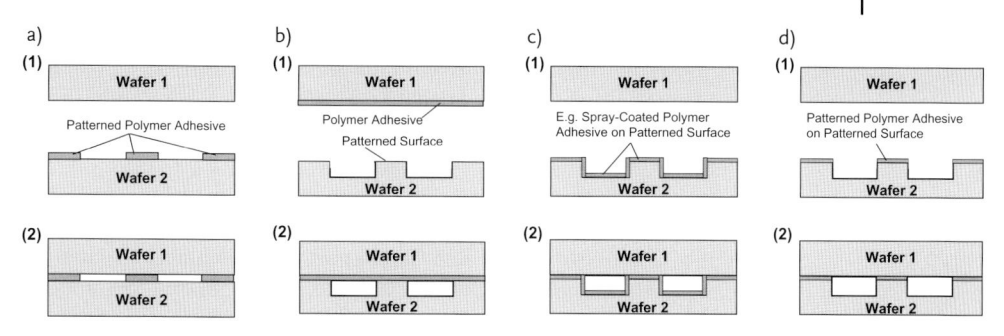

Figure 3.5 Examples for localized polymer adhesive wafer bonding with (a) a patterned polymer adhesive, (b) a patterned wafer surface, (c) a spray-coated polymer adhesive on a patterned wafer surface, and (d) a patterned polymer adhesive on a patterned wafer surface. (1) Before bonding; (2) after bonding.

one or both bonding surfaces are patterned using photolithographic techniques in combination with etching and/or deposition techniques to define the local areas that are to be bonded. The polymer adhesive is then deposited on the patterned surface and/or on the corresponding bonding surface, for example, by spray coating, spin coating, stamping, or vapor deposition of the polymer adhesive as shown in Figures 3.5b and c.

For localized bonding, generally the basic process scheme and bonding parameter influences apply as for wafer bonding with self-contained polymer adhesive layers as described in Section 3.3.1. It is important to realize that in localized wafer bonding, the effective bonding pressure equals the bonding force that is applied with the bond chucks to the wafer stack divided by the patterned bond area. This is different from bonding with self-contained intermediate polymer adhesive layers, where the bonding pressure equals the bonding force that is applied with the bond chucks to the wafer stack divided by the total wafer area. In localized polymer adhesive wafer bonding it is also important that the polymer adhesive remains sufficiently firm during the bonding process to retain the shape of the patterned structures but still sufficiently deforms to adapt to the wafer surfaces for complete bonding. If the polymer adhesive viscosity becomes too low, the polymer reflows during the bonding process, and consequently the patterned polymer structures lose their shape. As a result, the bonded wafer areas become larger and the gap height (as initially defined by the polymer thickness) between the two wafer surfaces decreases in an uncontrolled way [18]. For some polymer adhesives that can be patterned with lithographic techniques as shown in Figure 3.5a, the available process window of the lithographic patterning may be limiting for retaining sufficient reflow capabilities of the patterned polymer layers to achieve good bonding results. Broader and more stable process windows may typically be obtained when the local bonding areas at the wafer surfaces are patterned and subsequently a thin polymer adhesive layer is deposited on one or both of the wafer surfaces as shown in Figures 3.5b and c.

Other nonlithographic methods to pattern a polymer adhesive include spray coating of a liquid polymer with a shadow mask, local dispensing of a liquid polymer, or screen printing of a liquid polymer. The lamination of polymer sheets that are patterned by punching or cutting with a water jet or laser, is another suitable way to apply a polymer adhesive only on certain wafer areas. However, many of these methods have limitations concerning control of the polymer thickness and the smallest achievable feature sizes of the locally bonded and unbonded areas.

Localized polymer adhesive wafer bonding has also been combined with other localized bonding techniques such as direct metal or solder bonding in the same wafer bonding step [10]. In these techniques, the minimum required bonding temperature for one bond type (e.g., the direct bond) must not be higher than the maximum temperature stability of the second bond type (e.g., the polymer adhesive bond). These hybrid localized metal and polymer wafer bonding technologies are described in Chapter 11 of this book.

3.4
Wafer-to-Wafer Alignment in Polymer Adhesive Wafer Bonding

For many applications of polymer adhesive wafer bonding, accurate alignment between the bonded wafers is essential. For precise wafer-to-wafer alignment in wafer bonding, various techniques have been implemented, including wafer backside alignment with a digitized image, the SmartView® method, intersubstrate microscopy, infrared transmission microscopy, transparent wafer with optical microscopy, and through-wafer holes with optical microscopy [2]. However, polymer adhesive wafer bonding is based on reflow or deformation and subsequent hardening of the intermediate polymer adhesive, and thus the polymer reflow process can cause problems for the achievable post-bond wafer-to-wafer alignment accuracy. When the wafer stack is pressed together with the bond chucks during the bonding process, shear forces acting in parallel to the bond line almost inevitably occur. If the polymer adhesive attains a low-viscosity phase during the bonding process, these shear forces can result in the wafers shifting relative to each other and thereby very significantly reducing the post-bond wafer-to-wafer alignment accuracy. One possibility to reduce this effect would be to use fixed and well-aligned bond chucks that do not introduce shear forces that cause the wafers to shift relative to each other during the bonding process. However, current state-of-the-art wafer bonding equipment typically does not provide these features. Nonetheless, there are a number of alternative approaches available to prevent large alignment shifts.

One approach that can prevent or reduce bonding-induced alignment shifts in polymer adhesive wafer bonding is the use of polymer adhesives that do not reflow and transform into a low-viscosity phase during the bonding process [6]. This can be achieved by using a thermosetting polymer that is partially crosslinked before bonding to an extent that it does not significantly reflow during bonding or by using a thermosetting polymer that is heated during bonding below the melting

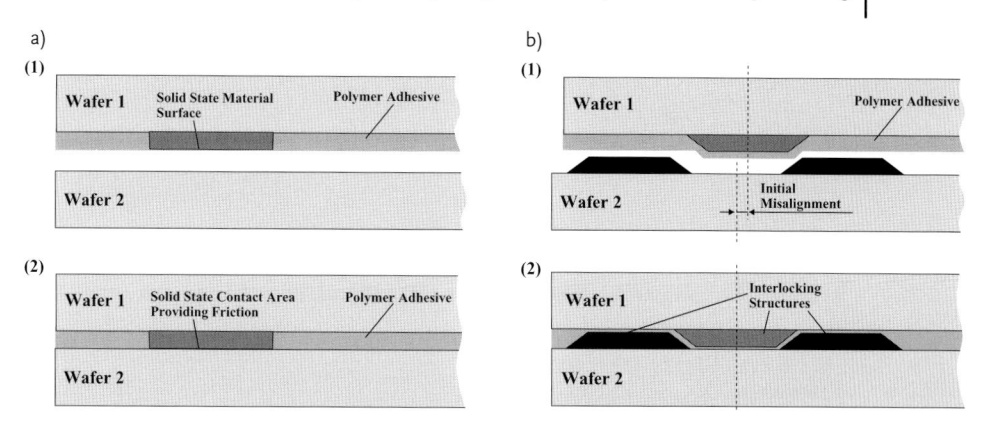

Figure 3.6 Techniques to improve obtainable wafer-to-wafer alignment in polymer adhesive wafer bonding (a) with frictional nonreflowable surface structures [45] and (b) with interlocking or keyed alignment structures [46]. (1) Before bonding; (2) after bonding.

point of the polymer. A disadvantage of this approach is that it is only practical for bonding wafers with low or no surface topographies where the polymer does not have to be redistributed significantly to prevent void formation. This approach is not suitable for bonding of wafers with high surface topographies.

A second approach that can prevent or reduce bonding-induced alignment shifts is to introduce surface structures on the wafers that are not covered with the polymer adhesive and that do not reflow during the bonding process as shown in Figure 3.6a. Such structures create frictional forces between the two wafer surfaces and can prevent the wafers from shifting relative to each other while the intermediate polymer is in a low-viscosity state in the bonding process [45]. These frictional structures may consist of metals, inorganic materials, or fully crosslinked thermosetting polymers that do not reflow during bonding.

A third approach, which is possible for submicrometer wafer-to-wafer alignment, is to introduce keyed alignment structures [46] that mechanically interlock the wafers and prevent them from shifting relative to each other or even provide a mechanical self-alignment function as shown in Figure 3.6b. Such corresponding interlocking structures may be etched into the wafer surfaces or deposited and patterned using dielectric materials, metals, or crosslinked thermosetting polymers. The placement accuracy of the corresponding interlocking structures on the wafer surfaces is only limited by the overlay accuracy of the photolithography and the etching processes used. Although added process steps are needed for keyed alignment during bonding, excellent post-bond wafer-to-wafer alignment accuracies of less than 1 µm have been reported [46].

When wafers are bonded that consist of dissimilar materials having different CTEs, additional problems for the obtainable post-bond wafer-to-wafer alignment accuracy can be expected. For a temperature increase of 100 °C, the difference in thermal expansion between two wafers consisting of dissimilar materials can be several tens of micrometers. If the wafers are joined and bonded to each other at an increased temperature, the difference between the thermal expansions of the two wafers from the temperature increase will result in a misalignment that is of the same order as the difference in thermal expansion of the two wafers. To accurately align and bond wafers that consist of dissimilar materials either their CTEs must match or the bonding must be performed near room temperature.

3.5
Examples for Polymer Adhesive Wafer Bonding Processes and Programs

In this section, process parameters and programs for wafer bonding with three different polymer adhesives are presented. The proposed program sequences and parameter sets represent possible processes that provide good bonding results. They are not optimized with respect to short bond cycle times. The programs should be viewed as examples that can serve as a starting point for the development of optimized, application-specific polymer adhesive wafer bonding programs. The program parameters may be adapted, the process sequence may be changed, and/or process steps may be added or removed. To counteract possible void formation at the bond interface, the indicated bonding pressure may generally be increased as long as the wafers do not break during bonding. When developing new or improved polymer adhesive wafer bonding process sequences and programs, it is useful to perform initial bonding experiments with glass wafers. This allows visual inspection of the bond interface through the glass wafers and observation of possible void formation.

3.5.1
Bonding with Thermosetting Polymers for Permanent Wafer Bonds (BCB) or for Temporary Wafer Bonds (mr-I 9000)

Dry-etch BCB is a thermosetting B-stage polymer that is compatible with semiconductor environments. BCB is a suitable polymer for permanent adhesive wafer bonds, which are very strong, resistant to various acids and solvents, and can withstand temperatures well above 200 °C. BCB layer thicknesses from below 0.3 μm to above 10 μm have been successfully used in adhesive wafer bonding [6, 29]. The excellent reflow capabilities of BCB during bonding as indicated in Figure 3.2a makes wafer bonding with BCB layers comparatively tolerant to topographies and particles at the wafer surfaces. BCB is supplied by Dow Chemical Company in the USA. Table 3.4 gives and Figure 3.7 depicts an example of a preparation sequence and a wafer bonding program for wafer bonding with dry-etch BCB as intermediate polymer adhesive, respectively. Soft-baked or partially cured BCB coatings can be stored for several days or weeks in particle-free environments before the actual wafer bonding is performed without compromising the bonding results.

The nano-imprint resist series mr-I 9000 is a thermosetting B-stage polymer that is compatible with semiconductor environments. It is a suitable polymer for

Table 3.4 Example of a preparation sequence for wafer bonding with thermosetting dry-etch BCB or mr-I 9000 as intermediate polymer adhesives.

No.	Process step	Comments
1	Clean wafers in an ultrasound bath with deionized water.	Alternative cleaning procedures to remove particles from the wafer surfaces may be used.
2	Dry wafers in a rinse and dry equipment.	Alternative procedures, such as drying at elevated temperatures in an oven, may be used.
3	Spin- or spray-coat thermosetting polymer precursor on the surface(s) of one or both wafers to be bonded.	The resulting polymer layer thickness depends on the viscosity of the polymer precursor in combination with the spin speed.
4	Place the wafer(s) with the polymer coating(s) for 2 min on a hotplate at 110 °C.	Evaporate solvents without crosslinking the thermosetting polymer (110 °C for 2 min is suitable for both BCB and mr-I 9000). Other temperatures and times may also be used.
5	Option 1. Place two wafers on the bond fixture with or without spacers separating the bond surfaces and transfer the fixture to the bond chamber.	Option 1. BCB layers reflow extremely well during bonding and thus can readily compensate for topographies or particles at the wafer surfaces. However, BCB also flows easily from areas with higher pressure to areas with lower pressure, leading to thickness nonuniformity after bonding. Layers of mr-I 9000 typically have well-balanced reflow capabilities.
	Option 2. Place wafer(s) with BCB coating(s) on a hotplate in a nonoxidizing atmosphere (e.g., nitrogen) for 30 min at 190 °C to partially crosslink the BCB. Then place two wafers on the bond fixture with or without spacers and transfer the fixture to the bond chamber.	Option 2. The partially crosslinked BCB layer reflows only marginally during bonding. Thus, well-defined BCB layer thicknesses after bonding can be achieved. However, the BCB layers cannot compensate large topographies or particles at the wafer surfaces. Different combinations of crosslinking temperatures and times can be chosen to provide different BCB reflow viscosities.

temporary adhesive wafer bonds that can be easily debonded or etched using, for example, oxygen plasma processes. Bonding with mr-I 9000 layers has been used in emerging heterogeneous micro- and nanosystem integration technologies. Wafer bonds with mr-I 9000 layers can withstand temperatures of around 100 °C and under certain conditions of up to 300 °C. Layer thicknesses of mr-I 9000 from 0.3 up to 5 µm have been successfully used in adhesive wafer bonding [28]. The well-balanced reflow capabilities of mr-I 9000 during bonding result in void-free

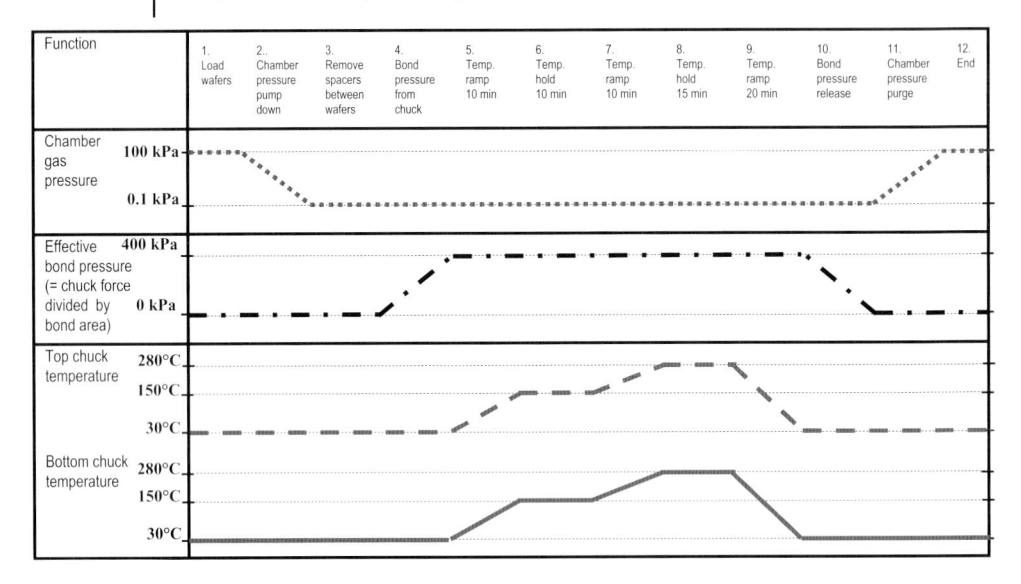

Figure 3.7 Example bonding program for wafer bonding with thermosetting dry-etch BCB as intermediate adhesive layer.

and uniform bonds, even if topographies are present at the wafer surfaces. Micro Resist Technology GmbH in Germany supplies mr-I 9000. The preparation sequence in Table 3.4 (option 1) can be used for wafer bonding with mr-I 9000 as polymer adhesive. For wafer bonding with mr-I 9000 layers the example bonding program steps in Figure 3.7 can be used when adjusting the temperature ramping steps 5 to 9 of the top and the bottom chuck to: step 5, temp. ramp 5 min to 110 °C; step 6, temp. hold 10 min at 110 °C; step 7, temp. ramp 20 min to 200 °C; step 8, temp. hold 30 min at 200 °C; step 9, temp. ramp 10 min to 30 °C. Soft-baked or partially cured mr-I 9000 coatings can be stored for several days or weeks in particle-free environments before the actual wafer bonding is performed without compromising the bonding results.

3.5.2
Bonding with Thermoplastic Polymer (HD-3007) for Temporary and Permanent Wafer Bonds

HD-3007 is a thermoplastic B-stage polyimide adhesive that is compatible with semiconductor environments. HD-3007 has been specially developed for polymer adhesive wafer bonding applications and is a suitable polymer for temporary and permanent adhesive wafer bonds. Softening and reflow of the thermoplastic HD-3007 layers are recognizable at around 200 °C and increase gradually with increasing temperature. Wafer bonds with HD-3007 layers can withstand temperatures of above 200 °C and they can be debonded by exposure to heat, solvents, or laser light [47]. Typical applications of wafer bonding with HD-3007 include three-

Table 3.5 Example of a preparation sequence for wafer bonding with HD-3007 as intermediate polymer adhesive.

No.	Process step	Comments
1	Clean wafers in an ultrasound bath with deionized water.	Alternative cleaning procedures that remove particles from the wafer surfaces may be used.
2	Dry wafers in a rinse and dry equipment.	Alternative drying procedures, including drying at elevated temperatures in an oven, may be used.
3	Spin- or spray-coat the HD-3007 precursor on the surface(s) of one or both wafers to be bonded.	The resulting HD-3007 layer thickness depends on the viscosity of the HD-3007 precursor in combination with the spin speed, with typical thicknesses between 2.5 and 5 µm.
4	Place the wafer(s) with the HD-3007 coating(s) for 90 s on a hotplate at 90 °C, followed by 90 s at 120 °C.	Soft-bake the HD-3007 layer(s) to remove solvents.
5	Place the wafer(s) with the HD-3007 coating(s) in an oven with a nitrogen or other nonoxidizing atmosphere and: Ramp at 5 °C min^{-1} to 200 °C and hold for 30 min. Ramp at 5 °C min^{-1} to 300 °C and hold for 60 min. Ramp at 10 °C min^{-1} down to 150 °C. Fast ramp down to 40 °C.	Cure the HD-3007 layer(s) and remove volatile substances emerging in the curing/imidization process. The curing/imidization step has to be done prior to bonding to prevent the volatile substances from getting trapped at the bond interface. The thermoplastic properties of HD-3007 are used in the wafer bonding process.
6	Place two wafers on the bond fixture with spacers separating the bond surfaces and transfer the fixture to the bond chamber.	

dimensional packaging applications and bonding of temporary handle wafers in, for example, wafer back grinding and thinning processes. HD-3007 is supplied by HD MicroSystems in the USA. Table 3.5 gives and Figure 3.8 depicts an example of a preparation sequence and a bonding program for wafer bonding with HD-3007 as intermediate polymer adhesive, respectively.

3.6
Summary and Conclusions

Wafer bonding with intermediate polymer adhesives is a generic and CMOS-compatible technology that provides unique possibilities for the fabrication,

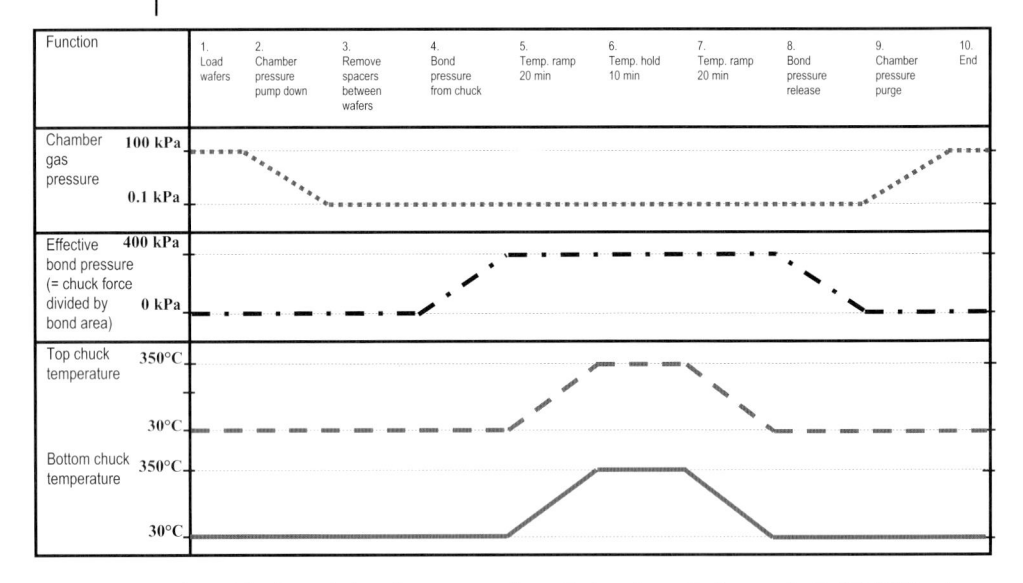

Figure 3.8 Example bonding program for wafer bonding with thermoplastic HD-3007 polymer as intermediate adhesive layer.

integration, and packaging of micro- and nanosystems, such as three-dimensional ICs, MEMS, MEMS/ICs, photonic, radiofrequency, and bio-MEMS. Process schemes and parameters for wafer bonding with self-contained intermediate polymer films and with patterned polymer adhesives (localized polymer adhesive wafer bonding) are readily available in the literature. Polymer adhesives that work specifically well for semiconductor wafer bonding applications include epoxy-like B-stage polymers (e.g., BCB, SU8, nano-imprint resists, and some negative photoresists) and most thermoplastic polymers (e.g., PMMA, nano-imprint resists, copolymers).

The main advantages of polymer adhesive wafer bonding are its insensitivity to surface topography, low bonding temperature, compatibility with standard CMOS IC wafer processing, and ability to join practically any kind of wafer materials. Polymer adhesive wafer bonding requires no special wafer surface treatments such as planarization. Structures and particles at wafer surfaces can be tolerated and compensated to some extent by the polymer adhesive. Polymer adhesive wafer bonding is a simple, robust, and low-cost bonding process.

References

1 Nobel, C. (1992) *Industrial Adhesives Handbook*, Casco Nobel, Fredensborg, Denmark.

2 Niklaus, F., Stemme, G., Lu, J.-Q., and Gutmann, R.J. (2006) Adhesive wafer bonding. *J. Appl. Phys.*, **99** (3), 031101.

3 Traeger, R.K. (1976) Hermeticity of polymeric lid sealant. Proceedings of the Electronic Components Conference, San Francisco, CA, pp. 361–367.

4 Stefan, P., Bioh, K., and James, L. (2008) Temporary bonding/debonding for ultrathin substrates. *Solid State Technol.*, **51**, 60–65.

5 Shuangwu, M.H., Pang, D.L.W., Nathapong, S., and Marimuthu, P. (2008) Temporary bonding of wafer carrier for 3D-wafer level packaging. Proceedings of the Electronic Packaging Technology Conference, Singapore, pp. 405–411.

6 Niklaus, F., Kumar, R.J., McMahon, J.J., Yu, J., Lu, J.-Q., Cale, T.S., and Gutmann, R.J. (2006) Adhesive wafer bonding using partially cured benzocyclobutene (BCB) for three-dimensional integration. *J. Electrochem. Soc.*, **153** (4), G291–G295.

7 Niklaus, F., Enoksson, P., Griss, P., Kälvesten, E., and Stemme, G. (2001) Low temperature wafer level transfer bonding. *IEEE J. Microelectromech. Syst.*, **10** (4), 525–531.

8 Niklaus, F., Kälvesten, E., and Stemme, G. (2001) Wafer-level membrane transfer bonding of polycrystalline silicon bolometers for use in infrared focal plane arrays. *J. Micromech. Microeng.*, **11**, 509–513.

9 Niklaus, F., Haasl, S., and Stemme, G. (2003) Arrays of monocrystalline silicon micromirrors fabricated using CMOS compatible transfer bonding. *IEEE J. Microelectromech. Syst.*, **12** (4), 465–469.

10 Despont, M., Drechsler, U., Yu, R., Pogge, H.B., and Vettiger, P. (2004) Wafer-scale microdevice transfer/interconnect: its application in an AFM-based data-storage system. *J. Microelectromech. Syst.*, **13** (6), 895–901.

11 Christiaens, I., Van Thourhout, D., and Baets, R. (2004) Low-power thermo-optic tuning of vertically coupled microring resonators. *Electron. Lett.*, **40** (9), 560–561.

12 Saharil, F., Wright, R.V., Rantakari, P., Kirby, P.B., Vähä-Heikkilä, T., Niklaus, F., Stemme, G., and Oberhammer, J. (2010) Low-temperature CMOS-compatible 3D-integration of monocrystalline-silicon based PZT RF MEMS switch actuators on RF substrates. Proceedings of MEMS, Hong Kong, China, pp. 47–50.

13 Takato, H. and Shimokawa, R. (2001) Thin-film silicon solar cells using an adhesive bonding technique. *IEEE Trans. Electron Dev.*, **48** (9), 2090–2094.

14 Matsuo, S., Tateno, K., Nakahara, T., and Kurokawa, T. (1997) Use of polyimide bonding for hybrid integration of a vertical cavity surface emitting laser on a silicon substrate. *Electron. Lett.*, **33** (13), 1148–1149.

15 Badihi, A. (1999) Shellcase ultrathin chip size package. Proceedings of Advanced Packaging Materials Conference: Processes, Properties and Interfaces, Braselton, GA, pp. 236–240.

16 Zoberbier, M., Hansen, S., Hennemeyer, M., Tönnies, D., Zoberbier, R., Brehm, M., Kraft, A., Eisner, M., and Völkel, R. (2009) Wafer level cameras – novel fabrication and packaging technologies. Proceedings of the International Image Sensor Workshop, Bergen, Norway.

17 Goetz, M. and Jones, C. (2002) Chip scale packaging techniques for RF SAW devices. IEEE Proceedings of the Electronics Manufacturing Technology Symposium, San Jose, CA, pp. 63–66.

18 Oberhammer, J., Niklaus, F., and Stemme, G. (2003) Selective wafer level adhesive bonding with benzocyclobutene for fabrication of cavities. *Sens Actuators A*, **105** (3), 297–304.

19 Oberhammer, J., Niklaus, F., and Stemme, G. (2004) Sealing of adhesive bonded devices on wafer-level. *Sens Actuators A*, **110** (1–3), 407–412.

20 Lapisa, M., Stemme, G., and Niklaus, F. (2011) Wafer-level heterogeneous integration for MOEMS, MEMS, and NEMS. *IEEE J Sel Top Quant Electron.*, **17** (3), 629–644.

21 Kazlas, P.T., Johnson, K.M., and McKnight, D.J. (1998) Miniature liquid-crystal-on-silicon display assembly. *Optics Letters*, **23** (12), 972–974.

22 Jackman, R.J., Floyd, T.M., Ghodssi, R., Schmidt, M.A., and Jensen, K.F. (2001) Microfluidic systems with on-line UV detection fabricated in photodefinable epoxy. *J. Micromech. Microeng.*, **11**, 263–269.

23 Tong, Q.-Y. and Gösele, U. (1999) *Semiconductor Wafer Bonding: Science and Technology*, John Wiley & Sons, Inc., New York.
24 Yacobi, B.G., Martin, S., Davis, K., Hudson, A., and Hubert, M. (2002) Adhesive bonding in microelectronics and photonics. *J. Appl. Phys.*, **91**, 6227–6262.
25 Alvino, W.M. (1995) *Plastics for Electronics: Materials, Properties, and Design*, McGraw-Hill, New York.
26 Flick, E.W. (1986) *Adhesives, Sealants and Coatings for the Electronic Industry*, Noyes Publications, Park Ridge, NJ.
27 Van der Groen, S., Rosmeulen, M., Baert, K., Jansen, P., and Deferm, L. (1997) Substrate bonding techniques for CMOS processed wafers. *J. Micromech. Microeng.*, **7**, 108–110.
28 Niklaus, F., Decharat, A., Forsberg, F., Roxhed, N., Lapisa, M., Populin, M., Zimmer, F., Lemm, J., and Stemme, G. (2009) Wafer bonding with nano-imprint resists as sacrificial adhesive for fabrication of silicon-on-integrated-circuit (SOIC) wafers in 3D integration of MEMS and ICs. *Sens Actuators A*, **154**, 180–186.
29 Niklaus, F., Enoksson, P., Kälvesten, E., and Stemme, G. (2001) Low temperature full wafer adhesive bonding. *J. Micromech. Microeng.*, **11** (2), 100–107.
30 Bilenberg, B., Nielsen, T., Clausen, B., and Kristensen, A. (2004) PMMA to SU-8 bonding for polymer based lab-on-a-chip systems with integrated optics. *J. Micromech. Microeng.*, **14** (6), 814–818.
31 Bhattacharya, S., Datta, A., Berg, J.M., and Gangopadhyay, S. (2005) Studies on surface wettability of poly(dimethyl) siloxane (PDMS) and glass under oxygen-plasma treatment and correlation with bond strength. *J. Microelectromech. Syst.*, **14** (3), 590–597.
32 Oh, K.W., Han, A., Bhansali, S., and Ahn, C.H. (2002) A low-temperature bonding technique using spin-on fluorocarbon polymers to assemble microsystems. *J. Micromech. Microeng.*, **12**, 187–191.
33 Shores, A.A. (1989) Thermoplastic films for adhesive bonding: hybrid microcircuit substrates. Proceedings of the Electronic Components Conference, Houston, TX, pp. 891–895.
34 Selby, J.C., Shannon, M.A., Xu, K., and Economy, J. (2001) Sub-micrometer solid-state adhesive bonding with aromatic thermosetting copolyesters for the assembly of polyimide membranes in silicon-based devices. *J. Micromech. Microeng.*, **11**, 672–685.
35 Noh, H., Moon, K., Cannon, A., Hesketh, P.J., and Wong, C.P. (2004) Wafer bonding using microwave heating of parylene intermediate layers. *J. Micromech. Microeng.*, **14** (4), 625–631.
36 Wang, X., Lu, L.-H., and Liu, C. (2001) Micromachining techniques for liquid crystal polymer. IEEE Proceedings of MEMS, Interlaken, Switzerland, pp. 126–130.
37 Nguyen, H., Patterson, P., Toshiyoshi, H., and Wu, M.C. (2000) A substrate-independent wafer transfer technique for surface-micromachined devices. Proceedings of MEMS, Miyazaki, Japan, pp. 628–632.
38 Glasgow, I.K., Beebe, D.J., and White, V.E. (1999) Design rules for polyimide solvent bonding. *J. Sens. Mater.*, **11**, 269–278.
39 Pham, N.P., Boellaard, E., Burghartz, J.N., and Sarro, P.M. (2004) Photoresist coating methods for the integration of novel 3-D RF microstructures. *J. Microelectromech. Syst.*, **13** (3), 491–499.
40 Limb, S.J., Labelle, C.B., Gleason, K.K., Edell, D.J., and Gleason, E.F. (1996) Growth of fluorocarbon polymer thin films with high CF_2 fractions and low dangling bond concentrations by thermal chemical vapor deposition. *Appl. Phys. Lett.*, **68** (20), 2810–2812.
41 Schift, H. (2008) Nanoimprint lithography: an old story in modern times? A review. *J. Vac. Sci. Technol.*, **B26** (2), 458–480.
42 Rowland, H.D., Sun, A.C., Schunk, P.R., and King, W.P. (2005) Impact of polymer film thickness and cavity size on polymer flow during embossing: toward process design rules for nanoimprint lithography. *J. Micromech. Microeng.*, **15**, 2414–2425.

43 Taklo, M.M.V., Bakke, T., Vogl, A., Wang, D.T., Niklaus, F., and Balgård, L. (2008) Vibration sensor for wireless condition monitoring. Proceedings of the Pan Pacific Microelectronics Symposium, Hawaii, USA.

44 Carlborg, C.F., Haraldsson, K.T., Cornaglia, M., Stemme, G., and van der Wijngaart, W. (2010) Large scale integrated 3D microfluidic networks through high yield fabrication of vertical vias in PDMS. Proceedings of MEMS, Hong Kong, China, pp. 240–243.

45 Niklaus, F., Enoksson, P., Kälvesten, E., and Stemme, G. (2003) A method to maintain wafer alignment precision during adhesive wafer bonding. *Sens. Actuators A*, **107** (3), 273–278.

46 Lee, S.H., Niklaus, F., Kumar, R.J., Li, H.-F., McMahon, J.J., Yu, J., Lu, J.-Q., Cale, T.S., and Gutmann, R.J. (2006) Fine keyed alignment and bonding for wafer-level 3D ICs. *MRS Symp. Proc.*, **914**, 433–438.

47 HD MicroSystems (2008) Process guide: HD-3007 polyimide adhesive, revision September 2008.

4
Anodic Bonding

Adriana Cozma Lapadatu and Kari Schjølberg-Henriksen

4.1
Introduction

The anodic bonding technique, also known as "field-assisted sealing" or "electrostatic bonding," was described for the first time by Wallis and Pomerantz in 1968 [1, 2]. The process has been demonstrated to be possible with a number of material combinations, but the pair that accounts for most applications is silicon and Pyrex glass. Joining silicon wafers with glass wafers by anodic bonding has proven to be a reliable method for packaging at the wafer level. Currently it is a well-established industrial technique that is reported to account for the majority of packaging applications for microelectromechanical systems (MEMS) [3].

Ceramics can also be used instead of glass [4]. Bonding of glass and ceramics can be done against metals [5], alloys, and other semiconductors, provided the thermal expansion coefficients of the two materials are closely matched. There are certain metals, such as aluminum and Kovar (a Fe–Ni–Co alloy), that because of the thermal mismatch cannot be sealed to glasses in their bulk form, but can be bonded as thin films or foils [2].

Anodic bonding is in essence a relatively simple process that provides high-quality hermetic seals. It involves bonding at elevated temperature, using the assistance of a strong electrostatic field. The bonding temperature is relatively low (300–450 °C for bonding silicon to borosilicate glass wafers), allowing the process to be performed at the end of the fabrication sequence. One of the main advantages of this process is that it can tolerate rougher surfaces than silicon direct bonding and does not impose the same very strict restrictions on the cleanliness of the bonding environment. On the other hand, it involves the use of alkali-containing glasses and it necessitates strong electric fields, which make it incompatible with some microelectronic devices. Another drawback is the thermal residual stress in the bonded structures, although the thermal expansion coefficients of silicon and glasses that are used for bonding are fairly close to each other.

Joining two silicon wafers by anodic bonding using a thin intermediate glass film, deposited by sputtering [6–9] or by vacuum evaporation [10, 11], has also attracted the attention of both academia and industry.

Handbook of Wafer Bonding, First Edition. Edited by Peter Ramm, James Jian-Qiang Lu, Maaike M.V. Taklo.
© 2012 Wiley-VCH Verlag GmbH & Co. KGaA. Published 2012 by Wiley-VCH Verlag GmbH & Co. KGaA.

As dimensions of devices decrease, while their density and complexity increase, the requirements that must be fulfilled by anodic bonding equipment are more and more strict. Accurate alignment with tolerances in the micrometer range is required, often for multiple wafer stacks. A high bonding yield is required and this depends on the ability to ensure a uniform heating and voltage distribution over the wafers. The need to bond in controlled environments (gas composition and pressure) adds to the challenges. With the transition of microsystems technology from the development stage to industrial production, a high throughput becomes necessary for cost-efficient production.

4.2
Mechanism of Anodic Bonding

4.2.1
Glass Polarization

Anodic bonding is an electrochemical process that relies on the polarization of alkali-containing glasses. It can be achieved by setting the two wafers to be bonded together on a hotplate (300–450 °C), so that the glass becomes sufficiently conductive, and applying a high DC voltage to the pair (400–1000 V) such that the glass is negative with respect to the silicon. The bonding temperature is well below the glass transition temperature (~600 °C) and thus there is no macroscopic deformation of the glass during the bonding process.

The phenomenon of glass polarization has been extensively studied, both experimentally and theoretically, under steady-state or slowly varying conditions [12–14]. When a DC voltage is applied to such a glass at elevated temperatures, the alkali cations are depleted from the vicinity of the anode and transported towards the cathode, as illustrated in Figure 4.1. For most of the glasses used for anodic bonding the conduction is primarily realized by Na^+ ions and to a very small extent by K^+ ions. As the positive alkali ions are displaced towards the cathode, they leave

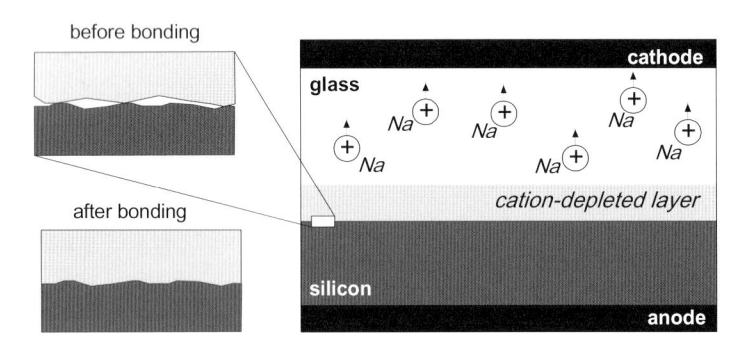

Figure 4.1 Mechanism of silicon–glass anodic bonding.

near the anode an alkali-depleted region. The depth of this region increases in time and depends on the temperature and the applied voltage [13]. The anode must neither provide nor accept mobile ions from the glass. If positive ions are supplied by the anode (called a nonblocking electrode in this case), no polarization occurs, the glass behaves like a resistor, and consequently bonding does not occur. Silver, for example, is a nonblocking anode [12].

Experimental evidence for the alkali-depleted regions has been obtained by different research groups, using ion scattering spectroscopy and electron microprobe analysis [13], secondary ion and secondary neutral mass spectroscopy [15], elastic recoil detection analysis [16], and transmission electron microscopy [17–19].

During the anodic bonding process, the cathode acts as a nonblocking electrode, i.e. it allows the sodium ions to leave the glass. They react with the surface water and the humidity of the air and form sodium hydroxide, a white substance that accumulates on the glass surface and on the cathode, around the points of their contact, partially etching the glass. Additionally, electrode erosion takes place, reducing its lifetime.

4.2.2
Achieving Intimate Contact

Because of surface roughness, the initial contact between silicon and glass occurs only at a few locations. The two wafers are separated over almost their entire area by a small gap (see the detail in Figure 4.1), as indicated by interference fringes present at the interface. Due to the fact that the rest of the glass is conductive, almost the entire applied voltage drops over the growing alkali-depleted layer and this gap. The resulting electric field is very high, of the order of several megavolts per centimeter. Since electrostatic forces of attraction are largest where the gap is narrowest, that is, around the periphery of the contact points, the nearby regions will be the next to be pulled together. In this way, the area of contact spreads outwards from the initial contact points across the entire wafer. This is clearly shown by the disappearance of the interference fringes from the interface. A large electrostatic field persists in the depletion layer, maintaining the parts under appreciable electrostatic pressure.

Planar cathodes are used in most applications because they create a more homogeneous electric field distribution and the bonding efficiency is improved considerably. Bonding starts simultaneously in more points on the wafer, which decreases the total bonding time. On the other hand, various bond fronts can lead to entrapment of gas at the interface and consequently to nonbonded areas. If channels are etched in one of the wafers, connected to the wafer edges, air can escape and this problem is avoided. If a point cathode is used, the bonding front spreads outwards from a spot beneath the electrode, preventing any air being trapped at the interface. However, using point electrodes is inefficient and less reliable, especially at low temperatures and voltages. The process time increases considerably and the quality of the seal degrades towards the edges of the wafer [20].

Experimental studies have shown that the temperature is a most important parameter determining intimate contact. The voltage required to achieve contact at constant time decreases with increasing temperature.

Since the electrostatic forces across any air gap are dependent on the square of the voltage and the inverse square of the gap depth, a low voltage and a high roughness will both inhibit intimate contact. Surface roughness, topography steps, and foreign particles on the mating surfaces must all be well controlled for good bonding. Foreign particles produce nonbonded regions around them that can jeopardize the hermeticity of the seal. The dimensions of these regions depend on the bonding temperature, the height of the particles, and the applied voltage. For small steps and low surface roughness, the intimate surface contact required for bonding can be achieved by local elastic, plastic, or viscous deformation of at least one of the surfaces [19, 21]. Due to the contribution of the electrostatic forces, anodic bonding is less sensitive to the imperfections of the two surfaces than direct bonding.

When an oxide is present at the interface, the bonding voltage is shared between the oxide and the interface gap and it is therefore more difficult to achieve intimate contact between the wafers. The bonding time increases dramatically with increasing oxide thickness and, if the oxide is too thick, bonding does not occur at all.

It has been reported that a hydrophilic silicon surface has a higher rate of formation of intimate contact than a hydrophobic surface [22] and that contact is achieved somewhat faster for p-type than for n-type silicon [23].

4.2.3
Interface Reactions

When silicon and glass are in intimate contact, chemical reactions take place at the interface, resulting in the oxidation of the silicon substrate and consequently in permanent bonds with the glass. The formation of the oxide layer is limited by the decreasing drift of charged species, caused by the decrease of the electric field strength as the thickness of the depleted layer increases.

Several published studies suggest that Si–O–Si bonds are formed at the silicon–glass interface. The source of the oxygen is considered to be nonbridging oxygen ions from the alkali-depleted layer that are displaced towards the anode [13], oxygen present on the silicon and glass surfaces especially in the form of hydroxyl groups, or oxygen from water dissociation that occurs in the "leached" superficial glass layer [16].

The formation of a thin oxide layer during bonding has been proven experimentally. Selectively etching away the glass that had been bonded to silicon and subsequent surface profiling revealed a 10 nm step between the previously bonded and nonbonded areas [24]. This indicates that silicon is consumed during anodic bonding and an oxide is formed. Elastic recoil detection analysis of anodically bonded silicon–glass pairs [16] showed a 4–8 nm shift of the oxygen profile towards the anode, suggesting that the oxidation of the anode material had taken place during glass polarization. This shift was not observed for thermal treatments without an applied voltage, demonstrating that anodic bonding is not a purely thermally activated process. Transmission electron microscopy together with energy loss spectroscopy or with marker experiments made it possible to demonstrate that an oxide layer of 5–20 nm was formed during silicon–glass anodic bonding [25].

4.3
Bonding Current

Monitoring the electrical current through the wafer pair is a method to follow the development of the bonding process. Initially, the bonding current increases sharply due to the increase in the contact area as the wafers are pulled into intimate contact by the electrostatic forces. Simultaneously a competing effect results from the fact that the depth of the cation-depleted layer increases, causing a decrease of the polarization current. The resulting bonding current reaches a peak, then gradually decreases until it reaches a relatively stable low value.

Figure 4.2 shows typical current–time characteristics, measured during bonding of silicon and glass pieces with an area of 1 cm^2 and a thickness of 0.5 mm [26]. The parameters are the applied voltage and the bonding temperature. The higher the temperature and/or the voltage the higher the peak current and the shorter the time required for complete bonding.

The glass thickness also affects the bonding time and the final result of the bonding process. A thicker glass means a higher electrical resistance, decreasing

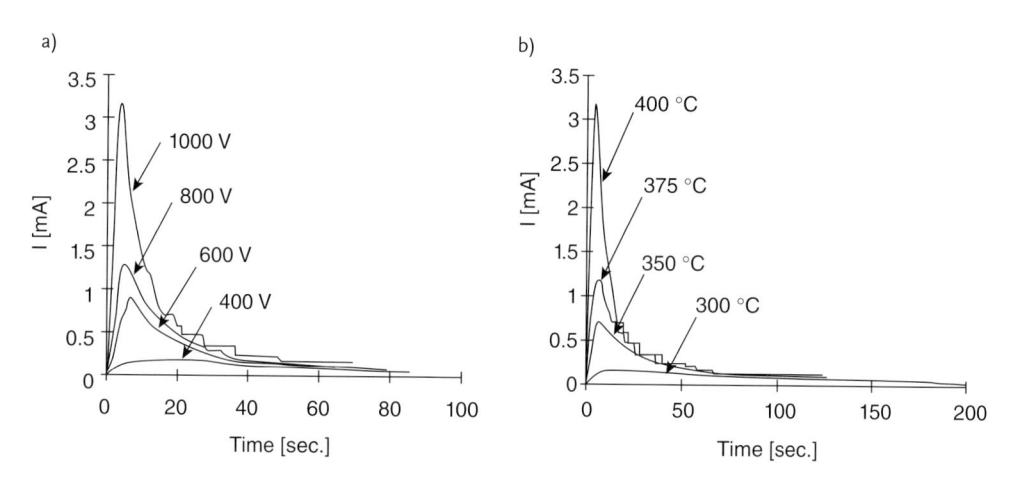

Figure 4.2 Current–time bonding characteristics: (a) variation of voltage at fixed temperature of 400 °C; (b) variation of temperature at fixed voltage of 1000 V.

the effective applied voltage to the glass–silicon interface and thus generating a weaker bond.

It has been reported [27] that bonding is complete for various experimental conditions at constant charge. The same authors point out, however, that it is very difficult to relate quantitatively the charge measured in the external circuit to the interfacial reaction kinetics, and ultimately to bond formation. Nevertheless, it has been observed experimentally that the magnitude of the total electrical charge as illustrated by the bonding current provides information about the quality and the strength of the bond [26, 28].

Bonding can also be performed in vacuum. Because of the poor heat transfer, the bonding is achieved much more slowly than at atmospheric pressure, a fact also illustrated by the appearance of the bonding current curve. At intermediate pressures electrical breakdown of the gas in the bonding chamber occurs (according to Paschen's law [29]), setting limits to the pressure range that can be used for bonding. The exact values of these limits depend to some extent on the configuration of the bonding equipment, the gas composition, and the particularities of the mechanical design of the structures being bonded.

Even when bonding is performed in high vacuum one has to take care to avoid plasma discharging that may occur in the residual gas in the sealed cavities. This residual gas is the result of gas release from the inner surfaces of the cavities and of the reactions that lead to bonding.

4.4
Glasses for Anodic Bonding

From the description of the bonding mechanism it is obvious that a first criterion for the choice of a glass suitable for anodic bonding is its content of alkali ions, so that the glass becomes sufficiently conductive at the bonding temperature. A second criterion is the degree of thermal mismatch between the glass and the anode material. Depending on the specific application, other characteristics may become important, such as a lower temperature at which the glass becomes conductive, optical properties, and resistance to specific chemicals.

Two of the most-used glass materials are Corning 7740 (Pyrex) and Schott Borofloat, which for the purposes of anodic bonding are in essence identical. Both these types are long-standing compositions, developed for other applications several decades ago. They were found suitable for anodic bonding due to their sodium content and the close match between their coefficients of thermal expansion and that of silicon.

HOYA's SD-2 glass was specially engineered for anodic bonding to silicon wafers. The curve representing its thermal expansion coefficient as a function of temperature matches closely that of crystalline silicon, in order to minimize the distortion or bowing effect caused by the thermal mismatch between the two wafers.

4.5
Characterization of Bond Quality

The evolution of the bonding current is a valuable source of information about the progress of the bonding process. In a typical processing sequence the bonding current, the temperature, the applied voltage, and the pressure in the bonding chamber are monitored continuously.

A high-quality seal is characterized by high mechanical strength and low residual stress in the bonded structures, and, in the case of hermetic seals, hermeticity over the product lifetime. Often a low, well-controlled gas pressure is required in the sealed cavities. These parameters are very difficult to measure quantitatively for product chips. The measurements are destructive and/or time consuming, often requiring special test structures. Characteristic values for a certain process can be obtained during product and process development.

The bond strength can be measured by fracture tests, such as pressure, pull, shear, or bending tests. The values obtained depend on the geometry of the test structures and on the method used to apply the force. Usually, they are found in the range 5–45 MPa [2, 30, 31].

It is a commonly observed phenomenon that cracking occurs in the glass, which is most often interpreted as the seal being stronger than the glass. According to the studies presented in [30], this behavior is rather a reflection of the asymmetric stress that develops during interfacial cracking between two dissimilar materials. The interface constitutes an irregularity and stress concentrations are unavoidable at irregularities. There are three main stress components that contribute to fracture: externally applied loads, thermal mismatch between the two dissimilar materials, and stress caused by compressed foreign particles at the interface.

The thermal residual stress generated during cooling from the bonding temperature to room temperature is another criterion for evaluating bond quality. It is caused by the difference between the thermal expansion coefficients of silicon and glass. The bonding-induced strain in silicon can be calculated from curvature measurements performed before and after the bonding process. They reveal that bonding at the lower end of the processing temperature range induces a compressive thermal residual stress in silicon. The higher process temperatures lead to tensile stress in silicon. For a given silicon thickness, there is for each thickness of glass an intermediate temperature that provides a no-stress bond. For a 0.5 mm thick glass samples that temperature was found to be 315 °C, while for 1.5 mm thick glass it was around 260 °C [26]. Bonding with thicker glass induces higher stress in silicon. Similar results were also obtained from Raman spectroscopy [26].

The hermeticity of the seal can be verified using several methods, their efficiency depending on the requirements of the specific application. Bombing the devices with gases like helium, argon, nitrogen, radioactive krypton, and neon has been reported for testing the hermeticity of vacuum-encapsulated devices. Long-term measurements of various parameters of bonded devices can reveal the increase of pressure in the cavities in the case of leaking seals. Measuring the pressure-dependent quality factor of resonating micromechanical devices is one of the most sensitive methods [32]. Note that after long-term exposure at temperatures above 300 °C, the diffusion of gas across 30–200 μm bonded regions and in the materials themselves can be substantial even if the seal is hermetic at room temperature [33].

More recently, an ultrafine leak test based on Q-factor monitoring was reported [34], the so-called neon ultrafine leak test, which has the potential to be used for in-line critical leak rate screening at the wafer level. It can only be applied for resonating devices.

Through consideration and experience, it has been found that in production a good method for screening dice with faulty silicon–glass bonding areas is visual inspection after dicing the wafer [35]. Nonbonded areas will appear as areas with significantly lighter gray color or, if the separation between glass and silicon is larger, colored interference fringes will appear.

4.6
Pressure Inside Vacuum-Sealed Cavities

Many micromechanical devices need a very low pressure in their cavities. For sealed absolute pressure sensors this avoids the problems induced by the temperature dependency of the gas in the reference cavity. For inertial sensors, the cavity pressure determines the damping and, thus, their dynamic behavior [36]. The high quality factors required for resonant structures for high resolution and wide dynamic range can only be obtained by reducing the air damping and thus the pressure [37]. In infrared sensors, thermal conductance between the sensing element and its package must be minimized for high sensitivity, which can be obtained by packaging in high vacuum. Vacuum encapsulation of radio-frequency switches is required to improve contact reliability and switching time.

Anodic bonding in high vacuum is possible, but the pressure in the cavities after bonding is higher than the pressure in the bonding chamber [38, 39]. The gases released from the inner surfaces of the cavities and those generated in the reactions that lead to bonding are trapped in the cavities and influence the performance of sealed devices. The residual gas pressure from anodic bonding of bulk micromachined sensors can be as low as 1 mbar [32], but also as high as 500 mbar for devices with very low cavity volume [38] for capacitive pressure sensors. Even for the same type of device, wafer-to-wafer and chip-to-chip variations between 0.5 and 20 mbar have been observed.

Reported methods to measure the residual pressure in anodically bonded cavities include measurements of membrane deflections [38–40], characterization of the temperature dependency of the output signal of pressure sensors [41], and measurements of the Q-factor of resonant structures [37, 42–46]. Of all these, the Q-factor method is recognized as the most reliable and sensitive method. At low pressures this parameter is strongly dependent in a very reproducible manner on the ambient pressure. The accuracy of the method based on membrane deflection was estimated to be in the range ±5 mbar [39], affected by the bonding-related thermal residual stress in the membranes. As for the characterization of the temperature dependency of the output signal of pressure sensors, reference test structures and/or reference measurements after puncturing the cavities are required in order to separate the influence of other parameters on the output signal, such as offset voltages caused by bonding-related residual stress or lithography inaccuracies.

For devices that require high-vacuum encapsulation, innovative techniques are required in order to solve the problems of residual gases. Examples reported in the literature are the use of getter materials and diffusion barriers to gas atoms [42, 47].

4.7
Effect of Anodic Bonding on Flexible Structures

An important issue for the design of micromechanical structures to be electrostatically bonded is how to avoid the bonding/sticking of the flexible silicon parts to the glass. For the usual dimensions of micromachined devices and small separation gaps, the electrostatic forces are high enough to produce collapse of the flexible parts onto the glass. Therefore, special precautions must be taken to avoid these problems. The alternatives include: increasing the gap between the silicon and glass to values for which the electrostatic attraction decreases to safe values; the use of shield electrodes that keep the silicon and the glass at the same potential in the regions not to be bonded; and the selective use of intermediate layers that prevent bonding [48]. The first of these solutions is not suitable for capacitive devices for example, since they require small gaps in order to provide acceptable mechanical sensitivities. When using a shield electrode, special care must be taken with its shape, because electrostatic forces develop in any region without shield and, depending on the opening size and on the flexibility of the silicon underneath, undesired bonding may still occur.

4.8
Electrical Degradation of Devices during Anodic Bonding

Some of the devices that are sealed by anodic bonding are purely mechanical structures, like microfluidic mixers and inkjet print heads. However, more often than not, the devices being sealed contain electrical components. These could be implanted resistors on a membrane, diodes, or capacitors. There are also some examples of monolithically integrated devices containing a sensor and its readout electronics on a chip being sealed by anodic bonding. When packaging devices with electrical components, two main concerns are commonly expressed. The first concern is whether sodium ions from the glass may contaminate the device. The second concern is whether the high electric field applied during anodic bonding may damage the device being encapsulated. The following subsections give an overview of research on these two issues.

4.8.1
Degradation by Sodium Contamination

Sodium is the main cause of instability in metal–oxide–semiconductor (MOS) devices [49, 50]. Sodium ions are charged and highly mobile in SiO_2. Once present in the SiO_2 on the surface of a device, Na^+ will cause changes in the surface potential that vary across the area of the device and with time. Therefore, the presence of sodium in the SiO_2 may change and destabilize the resistance of underlying implanted resistors. Sodium causes shift and drift of the threshold voltage of MOS transistors and may also change the leakage current and breakdown voltage of p–n-junction diodes [50]. Sodium neutralization methods and improved laboratory cleanliness and chemicals now permit the routine fabrication of stable MOS devices. Nevertheless, furnace tubes and processing in general are monitored closely to detect possible occurrence of ionic contamination, even in modern integrated circuit fabrication plants.

The glasses used for anodic bonding contain approximately 4 wt% sodium. Whenever anodic bonding and electrical components are mentioned together, it is emphasized that the sodium content in the glass is a potential problem [51–54]. During anodic bonding, the sodium ions in the glass move and a sodium-depleted zone is formed at the interface between the glass and the wafer to be bonded. The sodium ions are allowed to leave the glass, reacting with surface water to form NaOH at the cathode. Hence, sodium ions are indeed present and movable during the anodic bonding process.

Schjølberg-Henriksen et al. [55] assessed the ionic contamination of SiO_2 during anodic bonding. MOS capacitors with either a solid rectangular gate (SG) or a gate consisting of thin metal fingers (FG) were used as test structures. The capacitor wafers were bonded to Corning 7740 (Pyrex) glass wafers. Cavities enclosing the capacitors were made by sawing the glass or etching the silicon. The vertical distance between the gate metal and the ceiling of the glass cavity was either $d = 2\,\mu m$ or $d = 200\,\mu m$. In addition, some capacitors on each wafer were situated outside the bonded area. On one wafer, a 1000 Å thick film of plasma-enhanced chemical vapor deposited Si_3N_4 was deposited on top of the gate aluminum. The nitride layer covered the exposed oxide completely, but openings were made on top of the probing pads so that the capacitors could be measured. Figure 4.3 shows a schematic of the silicon–glass stack and the various capacitor locations.

In their work, Schjølberg-Henriksen et al. reported that anodic bonding caused sodium contamination of the SiO_2 situated both within and outside glass cavities. The increase of sodium ion concentration was of the order of $10^{13}\,cm^{-2}$ for capacitors situated outside the glass, about $10^{11}\,cm^{-2}$ for capacitors with $d = 2\,\mu m$, and about $10^{10}\,cm^{-2}$ for capacitors with $d = 200\,\mu m$. Thus, the contamination depended both on the location of the oxide and on the geometry of the glass cavity. Moreover, the aluminum gate was found to partly shield the oxide from contamination. The

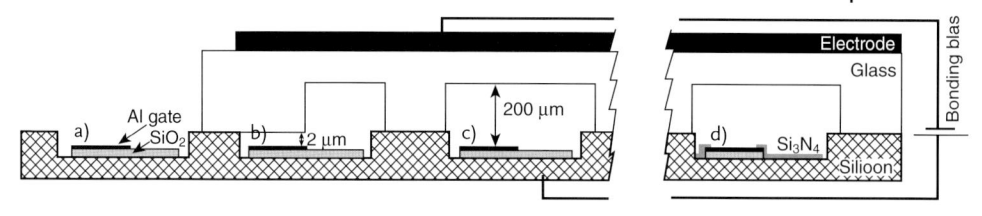

Figure 4.3 Silicon–glass stack with MOS capacitors. The hatched area is silicon, white is Pyrex glass, light gray is oxide, and black is the aluminum gate electrode. The plate electrode used during bonding is shown on top of the glass. The three different capacitor locations are (a) outside the glass, (b) within the glass with $d = 2\,\mu m$, and (c) within the glass with $d = 200\,\mu m$. (d) Capacitor with protective nitride coating and $d = 200\,\mu m$.

increase in sodium ion concentration was between 2 and 5 times larger for FG capacitors than for SG capacitors. This was attributed to mechanical shielding of the SiO$_2$ by the gate aluminum. The nitride film was found to be very beneficial in protecting the capacitors. The nitride prevented sodium contamination of the capacitors situated within glass cavities, and reduced the contamination of the capacitors situated outside the glass. The reported protective property of nitride agreed well with the work of Visser *et al.* [33], who found that a 1000 Å thick film of nitride served as diffusion barrier for sodium in a 3 μm thick film of sputtered glass.

4.8.2
Degradation by High Electric Fields

During anodic bonding, a high voltage is applied between the wafers that are to be bonded together. Electric fields will be present across cavities formed between the wafers, and the fields may damage devices situated in these cavities.

4.8.2.1 Shielding of Devices

Various solutions for shielding devices in cavities have been presented in the literature. In the accelerometer of Gianchandani *et al.* [56], a metal plane was deposited on the glass surface above CMOS circuits. During anodic bonding, the metal plane was ground, shielding the circuitry in the cavity from the high electric fields. In the devices of Matsumoto and Esashi [57, 58], the polysilicon gates of MOS transistors were connected to the silicon substrate during anodic bonding. It was reported that damage was avoided if the process time was kept below 30 min so that aluminum spiking did not occur. Chavan and Wise [59] combined the two protection methods during packaging of a monolithic pressure sensor. A metal pattern was deposited on the glass to block the electric field in the circuit area. In addition, the signal lines were shorted to the silicon bulk during anodic bonding by a jumper deposited after wafer probe testing. After bonding, the shorts were removed when the lightly doped bulk silicon was dissolved during ethylenediamine pyrocatechol etch.

It may not be necessary to shield the electronics completely from the electric field. Von Arx *et al.* [60] packaged electronics intended for invasive use in a custom-fabricated glass capsule. The capsule was anodically bonded to polysilicon, applying 2000 V at 320 °C for about 10 min. The cavity of the capsule was about 2 mm deep. According to the authors, this large distance prevented the electric field from becoming large enough to damage MOS devices.

For certain applications, shielding of the electronic components may not be required at all. Schabmueller *et al.* [61] presented a single-chip system for a polymerase chain reaction system with fiber optics. The diode detector of the system was situated in a chamber which was sealed by anodic bonding. The diode was degraded by the anodic bonding, but could still be used for light detection.

4.8.2.2 Mechanism of Degradation by Electric Fields

The design and requirements of a single-chip system decide if shielding of the electronics during anodic bonding is necessary. The shielding methods discussed above were found to provide sufficient protection for their respective circuits. However, a general understanding of the electrical effects during anodic bonding could be useful. With such an understanding, system-specific trial and error would be redundant.

A limited number of studies have been carried out on the electrical effects of anodic bonding. The silicon surface underneath the bonded interface has been investigated by Lapadatu and coworkers. Anodic bonding gave decreased resistivity, decreased breakdown voltage of p–n junctions, and increased leakage currents. The damage was attributed to ionic movements during bonding, the formation of a thin oxide layer, and related charging effects [62].

Decroux *et al.* [63] measured MOS capacitors situated in glass holes with different diameters. Even though the capacitors were situated outside the glass, significant effects of anodic bonding on the MOS structures were observed. Shifts in the flat-band voltage and a large increase in the interface trap density occurred during bonding in all the capacitors, regardless of hole diameter. The damage was suggested to result from ionization of the air in the hole, which enabled injection of electrons from the silicon substrate into the oxide. The damage was reported to be avoided by covering the capacitor with a metal layer that was connected via a diode to the substrate. The arrangement discharged the metal at high temperatures, that is, during bonding [64].

There have been a limited number of studies on the electrical effects of anodic bonding. Shirai and Esashi [65] and van der Groen *et al.* [66] monitored MOS structures situated within glass cavities. The structures were characterized before and after anodic bonding. In the work of van der Groen *et al.*, the MOS structures were severely degraded or even completely destroyed during anodic bonding. Those authors found that damage could be modeled by a parallel capacitor, but no further explanation of the origin of the parallel capacitor was suggested.

Shirai and Esashi found that bonding induced threshold voltage shifts of about 3 V for n-type MOSFETs and about 6 V for p-type MOSFETs. The shifts were not observed if the gates of the transistors were connected to the silicon substrate during bonding. They believed the shifts to be caused by negative-bias temperature instability (NBTI). NBTI is an oxide-degrading mechanism that is caused by the combination of applied electric field in the megavolt per centimeter range at temperatures above 250 °C. NBTI increases the fixed oxide charge, N_f, and the interface trap density, D_{it}, by equal amounts. A further description of NBTI is given in [67].

Schjølberg-Henriksen et al. [68] investigated the increases in oxide charge and interface trap density using MOS capacitors as test structures. The capacitors were situated outside the glass and within glass cavities as shown in Figure 4.3. D_{it} and N_{eff} were measured before and after bonding, where N_{eff} was defined as the effective oxide charge, consisting of the oxide fixed charge, the oxide trapped charge, and a contribution from the mobile ions present in the oxide. Within glass cavities, the capacitors with $d = 2\,\mu m$ had an increase in N_{eff} of about $4 \times 10^{11}\,cm^{-2}$. For the capacitors with $d = 10\,\mu m$ and $d = 200\,\mu m$, the increase was in the region of $10^{10}\,cm^{-2}$. D_{it} increased by quantities similar to the increases in N_{eff}. The increases in N_{eff} and D_{it} were proposed to be caused by NBTI. For capacitors situated outside the glass, the measurements were similar to those of Decroux et al. [63]. Instead of the explanation of Decroux et al. of electron injection, Schjølberg-Henriksen et al. suggested the damage was caused by sodium contamination originating from the bulk glass.

4.9
Bonding with Thin Films

As an alternative to bonding with bulk glass, silicon–silicon anodic bonding with a thin intermediate glass layer takes advantage of the relatively low temperature specific to anodic bonding and the low residual stress characteristic of the bonding of two wafers made of the same material. Because the glass layer is very thin (a few micrometers), its influence on the thermal behavior of the bonded ensemble is negligible for most applications.

The glass layer can be deposited either by sputtering [6–9] or evaporation [10, 11]. The former is a very slow process and requires a considerable time to deposit suitable glass layers. Although successful bonding with very thin glass layers (20 nm) at low voltages (10 V) has been reported in the literature [9], other authors noted that layers thinner than 0.5 μm do not bond [7, 8]. Evaporation yields deposition rates three orders of magnitude higher (up to $20\,nm\,s^{-1}$) compared to sputtering, making this process more suitable for production.

The bonding procedure with intermediate glass is basically the same as that described for bulk glass, with the mention that the negative voltage is applied to the glass-coated wafer. Nevertheless, some particularities arise from the configuration of the wafer pair, as described below.

Since at the bonding temperature silicon acts as a metal, an equipotential surface establishes immediately across the glass-coated silicon wafer, regardless the shape of the cathode of the bonding equipment, and therefore bonding begins simultaneously in many points. This is beneficial for the speed of the process, but may result in trapped gas at the interface and thus in nonbonded regions. With proper design, the otherwise trapped gas can escape through channels connected to the wafer edge.

As far as the bonding temperature is concerned, it has conflicting effects. On the one hand, the deformability of the glass layer increases with temperature, and hence its capability to overcome the imperfections of the surfaces [21]. The mobility of the sodium ions also increases with temperature, thus decreasing the glass resistivity and increasing the electrostatic forces developed between the two surfaces in contact. On the other hand, the breakdown voltage of the glass layer, and hence the maximum voltage that can be used for bonding, decreases as the temperature increases. Consequently, both the opposition to bonding and the bonding driving force are limited at higher temperatures.

One important problem encountered in bonding with sputtered layers is the production of a glass layer that can withstand the high electric field required for bonding. Point defects in the glass layer, hard foreign particles present at the interface, and small defects at the edges of the wafer originating from handling can cause the breakdown to occur at much smaller electric fields than for defect-free layers. An improvement can be obtained if the glass is deposited on top of a thermally grown oxide and if the wafer is thermally annealed prior to bonding. Such a treatment (e.g., 3.5 h at 550 °C in nitrogen) is also recommended for stress relief in the sputtered layers [9].

Because of the lower applied voltage, the electrostatic forces that develop during bonding are lower than in the case of conventional anodic bonding, making this process much more sensitive to surface imperfections.

Although there has been some time since this method was first proposed, there are still no reported industrial breakthrough for bonding with sputtered glass, probably because of the drawbacks mentioned.

The much higher glass deposition rates achievable by evaporation allow thicker and hence more breakdown-resistant glass layers to be obtained in shorter times. Glass layers of 5–7 µm in thickness were successfully used to bond two silicon wafers, with high yield at bonding temperatures greater than 350 °C. A high compressive stress (225 MPa) exists in the as-deposited layers, but it is reported that this stress can be relieved by annealing at 340 °C [69]. A bonding-induced, relatively small compressive stress has been noted. It is mainly influenced by the geometry and the applied force of the bonding electrodes.

4.10
Conclusions

Anodic bonding is by its physical and chemical nature a method with a high potential resulting in reliable hermetic seals. There are adverse effects acting on the bonded structures that must be taken into account and dealt with during the process and product development. This technique becomes then a reliable tool for the wafer-level bonding of MEMS structures, optics, fluidics, and bio-MEMS. The method is used in an increasing number of industrial products and it is also offered as part of industrial foundry services.

More work is needed to minimize the negative effects on mechanical flexible parts and to develop robust designs that limit ionic contamination and charging effects when bonding on and near electrically active devices.

References

1 Pomerantz, D.I. (1968) Anodic bonding. US Patent 3,397,278.
2 Wallis, G. and Pomerantz, D.I. (1969) Field-assisted glass–metal sealing. *J. Appl. Phys.*, **40**, 3946–3949.
3 Linder, P., Dragoi, V., Farrens, S., Glinsner, T., and Hangweier, P. (2004) Advanced techniques for 3D devices in wafer-bonding processes. *Solid State Technol.*, **47** (5), 55–58.
4 Gustafsson, K., Hok, B., Johansson, S., and Murray, T. (1988) Field-assisted bonding of a machinable ceramic to silicon and metals. *Sens. Mater.*, **2**, 65–72.
5 Pomerantz, D.I. (1968) Bonding electrically conductive metals to insulators. US Patent 3,417,459.
6 Brooks, A.D., Donovan, R.P., and Hardesty, C.A. (1972) Low-temperature electrostatic silicon-to-silicon seals using sputtered borosilicate glass. *J. Electrochem. Soc.*, **119** (4), 545–546.
7 Esashi, M., Nakano, A., Shoji, S., and Hebiguchi, H. (1990) Low-temperature silicon-to-silicon anodic bonding with intermediate low melting point glass. *Sens. Actuators A*, **23** (1–3), 931–934.
8 Hanneborg, A., Nese, M., Jakobsen, H., and Holm, R. (1992) Silicon-to-thin film anodic bonding. *J. Micromech. Microeng.*, **3**, 117–121.
9 Berenschot, J.W., Gardeniers, J.C.E., Lammerink, T.S.J., and Elwenspoek, M. (1994) New applications of RF sputtered glass films as protection and bonding layers in silicon micromachining. *Sens Actuators A*, **41–42**, 338–343.
10 de Reus, R. and Lindahl, M. (1997) Si-to-Si wafer bonding using evaporated glass. Technical Digest of the International Conference on Solid-State Sensors and Actuators (Transducers 97), Chicago, IL, 16–19 June 1997, pp. 661–664.
11 Choi, W.B., Ju, B.K., Lee, Y.H., Haskard, M.R., Sung, M.Y., and Oh, M.H. (1997) Anodic bonding technique under low temperature and low voltage using evaporated glass. *J. Vac. Sci. Tecnol.*, **B15**, 477–481.
12 Wallis, G. (1970) Direct-current polarization during field-assisted glass–metal sealing. *J. Am. Ceram. Soc.*, **53** (10), 563–567.
13 Carlson, D.E., Hang, K.W., and Stockdale, G.F. (1972) Electrode "polarization" in alkali-containing glasses. *J. Am. Ceram. Soc.*, **55**, 337–341.
14 Carlson, D.E. (1974) Ion depletion of glass at a blocking anode: I. Theory and experimental results for alkali silicate glasses. *J. Am. Ceram. Soc.*, **57** (7), 291–294.
15 Gossnik, G. (1978) SIMS analysis of a field-assisted glass-to-metal seal. *J. Am. Ceram. Soc.*, **61** (11–12), 539–540.
16 Nitzsche, P., Lange, K., Schmidt, B., Grigull, S., Kreissig, U., Thomas, B., and Hertog, K. (1998) Ion drift processes in Pyrex-type alkali-borosilicate glass during anodic bonding. *J. Electrochem. Soc.*, **145** (5), 1755–1762.
17 Van Helvoort, A.T.J., Knowles, K.M., and Fernie, J.A. (2003) Characterization of cation depletion in Pyrex during electrostatic bonding. *J. Electrocem. Soc.*, **150** (10), G624–G629.
18 Xing, Q.F., Yoshida, M., and Sasaki, G. (2002) TEM study of the interface of the anodic-bonded silicon/glass. *Scr. Mater.*, **47**, 577–582.
19 Knowles, K.M. and van Helvoort, A.T.J. (2006) Anodic bonding. *Int. Mater. Rev.*, **51** (5), 273–311.

20 Plaza, J.A., Esteve, J., and Lora-Tamayo, E. (1997) Nondestructive anodic bonding test. *J. Electrochem. Soc.*, **144** (5), L108.

21 Anthony, T.R. (1983) Anodic bonding of imperfect surfaces. *J. Appl. Phys.*, **54**, 2419–2428.

22 Lee, D.J., Ju, B.K., Jang, J., Lee, K.B., and Oh, M.H. (1999) Effects of a hydrophilic surface in anodic bonding. *J. Micromech. Microeng.*, **9**, 313–318.

23 Lee, T.M.H., Lee, D.H.Y., Liaw, C.Y.N., Lao, A.I.K., and Hsing, I.M. (2000) Detailed characterization of anodic bonding process between glass and thin-film coated silicon substrates. *Sens. Actuators A*, **86**, 103–107.

24 Baumann, H., Mack, S., and Münzel, H. (1995) Bonding of structured wafers, *3rd International Symposium on Semiconductor Wafer Bonding: Science, Technology and Applications*, vol. 95-7, Electrochemical Society, Pennington, NJ, p. 471.

25 Van Helvoort, A.T.J., Knowles, K.M., Holmestad, R., and Fernie, J.A. (2004) Anodic oxidation during electrostatic bonding. *Philos. Mag.*, **84**, 505–519.

26 Cozma, A. and Puers, R. (1995) Characterisation of the electrostatic bonding of silicon and Pyrex glass. *J. Micromech. Microeng.*, **5**, 98–102.

27 Albaugh, K.B. and Rasmussen, D.H. (1992) Rate processes during anodic bonding. *J. Am. Ceram. Soc.*, **75**, 2644–2648.

28 Rogers, T., Aitken, N., Stribley, K., and Boyd, J. (2005) Improvements in MEMS gyroscope production as a result of using in situ, aligned, current-limited anodic bonding. *Sens. Actuators A*, **123–124**, 106–110.

29 Paschen, F. (1889) *Ann. Physik*, **273** (5), 69–75.

30 Obermeier, E. (1995) Anodic wafer bonding, *3rd International Symposium on Semiconductor Wafer Bonding: Science, Technology and Applications*, vol. 95-7, Electrochemical Society, Pennington, NJ, p. 212.

31 Johansson, S., Gustafsson, K., and Schweith, J. (1988) Influence of bonded area ratio on the strength of FAB seals between silicon microstructures and glass. *Sens. Mater.*, **4**, 209–221.

32 Corman, T., Enoksson, P., and Stemme, G. (1998) Low-pressure-encapsulated resonant structures with integrated electrodes for electrostatic excitation and capacitive detection. *Sens. Actuators A*, **66** (1–3), 160–166.

33 Visser, M.M., Moe, S.T., and Hanneborg, A.B. (2001) Diffusion at anodically bonded interfaces. *J. Micromech. Microeng.*, **11**, 376–381.

34 Reinert, W. (2006) Industrial wafer level vacuum encapsulation of resonating MEMS devices. Symposium on Design, Test, Integration and Packaging of MEMS/MOEMS, DTIP 2006, Stresa, Lago Maggiore, Italy, 26–28 April 2006.

35 Jakobsen, H., Cozma Lapadatu, A., and Kittilsland, G. (2001) Anodic bonding for MEMS, *Semiconductor Wafer Bonding IV: Science, Technology and Applications*, vol. 97-36, Electrochemical Society, Pennington, NJ, pp. 243–254.

36 Minani, K., Moriuchi, T., and Esashi, M. (1995) Cavity pressure control for critical damping of packaged micro mechanical devices. Proceedings of the 8th International Conference on Solid-State Sensors and Actuators, Eurosensors IX, Stockholm, Sweden, 25–29 June 1995, p. 240.

37 Corman, T., Enoksson, P., and Stemme, G. (1997) Gas damping of ellectrostatically excited resonators. *Sens. Actuators A*, **61** (1–3), 249–255.

38 Henmi, H., Shoji, S., Shoji, Y., Yoshimi, K., and Esashi, M. (1994) Vacuum packaging for microsensors by glass–silicon anodic bonding. *Sens. Actuators A*, **43** (1–3), 243–248.

39 Mack, S., Baumann, H., Gösele, U., Werner, H., and Schögl, R. (1997) Analysis of bonding-related gas enclosure in micromachined cavities sealed by silicon wafer bonding. *J. Electrochem. Soc.*, **144** (3), 1106–1111.

40 Minami, K., Moriuchi, T., and Esashi, M. (1995) Cavity pressure control for critical damping of packaged micromechanical devices. Technical Digest of the 7th International Conference on Solid-State Sensors and Actuators (Transducers 95), Stockholm, Sweden, 25–29 June 1995, p. 240.

41 Gooch, R., Schimert, T., McCardel, W., Ritchey, B., Gilmour, D., and Koziarz, W. (1999) Wafer-level packaging for MEMS. *J. Vac. Sci. Tecnol*, **A17** (4), 2295–2299.

42 Sparks, D., Massoud-Ansari, S., and Najafi, N. (2004) Reliable vacuum packaging using NanoGetters and glass frit bonding. *Proc. SPIE*, **5343**, 70–78.

43 Legtenberg, R. and Tilmans, H.A.C. (1994) Electrostatically driven vacuum-encapsulated polysilicon resonators. Part I. Design and fabrication. *Sens. Actuators A*, **45** (1), 57–66.

44 Zook, J.D. and Herb, W.R. (1999) Polysilicon sealed vacuum cavities for microelectromechanical systems. *J. Vac. Sci. Technol*, **A17** (4), 2286–2294.

45 Lee, B., Seok, S., and Chun, K. (2003) A atudy on wafer level vacuum packaging for MEMS devices. *J. Micromech. Microeng.*, **13** (5), 663–669.

46 Cheng, Y.T., Hsu, W.T., Lin, L., Nguyen, C.T., and Najafi, K. (2001) Vacuum packaging technology using localized aluminium/silicon-to-glass bonding. Technical Digest of the 14th International Conference on Micro Electro Mechanical Systems, MEMS 2001, Interlaken, Switzerland, 21–25 January, pp. 18–21.

47 Moraja, M., Amiotti, M., and Kullberg, R. (2003) New getter configuration at wafer level for assuring long term stability of MEMS. *Proc. SPIE*, **4980** (260), 260–267.

48 Veenstra, T.T., Berenschot, J.W., Gardeniers, J.G.E., Sanders, R.G.P., Elwenspoek, M.C., and van den Berg, A. (2001) Use of selective anodic bonding to create micropump chambers with virtually no dead volume. *J. Electrochem. Soc.*, **148**, G68–G72.

49 Pierret, R.F. (1996) *Semiconductor Device Fundamentals*, Addison-Wesley, Reading, MA.

50 Nicollian, E.H. and Brews, J.R. (1982) *MOS (Metal-Oxide-Semiconductor) Physics and Technology*, John Wiley & Sons, Inc., New York.

51 Audet, S.A. and Edenfeld, K.M. (1997) Integrated sensor wafer-level packaging. Technical Digest of the 9th International Conference on Solid-State Sensors and Actuators, Chicago, IL, June 1997.

52 Lee, C., Huang, W.-F., and Shie, J.-S. (2000) Wafer bonding by low-temperature soldering. *Sens. Actuators A*, **85** (1–3), 330–334.

53 Lin, L. (2000) MEMS post-packaging by localized heating and bonding. *IEEE Trans. Electron Devices*, **23** (4), 608–616.

54 Wild, M.J., Gillner, A., and Poprawe, R. (2001) Locally selective bonding of silicon and glass with laser. *Sens. Actuators A*, **93** (1), 63–69.

55 Schjølberg-Henriksen, K., Jensen, G.U., Hanneborg, A., and Jakobsen, H. (2003) Sodium contamination of SiO_2 caused by anodic bonding. *J. Micromech. Microeng.*, **13** (6), 845–852.

56 Gianchandani, Y., Ma, K.J., and Najafi, K. (1995) A CMOS dissolved wafer process for integrated P++ micromechanical systems. Technical Digest of the 8th International Conference on Solid-State Sensors and Actuators, Stockholm, Sweden, June 1995.

57 Matsumoto, Y. and Esashi, M. (1993) Integrated silicon capacitive accelerometer with PLL servo technique. *Sens. Actuators A*, **A39** (3), 209–217.

58 Esashi, M. (1993) Micromachining for packaged sensors. Technical Digest of the 7th International Conference on Solid-State Sensors and Actuators, Yokohama, Japan, 7–10 June 1993.

59 Chavan, A.V. and Wise, K.D. (2002) A monolithic fully integrated vacuum-sealed CMOS pressure sensor. *IEEE Trans. Electron Devices*, **49** (1), 164–169.

60 von Arx, J., Ziaie, B., Dokmeci, M., and Najafi, K. (1995) Hermeticity testing of glass–silicon packages with on-chip feedthroughs. Technical Digest of the 8th International Conference on Solid-State Sensors and Actuators, Stockholm, Sweden, 25–29 June 1995.

61 Schabmueller, C.G.J., Pollard, J.R., Evans, A.G.R., Wilkinson, J.S., Ensell, G., and Brunnschweiler, A. (2000) Integrated diode detector and optical fibres for in-situ detection of DNA amplification within micromachined PCR chips. Technical Digest of the 11th Micromechanics Europe Workshop, Uppsala, Sweden, October 2000.

62 Lapadatu, A.C., Jakobsen, H., and Puers, R. (1999) The effects of anodic bonding on the electrical characteristics of the p–n junctions. Technical Digest of the 10th International Conference on Solid-State Sensors and Actuators, Sendai, Japan, 7–10 June 1999.

63 Decroux, M., van den Vlekkert, H.-H., and de Rooij, N.F. (1986) Glass Encapsulation of ChemFet's: a simultaneous solution for ChemFet packaging and ion-selective membrane fixation. Proceedings of the 2nd International Meeting on Chemical Sensors, Bordeaux, France, 7–10 July 1986.

64 van den Vlekkert, H.-H., Decroux, M., and de Rooij, N.F. (1987) Glass Encapsulation of chemical solid state sensors based on anodic bonding. Technical Digest of the 4th International Conference on Solid-State Sensors and Actuators, Tokyo, Japan, 2–5 June 1987.

65 Shirai, T. and Esashi, M. (1992) Circuit damage by anodic bonding (in Japanese), ST-92-7. JIEE Technical Meeting, pp. 9–17.

66 van der Groen, S., Rosmeulen, M., Jansen, P., Deferm, L., and Baert, K. (1996) Bonding techniques for single crystal TFT AMLCDs. *Proc. SPIE*, **2881**, 194–200.

67 Helms, C.R. and Poindexter, E.H. (1994) The silicon–silicon-dioxide system: its microstructure and imperfections. *Rep. Prog. Phys.*, **57** (8), 791–852.

68 Schjølberg-Henriksen, K., Jensen, G.U., Hanneborg, A., and Jakobsen, H. (2004) Anodic bonding for monolithically integrated MEMS. *Sens. Actuators A*, **114** (2–3), 332–339.

69 Weichel, S., de Reus, R., and Lindahl, M. (1998) Silicon-to-silicon wafer bonding using evaporated glass. *Sens. Actuators A*, **70** (1–2), 179–184.

B. Direct Wafer Bonding

5
Direct Wafer Bonding
Manfred Reiche and Ulrich Gösele

5.1
Introduction

The adhesion of flat, polished surfaces has been well known for a long time. It has been applied for centuries in the glass industry and is termed as *"Ansprengen"* or "binging into contact" [1]. The effect was studied first by Lord Rayleigh in 1936 [2] showing the presence of an interaction energy between two bonded glass plates which was quantified by tensile strength measurements. Bonding of semiconductor wafers was described by Wallis and Pomerantz in 1969 [3]. Here, silicon wafers were bonded to sodium-containing glass wafers at high temperatures (500 °C) and by applying an electric field. This process is known as anodic bonding. In contrast, semiconductor wafer direct bonding (SWDB), or fusion bonding, describes a method to join mirror-polished semiconductor wafers at room temperature without the addition of any glue or external forces. The bonding of silicon wafers was first reported in the mid-1980s. Lasky [4] described the bonding of two oxidized silicon wafers to form silicon-on-insulator (SOI) substrates. At the same time, Shimbo *et al.* [5] analyzed the bonding behavior of non-oxidized silicon wafers forming an epitaxial substrate for power devices. Since this time, numerous applications of SWDB have been published. In 1988, Petersen *et al.* [6] reported the formation of pressure sensors by bonding of structured silicon wafers, which initiated modern silicon bulk micromechanics. Furthermore, SWDB has also been applied to join wafers of other semiconductor materials such as III–V and II–VI compounds, or to combine different materials. Today, SWDB is a key process in the mass production of advanced substrates for integrated devices (SOI), and in the production of three-dimensional microelectromechanical systems (MEMS) [7–9]. The realization of compliant or engineered substrates is also an important issue for further applications [10].

5.2
Surface Chemistry and Physics

SWDB requires wafers with a high degree of flatness, parallelism, and smoothness. Also clean surfaces are necessary which are free of particulate, organic, and metallic contaminations. This is important because the surface cleanliness has a direct effect on both the structural and electrical properties of the bonding interface as well as on the resulting electrical properties of the bonded material. After cleaning, activation of the surfaces is required prior to bonding. Then the two mating wafers are brought together face to face in air at room temperature. The top wafer floats on the other due to the presence of a thin cushion of air between both wafers. When an external pressure is applied onto a small part of the pair to push out the intermediate air, a bond is formed by surface attraction forces between the wafers at this location (Figure 5.1). For wafer bonding, the interaction between the two surfaces is important. The total energy or adhesion energy W of two planar surfaces at a distance D apart is given by [11]

$$W = \frac{A_{132}}{12\pi}\left[\frac{1}{D_0^2} - \frac{1}{D^2}\right] = \frac{A_{132}}{12\pi D_0^2}\left[1 - \frac{D_0^2}{D^2}\right] \tag{5.1}$$

per unit area, where D_0 is the interatomic distance and

$$A_{132} = \frac{3kT}{2}\sum_{n=0}^{\infty}\sum_{s=1}^{\infty}\frac{(\Delta_{13} - \Delta_{23})^s}{s^2} \tag{5.2}$$

is the Hamaker constant for surface 1 interacting with surface 2 through medium 3 [12]. Δ_{13} and Δ_{23} are the differences in the dielectric response of the three materials, T is the temperature, and k is Boltzmann's constant. The Hamaker constant for the interaction of two amorphous silica surfaces across water was measured as $A_{132} = 1.69 \times 10^{-20}$ J at 20 °C [13].

At $D = D_0$ (both surfaces are in contact) Eq. (5.1) gives $W = 0$, while for $D = \infty$ (two isolated surfaces) one has

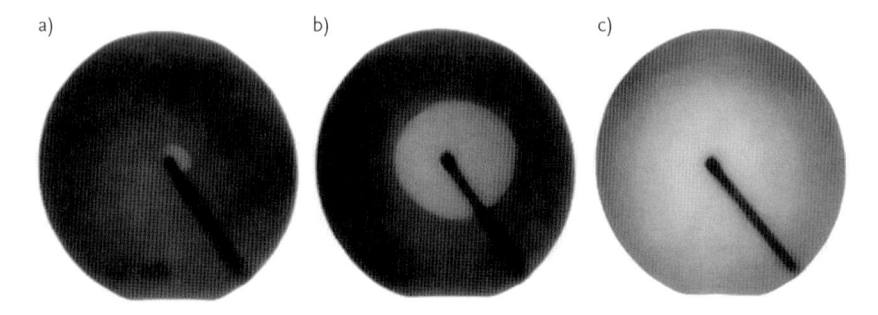

Figure 5.1 Infrared microscope images of the propagation of a bond wave [7]. (a) Formation of a locally bonded area by applying an external pressure. Propagation of the bond wave (b) after 3 s and (c) after 5 s.

$$W = \frac{A_{132}}{12\pi D_0^2} = 2\gamma \qquad (5.3)$$

or

$$\gamma = \frac{A_{132}}{24\pi D_0^2} \qquad (5.4)$$

In other words, the surface energy γ equals half the energy needed to separate two flat surfaces from contact to infinity; i.e., it is half the adhesion energy. Adhesion is caused by different forces acting at an interface. Most important for wafer bonding are the following.

- Capillary force.
- Electrostatic force initiated by Coulomb interaction between charged objects or from the contact potential between two surfaces caused by differences in the local energy states and electron work functions.
- Van der Waals force resulting from the interactions between atomic or molecular oscillating or rotating electrical dipoles within the interacting media. There are three types of interactions [14]:
  - the Keesom force which is the interaction between two permanent dipoles,
  - the Debye force which is the interaction between one permanent dipole and one induced dipole, and
  - the London or dispersion force which is the interaction between two induced dipoles.
- Solid bridging caused by impurities.
- Hydrogen bonding between OH groups as the separation between the surfaces becomes small.

Measurements on silicon microstructures showed that capillary force dominates [15–17]. It is about $2 \times 10^2\,\mu\text{N}\,\mu\text{m}^{-2}$ at a separation distance of two smooth silicon surfaces of 1 nm. Increasing the distance to 10 nm reduces the capillary force to about $5\,\mu\text{N}$. In addition, electrostatic forces, van der Waals forces, and hydrogen bonding are about one order of magnitude lower. For short distances between both surfaces ($D \approx 1\,\text{nm}$) hydrogen bonding and van der Waals forces are about $10\,\mu\text{N}\,\mu\text{m}^{-2}$, while electrostatic forces reach values of about $5\,\mu\text{N}$. All these forces act only over short ranges and the effect depends on the specific surface conditions. Most important is the surface roughness. Several models have been proposed to predict the effect of roughness on adhesion between two surfaces. These treatments have considered Hertzian, Johnson–Kendal–Roberts, and Deryagin–Muller–Toporov contacts [18].

Also, the chemistry of species on the silicon surfaces affects the various forces. Silicon surfaces are covered with a thin oxide layer under room temperature conditions. X-ray photoelectron (XPS) and high-resolution electron energy loss spectroscopy proved the existence of a large number of singular and associated OH groups causing the hydrophilicity of such surfaces [19]. Detailed investigations revealed changes in the angle of the Si–O–Si bridge indicating structural changes of different oxides and a dependence on the storage time.

Hydrophobic surfaces are obtained by removing the oxide, such surfaces being mainly characterized by the presence of Si–H groups. Because the oxide is usually removed by hydrofluoric acid (HF), Si–F groups are also obtained. Ermolieff *et al.* [20] correlated the presence of Si–F bonds with the amount of remaining oxide (indicated by the presence of O–Si–F bonds). The transition from hydrophilic to hydrophobic surface conditions is assumed to be at a concentration of 25% Si–F bonds. Furthermore, the pH and the concentration of the etch solution also affect the structural and chemical nature on the silicon surface [22, 21].

Models of the atomic mechanisms occurring at the interface during wafer bonding have been developed based on the different nature of hydrophilic and hydrophobic surfaces.

5.3
Wafer Bonding Techniques

5.3.1
Hydrophilic Wafer Bonding

Wafer bonding using hydrophilic surfaces is the most commonly used technique. Silicon surfaces are covered with an oxide layer under room temperature conditions. If oxidized surfaces are bonded an oxide layer results at the interface (Figure 5.2).

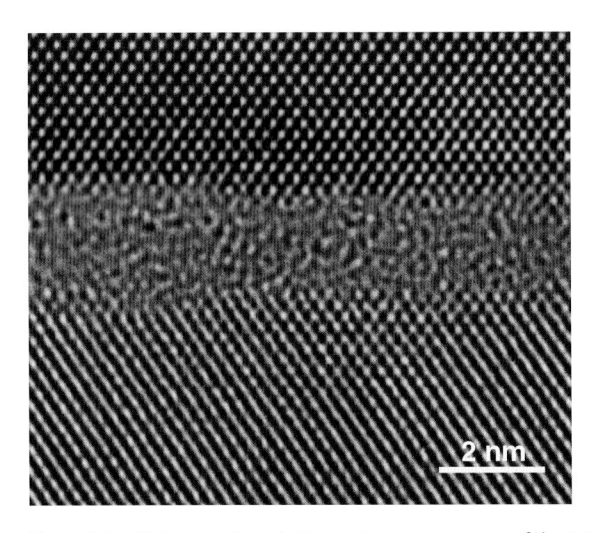

Figure 5.2 High-resolution electron microscope image of the interface of bonded hydrophilic wafers.

A model for hydrophilic wafer bonding was first described by Stengl et al. [23] using the analogy of surface chemistry of silica and oxidized silicon. Based on results of infrared spectroscopy, a three-dimensional hydrogen bonded network of water molecules is assumed. The water is primarily bonded via Si–OH groups on the silica surface. During heating above 180 °C the adsorbed water molecules desorb under atmospheric pressure leaving a hydroxylated silica surface, on which most of the SiO groups are linked via hydrogen atoms. OH groups are bonded more stably with increasing temperature. Using experimental data of Maszara et al. [24] the bond energy γ_B between two hydrophilic silicon surfaces is calculated by

$$\gamma_B = (1960 - \gamma_S)(1 - e^{-k_2 t}) + \gamma_S \tag{5.5}$$

where t is the time and k_2 a rate constant describing the time dependence of the reaction SiOH:SiOH → SiOSi + H$_2$O. Parameter γ_S is given by

$$\gamma_S = 630 \times \exp\left(\frac{-E_A}{kT}\right) \tag{5.6}$$

with $E_A = 0.05$ eV. For room temperature bonding, Eq. (5.5) results in $\gamma_B \approx 60$ mJ m^{-2}, which is about twice the value of the surface energy calculated from Eq. (5.4).

The model was further developed by Tong and Gösele [7]. They proposed that for room temperature conditions, chains consisting of three or more hydrogen-bonded water molecules bridge the interface. This is based on the fact that hydrogen-bonded water triplets are more stable than single water molecules or dimers. They also pointed out that two main types of silanols are present on the oxide surface: singular silanols (Si–OH) and associated or vicinal silanols (Si–OH–O–Si). Assuming by analogy with silica a maximum concentration of four to six silanol groups per square nanometer and all groups adsorb water molecules that are connected by hydrogen bonds across the two surfaces, the bond energy was estimated to be

$$\gamma_B = \frac{1}{2}(2 d_{OHi} E_{hi} + d_{OHH} E_{hH}) \tag{5.7}$$

where d_{OHi}, d_{OHH} and E_{hi}, E_{hH} are the surface density of silanol groups and hydrogen bond energy of the isolated and associated silanol groups, respectively. Using $d_{OHi} = 1.4 \times 10^{14}$ cm^{-2}, $d_{OHH} = 1.4 \times 10^{14}$ cm^{-2}, $E_{hi} = 7 \times 10^{-17}$ mJ per bond, and $E_{hH} = 4.2 \times 10^{-17}$ mJ per bond, the bond energy of room temperature bonded wafers is calculated as

$$\gamma_B = 165 \text{ mJ m}^{-2} \tag{5.8}$$

which is about a factor of two higher than measured values. The difference was explained by polymerization processes of silanol groups at room temperature [7].

Another interpretation based on molecular dynamic simulations was given by Litton and Garofalini [25]. They discussed the presence of an additional type of silanol, the geminal silanol (Si–(OH)$_2$) found by NMR analysis on silica gel

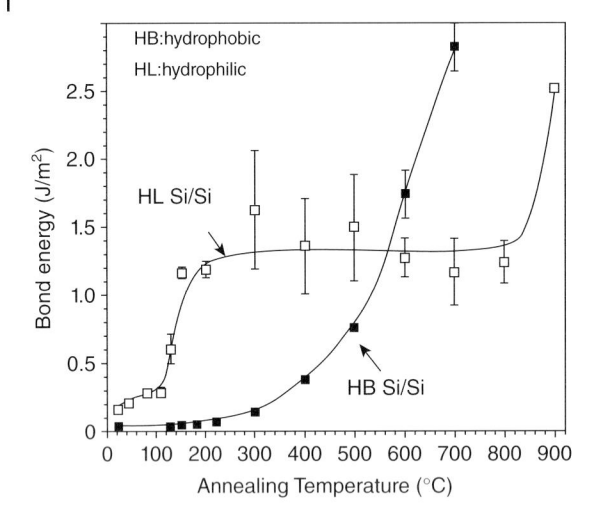

Figure 5.3 Dependence of bond energy on annealing temperature for hydrophilic and hydrophobic bonded wafers.

surfaces. It was also assumed that the mean siloxane (Si–O–Si) bond angle is lower (~130°) for thin oxides prepared by wet chemical cleaning than for bulk SiO_2 (~144°). This indicates that the oxide is strained due to the Si–SiO_2 interface.

Annealing after the initial bonding process at room temperature results in changes of the interface chemistry. Measurements of the interface energy of bonded hydrophilic wafers show a different behavior for different temperature ranges (Figure 5.3) [26]. The calculated activation energies [7] refer to

- the rearrangement of interface water at temperatures below 110 °C,
- polymerization processes of silanol groups across the interface at 110 °C $\leq T \leq 150$ °C,
- a temperature range between 150 and 800 °C which is characterized by an almost constant bond energy, and
- complete bonding via oxide at $T > 800$ °C.

The bond energy does not vary for different hydrophilic surface conditions, that is, bonded hydrophilic Si–Si, Si–SiO_2, and SiO_2–SiO_2 pairs result in the same values of γ_B.

5.3.2
Hydrophobic Wafer Bonding

When the oxide layer from a crystalline silicon substrate is removed with HF, a hydrophobic surface with unique properties is obtained, that is, having a good

Figure 5.4 High-resolution electron microscope image of the interface of bonded hydrophobic wafers.

resistance to chemical attack and a low surface recombination velocity, which means a surface with a very low density of surface states.

The etching of the oxide is assumed to be a two-step process. First, most of the oxide layer is rapidly dissolved in HF, forming SiF_6^{2-} ions in solution. Second, anodic dissolution of the last monolayer of oxidized silicon (Si^{n+} with $n = 1, 2, 3$) occurs, resulting in a hydrogen-passivated surface. The dominant species are dihydride (Si–H$_2$) for Si(100) surfaces and monohydride (Si–H) for Si(111) surfaces [27].

Bonded hydrophobic wafers are characterized by interfaces completely different from those of bonded hydrophilic wafers (Figure 5.4). The removal of the oxide results, as in the case of bicrystals, in the two silicon lattices being in contact. Crystal defects (dislocations) are generated forming a two-dimensional network in order to match both crystal lattices. The dislocation structure depends on the misorientation. In general, the twist component causes a network of pure screw dislocations, while the tilt component is compensated by a periodic array of 60° dislocations [28].

The first detailed analyses of interfaces of bonded hydrophobic wafers were carried out by Bengtsson and Engström in 1989 [29]. A first concept for hydrophobic wafer bonding was presented by Bäcklund and coworkers [30, 31] suggesting van der Waals forces as the origin of the attraction forces. Further investigations assumed the formation of hydrogen bonds via Si–F groups on the hydrophobic surface [26]. The surface energy was estimated using the equation

$$\gamma_S = \frac{1}{2}(2d_{Si\text{-}F}E_{hHF}) \tag{5.9}$$

where $d_{Si\text{-}F}$ is the surface density of Si–F bonds and E_{hHF} is the lowest bond energy of the hydrogen bonded H–F cluster across the two mating surfaces [7]. Using $d_{Si\text{-}F} = 10^{14}\,cm^{-2}$ and $E_{hHF} = 25.1\,kJ\,mol^{-1}$, the bond energy was calculated to be $\gamma_B \leq 42\,mJ\,m^{-2}$, which is in accordance with experimentally measured data. Analyses of HF-treated surfaces and interfaces of bonded hydrophobic wafers proved the existence of fluorine; the main species, however, is hydrogen [19, 20, 27, 32].

This means that hydrogen bonds like Si–H⋯H–Si are probably more favored. The contribution of the different hydrogen bonds depends on the pretreatment, that is, if the hydrophobization is carried out using diluted HF solutions, buffered HF solutions (HF/NH$_4$F, etc.), or plasma etch techniques [33].

The behavior of the interface energy as a function of the annealing temperature is quite different for bonded hydrophobic wafers (Figure 5.3) [26]. The interface energy is nearly constant for annealing temperatures up to 150 °C; at higher temperatures γ_B increases. There are, however, two different regimes. For 150 °C ≤ T ≤ 300 °C the increase of the interface energy is characterized by an activation energy of 0.21 eV, while an activation energy of 0.36 eV was determined for annealing at higher temperatures [7]. Both activation energies correlate to different interface processes. There is a relation to the existence of Si–CH$_x$ groups (stable up to about 400 °C) and Si–H groups detected up to about 600 °C on hydrophobic silicon surfaces [19].

5.3.3
Low-Temperature Wafer Bonding

Annealing at high temperatures is required after the initial bonding at room temperature for both hydrophilic and hydrophobic surface conditions. The reason for this is the generation of sufficiently high interface energies by transformation of weaker bonds into covalent bonds via the interface. Typical annealing temperatures are above 800 °C. An increasing number of applications, however, require temperatures below 500 °C. This is especially true for the integration of fully processed device wafers reducing the thermal budget to temperatures below 450 °C. Because the interface energy is insufficiently low for bonding using standard conditions up to these temperatures, modified techniques especially of wafer bonding under hydrophilic conditions have been developed. The key issue for low-temperature wafer bonding is the modification of the wafer surfaces either due to increasing the hydrophilicity, which increases the number of hydrogen bonds via the interface, or due to the generation of new types of chemical bonds stable already at lower temperatures. This includes the application of different thin layers, such as patternable materials (epoxy-based negative resist (SU-8), benzocyclobutene-based polymers, or other types of photoresist) [34, 35], low-temperature oxide layers (tetraethylorthosilicate (TEOS), spin-on glass, etc.) [36, 37], and layers deposited by plasma-enhanced chemical vapor deposition (PE-CVD; amorphous silicon, oxide, nitride) [38]. Alternative techniques use the activation of the wafer surface before bonding. For example, wet chemical processes have been applied resulting in an increase of the interface energy by a factor of about two [7, 39]. Higher interface energies have been obtained by surface activation using plasma processes. Especially, treatments in an oxygen plasma are most common where different plasma sources were applied (reactive ion etching, inductively coupled plasma, electron cyclotron resonance plasma, microwaves) [40–43]. The increasing interface energy of plasma-treated surfaces was discussed in terms of

- an increasing surface roughness, which increases the total interface area;

- an increasing oxide thickness, combined with a higher porosity of the amorphous layer resulting in an increasing incorporation of additional water molecules; or

- an accelerated transformation of silanol to siloxane bonds.

Treatments in other, non-oxygen-containing plasma environments (Ar, N_2, Cl_2), however, resulted in an analogous increase of the interface energy [44, 45]. This means that other reasons are probably more important. Analyses of dielectric barrier discharges (DBDs) have shown that the activation effect of semiconductor surfaces is mainly caused by electrons and UV radiation instead of ions [46]. The effect of UV radiation is an additional cleaning of the surfaces by cracking of CH_x compounds, while electrons generate surface-active sites. XPS investigations of DBD-treated germanium wafers, for instance, showed the reduction of the Ge [2+] and accelerated formation of the Ge [4+] oxidation states by sp^3 hybridization in consequence of the interaction with high-energy electrons generated in the discharge [47]. This causes more active sites to bond OH groups on the surface, or could favor the preferred formation of Ge–O–Ge bonds even at low temperatures.

5.3.4
Wafer Bonding in Ultrahigh Vacuum

For a number of applications, even the bonding of dissimilar materials, it would be desirable to be able to reach the bulk bonding strength by room temperature bonding without the requirement of an additional heating step. In this context, the question arises as to whether it is possible to bring the flat and clean surfaces of two wafers without any adsorbed foreign atoms or molecules together at room temperature so that covalent bonds will form directly at room temperature. One possible solution is to do the bonding in ultrahigh vacuum (UHV).

Haneman et al. [48] showed that the rebonding of the surfaces of partially cleaved germanium crystals in UHV appears to lead to covalent bonds. For bonding of silicon wafers, the surfaces were cleaned, hydrophobized by either by dipping in diluted HF [49] or cleaning in a hydrogen-containing plasma [50], and transferred into a UHV chamber [51]. Upon annealing to about 600 °C the hydrogen is desorbed and the silicon surface reconstructs into Si(100)-2 × 1 via formation of Si–Si dimers which give a more stable configuration with only one dangling bond per surface atom. For Si(111) surfaces, a (7 × 7) reconstruction is favored [52]. The energy of a reconstructed Si–Si bond is still higher than that of an undisturbed bond because of the strong bending. For sufficiently small separation between surfaces, the electronic surface states overlap and an "adhesive avalanche" [53] occurs via reconstruction breaking and covalent bonding as predicted by molecular dynamics simulations [54, 55]. The energy released during the bonding process is spent on rearranging surface atoms. Even if the two wafers are bonded in such a manner that their crystallographic orientations are perfectly aligned, the interface does not resemble a perfect structure due to incomplete bonding arising from atomic steps and fast energy dissipation. In real cases, as for hydrophobic wafer bonding, there is always a tilt and twist misorientation between the bonded wafers which makes the rearrangement less perfect, resembling a grain boundary at the bonded interface. The atoms at the interface between the two crystals have arrangements different from those in the bulk material, with important consequences for the electronic properties. In either case, the bonding energy is above $2\,\text{J}\,\text{m}^{-2}$ directly after room temperature bonding [56].

Wafer bonding in UHV has been applied to bond numerous dissimilar materials such as GaAs–Si, InP–Si, LiNbO$_3$–Si, LiTaO$_3$–Si, or silicon to platinum films. Most of these applications use surface cleaning and activation by an argon beam [57–59].

5.4
Properties of Bonded Interfaces

Wafer bonding of hydrophilic surfaces results in the presence of an oxide layer in the interface. The thickness of the oxide layer is only a few nanometers if native oxides are present (Figure 5.2). Most applications, however, use oxides of different thickness grown by thermal wet or dry oxidation, or low-temperature oxides such as TEOS and PE-CVD and high-density plasma oxides. The mechanical properties of bonded wafer pairs (stress) and the electrical properties depend on the applied oxide layer. The latter are especially important in the production of advanced substrates (SOI).

Infrared spectroscopy (attenuated total reflectance spectroscopy) and secondary ion mass spectroscopy measurements revealed that there are hardly any contaminants at the bonded interfaces [32]. Only oxygen (or Si–O vibration modes) is detected at the interface of bonded hydrophilic wafer pairs after annealing at elevated temperatures (>900 °C). At lower temperatures, a certain amount of OH is present at the bonding interface. Typical breakdown voltages of more than $7 \times 10^7\,\text{V}\,\text{cm}^{-1}$ are measured at room temperature on bonded p–n wafers dropping only slightly if the measuring temperature increases to 150 °C. A model describing the electrical properties of interfaces of bonded hydrophilic wafers was published by Bengtsson and Engström [29] referring to interface state densities of about $10^{11}\,\text{cm}^{-2}$. More recent measurements proved that interface state densities of $10^{10}\,\text{cm}^{-2}$ or below can be obtained.

Defects in the interfaces of bonded wafers are obtained either by flatness variations or particles enclosed in the interface.

There are numerous theoretical studies of the conditions under which a gap separating two wafers will prevent bonding [7]. According to Yu and Suo [60], a relation exists between the distance of both wafers R (as a consequence of the flatness) and the lateral dimension of the gap l. Assuming the thickness of both wafers is d and $l > 4d$, complete bonding is obtained if

$$R < \frac{l^2}{\sqrt{2E'd^3/3\gamma}} \tag{5.10}$$

where $E' = E/(1-v^2)$, E being Young's modulus and v Poisson's ratio, and γ is the surface energy. In cases where $l < 4d$, the condition of gap closing is

$$R < 5.1 \left(\frac{l\gamma}{E'} \right)^{1/2} \tag{5.11}$$

An expression of the unbonded area caused by particles of radius r_p was derived in [7]. Assuming the same thickness of both wafers and that the particle is incompressible, the unbonded area is calculated as

$$R = \left(\frac{2}{3} E' \frac{d^3}{\gamma} \right)^{1/4} r_p^{1/2} \tag{5.12}$$

Unbonded areas caused by an insufficient surface flatness or by particles appear immediately after bonding at room temperature. However, subsequent annealing after the bonding can also cause the generation of unbonded areas. These interface defects are generally known as bubbles and are observed mainly during annealing at temperatures below 900 °C [61]. Based on the assumption that hydrocarbons and hydrogen play an important role, Mitani and Gösele [62] presented a thermodynamic model of bubble formation. They assumed also two different sets of bubbles according to the detection limits of the applied analysis methods (larger bubbles (diameter > 1 mm) detected by infrared microscopy, and smaller bubbles (diameter < 500 µm) detected by X-ray topography). The critical radius for large bubbles was determined to be

$$r_{\text{crit}}(t) = \left[\frac{16\gamma E d^3 x_{\text{ox}}^2}{9\alpha(1-v^2)B^2C^2} \right]^{1/4} \propto \left(\frac{x_{\text{ox}}}{C} \right)^{1/2} \tag{5.13}$$

where α is a geometrical factor ($1/3 \leq \alpha \leq 1/2$), B a proportionality constant between the pressure inside of the bubble and areal contamination concentration C, and x_{ox} is the oxide thickness. This equation indicates that bubbles can more easily be prevented by thick oxide layers.

Using analysis methods with higher resolution, bubble formation was recently studied by statistical methods [46]. Measurements of bubble sizes and distances, and their distribution functions proved numerous individual processes of nucleation, growth, coalescence, and dissolution. All processes are controlled by diffusion and follow the thermodynamic model of steady-state nucleation and growth. The dominance of the individual processes depends on the temperature and time of the annealing. Increasing the annealing time results in the formation of a second set of bubbles which can probably be correlated with previously described results [62].

Bonded hydrophobic wafers are characterized by completely different interfaces. The removal of the oxide results, as in the case of bicrystals, in two silicon lattices being in contact. Crystal defects (dislocations) are generated forming a two-dimensional network in order to match both crystal lattices. Bonding of Si(100) wafers causes a Σ1 (100) small angle grain boundary characterized by a square-like mesh of screw dislocations expected from theory [63]. These dislocations are formed by the rotational misfit (twist) between both crystal lattices. There is, however, an additional tilt component caused by the deviation on the [001] axis of real wafers (cutoff). The tilt component is compensated by a periodic array of 60° dislocations. The spacings between dislocations S in both networks are indirectly proportional to the misalignment angle and are given by

$$S_{twist} = \frac{a}{2\sqrt{2} \times \sin(\vartheta_{twist}/2)} \quad (5.14)$$

for the screw dislocation network. On the other hand, the relation between dislocation distance and tilt angle of the network formed by 60° dislocations is

$$S_{tilt} = \frac{a}{\sqrt{2} \times \tan \vartheta_{tilt}} \quad (5.15)$$

In both equations a is the lattice constant ($a = 0.543$ nm for silicon) and ϑ_{twist} and ϑ_{tilt} are the angles of misorientation of the twist and tilt component, respectively.

Both dislocation fractions were investigated for hydrophobic wafers bonded under environmental conditions [64–66] and under UHV conditions [56, 67]. There have been different observations for wafer pairs bonded under UHV conditions. Some reports present dislocation networks already after bonding at room temperature [68], while dislocation networks are observed by other authors only after a subsequent annealing above 800 °C [56]. The reason probably is that the boundary is unrelaxed after bonding at room temperature, while relaxation occurs by energetic reasons only at higher temperatures [69]. Bonded interfaces show nanometer-sized shallow voids at temperatures below 800 °C. The density and size of these voids depend on the temperature relating to accommodation processes. Analogous analyses of dislocation networks are reported for hydrophobic wafers bonded under environmental conditions only after annealing above 800 °C. The reason is the low bonding strength at lower temperatures causing debonding during sample preparation for transmission electron microscopy.

Dislocations in the interface of bonded wafers possess numerous remarkable properties which may be used for various applications [70, 71]. The electrical properties of bonded hydrophobic silicon wafers were studied by Bengtsson et al. [72]. Measurements of the capacitance–voltage (C–V) characteristics of bonded unipolar wafers were interpreted on the assumption of two distributions of interface states, one of acceptors and one of donors, causing a potential barrier at the bonded interface. The origin of the interface states was assumed to be impurities and crystal defects. More recent analyses using the electron beam-induced current technique proved barrier heights generally smaller than 100 meV for different types of bonded hydrophobic wafers [73]. The concentration of deep levels along the interface was determined to be of the order of 10^5 cm^{-1}. Low concentrations of deep levels at the interface are also the reason for low dark currents, improved C–V characteristics, and fast rise times of p–i–n diodes prepared on bonded hydrophobic wafers [74]. Furthermore, current–voltage measurements of thin hydrophobic bonded layers on SOI substrates containing dislocations proved an increase of the lateral interface conductivity by six orders of magnitude compared to layers without dislocations [75]. The reason is that dislocations form channels of higher conductance in the silicon layer resulting in a higher carrier concentration in the channel but increase also the effective carrier mobility, which is consistent with quasi-ballistic transport along dislocations. Measurements of metal–oxide–semiconductor field-effect-transistors (MOSFETs) indicate an increase of the drain current (I_D) of about one order of magnitude even at low gate voltages required for future high-performance devices [73]. The analyses were carried out on MOSFETs containing only a small number of dislocations in the channel making it possible to extrapolate a current of $I_D \approx 5 \times 10^{-2}$ A for a MOSFET having only one dislocation in the channel. This corresponds to a current density of about $500\,\mathrm{A\,cm^{-1}}$.

5.5
Applications of Wafer Bonding

Numerous publications concerning the applications of SWDB have appeared during the last two decades. These applications cover all areas of modern micro- and nanoelectronics as well as MEMS. The following summarizes only some applications especially important for future developments.

5.5.1
Advanced Substrates for Microelectronics

SOI technology has appeared as an effective means of extending CMOS scaling for mainstream high-performance or low-power applications. The combination of wafer bonding and layer transfer by hydrogen-induced splitting is the smartest and most effective method of producing SOI wafers. If hydrogen ions are implanted at high doses (>$5 \times 10^{16}\,\mathrm{cm^{-2}}$) platelet-like planar defects (microcavities) are generated on {100} and {111} planes near the implantation depth R_p. Some hydrogen ions bond to the dangling silicon bonds in the microcavities, while others fill these voids. If such an ion-implanted wafer is heated to 400–500 °C, more hydrogen segregates into the voids in the form of molecular hydrogen, the pressure builds up to the point of fracture, and blistering is obtained. Blistering phenomena caused by surface bombardment with hydrogen or inert gases (helium) have been seen in the past, and much effort was focused on preventing them. Bruel [76–78] realized that the deleterious effect could be harnessed to accomplish a weakened plane or zone that makes it possible to obtain a controlled cut through the crystalline lattice. The key is to introduce a stiffener layer (another wafer bonded to the surface of the implanted wafer) that prevents blistering and redirects the pressure that builds up in microcavities in a lateral direction. The process is generally known as Smart Cut® [78]. Compared to conventional layer transfer methods which employ wafer bonding and polishing/etching thinning techniques, e.g., bond-and-etched back SOI, ion implantation-induced layer splitting has several advantages. First, a high thickness uniformity of the transferred layer is guaranteed by the ion implantation process which allows the control of R_p (determining the thickness of the transferred layer) within a few percent over the whole wafer area. Second, the implanted wafer can be reused for the next cycle after a short polishing step.

The thickness control and thickness uniformity are important issues for the application of SOI wafers in high-performance device fabrication. Today's commercial SOI wafers for partially depleted CMOS applications are characterized by device layer thicknesses ranging from 90 nm down to 30 nm at a thickness uniformity of ±3 nm. The wafer diameter is 300 mm. In addition, the thickness of the device layer for fully depleted CMOS is below 20 nm. The thickness uniformity here is ±1 nm.

Besides reductions in device layer thickness, SOI material with ultrathin buried oxide (UT-BOX) layers is important for some applications. For very low-power devices, for instance, UT-BOX layers offer the possibility to easily form buried n- and p-regions in the handle wafer, which can be used as back gates. By applying a back bias, the off current is reduced, while in the forward bias mode the threshold voltage is lowered resulting in a current drive increase [79]. Another advantage of UT-BOX SOI is the reduction of local MOSFET self-heating. An improvement by a factor of three in thermal conductance is obtained by reducing the UT-BOX layer thickness from 150 to 20 nm [80].

Strain engineering and material innovations have been identified as the main contributors to the continued performance improvement in CMOS devices. Besides SOI, significant improvements of performance are obtained by increased carrier mobility which has been reported for devices fabricated on strained silicon layers. Combining the advantages of SOI and strained silicon results in strained silicon-on-insulator (SSOI) substrates merging the properties of both materials [81]. For the fabrication of SSOI wafers, strained silicon layers are grown on a relaxed SiGe virtual substrate and are then transferred to oxidized silicon handle wafers by a combination of wafer direct bonding and layer transfer by hydrogen-induced layer splitting. The strain in the silicon layer grown on a relaxed SiGe buffer is induced by the lattice mismatch between silicon and SiGe. Because the lattice parameter of $Si_{1-x}Ge_x$ ($0 \leq x \leq 1$) alloys varies between 0.5431 nm for silicon ($x = 0$) and 0.5657 nm for germanium ($x = 1$), tensile strain is induced in the silicon layer. The strain is generally biaxial.

The combination of different materials to form compliant substrates is another important application for semiconductor wafer bonding. For instance, the significantly larger bulk mobilities for both electrons (3900 versus 1400 $cm^2 V^{-1} s^{-1}$) and holes (1900 versus 500 $cm^2 V^{-1} s^{-1}$) for germanium over silicon have led to a resurgence of interest in germanium. Single-crystalline germanium substrates are available up to 300 mm in diameter, but there are some significant disadvantages (extremely high material costs, higher density of germanium (5.32 $g cm^{-3}$) compared to silicon (2.33 $g cm^{-3}$) resulting in an about 2.3 times higher mass of a germanium wafer) making germanium wafers ineffective in mass production. Therefore germanium layers on silicon or germanium on insulator (GOI) are alternatives. Specific techniques have been applied to bond germanium wafers with or without an oxide layer in between on silicon substrates [47]. An alternative way to fabricate such compliant substrates is the transfer of thin layers (germanium) to a silicon substrate instead of the bonding of bulk wafers. Wafer bonding techniques have recently been published using combinations of wafer bonding and hydrogen-induced layer splitting [82–84]. These techniques apply thin germanium layers produced either by hydrogen implantation into germanium wafers [82, 83] or, in a smarter and more competitive way, by transferring of germanium layers grown by CVD methods on silicon substrates [84].

5.5.2
MEMS and Nanoelectromechanical Systems

SWDB is also an ideal technique for combining different structures to form functionalized elements in a three-dimensional arrangement. The better understanding of the fundamental principles of physical and chemical processes on surfaces and bonded interfaces, and especially the development of planarization techniques, results in an increasing number of applications of various material combinations [85, 86]. Wafer bonding has now been applied in research into and production of nearly all types of MEMS including mechanical, fluidic, power, optical, medical, and biological microsystems. Much progress has been made during the last few years in wafer bonding of numerous materials including III–V and II–VI compounds, SiC, and complex oxides ($LiNbO_3$, $SrTiO_3$, etc.). Current reviews are presented, for instance, in [9, 87].

The combination of electronic components with other functions to realize systems-on-a-chip is another application for wafer bonding. Today's systems-on-a-chip already contain digital, analogue, mixed-signal, and radio-frequency functions but future applications need also the integration of optoelectronic components (e.g., for optical data communication) or other functionalities.

5.6
Conclusions

SWDB offers a new degree of freedom in the design of material combinations without the common restrictions of the structure (amorphous, polycrystalline, orientation, lattice constant) of the materials to be bonded. It is already established for industrial fabrication of advanced substrates (SOI) and is a key technology for further developments in this area (SSOI, GOI).

The basic processes of semiconductor wafer bonding are well understood for silicon. These include the chemical and physical processes on silicon surfaces and bonded interfaces, the mechanical properties of bonded structures, the effect of thermal treatments after the bonding process, and the behavior of bonded wafers during subsequent device process steps. Further research, however, is required to realize SOI substrates and other engineered substrates with ultrathin device layers (as for fully depleted devices), the preparation of other dissimilar substrates, and future applications of MEMS.

References

1 Biermann, U.K.P., van Gorkum, A.A., and Pals, J.A. (1995) Direct bonding: from an optical technology to a broad research topic. *Philips J. Res.*, **49**, 1–10.
2 Lord Rayleigh (1936) A study of glass surfaces in optical contact. *Proc. R. Soc. Lond. A*, **156**, 326–349.
3 Wallis, G. and Pomerantz, D.I. (1969) Field assisted glass–metal sealing. *J. Appl. Phys.*, **40**, 3946–3949.
4 Lasky, J.B. (1986) Wafer bonding for silicon-on-insulator technologies. *Appl. Phys. Lett.*, **48**, 78–80.
5 Shimbo, M., Furukawa, K., Fukuda, K., and Tanzawa, K. (1986) Silicon-to-silicon direct bonding method. *J. Appl. Phys.*, **60**, 2987–2989.
6 Petersen, K., Barth, P., Poydock, J., Brown, J., Mallon, J., and Bryzek, J. (1988) Silicon fusion bonding for pressure sensors, in *Solid-State Sensors and Actuators Workshop Technical Digest*, IEEE, New York, pp. 144–147.
7 Tong, Q.-Y. and Gösele, U. (1999) *Semiconductor Wafer Bonding*, John Wiley & Sons, Inc., New York.
8 Iyer, S.S. and Auberton-Herve, A.J. (eds) (2002) *Silicon Wafer Bonding Technology for VLSI and MEMS Application*, IEEE, Piscataway, NJ.
9 Alexe, M. and Gösele, U. (2004) *Wafer Bonding: Applications and Technology*, Springer Verlag, Heidelberg.
10 Suga, T., Baumgart, H., Hobart, K.D., Bagdahn, J., Colinge, C., and Moriceau, H. (eds) (2008) *Semiconductor Wafer Bonding 10: Science, Technology, and Applications*, Electrochemical Society, Pennington, NJ.
11 Israelachvili, N. (2006) *Intermolecular and Surface Forces*, Academic Press, London, p. 201.
12 Eichenlaub, S., Chan, C., and Beaudoin, S.P. (2002) Hamaker constants in integrated circuit metallization. *J. Colloid Interface Sci.*, **248**, 389–397.
13 Butt, H.J., Graf, K., and Kappl, M. (2004) *Physics and Chemistry of Interfaces*, Wiley-VCH Verlag GmbH, Weinheim, p. 89.
14 Bergström, L. (1997) Hamaker constants of inorganic materials. *Adv. Colloid Interface Sci.*, **70**, 125–169.
15 Maboudian, R. and Howe, R.T. (1997) Adhesion in surface micromechanical structures. *J. Vac. Sci. Technol. B*, **15**, 1–20.
16 Legtenberg, R., Tilmans, H.A.C., Elders, J., and Elwenspoek, M. (1994) Stiction of surface micromachined structures after rinsing and drying: model and investigation of adhesion mechanisms. *Sens. Actuators A*, **43**, 230–238.
17 Jones, R., Pollock, H.M., Cleaver, J.A.S., and Hodges, C.S. (2002) Adhesion forces between glass and silicon surfaces in air studied by AFM: effects of relative humidity, particle size, roughness, and surface treatment. *Langmuir*, **18**, 8045–8055.
18 Maugis, D. (1996) On the contact and adhesion of rough surfaces. *J. Adhes. Sci. Technol.*, **10**, 161–175.
19 Grundner, M. and Jacob, H. (1986) Investigations of hydrophilic and hydrophobic silicon (100) wafer surfaces by X-ray photoelectron and high-resolution electron energy loss spectroscopy. *Appl. Phys. A*, **39**, 73–82.
20 Ermolieff, A., Martin, F., Amouroux, A., Marthon, S., and Westendorp, J.F.M. (1991) Surface composition analysis of HF vapour-cleaned silicon by X-ray photoelectron spectroscopy. *Semicond. Sci. Technol.*, **6**, 98–102.
21 Tong, Q.-Y., Lee, T.-H., Gösele, U., Reiche, M., Ramm, J., and Beck, E.

(1997) The role of surface chemistry in bonding of standard silicon wafers. *J. Electrochem. Soc.*, **144**, 384–389.

22. Carroll, M.S., Sturm, J.C., and Yang, M. (2000) Low-temperature preparation of oxygen- and carbon-free silicon and silicon–germanium surfaces for silicon and silicon–germanium epitaxial growth by rapid thermal chemical vapor deposition. *J. Electrochem. Soc.*, **147**, 4652–4659.

23. Stengl, R., Tan, T., and Gösele, U. (1989) A model for silicon wafer bonding process. *Jpn. J. Appl. Phys.*, **28**, 1735–1741.

24. Maszara, W.P., Goetz, G., Caviglia, A., and McKitterick, J.B. (1988) Bonding of silicon wafers for silicon-on-insulator. *J. Appl. Phys.*, **64**, 4943–4950.

25. Litton, D.A. and Garofalini, S.H. (2001) Modeling of hydrophilic wafer bonding by molecular dynamics simulations. *J. Appl. Phys.*, **89**, 6013–6023.

26. Tong, Q.-Y., Schmidt, E., Gösele, U., and Reiche, M. (1994) Hydrophobic silicon wafer bonding. *Appl. Phys. Lett.*, **64**, 625–627.

27. Trucks, G.W., Raghavachari, K., Higashi, G.S., and Chabal, Y.J. (1990) Mechanism of HF etching of silicon surfaces: a theoretical understanding of hydrogen passivation. *Phys. Rev. Lett.*, **65**, 504–507.

28. Reiche, M. (2008) Dislocation networks formed by silicon wafer direct bonding. *Mater. Sci. Forum*, **590**, 57–78.

29. Bengtsson, S. and Engström, O. (1989) Interface charge control of directly bonded silicon structures. *J. Appl. Phys.*, **66**, 1231–1239.

30. Bäcklund, Y., Ljungberg, K., and Söderbärg, A. (1992) A suggested mechanism for silicon direct bonding from studying hydrophilic and hydrophobic surfaces. *J. Micromech. Microeng.*, **2**, 158–160.

31. Ljungberg, K., Söderbärg, A., and Bäcklund, Y. (1993) Spontaneous bonding of hydrophobic silicon surfaces. *Appl. Phys. Lett.*, **62**, 1362–1364.

32. Reiche, M., Hopfe, S., Gösele, U., Strutzberg, H., and Tong, Q.-Y. (1996) Characterization of interfaces of directly bonded silicon wafers: a comparative study of secondary ion mass spectroscopy, multiple internal reflection spectroscopy, and transmission electron microscopy. *Jpn. J. Appl. Phys.*, **35**, 2102–2107.

33. Reiche, M., Gösele, U., and Wiegand, M. (2000) Modification of Si(100) surfaces by SF_6 plasma etching – application to wafer direct bonding. *Cryst. Res. Technol.*, **35**, 807–821.

34. Pan, C.T., Yang, H., Shen, S.C., Chou, M.C., and Chou, H.P. (2002) A low-temperature wafer bonding technique using patternable materials. *J. Micromech. Microeng.*, **12**, 611–615.

35. Li, S., Freidhoff, C.B., Young, R.M., and Ghodssi, R. (2003) Fabrication of micronozzles using low-temperature wafer-level bonding with SU-8. *J. Micromech. Microeng.*, **13**, 732–738.

36. Steinkirchner, J., Martini, T., Reiche, M., Kästner, G., and Gösele, U. (1995) Silicon wafer bonding via designed monolayers. *Adv. Mater.*, **7**, 662–665.

37. Alexe, M., Dragoi, V., Reiche, M., and Gösele, U. (2000) Low temperature GaAs/Si direct wafer bonding. *Electron. Lett.*, **36**, 677–678.

38. Wiegand, M., Reiche, M., Gösele, U., Gutjahr, K., Stolze, D., Longwitz, R., and Hiller, E. (2000) Wafer bonding of silicon wafers covered with various surface layers. *Sens. Actuators*, **86**, 91–95.

39. Esser, R.H., Hobart, K.D., and Kub, F.J. (2003) Improved low-temperature Si–Si hydrophilic wafer bonding. *J. Electrochem. Soc.*, **150**, G228–G231.

40. Bengtsson, S. and Amirfeiz, P. (2000) Room temperature wafer bonding of silicon, oxidized silicon, and crystalline quartz. *J. Electron. Mater.*, **29**, 909–915.

41. Farrens, S.N., Dekker, J.R., Smith, J.K., and Roberds, B.E. (1995) Chemical free room temperature wafer to wafer direct bonding. *J. Electrochem. Soc.*, **142**, 3949–3955.

42. Kissinger, G. and Kissinger, W. (1993) Void-free silicon-wafer-bond strengthening in the 200–400 °C range. *Sens. Actuators A*, **36**, 149–156.

43. Wiegand, M., Reiche, M., and Gösele, U. (2000) Time-dependent surface properties and wafer bonding of O_2-plasma-treated

silicon (100) surfaces. *J. Electrochem. Soc.*, **147**, 2734–2740.

44 Reiche, M., Wiegand, M., and Dragoi, V. (1999) Plasma activation for low-temperature wafer direct bonding, in *Semiconductor Wafer Bonding: Science, Technology and Applications V*, vol. 99-35 (eds C.E. Hunt, H. Baumgart, U. Gösele, and T. Abe), Electrochemical Society, Pennington, NJ, pp. 292–301.

45 Suni, T., Henttinen, K., Suni, I., and Mäkinen, J. (2002) Effects of plasma activation on hydrophilic bonding of Si and SiO_2. *J. Electrochem. Soc.*, **149**, G348–G351.

46 Reiche, M., Radu, I., Gabriel, M., Zoberbier, M., Hansen, S., and Eichler, M. (2005) Mechanisms of low-temperature wafer bonding, in *Semiconductor Wafer Bonding VIII: Science, Technology, and Applications*, vol. 2005-02 (eds K.D. Hobart, S. Bengtsson, H. Baumgart, T. Suga, and C.E. Hunt), Electrochemical Society, Pennington, NJ, pp. 326–337.

47 Radu, I., Reiche, M., Zoberbier, M., Gabriel, M., and Gösele, U. (2005) Low-temperature wafer bonding via DBD surface activation, in *Semiconductor Wafer Bonding VIII: Science, Technology, and Applications*, vol. 2005-02 (eds K.D. Hobart, S. Bengtsson, H. Baumgart, T. Suga, and C.E. Hunt), Electrochemical Society, Pennington, NJ, pp. 295–302.

48 Haneman, D., Roots, W.D., and Grant, J.T.P. (1967) Atomic mating of germanium surfaces. *J. Appl. Phys.*, **38**, 2203–2212.

49 Fenner, D.B., Biegelsen, D.K., and Bringans, R.D. (1989) Silicon surface passivation by hydrogen termination: a comparative study of preparation methods. *J. Appl. Phys.*, **66**, 419–424.

50 Ramm, J., Beck, E., Zueger, A., Dommann, A., and Pixley, R.E. (1993) Hydrogen cleaning of silicon wafers. Investigation of the wafer surface after plasma treatment. *Thin Solid Films*, **228**, 23–26.

51 Gösele, U., Stenzel, H., Martini, T., Steinkirchner, J., Conrad, D., and Scheerschmidt, K. (1995) Self-propagating room-temperature silicon wafer bonding in ultrahigh vacuum. *Appl. Phys. Lett.*, **67**, 3614–3616.

52 Boland, J.J. (1991) The importance of structure and bonding in semiconductor surface chemistry: hydrogen on the Si(111)-7 × 7 surface. *Surf. Sci.*, **244**, 1–14.

53 Nelson, J.S., Dodson, B.W., and Taylor, P.A. (1992) Adhesive avalanche in covalently bonded materials. *Phys. Rev. B*, **45**, 4439–4444.

54 Conrad, D., Scheerschmidt, K., and Gösele, U. (1996) Molecular dynamics simulations of silicon wafer bonding. *Appl. Phys. A*, **62**, 7–12.

55 Kubair, D.V., Cole, D.J., Ciacchi, L.C., and Spearing, S.M. (2009) Multiscale mechanics modeling of direct silicon wafer bonding. *Scr. Mater.*, **60**, 1125–1128.

56 Reznicek, A., Scholz, R., Senz, S., and Gösele, U. (2003) Comparative TEM study of bonded silicon/silicon interfaces fabricated by hydrophilic, hydrophobic, and UHV wafer bonding. *Mater. Chem. Phys.*, **81**, 277–280.

57 Takagi, H., Maeda, R., Hosoda, N., Chung, T.R., and Suga, T. (1999) Room-temperature wafer bonding using Ar beam surface activation, in *Semiconductor Wafer Bonding: Science, Technology, and Applications V*, vol. 99-35 (eds C.E. Hunt, H. Baumgart, U. Gösele, and T. Abe), Electrochemical Society, Pennington, NJ, pp. 265–274.

58 Takagi, H., Maeda, R., and Suga, T. (2001) Room-temperature wafer bonding of Si to $LiNbO_3$, $LiTaO_3$ and $Gd_3Ga_5O_{12}$ by Ar-beam surface activation. *J. Micromech. Microeng.*, **11**, 348–352.

59 Takagi, H., Maeda, R., Hosoda, N., and Suga, T. (1999) Room-temperature bonding of Si wafers to Pt films on SiO_2 or $LiNbO_3$ substrates using Ar-beam surface activation. *Jpn. J. Appl. Phys.*, **38**, L1559–L1561.

60 Yu, H.H., and Suo, Z. (1998) A model of wafer bonding by elastic accommodation. *J. Mech. Phys. Solids*, **46**, 829–844.

61 Mitani, K., Lehmann, V., Stengl, R., Feijoo, D., Gösele, U., and Massoud, H.Z. (1991) Causes and prevention of temperature-dependent bubbles in silicon

wafer bonding. *Jpn. J. App. Phys.*, **30**, 615–622.

62 Mitani, K. and Gösele, U. (1992) Formation of interface bubbles in bonded silicon wafers: a thermodynamic model. *Appl. Phys. A*, **54**, 543–552.

63 Bollmann, W. (1970) *Crystal Defects and Crystalline Interfaces*, Springer Verlag, New York.

64 Gafiteanu, R., Chevacharoenkul, S., Goesele, U.M., and Tan, T.Y. (1993) Twist boundaries in silicon. A model system. *Inst. Phys. Conf. Ser.*, **134**, 87–90.

65 Benamara, M., Rocher, A., Sopéna, P., Claverie, A., Laporte, A., Sarrabayrouse, G., Lescouzères, L., and Peyre-Lavigne, A. (1996) Structural and electrical investigations of silicon wafer bonding interfaces. *Mater. Sci. Eng. B*, **42**, 164–167.

66 Wilhelm, T., Mchedlidze, T., Yu, X., Arguirov, T., Kittler, M., and Reiche, M. (2008) Regular dislocation networks in silicon, part 1: structure. *Solid State Phenom.*, **131–133**, 571–578.

67 Akatsu, T., Scholz, R., and Gösele, U. (2004) Dislocation structure in low-angle interfaces between bonded Si(001) wafers. *J. Mater. Sci.*, **39**, 3031–3039.

68 Plößl, A., Scholz, R., and Akatsu, T. (1999) Room temperature covalent bonding: effect on interfacial properties, in *Semiconductor Wafer Bonding: Science, Technology, and Applications V*, vol. 99-35 (eds C.E. Hunt, H. Baumgart, U. Gösele, and T. Abe), Electrochemical Society, Pennington, NJ, pp. 232–243.

69 Schober, T. and Balluffi, R.W. (1970) Quantitative observation of misfit dislocation arrays in low and high angle twist grain boundaries. *Philos. Mag.*, **21**, 109–123.

70 Kittler, M., Yu, X., Mchedlidze, T., Arguirov, T., Vyvenko, O.F., Seifert, W., Reiche, M., Wilhelm, T., Seibt, M., Voß, O., Wolff, A., and Fritzsche, W. (2007) Regular dislocation networks in silicon as a tool for nanostructure devices used in optics, biology, and electronics. *Small*, **3**, 964–973.

71 Kittler, M. and Reiche, M. (2009) Dislocations as active components in novel silicon devices. *Adv. Eng. Mater.*, **11**, 249–258.

72 Bengtsson, S., Andersson, G.I., Andersson, M.O., and Engström, O. (1992) The bonded unipolar silicon–silicon junction. *J. Appl. Phys.*, **72**, 124–140.

73 Yu, X., Arguirov, T., Kittler, M., Seifert, W., Ratzke, M., and Reiche, M. (2006) Properties of dislocation networks formed by Si wafer direct bonding. *Mat. Sci. Semicond. Proc.*, **9**, 96–101.

74 Reiche, M., Hiller, E., and Stolze, D. (2002) New substrates for MEMS, in *Proceedings of IEEE Sensors 2002*, vol. 1, IEEE, Piscataway, NJ, pp. 607–612.

75 Reiche, M., Kittler, M., Buca, D., Hähnel, A., Zhao, Q.-T., Mantl, S., and Gösele, U. (2010) Dislocation-based Si nanodevices. *Jpn. J. Appl. Phys.*, **49**, 04DJ02.

76 Bruel, M. (1994) Process for the production of thin semiconductor material films. US Patent 5,374,564.

77 Bruel, M. (1995) silicon on insulator material technology. *Electron. Lett.*, **31**, 1201–1202.

78 Bruel, M. (1996) Application of hydrogen ion beams to silicon on insulator material technology. *Nucl. Instrum. Meth. Phys. Res. B*, **108**, 313–319.

79 Tsuchiya, R., Horiuchi, M., Kimura, S., Yamaoka, M., Kawahara, T., Maegawa, S., Ipposhi, T., Ohji, Y., and Matsuoka, H. (2004) Silicon on thin BOX: a new paradigm of the CMOSFET for low-power and high-performance application featuring wide-range back-bias control, in *IEEE International Electron Devices Meeting Technical Digest*, IEEE, Piscataway, NJ, pp. 631–634.

80 Bresson, N., Cristoloveanu, S., Oshima, K., Mazure, C., Letertre, F., and Iwai, H. (2004) Alternative dielectrics for advanced SOI MOSFETs: thermal properties and short channel effects, in *Proceedings of the IEEE SOI Conference*, IEEE, Piscataway, NJ, pp. 62–64.

81 Reiche, M., Moutanabbir, O., Hoentschel, J., Gösele, U., Flachowsky, S., and Horstmann, M. (2010) Strained silicon devices. *Solid State Phenom.*, **156–158**, 61–68.

82 Tracy, C.J., Fejes, P., Theodore, N.D., Maniar, P., Johnson, E., Lamm, A.J., Paler, A.M., Malik, I.J., and Ong, P. (2004) Germanium-on-insulator substrates by wafer bonding. *J. Electron. Mater.*, **33**, 886–892.

83 Chao, Y.-L., Scholz, R., Reiche, M., Gösele, U., and Woo, J.C.S. (2006) Characteristics of germanium-on-insulators fabricated by wafer bonding and hydrogen-induced layer splitting. *Jpn. J. Appl. Phys.*, **45**, 8565–8570.

84 Akatsu, T., Deguet, C., Sanchez, L., Allibert, F., Rouchon, D., Signamarcheix, T., Richtarch, C., Boussagol, A., Loup, V., Mazen, F., Hartmann, J.M., Campidelli, Y., Clavelier, L., Letertre, F., Kernevez, N., and Mazure, C. (2006) Germanium-on-insulator (GeOI) substrates – a novel engineered substrate for future high performance devices. *Mater. Sci. Semicond. Process.*, **9**, 444–448.

85 Haisma, J., Spierings, G.A.C.M., Michielsen, T.M., and Adema, C.L. (1995) Surface preparation and phenomenological aspects of direct bonding. *Philips J. Res.*, **49**, 23–46.

86 Haisma, J. and Spierings, G.A.C.M. (2002) Contact bonding, including direct-bonding in a historical and recent context of materials science and technology, physics and chemistry. *Mater. Sci. Eng. R*, **37**, 1–60.

87 Herrick, K.J., Kazior, T.E., Laroche, J., Liu, A.W.K., Lubyshev, D., Fastenau, J.M., Urteaga, M., Ha, W., Bergman, J., Brar, B., Bulsara, M.T., Fitzgerald, E.A., Clark, D., Smith, D., Thompson, R.F., Daval, N., and Celler, G.K. (2008) Engineering substrates for 3D integration of III-V and CMOS, in *Semiconductor Wafer Bonding 10: Science, Technology, and Applications*, vol. 16 (eds T. Suga, J. Bagdahn, H. Baumgart, C. Colinge, K.D. Hobart, and H. Moriceau), Electrochemical Society, Pennington, NJ, pp. 227–234.

6
Plasma-Activated Bonding

Maik Wiemer, Dirk Wuensch, Joerg Braeuer, and Thomas Gessner

6.1
Introduction

Several years ago, low-temperature wafer bonding was implemented industrially. In particular, direct wafer bonding or short direct bonding has a major advantage in microelectromechanical systems (MEMS) packaging because no additional intermediate layers, like in eutectic or adhesive bonding, are required. Direct bonding uses the formation of covalent bonds caused by the atomic contact of two clean and smooth surfaces and subsequently annealing at temperatures above 800 °C [1, 2]. As a result of system integration and enhanced packaging density it is essential to reduce these relatively high annealing temperatures. One possibility to reduce these process temperatures is the implementation of plasma activating procedures prior to bonding. Therefore, the bond technique here presented is called plasma-activated bonding (PAB). This activation process is generally employed because various suppliers of wafer bond equipment have offered special plasma-activation tools for several years.

In principle, two solids, regardless of which materials they are, can be bonded using PAB subject to the condition that their surfaces are flat and clean. The implementation of the PAB process requires a special high surface quality of the process wafers. The most important parameters are surface roughness, wafer flatness, and surface contamination caused by technology process steps before the bonding step. Typical values of 4-inch or 6-inch wafers are: surface roughness (root mean square) less than 0.5 nm (measured area $10 \times 10\,\mu m^2$), flatness less than 10 μm, and total thickness variation less than 2 μm [3–5]. In practice, it is expensive and associated with much technical effort to create a smooth and flat bonding surface [3]. Especially, the PAB of silicon is an area that has become of considerable scientific and commercial interest. Many researchers have investigated this bond technique and developed various theoretical models and proved these models with several experiments [2, 4, 6–13].

The system integration and packaging of microelectronic, micromechanical, and microanalysis systems are increasingly affected by three-dimensional chip stacking [14]. The fabrication of bonding on the wafer or chip scale plays a major

Handbook of Wafer Bonding, First Edition. Edited by Peter Ramm, James Jian-Qiang Lu, Maaike M.V. Taklo.
© 2012 Wiley-VCH Verlag GmbH & Co. KGaA. Published 2012 by Wiley-VCH Verlag GmbH & Co. KGaA.

role therein [15]. Different functionalities often require the implementation of heterogeneous materials with a large mismatch in coefficient of thermal expansion (CTE) or materials with a low melting point. Therefore the process temperature during wafer bonding is often the limiting factor for the development of MEMS.

The theory of direct bonding and the mechanism of plasma activation processes are presented in this chapter. Furthermore, a classification of PAB is given. Based on this classification, the principle of low- and atmospheric-pressure plasma activation of materials used in the direct bonding process is explained. The process flow of PAB and the characterization of the bonds produced are also introduced. The last section focuses on some applications for PAB.

6.2
Theory

This section describes briefly the theory of direct bonding, the mechanism of plasma activation processes, and the physical definition of plasma to enable an understanding of the PAB technique.

6.2.1
(Silicon) Direct Bonding

Two different methods are known for the direct bonding process. One is hydrophobic direct bonding. The other, and more interesting for plasma activation, is hydrophilic direct bonding. To understand the mechanisms of plasma treatment of a silicon surface, it is necessary to explain the procedure of hydrophilic direct bonding. The process includes pre-bonding at room temperature and annealing at high temperatures in order to develop strong covalent Si–Si bonds.

In the case of hydrophilic direct bonding, the bond process at room temperature occurs after contact of the wafers by hydrogen bonds between the chemisorbed water molecules located on the opposing wafer surfaces [3, 5, 16]. Typically, external pressure is applied to facilitate contact of the wafer surfaces. The bond strength particularly depends on the number of hydrogen bonds and therefore on the number of silanol groups and water molecules. A subsequent heat treatment (annealing) takes place, to increase the strength of the bonded wafers. Thereby the water molecules diffuse out either along the interface to the outside or through the native oxide to the bulk silicon. The wafers move closer together and form hydrogen bonds between the Si–OH groups. With increasing temperature the opposing silanol groups can react with each other to form covalent siloxane bonds (Si–O–Si) [1–3]. This reaction is the following:

$$Si-OH + OH-Si \rightarrow Si-O-Si + H_2O \tag{6.1}$$

The reasons for significant bond strengths on heat treatment at temperatures above 800 °C are enhanced water molecule diffusion and increased viscous flow of the materials. Especially, the increased viscous flow leads to greater surface contact between the wafers. In summary, a high bond strength depends on the cleanness of the silicon surface, the number of hydrogen bonds formed, and the diffusion of water molecules out of the interface during annealing at high temperatures [2, 3].

6.2.2
Mechanisms of Plasma on Silicon Surfaces

Tan *et al.* described the surface and interface effects caused by plasma treatments, which result in an increase of the bond strength of a silicon wafer compound at low temperatures [17]. Those authors propose four mechanisms:

1) **Removal of contaminants on the wafer surface.** The plasma treatment removes organic contaminants and adsorbates from the substrate surface. These contaminations increase the surface potential barrier and hinder the formation of strong bonds. The activation process cleans the surface und produces more dangling bonds which can enhance the surface energy.

2) **Increase in the number of silanol groups.** The plasma initiates a modification of the surface. Hence, the number of silanol groups on the silicon wafer is increased. As described in the process of hydrophilic direct wafer bonding at the beginning of this chapter, the bond strength of a wafer compound is correlated with the number of siloxane bonds formed, which are generated by the silanol groups on the wafer surface via the chemical reaction shown in Eq. (6.1) [1]. However, the work of Wiegand[2] shows that the density of the OH groups is almost the same as that without plasma treatment. With a method called power spectral density analysis and with the help of atomic force microscopy, he represented each surface's feature frequency or wavelength and verified that the total surface area increases after the plasma treatment. Therefore, the total number of silanol groups on the wafer surface increases, whereas the surface density of the silanol groups remains unchanged. Thus, more silanol groups can react with each other and the silicon surfaces are bonded via an increased number of oxygen bonds. So, more siloxane bonds are formed via the chemical reaction of Eq. (6.1) and the bond strength is finally increased. The increasing amount of silanol groups on the wafer surface is also mentioned in [6–8].

3) **Improvement of the diffusivity of water and gas trapped at the interface.** Plasma activation ensures an increased diffusivity and thus enhances the evacuation of water and gas trapped at the interface. Consequently, the reversible chemical reaction of Eq. (6.1) will intensify, so that the bond strength is significantly increased at lower temperatures.

4) **Enhancement of viscous flow.** A viscous flow of the surface layer at temperatures above 800 °C initiates the mechanism of hydrophilic direct bonding to complete bond formation of the wafer pair. Accordingly, it is necessary that

this viscous flow of the surface layer occurs at lower temperatures. A SiO_2 layer, for example, is damaged by the plasma process and it becomes porous [18]. Thus, the layer contains more water than it does before the plasma treatment, so that the viscosity is much reduced and a viscous flow might takes place at lower temperatures. Therefore, a larger contact area can now be available, so that the bond strength is increased.

To sum up, plasma activation enhances the bond strength in comparison to that of conventional direct bonding. This increase in bond strength using plasma treatment is based on cleaning of the wafer surface, an increase in the number of silanol groups, enhanced diffusivity, and lower viscosity of the oxide.

6.2.3
Physical Definition of a Plasma

A physical plasma is an ionized gas with a neutral electrical charge. That means that the density of positive and negative charges is equal [1]. In principle, a plasma consists of electrons, ions, radicals, and photons, for example ultraviolet (UV) radiation [19]. The electrons that are required to ignite the plasma are produced using a high-frequency (HF) alternating electrical field. This field is applied between two electrodes in the reactor chamber. In that field electrons are accelerated and collide with the plasma gas molecules. During this collision more and more electrons are ejected from the gas molecules due to the high kinetic energy of the accelerated electrons. These generated electrons lead to the creation of positively charged ions and radicals. The ions are also accelerated in the HF field. Because of the much higher mass of the ions compared to the electrons, the ions are not able to follow the fast alternating field. This is the reason why ions are accelerated in one direction, that is, towards the wafer. The high kinetic energy of the ions leads to an ion bombardment on the wafer surface and so the activation takes place [20].

6.3
Classification of PAB

The classification of the plasma types for PAB is quite challenging. There are several techniques to achieve plasma activation. These techniques can be subdivided in terms of the gases that are used in the process (oxygen, nitrogen, hydrogen, ammonia, argon, etc.), the atmosphere (low pressure, atmospheric pressure), and the plasma mode (reactive ion etching (RIE), inductively coupled plasma RIE (ICP-RIE), etc.) [21]. The background of the classification for the PAB used herein is the order of the technical development as well as the main parameter "pressure" of the plasma activation for bonding. Figure 6.1 represents this classification.

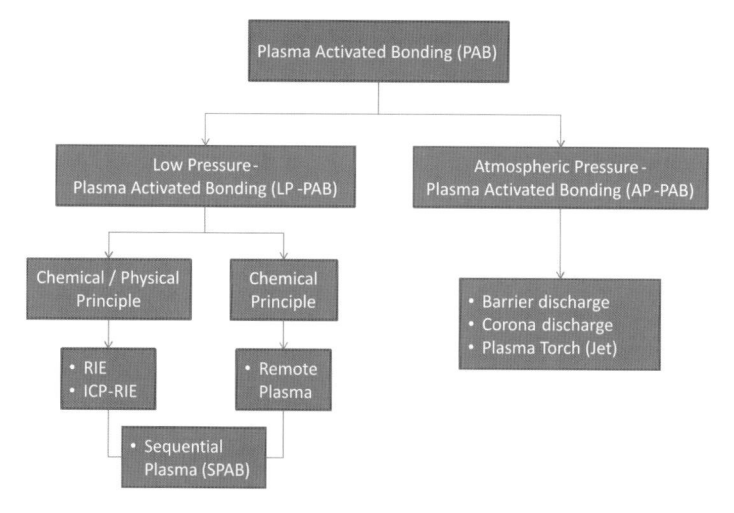

Figure 6.1 Classification of PAB.

6.3.1
Low-Pressure PAB

Low-pressure plasma systems operate in a vacuum, which typically ranges from 0.1 to 100 Pa with a continuous gas flow into the reactor. As mentioned before, there are a multiplicity of plasma modes that can be applied to activate the wafer surface. A commonly used plasma mode is RIE [4, 10, 22]. This RIE mode is usually adopted in dry etching processes. By reducing the plasma parameters, such as the HF power, it can also be used for surface activation. Also a variation of RIE, i.e., ICP-RIE, can be used in the wafer activation process [23].

In both RIE and ICP-RIE plasmas, ions and radicals interact with the surface. In contrast to these chemical/physical methods, another treatment is the so-called remote plasma which only uses chemical components. For the remote plasma, mostly neutral radicals react during the treatment. The advantage of this process is that ion bombardment is avoided so that the surfaces are less damaged. Longer plasma exposure times are possible in comparison to the RIE and ICP-RIE modes mentioned above [11, 24]. A special procedure for the remote plasma is the sequential plasma mode [12, 13]. In this special method the wafers are first activated with a short RIE plasma treatment, directly followed by a radical treatment [12]. Both plasma modes are applied in the same reactor chamber, so that no wafer handling between the two steps is necessary. An additional microwave source and an ion-trapping metal plate are used in the plasma reactor to generate the radicals. After the ion bombardment the radicals are separated from the ions by a trapping plate. The effect of the plasma on the wafer surfaces changes from chemical/physical plasma treatment to a chemical treatment caused by reactions between the radicals and the atoms on the surface [25].

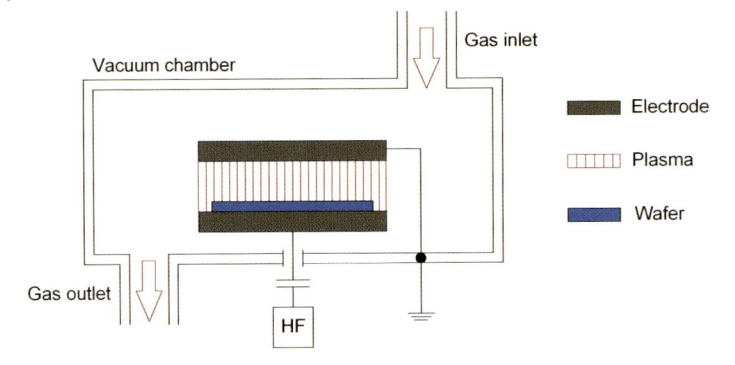

Figure 6.2 Schematic of a plasma reactor for low-pressure plasma activation.

Figure 6.2 shows a typical plasma reactor for low-pressure activation. Low-pressure activation requires a vacuum, additional process gases, and a HF electric field between the two electrodes.

The substrate exposed to the plasma is activated by ion bombardment and chemical reactions caused by radicals. During the period of positive voltage at the HF electrode the electrons move to this electrode. Due to the work function the electrons are not able to leave the electrode within the positive half-wave of the applied voltage. Therefore this electrode is negatively charged up to typically 1000 V (bias voltage). The massive ions are not able to follow the HF field so that they move to the negatively charged electrode where the wafer is placed. In most cases an operation frequency of 13.56 MHz is used [20, 26]. In contrast to conventionally used bias voltage for etching processes, the ions in plasma activation are only accelerated to approximately 100 V [4]. Beside the process gas used, the most important parameters for plasma activation are generator power, chamber pressure, operating distance, process time, and gas flow.

6.3.2
Atmospheric-Pressure PAB

Atmospheric-pressure PAB is a method to ignite plasma without using a low-pressure environment. That means that no expensive equipment for vacuum generation is needed [27].

Gabriel et al. reported that there is an important difference between treatments using low- and atmospheric-pressure plasmas [27]. For a plasma process at atmospheric pressure the interaction of ions with the substrate is negligible. In contrast, the low-pressure plasma treatment is dominated by physical processes with highly energetic ions and electrons on the substrate surface. The lifetime of ions in atmospheric-pressure plasma is below 100 μs. So there is almost no influence on the substrate surface [27, 28]. This is the reason why the surface processes using an atmospheric-pressure plasma can be described by the intense UV radiation,

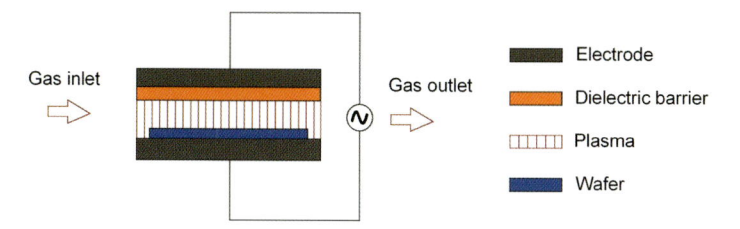

Figure 6.3 Schematic of atmospheric-pressure plasma activation.

which is generated by the dissociation of ions, the chemically active radicals, and the high concentration of electrons (about 10^{14}–10^{15} cm^{-3}) [27, 28].

With an atmospheric-pressure plasma it is possible to ignite the plasma both at specific local areas and the whole surface of the substrate. To generate stable plasma at atmospheric pressure, a method different from that of low-pressure plasma activation is required. Between two electrodes the plasma gas is ignited via an alternating voltage. To avoid sparks a dielectric has to be fixed to one or both of the electrodes. This mode is known as the dielectric barrier discharge (DBD) [9, 27–29]. Originally the DBD, also called silent discharges, was developed by Werner von Siemens (1857) to create ozone. For the DBD the shape of the electrode has to be similar to the substrate geometry to cover the entire surface [30].

The principle of the discharge with one dielectric barrier is shown in Figure 6.3. The electrical discharge between two metallic electrodes is separated by an isolator barrier. As a result of the discharge at atmospheric pressure a plasma is created in the space between the electrode and the isolator barrier.

The DBD uses alternating high voltage (1–10 kV) with frequencies of 0.5–500 kHz. Typical operating distance between the barrier and the ground electrode varies from 0.1 mm to several centimeters. Dielectric materials are often glass, quartz, ceramics, or other materials with high breakdown voltage [31, 32].

As mentioned before, the DBD enables local surface activation. Eichler *et al.* call this method "plasma printing" [29]. An additional mode of local atmospheric-pressure plasma activation is the plasma torch. The plasma torch can be used like a sequential writing method. In this mode the plasma is stable on the top of the nozzle and each geometry can be activated [33, 34].

6.4 Procedure of PAB

6.4.1 Process Flow

A schematic of a typical PAB process flow is shown in Figure 6.4. First, the wafers are often cleaned in deionized water or hydrogen peroxide-based solution (RCA,

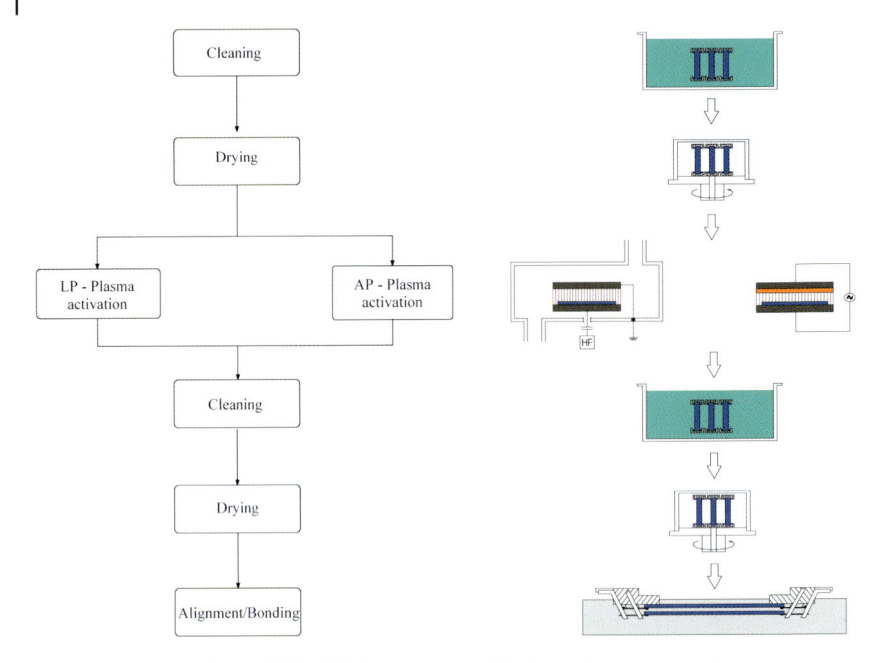

Figure 6.4 Process flow of PAB. (LP, low pressure; AP, atmospheric pressure.)

Piranha). This pretreatment reduces the organic and inorganic contaminations of the surface and creates hydrophilic surfaces. Thereafter the wafers are dried, for example in a spin dryer, and subsequently treated by plasma exposure. After plasma treatment the wafers are dipped in deionized water or hydrogen peroxide-based solution and dried. However, the wet chemical cleaning or the water dip is not essential for a PAB process. The next step is the pre-bonding procedure. Conventionally used bond equipment can be applied, which allows the alignment and the bonding of the wafers as well as the monitoring of all bonding parameters, such as temperature, time, and pressure, during the process. After the pre-bonding process at room temperature an annealing step can be optionally used to increase the bond strength.

6.4.2
Characterization Techniques

The quality of a wafer bond can be described by defect rate, hermiticity, bonding strength, and induced stresses. A number of techniques can be used to characterize a bond. The most important physical parameter of a plasma-activated bond is the bond strength. There are various measurement techniques to characterize the bond strength. A frequently used method is the crack-opening method, which is shown schematically in Figure 6.5a. With this technique a razor blade (thickness

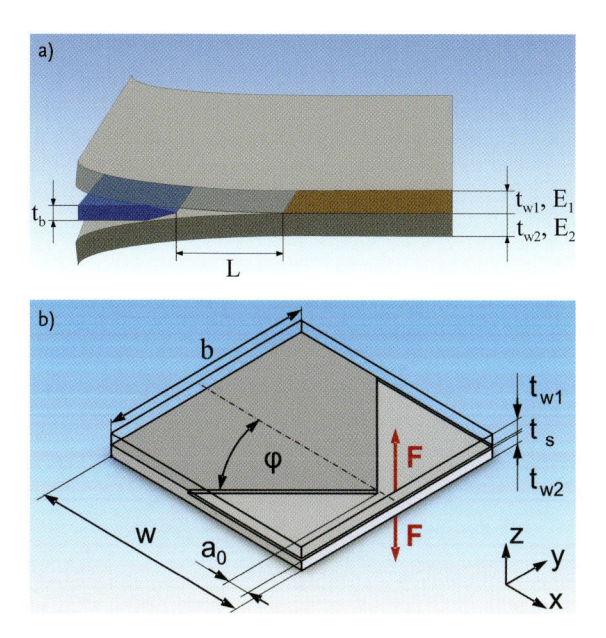

Figure 6.5 Schematics of test methods: (a) crack-opening method; (b) micro-chevron test.

t_b) separates the stacks into two parts (thicknesses t_{w1}, t_{w2}; Young's moduli E_1, E_2). The inserted blade creates two new surfaces and forms the equilibrium crack length L. The bond strength is defined by the surface energy γ and is given by

$$\gamma = \frac{3t_b^2 E_1 t_{w1}^3 E_2 t_{w2}^3}{16L^4(E_1 t_{w1}^3 + E_2 t_{w2}^3)} \tag{6.2}$$

Besides the crack-opening method, tensile or blister tests can be applied for these investigations [3, 35].

The crack-opening method provides no significant bond strengths due to fact that the crack length L is determined very subjectively by the operator. This inherent measurement uncertainty makes it very hard to compare published results between each other.

A much more accurate method to determine the bond strength is the so-called micro-chevron test (MC test). The MC test is a specially adapted method for microsystems technology [36, 37]. In this case the bond strength is determined by the fracture toughness k_{ic}. However, this test requires special sample preparation in an etch process to produce the required geometries (see also Figure 6.5b). Due to the special characteristics of the test, the crack length is not determined experimentally. The maximum force F_{max} is measured in a tensile test. The fracture toughness is calculated using the maximum tensile force, the geometrical parameters (width w, length b), and a specific geometry value y_{min}, which is specified in numerical analysis [36, 38]. The fracture toughness is determined as

$$k_{ic} = \frac{F_{max}}{b\sqrt{w}} Y_{min} \qquad (6.3)$$

Besides the characterization of the bond strength, hermeticity is an important aspect for MEMS applications. Visser *et al.* [39] performed leak tests with helium for plasma-activated bonds according to MIL-STD 883. Both the PAB and the direct bonded (annealed at 1100 °C) specimen showed a leak rate of less than 5.0×10^{-8} mbar l s^{-1}. Due to the small cavities (typical volume of 0.1 mm^3) of MEMS, ultralow leak rates (10^{-16} mbar l s^{-1}) must be realized. This ultrahigh resolution requires much more sensitive leakage measurements, so that the conventionally used methods like the helium leak tests are not adequate. Millar and Desmulliez described in a review ultralow leak rate detection methods. These methods ensure accurate hermeticity testing of MEMS [40].

6.4.3
Experiments and Results

Single-side-polished 4-inch silicon wafers with a thickness of 525 μm were used. All wafers were p-doped and (100)-orientated. The process flow of the following wafer preparation is based on Figure 6.4. First, all wafers were cleaned, using the RCA procedure. In the next step a low-pressure RIE plasma reactor was applied to activate the wafer surfaces (Figure 6.2). For plasma treatment duration, an optimum of 5 min was determined, which was used in the subsequent experiments. Process gases oxygen and nitrogen were used during the activation process. This activation process was followed by rinsing the wafers for 10 min in deionized water. After the pre-bonding process at room temperature, the wafers were annealed at various temperatures, which ranged from 40 to 200 °C. After dicing into single chevron test structures, the bond strength of the silicon–silicon interface was characterized using the MC test. As seen in Figure 6.6a, for all plasma-treated wafer stacks, the fracture toughness is two or three times higher than that for the nonactivated reference wafer stack [41].

Eichler *et al.* [9, 30] reported a dielectric barrier discharge at atmospheric pressure for low-temperature direct silicon wafer bonding (Figure 6.3). The measurement of the surface energy with a modified crack-opening method is realized *in situ* during annealing from room temperature to 200 °C. Figure 6.6b presents these results against the process gas variation. The experiments show that at annealing temperatures below 100 °C the surface energies are two to three times higher compared to the RCA-cleaned reference silicon wafer stack. By annealing the wafers up to 100 °C and treating with synthetic air (80 vol.% N_2 + 20 vol.% O_2) or nitrogen plasma a linear increase of the surface energy can be observed. The highest surface energy at 100 °C can be achieved with a synthetic air (N_2/O_2) treatment. The treatment with the process gas oxygen shows no significant increase below 100 °C. By annealing at temperatures between 100 and 200 °C the surface energy of the oxygen-treated surfaces rises strongly. After annealing at 200 °C the oxygen plasma-treated wafers achieve the highest final value among all process gases [9, 30].

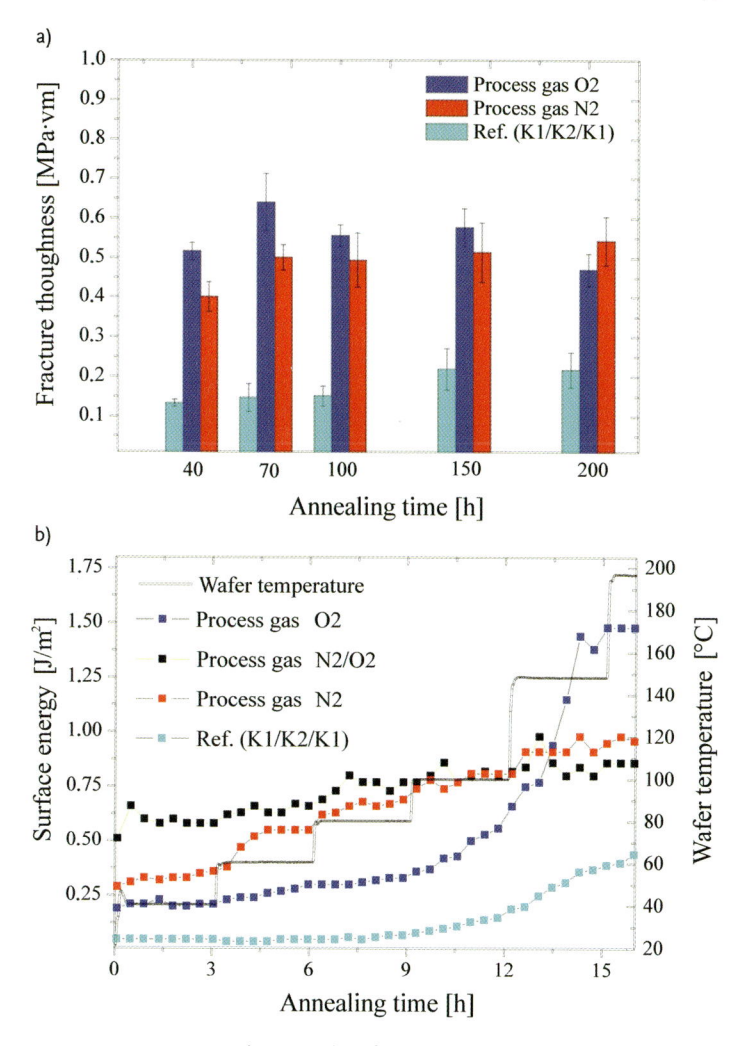

Figure 6.6 Fracture toughness and surface energy during annealing for various activated silicon wafer stacks: (a) low-pressure PAB; (b) atmospheric-pressure PAB.

6.5
Applications for PAB

Some selected applications are discussed in the following subsections. Some sensors are introduced: a pressure sensor and an image sensor. An optical microsystem and microfluidic packaging are also presented. At the end of this section the compatibility of PAB with the complementary metal–oxide–semiconductor (CMOS) standard process is illustrated. The applications described here are each representative for one of the different plasma modes mentioned above, for example, ICP-RIE, DBD, and sequential plasma treatment.

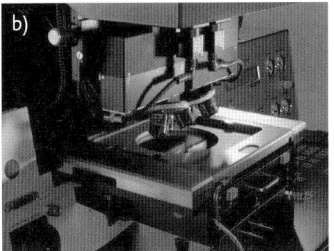

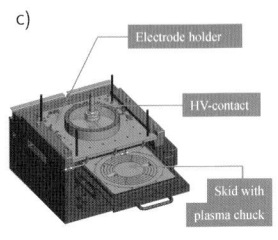

Figure 6.7 (a) Pressure sensor (in cooperation with EPCOS AG). (b) Mask aligner. (c) Tool for an atmospheric-pressure plasma activation system (upgradable in the SUSS Mask Aligner) [42].

6.5.1
Pressure Sensor

Figure 6.7a shows a pressure sensor (EPCOS AG) used as a demonstrator for atmospheric-pressure PAB. Here, a DBD activation of silicon wafers was applied at the Fraunhofer Institute for Surface Engineering and Thin Films. A grounded chuck acting as wafer carrier and an indium tin oxide (ITO)-coated high-voltage glass electrode are used for the activation equipment. In this case the ITO layer is the real electrode and the 750 μm thick glass substrate is the dielectric barrier. The gas-flushed gap between the substrate surface and the counter electrode has an operating distance of 500 μm. The discharge is powered by a corona generator which operates at a frequency of 28.5 Hz. The transfer of this technique within a mask aligner is currently planned (Figures 6.7b and c).

Before low-temperature bonding of the pressure sensor at the Fraunhofer Research Institution for Electronic Nano Systems, the ideal plasma parameters have to be determined. Hence, the process gases nitrogen and synthetic air (80 vol.% N_2 + 20 vol.% O_2) and a process time of 40 s were chosen. Before and after the atmospheric-pressure plasma treatment a cleaning step using deionized water was applied [30]. To increase the bonding strength an annealing at 350 °C for 6 h in nitrogen atmosphere was carried out. After dicing the sensors were characterized by burst tests. The bond strength of a synthetic air plasma-activated specimen was two times higher than that of nonactivated ones [42, 43].

6.5.2
Optical Microsystem

The development of a high-density optical microsystem applied to laser Doppler velocimeter sensors is presented by Higurashi and coworkers [39, 44]. Here, a two-step PAB technique is used as an alternative to eutectic AuSn or Au–Au thermocompression bonding.

Due to the high process temperatures (above 300 °C) of AuSn and Au–Au thermocompression bonding, Higurashi and coworkers developed an Au–Au surface-activated bonding technique. To clean the surfaces, low-pressure plasma activation with argon as process gas is applied. Subsequently Au–Au chip bonding takes place by contacting the substrates in ambient air for 30 s at temperatures below 250 °C [45]. Higurashi [44] asserted that an Au–Au bonded laser diode did not show any significant changes in electrical and optical performance compared with an unbonded one. After the first bonding step a further surface-activated bonding step is introduced to fabricate the microsystem (Figure 6.8). In this step, the optical bench including the bonded laser diode, micromirrors, and electrodes onto patterned silicon and a glass substrate including a photodiode, electrical interconnections, and microlenses are bonded at 150 °C.

6.5.3
Microfluidics Packaging

Howlader *et al.* developed a room temperature PAB technique using sequential plasma activation [25]. At first the wafers are exposed to an oxygen RIE treatment. After this process a microwave nitrogen radical treatment is applied. By using the sequential plasma activation as an *in situ* process, contamination is negligible. The implemented ion trap has two main functions. First, in the RIE mode the ion trap works as the counter electrode. Second, during the microwave treatment the trap operates as a filter to separate the uncharged radicals. The typical parameters for the oxygen RIE treatment are 13.56 MHz, 30 Pa chamber pressure, 200 W generator power, and 60 s process time. For the microwave plasma typical parameters are 2.45 GHz, 100 Pa chamber pressure, and 2000 W generator power. After the plasma activation the wafers were pre-bonded and kept in atmospheric air. Howlader *et al.* characterizes the bond strength by tensile testing. With this sequential plasma activation bond strengths similar to those of the bulk material (silicon or glass) could be achieved. To test the hermiticity a krypton-85 leak test was applied. The sealed cavities were placed in a high-pressure chamber with radioisotope krypton-85 gas mixture. To measure the radiation coming out of the specimen a scintillation counter was used. With this method the leak rate was determined to be lower than $1 \times 10^{-9}\,\text{Pa}\,\text{m}^3\,\text{s}^{-1}$.

Howlader *et al.* used the technique described above to produce a glass–glass microfluidic device. Based on the high bonding strength, and the hermetic sealing behavior, this technique can be used for such microfluidic applications. Furthermore, the low bonding temperature guarantees the survival of living cells in the microchip.

6.5.4
Backside-Illuminated CMOS Image Sensor

Optical devices such as image sensors are another application for PAB. Dragoi *et al.* [46] bonded a fully processed CMOS wafer to a blank one using PAB. First, the CMOS wafer was polished to achieve a flat surface which corresponds to the requirements of direct bonding. After that, plasma-enhanced chemical vapor

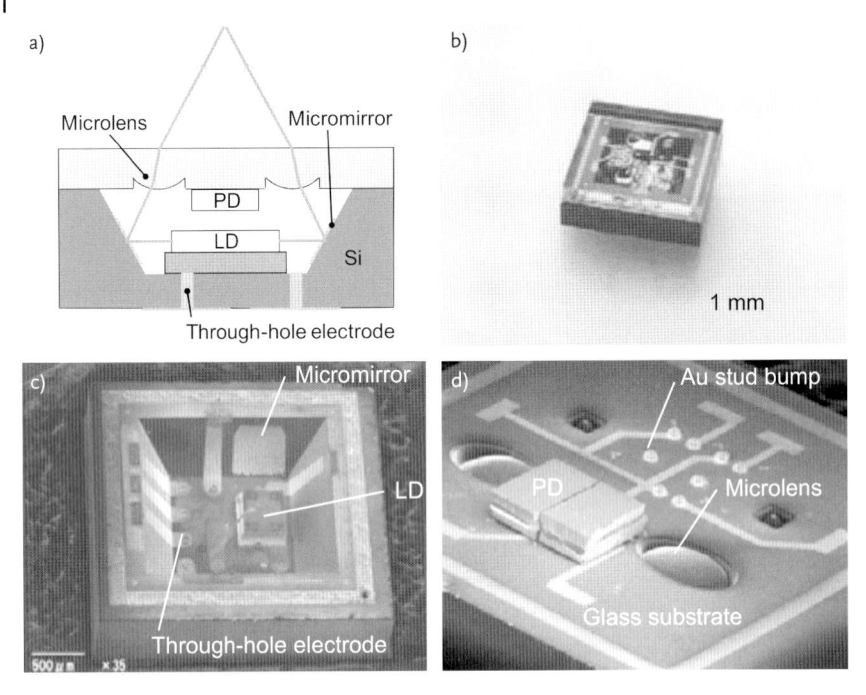

Figure 6.8 High-density optical microsystem: (a) cross-sectional assembly; (b) fabricated system; (c) integrated laser diode (LD) on optical bench; (d) integrated photodiode (PD) on Pyrex glass [44].

deposition was applied to form a silicon dioxide coating on both wafers. Subsequently an oxide layer densification was used to remove gas from the oxide layer. The oxide was then polished using standard chemical mechanical polishing (CMP). In the second step both wafers were plasma-activated and aligned by edge-to-edge alignment and optical alignment followed by spontaneous pre-bond at room temperature. After pre-bonding, the wafer stacks were annealed at 200–400 °C. The CMOS substrate was back-thinned by grinding and etching. The final technology step for processing the CMOS image sensor was adhesive wafer bonding of the optical elements like filters, lenses, and spacers.

6.5.5
CMOS Compatibility of Low-Pressure PAB

Sanz-Velasco et al. [47] determined the compatibility to the CMOS process with an oxygen plasma treatment prior to direct bonding. This is difficult because the impact of the plasma treatment on the electrical properties of the bonded interface and on the reliability of standard electronics is unclear. The plasma can have an influence on the components by contaminations or radiation damage. Here, contamination can be induced by the deposition of nonvolatile contaminants from the sputtering procedures or mobile ion contamination. The potential difference between the plasma and the wafer and the plasma and the chamber walls justifies this sputtering procedure. For RIE, the wafer is exposed to a bombardment by energetic ions, electrons, and photons, for example UV radiation. The photons can cause radiation damage in the materials. Such damage can be in the form of electron traps in gate oxides which are the reason for shifts in the threshold voltages of MOS devices. Atoms in the surface can also be displaced and implanted. Furthermore, destruction of the gate oxide can occur.

In order to test CMOS compatibility, Sanz-Velasco et al. investigated the influence of plasma activation on application-specific integrated circuits. For this they manufactured test structures using the CMOS CYE 0.8 μm process from Austria Mikro Systeme International AG. According to the CYE process a top passivation layer, consisting of 1 μm silicon oxynitride and 1 μm silicon nitride, is superimposed on the electronics to protect them during the plasma treatment. Only the contact pads are treated. ICP-RIE was used with a gas pressure of 40 mtorr, a radio-frequency power of 15 W, and a coil power of 800 W to expose the components for 30 s to the oxygen plasma [47].

To characterize the various influences, Sanz-Velasco et al. measured the $I-V$ characteristics of MOS transistors, the $C-V$ characteristics of MOS gate capacitances, and the frequency of oscillators prior and after the plasma treatment. All measurements showed no significant change. Thus, the authors concluded that the oxygen plasma treatment in ICP-RIE can be used for CMOS applications.

6.6
Conclusion

PAB is a fully CMOS-compatible technology that provides unique possibilities for fabrication and improvement of three-dimensional integrated microsystems. Process flow, limitations, and parameters for PAB at low and atmospheric pressure with sufficient bond strength, packaging reliability, and defect-free bonding interfaces are readily available in the literature. The main advantages of PAB are the low bonding temperature, the compatibility with standard integrated wafer processing, and the wafer bonding of heterogeneous materials with a high CTE mismatch or temperature-sensitive materials. However, PAB requires special wafer surface preparation such as CMP to reduce surface roughness, wafer flatness, and minimal surface contamination. For defect-free PAB, cleanliness prior to bonding should be at a very high level. In conclusion, PAB is a well-known technology and one possibility for three-dimensional wafer stacking used in MEMS packaging.

References

1 Stengl, R., Tan, T., and Gösele, U. (1989) A model for the silicon wafer bonding process. *Jpn. J. Appl. Phys.*, **28**, 1735.

2 Wiegand, M. (2001) Auswirkungen einer Plasmabehandlung auf die Eigenschaften des Niedertemperatur-Waferbondens

monokristalliner Siliziumoberflächen. PhD thesis, Martin-Luther-Universität Halle-Wittenberg.
3 Tong, Q., Goesele, U., and Society, E. (1999) *Semiconductor Wafer Bonding: Science and Technology*, John Wiley & Sons, Inc., New York.
4 Wiemer, M., Otto, T., Gessner, T., Hiller, K., Kapser, K., Seidel, H., Bagdahn, J., and Petzold, M. (2001) Implementation of a low temperature wafer bonding process for acceleration sensors. *MRS Symp. Proc.*, **682**, N4.6.
5 Plößl, A. and Kräuter, G. (1999) Wafer direct bonding: tailoring adhesion between brittle materials. *Mater. Sci. Eng.*, **R25**, 1.
6 Kissinger, G. and Kissinger, W. (1993) Void-free silicon-wafer-bond strengthening in the 200–400 °C range. *Sens. Actuators A*, **36**, 149.
7 Farrens, S.N., Dekker, J.R., Smith, J.K., and Roberds, B.E. (1995) Chemical free room temperature wafer to wafer direct bonding. *J. Electrochem. Soc.*, **142**, 3949.
8 Reiche, M., Gutjahr, K., Stolze, D., Burczyk, D., and Petzold, M. (1997) The effect of plasma pretreatment on the Si/Si bonding behavior. *Electrochem. Soc. Proc.*, **97-36**, 437.
9 Eichler, M., Michel, B., Hennecke, P., and Klages, C. (2009) Effects on silanol condensation during low temperature silicon fusion bonding. *J. Electrochem. Soc.*, **156**, H786.
10 Suni, T., Henttinen, K., Suni, I., and Mäkinen, J. (2002) Effects of plasma activation on hydrophilic bonding of Si and SiO. *J. Electrochem. Soc.*, **149**, G348.
11 Byun, K.Y., Ferain, I., and Colinge, C. (2010) Effect of free radical activation for low temperature si to si wafer bonding. *J. Electrochem. Soc.*, **157** (1), H109.
12 Wang, C., Higurashi, E., and Suga, T. (2008) Void-free room-temperature silicon wafer direct bonding using sequential plasma activation. *Jpn. J. Appl. Phys.*, **47**, 2526.
13 Suga, T., Kim, T.H., and Howlander, M.M.R. (2004) Combined process for wafer direct bonding by means of the surface activation method. Electronic Components and Technology Conference.
14 Tsang, C.K., Andry, P.S., Sprogis, E.J., Patel, C.S., Webb, B.C., Manzer, D.G., and Knickerbocker, J.U. (2006) CMOS-compatible through silicon vias for 3D process integration. *MRS Symp. Proc.*, **970**, 0970-Y01-01.
15 Hofmann, L., Kuechler, M., Gumprecht, T., Ecke, R., Schulz, S.E., and Gessner, T. (2008) Investigations on via geometry and wetting behavior for the filling of through silicon vias by copper electro depostition. MRS Conference Proceedings, AMC XXIV, p. 623.
16 Suni, T. (2006) Direct wafer bonding for MEMS and microelectronics. PhD thesis, Helsinki University of Technology.
17 Tan, C.M., Yu, W., and Wie, J. (2006) Comparison of medium-vacuum and plasma-activated low-temperature wafer bonding. *Appl. Phys. Lett.*, **88**, 114102.
18 Amirfeiz, P., Bengtsson, S., Bergh, M., Zanghellini, E., and Borjesson, B.L. (2000) Formation of silicon structures by plasma activated wafer bonding. *J. Electrochem. Soc.*, **147**, 2693.
19 Moriceau, H., Rieutord, F., Morales, C., and Sartori, S. (2005) A surface plasma activation before direct wafer bonding: a short review and recent results. Proceedings of the International Symposium on Semiconductor Wafer Bonding VIII: Science, Technology, and Applications, p. 34.
20 Hoppe, B. (1998) *Mikroelektronik 2*, Vogel Verlag und Druck, Würzburg.
21 Moriceau, H., Rieutord, F., Fournel, F. et al. (2009) Low temperature plasma activated direct bonding. WaferBond '09, Book of Abstracts, pp. 11–12.
22 Doll, A., Goldschmidtboeing, F., and Woias, P. (2004) Low temperature plasma-assisted wafer bonding and bond-interface stress characterization, in *17th IEEE International Conference on MEMS*, IEEE, New York, pp. 665–668.
23 Zhang, X. and Raskin, J. (2005) Low-temperature wafer bonding: a study of void formation and influence on bonding strength. *J. Microelectromech. Syst.*, **14** (2), 368.

24 Belford, R. and Sood, S. (2009) Surface activation using remote plasma for silicon to quartz wafer bonding. *Microsystem Technol.*, **15** (3), 407.

25 Howlader, M.M.R., Suehara, S., Takagi, H., Kim, T.H., Maeda, R., and Suga, T. (2006) Room-temperature microfluidics packaging using sequential plasma activation process. *IEEE Trans. Adv. Packaging*, **29** (3), 448.

26 Janzen, G. (1992) *Plasmatechnik Grundlagen, Anwendungen, Diagnostik*, Hüthig Buch Verlag, Heidelberg.

27 Gabriel, M., Johnson, B., Suss, R., Reiche, M., and Eichler, M. (2006) Wafer direct bonding with ambient pressure plasma activation. *Microsystem Technol.*, **12** (5), 397.

28 Eliasson, B. and Kogelschatz, U. (1991) Modeling and applications of silent discharge plasmas. *IEEE Trans. Plasma Sci.*, **19** (2), 309.

29 Eichler, M., Michel, B., Gabriel, M., and Klages, C.-P. (2009) Einsatz von Mikroplasmen für die Herstellung von Silizium-mehrlagenaufbauten. *Vakuum in Forschung und Praxis*, **21** (4), 6.

30 Eichler, M., Michel, B., Thomas, M., Gabriel, M., and Klages, C.-P. (2008) Atmospheric-pressure plasma pretreatment for direct bonding of silicon wafers at low temperatures. *Surf. Coat. Technol.*, **203**, 826.

31 Samson, J. and Ederer, D. (1998) *Experimental Methods in the Physical Sciences*, Academic Press, London.

32 Fridman, A. and Cho, Y.I. (2007) *Transport Phenomena in Plasma*, Academic Press/Elsevier, London.

33 Tendero, C., Tixier, C., Tristant, P., Desmaison, J., and Leprince, P. (2006) Atmospheric pressure plasmas: a review. *Spectrochim. Acta B*, **61** (1), 2.

34 Kim, M.C., Yang, S.H., Boo, J.-H., and Han, J.G. (2003) Surface treatment of metals using an atmospheric pressure plasma jet and their surface characteristics. *Surf. Coat. Technol.*, **174**, 839.

35 Maszara, W.P., Goetz, G., Cavailia, A., and McKitterick, J.B. (1988) Bonding of silicon wafers for silicon-on-insulator. *J. Appl. Phys.*, **64**, 4943.

36 Bagdahn, J., Plößl, A., Wiemer, M., and Petzold, M. (2003) Measurement of the local strength distribution of directly bonded silicon wafers using the micro-chevron-test. Proceedings of the 5th International Symposium on Semiconductor Wafer Bonding, PV 99-35, pp. 218–223.

37 Petzold, M., Knoll, H., and Bagdahn, J. (2001) Strength assessment of wafer-bonded micromechanical components using the micro-chevron-test. *Proc. SPIE*, **4558**, 133.

38 Vogel, K., Shaporin, A., Wünsch, D., and Billep, D. (2009) Crack propagation in micro-chevron-test samples of direct bonded wafers. 26th Danubia-Adria Symposium on Advances Experimental Mechanics.

39 Visser, M.M., Weichel, S., de Reus, R., and Hanneborg, A.B. (2002) Strength and leak testing of plasma activated bonded interfaces. *Sens. Actuators A*, **97–98**, 434.

40 Millar, S. and Desmulliez, M. (2009) MEMS ultra low leak detection methods: a review. *Sens. Rev.*, **29** (4), 339.

41 Wünsch, D., Müller, B., Wiemer, M., Besser, J., and Gessner, T. (2010) Aktivierung mittels Niederdruckplasma zur Herstellung von Si-Verbunden im Niedertemperatur-Bereich und deren Charakterisierung mittels Mikro Chevron Test. 2nd GMM Workshop Technologien und Werkstoffe der Mikrosystem- und Nanotechnik.

42 Wünsch, D., Wiemer, M., Cabriel, M., and Gessner, T. (2010) Low temperature wafer bonding for microsystems using dielectric barrier discharge. MST News, no.1/10, pp. 24–25.

43 Wiemer, M., Wünsch, D., Bräuer, J., Eichler, M., Hennecke, P., and Gessner, T. (2009) Low temperature bonding of hetero-materials using ambient pressure plasma activation. WaferBond '09, Book of Abstracts, pp. 73–74.

44 Higurashi, E. (2008) Low temperature bonding for optical microsystems application. *Electrochem. Soc. Trans.*, **16** (8), 93.

45 Higurashi, E., Imamura, T., Suga, T., and Sawada, R. (2007) Low-temperature bonding of laser diode chips on silicon

substrates using plasma activation of Au films. *IEEE Photon. Technol. Lett.*, **19** (24), 1994.

46 Dragoi, V., Mittendorfer, G., Wagenleitner, T., Lindner, P., and Wimplinger, M. (2009) Wafer bonding for backside illuminated CMOS image sensors. WaferBond '09, Book of Abstracts, pp. 121–122.

47 Sanz-Velasco, A., Bring, M., Rödjegård, H., Andersson, G.I., and Enoksson, P. (2006) Low temperature plasma-assisted-wafer-bonding for MEMS. *Electrochem. Soc. Trans.*, **3** (6), 355.

C. Metal Bonding

7
Au/Sn Solder

Hermann Oppermann and Matthias Hutter

7.1
Introduction

Eutectic soldering is in many cases more attractive over other bonding methods such as anodic bonding, fusion bonding, thermocompression bonding, or adhesive joining. The main reason is that the liquid solder wets the metal layers and due to the low viscosity it spreads easily over the bonding area. In this way it can accommodate higher surface topography and nonplanarity than other bonding methods.

Single components can be placed on a substrate simply by a pick-and-place procedure followed by a collective reflow process, which is favored for low-cost bonding. As this method does not require any force during bonding it is advantageous for soldering of sensitive components with membranes or other fragile features as well as for three-dimensional stacking. However, the solder height must be sufficient to allow for collapse during wetting and to account for the formation of high-melting intermetallic compounds at the solder interface. If a tool is used during bonding to apply a mechanical load, the method is called thermode bonding. The bond tool is usually heated when the force is applied. In order to avoid the solder being squeezed out of the bonding interface a thin solder layer is chosen. Both methods, reflow and thermode bonding, are described in Sections 7.3 and 7.4, respectively.

Most eutectic solders used in board assembly, die attach, and flip chip have a tin-rich composition. They are quite ductile showing a large plastic elongation before fracture. Due to the low melting point the maximum operational temperature is limited to 150 °C or lower. For higher temperatures, gold-rich solders like AuSn20, AuSi3, AuGe12, or AuIn28 are potential candidates (Table 7.1).

AuSn20 consists of two intermetallic compounds, whereas AuSi3 and AuGe12 are formed by the pure elementary phases. With increasing soldering temperature the microelectronic components might degrade faster due to higher diffusion rates. If there is a difference in the coefficient of thermal expansion an increase in the eutectic temperature will also result in larger thermomechanical stresses.

Handbook of Wafer Bonding, First Edition. Edited by Peter Ramm, James Jian-Qiang Lu, Maaike M.V. Taklo.
© 2012 Wiley-VCH Verlag GmbH & Co. KGaA. Published 2012 by Wiley-VCH Verlag GmbH & Co. KGaA.

Table 7.1 Gold-rich solder alloys.

Gold-rich eutectic solder alloys	Eutectic temperature (°C)	Components of microstructure
AuSn20	278	AuSn (δ) + Au$_5$Sn (ζ)
AuGe12	356	Au + Ge
AuSi3	363	Au + Si
AuIn28	450	AuIn + Au$_7$In$_3$ (γ')

These are the main reasons why AuSn20 as the solder alloy with the lowest melting point is used more often than other gold-rich solder alloys.

In photonics integration Au/Sn solder plays a unique role [1]. It is used as a fluxless process which is mandatory in order to avoid any contamination of optical surfaces. Post-cleaning processes are not considered as the results are often not satisfactory and are difficult to monitor. For optical alignment a post-bonding accuracy of 1 μm must be achieved. This requires a precise thermode bonder to hold the chip in place during solidification, or the self-alignment capability is used which is provided by the liquid solder. As the eutectic phase consists of a mixture of two intermetallic compounds, AuSn (δ) and Au$_5$Sn (ζ), the solder is quite hard and has a high creep resistance. These properties are advantageous for keeping the position stable during operational lifetime for a long period.

For hermetic encapsulation and especially for vacuum enclosure, a fluxless process is required which is applicable for Au/Sn solder. If a flux were to be used, flux residuals would outgas and corrode components within the cavity. Thermal management demands thin bondlines of a few micrometers and high thermal conductivity which have been achieved using Au/Sn. The solder is therefore used for high-power laser bars, radio-frequency power amplifiers, or high-power discrete devices in leadframe packages. Medical devices and some sensors have to operate in corrosive environments; Au/Sn solder is a very noble alloy, which withstands many types of corrosive attacks.

7.2
Au/Sn Solder Alloy

Au/Sn metallurgy involves two eutectic compositions, but only the gold-rich eutectic with 20 wt% tin is reliable. The tin-rich eutectic consists of brittle phases like AuSn$_4$ (η) and AuSn$_2$ (ε), which are prone to crack propagation at lower stress levels. We are therefore looking only at compositions of final interconnect above 50 at% gold.

The solder can be applied as preforms for die bonding. For thin solder layers, evaporation and sputtering are also possible and could be combined with liftoff techniques for patterning. We have developed an electroplating process: plating

7.2 Au/Sn Solder Alloy

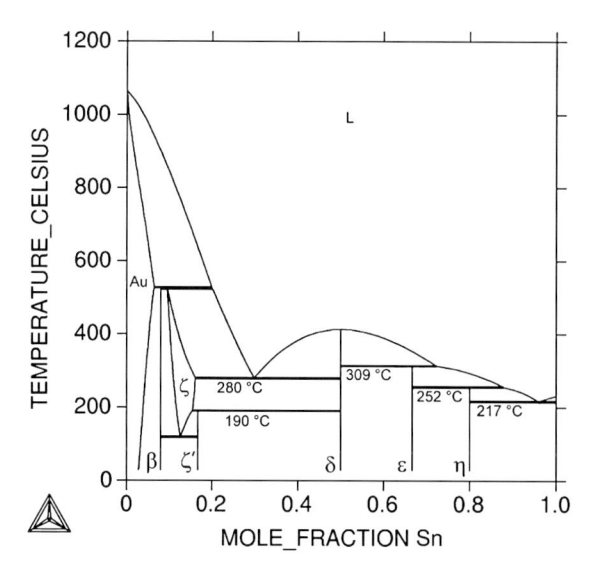

Figure 7.1 Au–Sn binary phase diagram [2].

gold first followed by tin. The solder alloy is formed in a post-processing step: either by bump reflow at the wafer level or after flip chip placement during reflow bonding. This results in a mature process window and provides a larger degree of freedom in the bump design compared to alloy plating.

In the standard bump reflow of a metal stack the temperature is increased continuously at approximately 5 to 10 K min^{-1} while the reactions between the phases occur according to the equilibrium phase diagram shown in Figure 7.1. Using higher temperature ramps (some flip chip bonders reach 500 to 3000 K s^{-1} and some laser bonders even more) will not allow the composition to reach the equilibrium, as limitations in phase nucleation rate, diffusion rate, and melt viscosity will retard the reactions. For these reasons a bump reflow prior to bonding will allow an easy control of phase transformations. According to calorimetric measurements (Figure 7.2) a eutectic reaction takes place at 217 °C: Sn + η-(AuSn$_4$) ↔ L$_1$; at 252 °C a peritectic reaction results in η-(AuSn$_4$) + L ↔ ε-(AuSn$_2$). As there is an excess of gold plated, the liquid phase L reacts with gold by diffusion-driven growth of intermetallic compounds and the liquid L disappears. Finally on reaching a temperature of 280 °C the second eutectic reaction forms the solder: δ-(AuSn) + ζ-(Au$_5$Sn) ↔ L$_2$. This solder composition is frequently used for bonding and is also called AuSn20. The results are shown in Figure 7.3. These reactions have been studied by calorimetric measurements [1, 3] and a cross-sectional analysis of intermediate stages has been published [4].

In the phase equilibrium diagram a high-temperature ζ phase and a low-temperature ζ' phase occur. The disordered high-temperature ζ phase has an extended range of solubility for tin. The crystal structure is close-packed hexagonal and with increasing tin concentration the lattice parameters increase as well. The

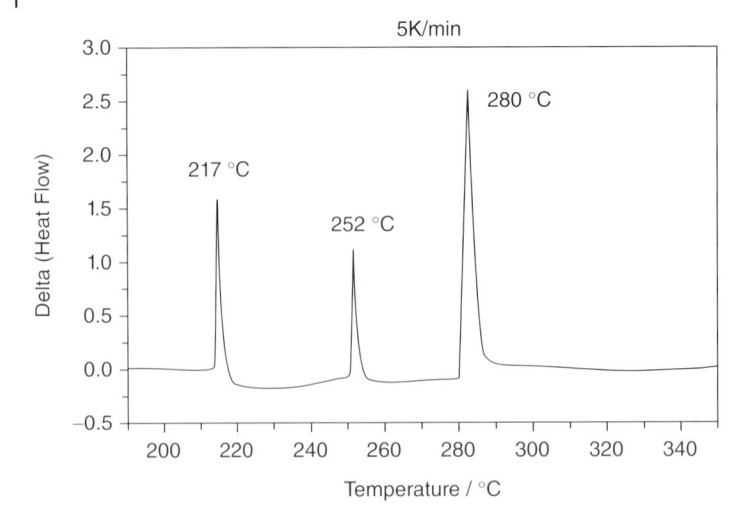

Figure 7.2 Calorimetric study of bump reflow.

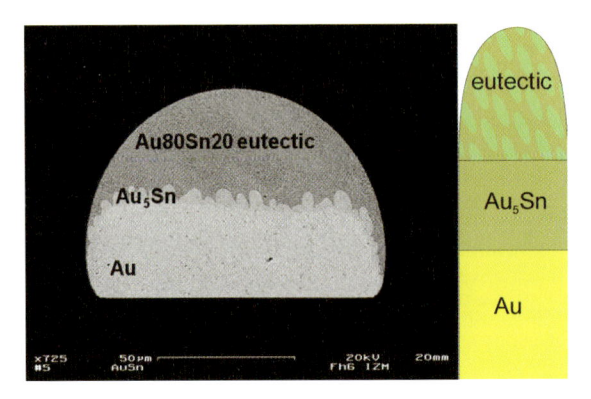

Figure 7.3 SEM image of AuSn solder after bump reflow. The bump consists of a gold socket or pillar, a solder cap of eutectic composition (AuSn20), and an intermetallic compound layer (Au_5Sn or ζ') in between.

ordered ζ' phase forms a superstructure and has a very narrow solubility range. The disordered ζ phase with the highest solubility of tin as formed during eutectic solidification transforms into the ordered ζ' phase at 190 °C. With increased gold concentration – as typically formed during diffusional growth in contact with a gold layer – the phase transformation is shifted to lower temperatures. The order–disorder phase transformations are evident in calorimetric studies as can be seen in Figure 7.4.

In Figure 7.5 all intermediate stacks formed between gold and tin layers are summarized. The as-plated condition ("ap") is somehow virtual, as the gold starts immediately to diffuse into the plated tin layer and transforms into intermetallic compounds at the interface. When the wafer with the layer stack is removed from

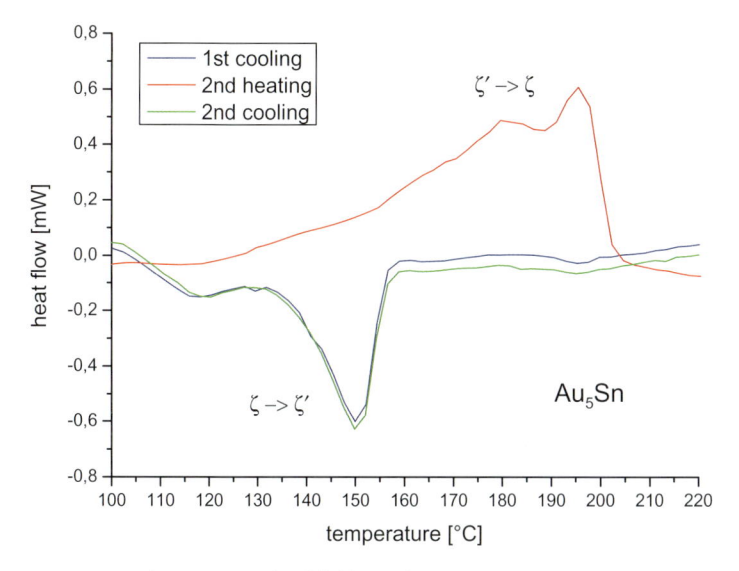

Figure 7.4 Calorimetric study of ζ–ζ′ transformation.

Figure 7.5 Schematic of metal stacks formed with different temperature treatments before bonding: ap, as plated; st, stored at room temperature; rl, reflow at lower temperature (217 °C < T_R < 280 °C); rm, similar to rl, but forming a metastable eutectic; p1 to p4, pretreatment with increasing temperature (200 °C < T_P < 280 °C) or time; pi, pretreatment with an inversion layer; rh, reflow at high temperature (T_R > 280 °C).

the electrolytic bath intermetallic compounds $AuSn_2$ (ε) and AuSn (δ) are already formed and they grow further even at room temperature ("st" in Figure 7.5; Figure 7.6). The $AuSn_4$ (η) phase does not form a complete layer between tin and $AuSn_2$ (ε), but single grains are formed there. This reflects that the $AuSn_4$ (η) phase is not stable at lower temperature.

At temperatures above 217 °C and below 280 °C the Au + Sn metal stack can be reflowed following the tin-rich eutectic. If the temperature rise stops below the

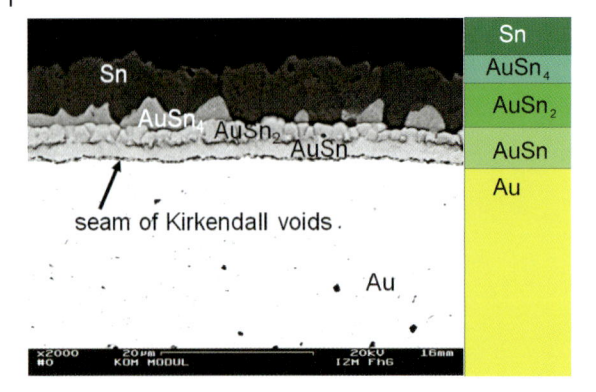

Figure 7.6 SEM image showing layers of gold, tin, and intermetallic compounds after room temperature storage ("st" in Figure 7.5).

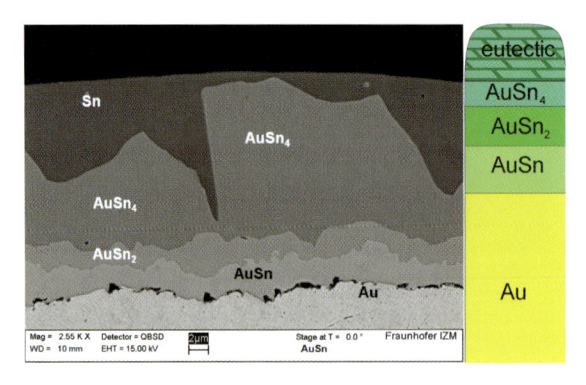

Figure 7.7 SEM image of AuSn bump after heating to 240 °C ("rl" in Figure 7.5).

peritectic reaction at 252 °C, after cooling the tin-rich eutectic is formed between tin and AuSn$_4$ (η). In this case a stabile eutectic reaction takes place but an abnormal microstructure is generated as shown in Figure 7.7. The typical two-phase region is not formed, because nucleation of new AuSn$_4$ (η) phase is suppressed and the AuSn$_4$ (η) layer grows during solidification and tin solidifies as it remains.

On increasing the temperature above the peritectic temperature of 252 °C, all AuSn$_4$ (η)transforms into liquid and AuSn$_2$ (ε). During solidification the nucleation of AuSn$_4$ (η) is still suppressed and as there is no remaining η phase to grow, another eutectic reaction between tin and AuSn$_2$(ε) occurs, which is not part of the equilibrium phase diagram. This reaction is metastable and a lower eutectic temperature of 207 °C was found. In Figure 7.8 the metal stack was heated to 255 °C, just above the peritectic reaction and quenched to prevent the gold from leaving as AuSn$_2$ (ε) directly on the already existing phase. It is clear that the metastable eutectic composition forms with a normal microstructure. When the

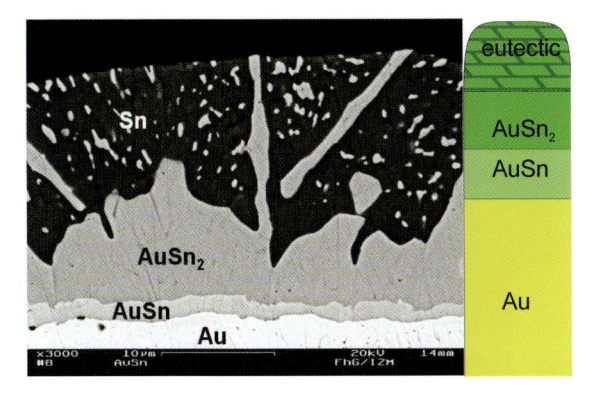

Figure 7.8 SEM image of AuSn bump after heating to 255 °C ("rm" in Figure 7.5).

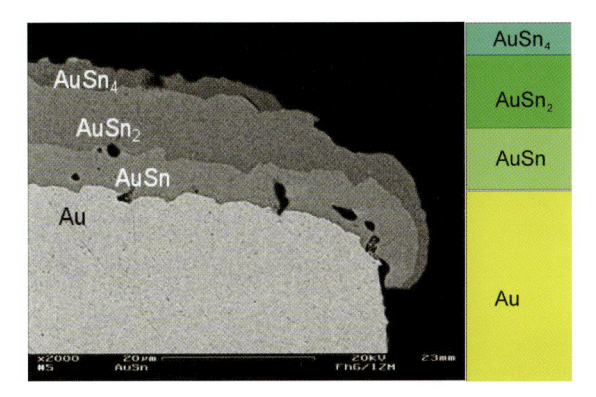

Figure 7.9 SEM image of AuSn bump layer of gold and tin after preconditioning ("p1" in Figure 7.5); no tin remaining.

solder melt is cooled more slowly, the remaining $AuSn_2$ (ε) layer grows leaving large tin grains, which can be described as metastable eutectic with abnormal microstructure.

Preconditioning has some advantages as this transforms the tin on top of the bump into intermetallic compounds. These intermetallic compounds have slower growth rates of oxide layers than pure tin. For flip chip assembly this is helpful if no flux should be used. Of course, the bump reflow above 280 °C also provides intermetallic compounds on the outer surface as shown in Figure 7.3. But for smaller bumps and thin tin layers, bump reflow is not an option as most of the tin transforms completely into high-melting ζ phase.

It should be mentioned that preconditioning does not have an effect on the ability to form the high-melting eutectic. Quite the contrary, the preconditioning increases the oxidation resistance and improves the arrangement of Kirkendall voids along the Au–Sn interface. In Figure 7.9 tin has been transformed

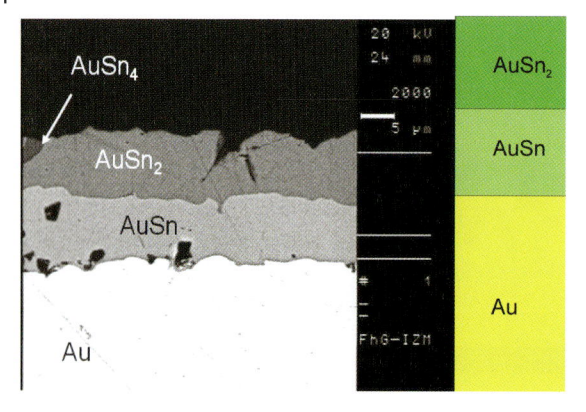

Figure 7.10 SEM image of AuSn bump layer of gold and tin after preconditioning ("p2" in Figure 7.5), with AuSn (δ) and AuSn$_2$ (ε); almost no AuSn$_4$ (η) remaining.

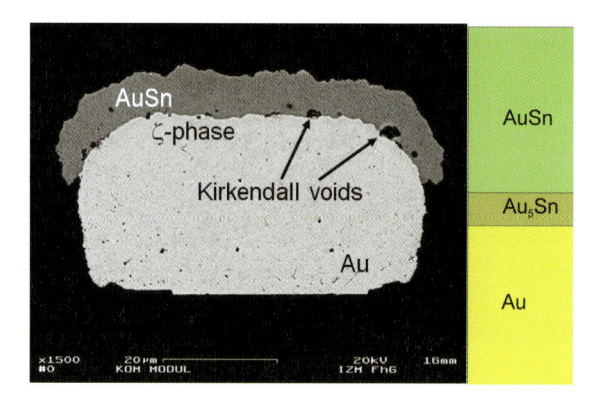

Figure 7.11 SEM image of AuSn bump layer of gold and tin after preconditioning ("p4" in Figure 7.5), with AuSn (δ) and Au$_5$Sn (ζ'), which is hardly distinguishable from gold in the SEM image.

completely and, on top of gold, layers of δ-(AuSn), ε-(AuSn$_2$), and η-(AuSn$_4$) are visible. With progression of preconditioning in Figure 7.10 AuSn$_4$ (η) disappears and in Figure 7.11 even the AuSn$_2$ (ε) is gone. In the late stage, Au$_5$Sn (ζ') appears between gold and δ phase. Kinetic studies have shown that the phase growth of η, ε, and δ phases is very fast and can be described by the diffusion of gold through the intermetallic compound layers. In contrast, it has been found that the growth rate of ζ phase is very slow and is based on the diffusion of tin through the ζ layer [5].

Using preconditioning assisted by a chemical reaction an inverse layer sequence with ζ' phase on top was achieved, which should further improve the corrosion resistance and wetting performance. During preconditioning a δ phase layer is formed on top of the bump. Further on at the outer surface tin is oxidized and

Figure 7.12 Au/Sn on nickel barrier, preconditioned into an inverse layer sequence ("pi" in Figure 7.5): Au/Au$_5$Sn (ζ′)/AuSn (δ)/Au$_5$Sn (ζ′).

evaporates as a volatile compound. The enrichment of gold at the surface leads to the formation of ζ′ phase on top. In Figure 7.12 the inverse layer sequence is clearly visible.

From all the different Au/Sn solder bump variations that can be achieved by modifying the metallurgical path, one can select the most appropriately adjusted bump to improve the bonding processes. It is possible to find a suitable process to be used for flip chip reflow soldering with or without self-alignment, for thermode bonding, for hermetic lid soldering, or for die attach.

7.3
Reflow Soldering

For reflow soldering, Au/Sn bumps are manufactured by electroplating of gold and tin in successive process steps. For the electrical connection of the bumps during electroplating a metallization consisting of TiW:N and gold is sputter-deposited on the wafer prior to the galvanic deposition of gold and tin. Even though Au/Sn solder neither wets the TiW:N layer nor reacts with it, it is used as a diffusion barrier and adhesion layer because it can be easily and selectively etched after electroplating. However, when using TiW:N the bump has to be composed in such a way that the total composition of the bump is richer in gold than the AuSn20 eutectic; that is, a solid gold layer has to remain beneath the eutectic cap after reflow to avoid dewetting from TiW:N. Scanning electron microscopy (SEM) images of cross sections of Au/Sn bumps are shown as-plated in Figure 7.6, after aging at 200 °C for 4 h in Figure 7.11, and after reflow in Figure 7.3. The electroplating process that has been described in [6] is followed by a reflow in liquid medium, which is necessary to form the eutectic solder cap on top of the bumps. Both electroplating and reflow are wafer-scale processes. After the reflow process is accomplished, these Au/Sn bumps consist of a gold layer with a eutectic solder cap on top (80 wt% gold and 20 wt% tin) showing a typical eutectic microstructure.

Between the eutectic cap and the remaining gold layer the intermetallic Au_5Sn phase forms during reflow, showing a typical dendritic structure (Figure 7.3).

As the temperature increases, tin is transformed by the growth of intermetallic compounds. At 280 °C the gold-rich eutectic reaction takes place. But if the tin transformation is completed, the liquid gold-rich and the liquid tin-rich phases come into contact; excessive mixing heat is generated and the turbulent flow can dissolve the gold irregularly. To ensure a satisfactory reflow result, the bumps are preconditioned at elevated temperature, that is, at 200 °C, for some minutes up to a few hours depending on the tin thickness, prior to the reflow. By applying this pretreatment the phase transformation during the reflow process can be controlled very well, that is, the gold layer is consumed evenly. Investigations regarding the phase transformations during the reflow are reported in detail elsewhere [4, 7]. Immediately after electroplating the formation and growth of tin-rich intermetallic phases starts accompanied by the creation of Kirkendall voids due to the fact that the tin-rich intermetallic phases $AuSn_4$, $AuSn_2$, and $AuSn$ grow predominantly driven by the diffusion of gold atoms rather than by tin atoms. As the Kirkendall voids form a continuous layer when stored at room temperature (Figure 7.6), tin oxides can form inside the voids. Shortly after the electroplating process, an annealing step at elevated temperature is carried out which prevents the formation of an open porous layer of Kirkendall voids. During the annealing step only a few larger but isolated voids form rather than a seaming layer of many small voids.

The volume and shape of the Au/Sn solder, that is, the height of the total bump and the thickness of the eutectic cap as well as the height of the remaining gold layer beneath the eutectic solder cap, have to be adjusted such that nonplanarity of die and substrate is compensated so that every solder bump comes into contact with its counterpart pad during flip chip assembly. In previous work it has been shown that it is possible to predict the shape of the Au/Sn bumps using the volume equation of cut spheres [7, 8].

The flip chip assembly is done using a pick-and-place and reflow process. During the reflow soldering the self-alignment mechanism can be used to let the chip move to its final position on the substrate [4, 9].

Experiments with test substrates have revealed that the concept of using the self-alignment effect in combination with mechanical stops is viable to achieve the highest alignment accuracy [3, 10, 11]. Silicon test vehicles were fabricated using reactive ion etching. They were designed such that two corresponding dies with cavities and stops could be placed on each other so that the bumps on the chip and pads on the substrate are intentionally misaligned before the reflow is carried out. During flip chip soldering the solder melts and wets the substrate pads, that is, the solder bumps collapse and the chip starts moving vertically towards the substrate until it stops due to those mechanical stops acting as spacers. The surface tension force of the misaligned molten solder also makes the chip simultaneously move horizontally in the other two directions until it is stopped by the mechanical stops. The restoring forces can be calculated using the equations shown in Figure 7.13. The surface tension of AuSn20 solder is $0.6\,\mathrm{N\,m^{-1}}$.

7.3 Reflow Soldering

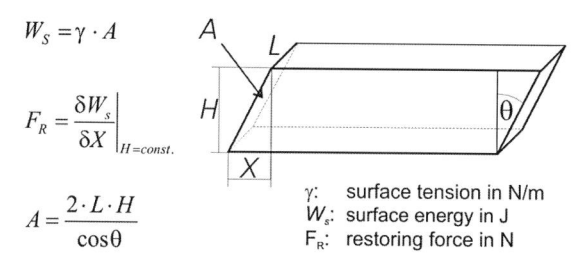

$$W_S = \gamma \cdot A$$

$$F_R = \left.\frac{\delta W_s}{\delta X}\right|_{H=const.}$$

$$A = \frac{2 \cdot L \cdot H}{\cos\theta}$$

γ: surface tension in N/m
W_s: surface energy in J
F_R: restoring force in N

Figure 7.13 Calculation of the restoring forces during self-alignment.

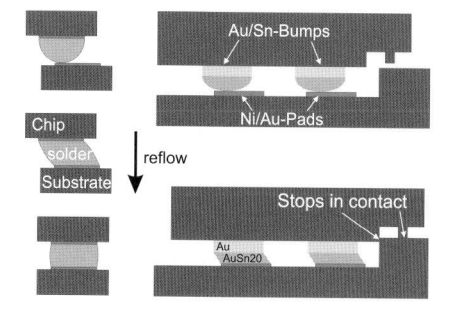

Figure 7.14 Schematic of the self-alignment mechanism. On the left-hand side a bump is shown after pick and place, during reflow, and after self-alignment has taken place.

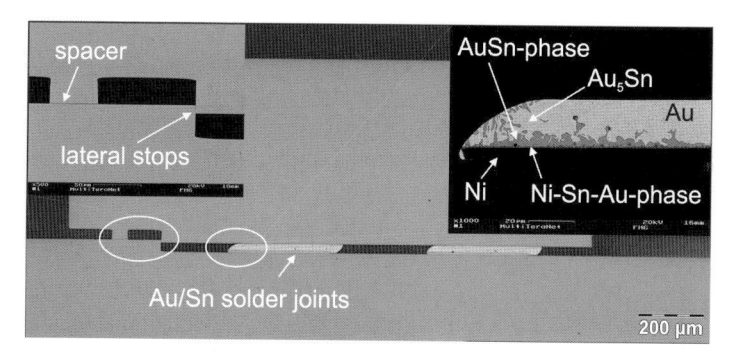

Figure 7.15 Cross-sectional images of flip chip assembly after self-alignment has taken place.

A schematic of the test vehicles after pick and place and after reflow is shown in Figure 7.14. As the restoring forces during self-alignment increase with increasing misalignment the pads on the substrate and bumps on the chip are still purposely misaligned to a certain extent (30 μm) when the stops are in contact with each other. This prevents the chip from stopping its movement just before it has reached its final position. In other words, the method employed uses misaligned solder joints and mechanical stops.

In Figure 7.15 a cross-sectional light microscopy image and a SEM image are shown revealing the results of the flip chip experiments. The stand-off and the etched silicon structure for lateral alignment of the chip are in direct contact with their counterparts on the substrate side. The chip moves until it is in contact with the mechanical stops.

Looking at the flip chip solder joints, a misalignment of 30 μm between the bump gold layer and the nickel pad on the substrate is visible although the stops are in direct contact with each other. Also, the intermetallic phases that form during soldering are visible. Due to the nickel which is dissolved in the solder to a certain extent the microstructure becomes coarse compared to the eutectic microstructure of the Au/Sn solder bumps. The solder joints consist of the intermetallic phases AuSn (δ) and Au_5Sn (ζ′). Between the nickel and the Au/Sn solder a ternary Ni–Au–Sn phase has formed, presumably the $Ni_3Sn_2(Au)$ phase.

7.4
Thermode Soldering

Unlike in reflow soldering, during thermode soldering either the bonding force or solder height is controlled. On the one hand, bending of very thin dies or substrates can thus be compensated. On the other hand, due to the pressure applied the Au/Sn solder layer has to be very thin in order to prevent squeezing of the molten solder out of the gap and to avoid excessive solder flow and shorts between neighboring joints [12]. Au/Sn bumps with excess gold, however, cannot be made too thin, otherwise the solder would transform into the intermetallic Au_5Sn (ζ′) phase and would isothermally solidify as the Au_5Sn (ζ) phase has a melting temperature of more than 500 °C. Typical soldering temperatures for AuSn20 solder are between 300 and 330 °C. As mentioned above, for a bump that is composed of the eutectic AuSn20, that is, has no excess gold, an under-bump metallization (UBM) has to be used, which reacts with the molten solder, that is, is wetted by the solder, but at the same time acts as an effective diffusion barrier [13]. From the literature it is known that TiPtAu is a suitable metallization for eutectic AuSn20 solder. However, the nonwettable TiW:N metallization, on which the electrodeposition of Au/Sn is based, cannot easily be replaced by titanium and platinum because both metals cannot be selectively wet-etched. For this reason a new deposition process had to be developed combining an electroplating process and liftoff technology. The process flow of this technology is shown in Figure 7.16. First, a TiW:N layer is sputter-deposited as an adhesion layer and diffusion barrier followed by sputter-deposition of gold. The gold layer is necessary for the electrical connection during electroplating because the electrical resistance of the TiW:N is too high. In a first lithographic step the bump area is opened and the gold is etched away there. After etching of the gold, a second lithography is done for the liftoff process. Then titanium, platinum, and gold layers are sputter-deposited and the liftoff process is carried out. A third lithography is followed and gold and tin are electrodeposited. Finally, the photoresist is removed and the gold layer and the TiW:N layer are selectively etched. This process flow results in a layer buildup of TiW:N/Ti/Pt/Au/Au(ed)/Sn(ed). The first gold etch removes the gold between the

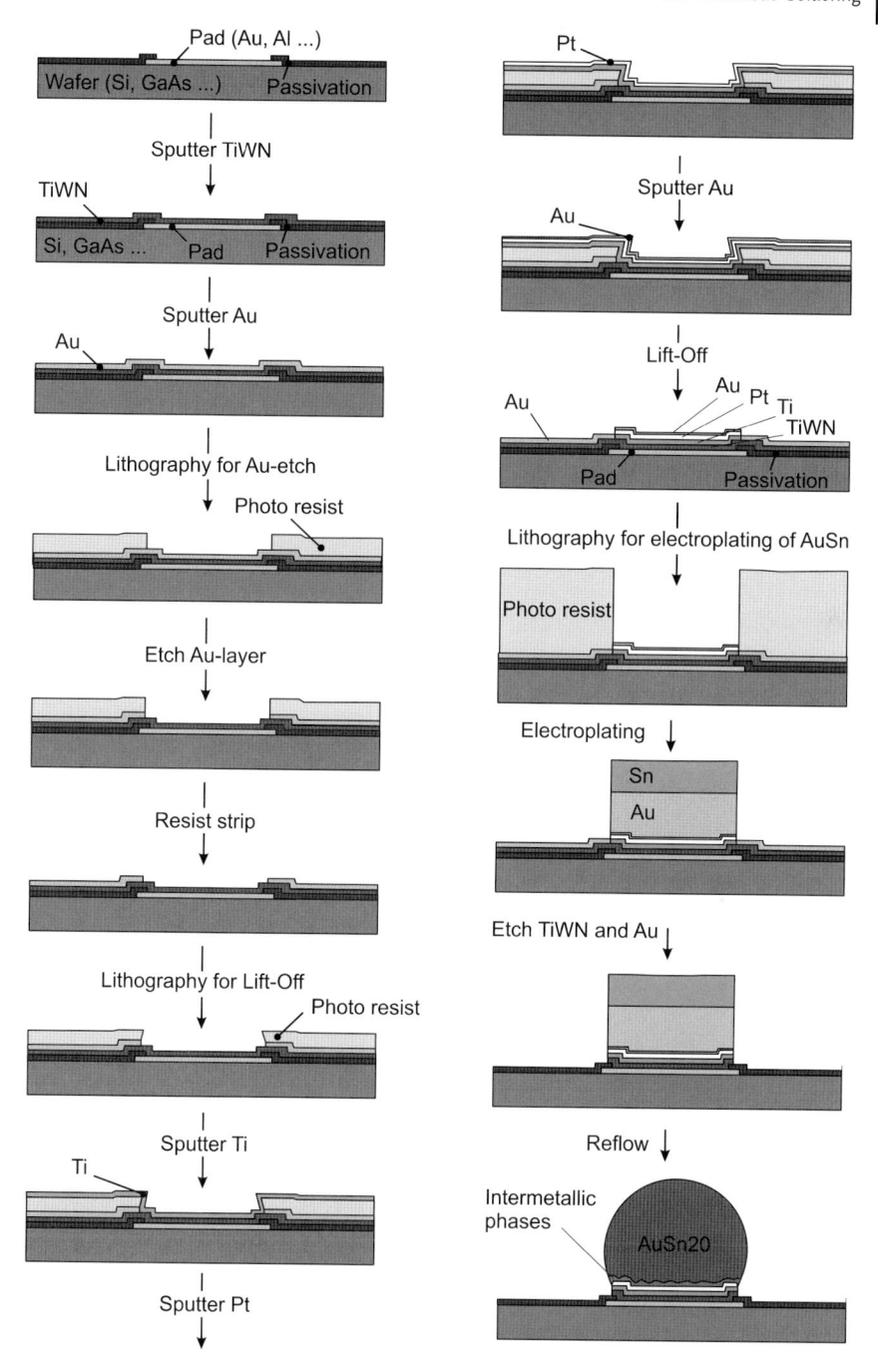

Figure 7.16 Electroplating process for eutectic AuSn20 solder on a TiPtAu metallization.

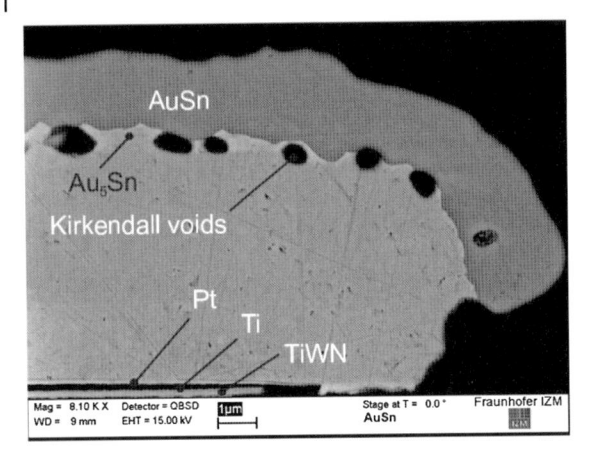

Figure 7.17 Cross-sectional SEM image of a Au/Sn bump on a TiW:N/Ti/Pt/Au metallization.

TiW and titanium layer. This is necessary in order to avoid the Au/Sn solder leaching out the gold within the stack and thus lifting the UBM.

After the electroplating is accomplished, the bumps are subjected to a preconditioning process, that is, they are annealed at temperatures above 200 °C for a certain period of time dependent on the thickness of a tin layer deposited in an activated atmosphere that prevents the formation of tin oxides on the bump surface. A bump of a total height of 8 µm after annealing is shown in Figure 7.17. The UBM consisting of TiW:N, titanium and platinum can be clearly seen. The tin layer has completely transformed into the AuSn (δ) and the Au$_5$Sn (ζ') phases. During annealing the Kirkendall voids do not disappear but they grow according to Oswald ripening so that there is no open porosity that could lead to inner oxidation of the voids.

During soldering such bumps, the Au/Sn solder melts totally and reacts with the metallization of the test substrate, which consists also of titanium, platinum, and gold. On both sides the platinum layer is totally dissolved in the liquid Au/Sn solder whereas the titanium layer is hardly dissolved in the liquid Au/Sn solder, but a reaction takes place between the gold and the titanium, forming the Au$_4$Ti phase. In Figure 7.18 a Au/Sn solder joint on Ti/Pt/Au metallization is shown. The bumps used for this joint consisted of a 2 µm thick gold layer followed by a 2 µm thick tin layer. Figure 7.19 shows the interface between the Au/Sn solder and the titanium layer at a much higher magnification. The platinum layer has been dissolved into the solder.

7.5
Aspects of Three-Dimensional Integration and Wafer-Level Assembly

In wafer-level integration, soldering based on Au/Sn is one of the promising technologies due to some unique properties:

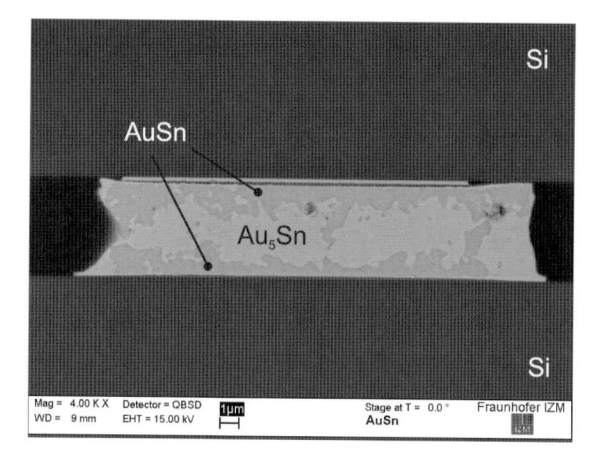

Figure 7.18 AuSn solder joint on Ti/Pt/Au metallization on chip and substrate side (both silicon).

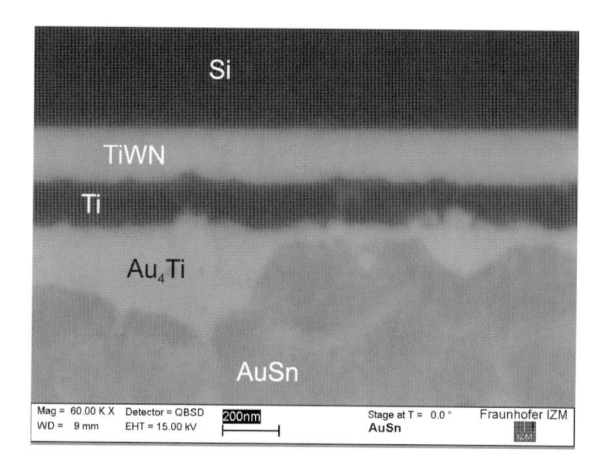

Figure 7.19 Cross-sectional SEM image showing in more detail the UBM and the interface between titanium and Au/Sn solder.

- fluxless bonding approach is possible;
- excellent wetting behavior;
- relatively low soldering temperature of around 300 °C;
- high operating temperature above 200 °C possible;
- enables transient liquid-phase bonding with re-melting at 512 °C;
- high creep resistance and yield strength;
- excellent corrosion resistance.

Au/Sn solder is used for light-emitting diode assembly at the wafer level [14] or the assembly of silicon lids in typical microelectromechanical systems (MEMS) applications [15–17]. A cross section of a Au/Sn solder frame is shown in Figure 7.20.

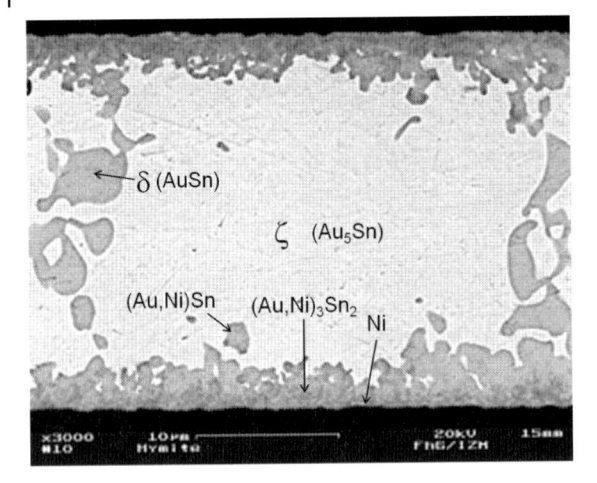

Figure 7.20 SEM image of silicon lid assembly at the wafer level using Au/Sn solder frame, nickel barrier layer, high-temperature storage at 200 °C for 500 h.

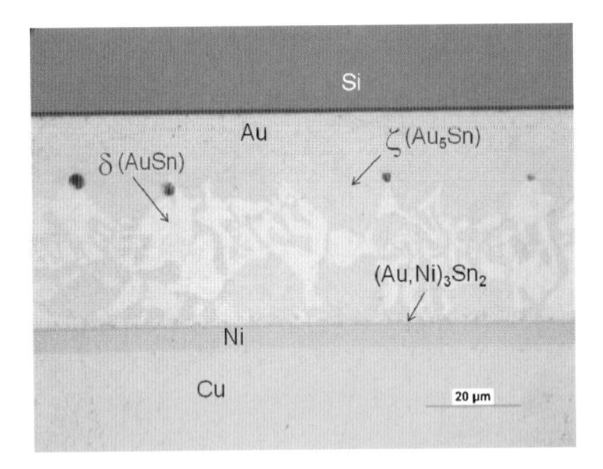

Figure 7.21 Light microscopy image of a silicon lid on an infrared imager chip sealed hermetically with a solder frame consisting of Cu/Ni/AuSn20/Au.

Thicker frames are plated with Cu/Ni/Au in order to increase the height. Au/Sn is plated to the silicon lid that is assembled at the wafer level by pick and place and reflow soldering under vacuum. Figure 7.21 shows the stack after hermetic soldering and consists of Cu/Ni/AuSn20/Au. For getter activation, the device must stay at 300 °C for 30 min. As can be observed in Figure 7.22, the eutectic composition changes to $(Ni,Au)_3Sn_2$, $(Au,Ni)Sn$ (δ) phase, and Au_5Sn (ζ') phase. Residual gas analysis shows the effectiveness of this procedure and indicates that Au/Sn solder is suitable for high-temperature applications.

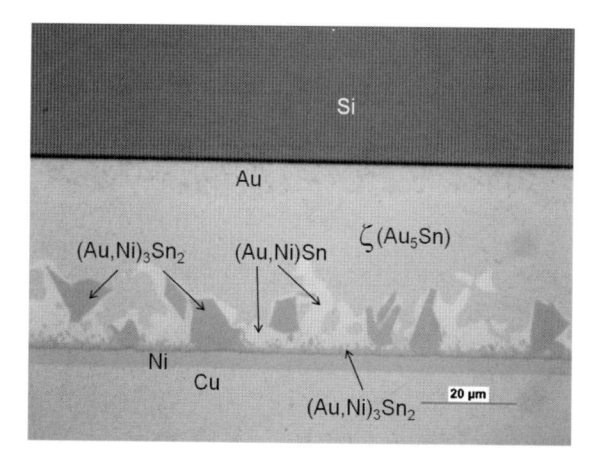

Figure 7.22 Hermetically sealed silicon lid on infrared imager chip after getter activation at 300 °C for 30 min.

Chip-to-wafer thermode bonding requires thin solder layers to avoid shortage between the contacts due to squeezing out of excess solder. The complete transformation of solder into high-melting intermetallic compound has the advantage that one device can be soldered next to another without re-melting the previous one [18]. This transient liquid-phase bonding where the solder is transformed into ζ' phase allows a sequential bonding at the wafer level. For this approach the tin is deposited at the device whereas the gold is provided at the bottom wafer.

A parallel bonding approach is achieved by first placing the devices with an intermediate bond on a handling wafer, creating a reconstructed wafer. The reconstructed wafer is bonded to the second wafer using Au/Sn solder. The high yield strength of Au/Sn allows for the handling wafer to be removed without distortion of the soldered interconnect.

A combination could be used for multiple three-dimensinal stacking: bonding a first device layer in a parallel process to the bottom wafer, sequentially followed by the next device layers [19].

For wafer-to-wafer bonding mechanical stops in the height are helpful to define the gap during application of pressure. This has been demonstrated with lid wafers on MEMS structures (Figure 7.23) as well as with chip-to-wafer bonding using reconstruction on handling wafers and performing wafer-to-wafer bonding.

7.6
Summary and Conclusions

Electroplating of gold and tin is a very economic deposition method. The TiW:N UBM is sputtered and can be removed easily by wet-etching processes. The

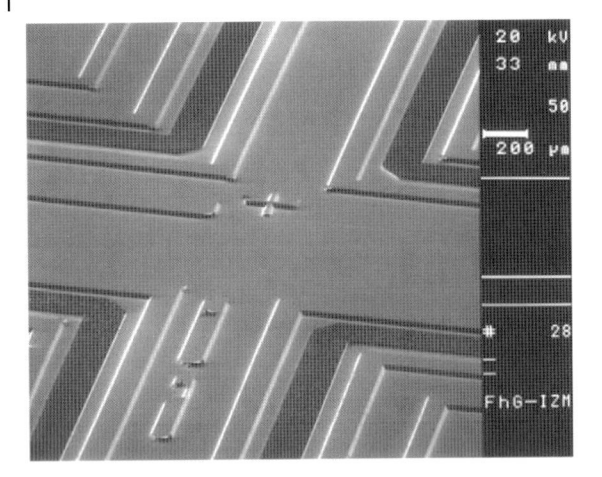

Figure 7.23 Au/Sn solder frame plated into a recess in a silicon lid.

dewetting characteristic of TiW implies an excess gold layer beyond the Au/Sn solder. If Au/Sn solder of eutectic composition without excess gold is required, a Ti/Pt/Au layer is sputtered and a combination of liftoff processes and wet etching is therefore introduced.

The Au/Sn metallurgy is complex due to the large number of intermetallic compounds and the different transformations taking place during reflow and bonding. But the reactions can be well described and are fully comprehensible.

It has been shown that Au/Sn solder has superior properties and the variation in metallurgy allows for a wide range of different process flows and adaptation to various applications. Reflow soldering using the self-alignment capability of the solder with mechanical stops has been shown as well as thermode bonding including transient liquid-phase soldering. The next challenge for wafer-level packaging will be self-assembly. Due to the superior properties we believe that Au/Sn solder will play a certain role there.

References

1 Oppermann, H. (2005) Solder interconnects: the role of AuSn solder in packaging, in *Materials for Information Technology: Devices, Interconnects and Packaging* (eds E. Zschech, C. Whelan, and T. Mikolajick), Springer Verlag, pp. 377–390.

2 TCS (1999) TCS Alloys Mobility Database, v. 2.0 (provided by Thermo-Calc Software). Database: AuSn.

3 Hutter, M., Oppermann, H., Engelmann, G., Dietrich, L., and Reichl, H. (2006) Precise flip chip assembly using electroplated AuSn20 and SnAg3.5 solder. Proceedings of the 56th Electronic Components and Technology Conference, San Diego, CA, 30 May–2 June 2006.

4 Kallmayer, C., Oppermann, H., Kloeser, J., Zakel, E., and Reichl, H. (1995)

Experimental results on the self-alignment process using Au/Sn metallurgy and on the growth of the ζ-phase during the reflow. Proceedings of the 1995 International Flip Chip, Ball Grid Array, TAB and Advanced Packaging Symposium (ITAP '95), San Jose, CA, February 1995, pp. 225–236.

5 Hutter, M. (2009) Verbindungstechnik Höchster Zuverlässigkeit Für Optoelektronische Komponenten. Doctoral thesis, Technische Universität Berlin.

6 Dietrich, L., Engelmann, G., Ehrmann, O., and Reichl, H. (1998) Gold and gold–tin wafer bumping by electrochemical deposition for flip chip and TAB. 3rd European Conference on Electronic Packaging Technology (EuPac'98), Nuremberg, Germany, 15–17 June 1998.

7 Hutter, M., Oppermann, H., Engelmann, G., Wolf, J., Ehrmann, O., Aschenbrenner, R., and Reichl, H. (2002) Calculation of shape and experimental creation of AuSn solder bumps for flip chip applications. Proceedings of the 52nd Electronic Components and Technology Conference, San Diego, CA.

8 Hutter, M., Oppermann, H., Engelmann, G., Wolf, J., Ehrmann, O., Aschenbrenner, R., and Reichl, H. (2002) Process control of the reflow of AuSn bumps. 6th VLSI Packaging Workshop of Japan, Kyoto Research Park, 12–14 November 2002, pp. 39–42.

9 Oppermann, H., Zakel, E., Engelmann, G., and Reichl, H. (1994) Investigation of self-alignment during flip-chip assembly using eutectic gold–tin metallurgy. 4th International Conference and Exhibition on Micro Electro, Opto and Mechanical Systems and Components, MST '94.

10 Hutter, M., Oppermann, H., Engelmann, G., and Reichl, H. (2004) High precision passive alignment flip chip assembly using self alignment and micromechanical stops. Proceedings of the 6th Electronics Packaging Technology Conference, EPTC 2004, Singapore, 8–10. December 2004, pp. 385–389.

11 Oppermann, H., Hutter, M., Engelmann, G., and Reichl, H. (2005) Passive alignment flip chip assembly using surface tension of liquid solder and micromechanical stops. *SPIE Proc. Ser.*, **5716**, 19–25.

12 Elger, G., Voigt, J., and Oppermann, H. (2001)Application of flip-chip bonders in AuSn solder processes to achieve high after bonding accuracy for optoelectronic modules. 14th Annual Meeting of the IEEE Lasers and Electro-Optics Society, San Diego, CA, 11–15 November 2001, pp. 437–438.

13 Anhöck, S., Oppermann, H., Kallmayer, C., Aschenbrenner, R., Thomas, L., and Reichl, H. (1998) Investigations of AuSn alloys on different end-metallizations for high temperature applications. 22nd IEEE/CPMT International Electronics Manufacturing Technology Symposium, IEMT Europe 98, pp. 156–165.

14 Elger, G., Jordan, R., Suchodoletz, M., and Oppermann, H. (2002) Development of an low cost wafer level flip chip assembly process for high brightness LEDs using the AuSn metallurgy. 35th International Symposium on Microelectronics, Denver, CO, 4–6 September 2002, pp. 199–204.

15 Rogge, B., Moser, D., Oppermann, H., Paul, O., and Baltes, H. (1998) Solder-bonded micromachined capacitive pressure sensors. *Proc. SPIE*, **3514**, 307–315.

16 Hutter, M., Suchodoletz, M., Oppermann, H., Engelmann, G., Ehrmann, O., and Reichl, H. (2004) Electroplated AuSn solder for flip chip assembly and hermetic sealing. 7th VLSI Packaging Workshop of Japan, Kyoto, 30 November–2 December 2004.

17 Wilke, M., Töpper, M., Kim, S., Klein, M., Drüe, K.-H., Müller, J., Wiemer, M., Glaw, V., Oppermann, H., Solzbacher, F., and Reichl, H. (2006) Development of a capping process based on Si and LTCC for a wireless neuroprosthetic implant. NIH/NINDS Neural Interfaces Workshop, Bethesda, MD, 21–23 August 2006.

18 Chu, K.-M., Lee, J.-S., Oppermann, H., Engelmann, G., Wolf, J., Reichl, H., and Jeon, D.Y. (2006) Multiple flip-chip assembly for hybrid compact optoelectronic modules using

electroplated AuSn solder bumps. Proceedings of the 39th International Symposium on Microelectronics (IMAPS 2006), San Diego, CA, 8–12 October 2006.

19 Brunschwiler, T., Rothuizen, H., Kloter, U., Reichl, H., Wunderle, B., Oppermann, H., and Michel, B. (2008) Forced convective interlayer cooling in vertically integrated packages. ITHERM2008, Lake Buena Vista, FL, 28–31 May 2008, pp. 1114–1125.

8
Eutectic Au–In Bonding

Mitsumasa Koyanagi and Makoto Motoyoshi

8.1
Introduction

A serious problem for large-scale integration (LSI) is that the signal propagation delay and the power consumption by the interconnections increase significantly as the transistor count increases. In addition, I/O circuits in LSI tend to consume more power due to the need to rapidly drive the output pins and the external wiring with large capacitances and inductances in package and printed circuit boards. As a result, it becomes increasingly difficult to achieve high performance and low power consumption in LSIs. To overcome these problems, the wiring length, the chip size, and the pin capacitance have to be reduced. Three-dimensional (3D) LSI satisfies these requirements and has thus been attracting considerable attention. One can easily reduce the wiring length, the chip size, and the pin capacitance by employing 3D LSIs and consequently one can increase the signal processing speed and decrease the power consumption [1–3]. Such 3D LSIs are also useful for increasing the wiring connectivity within a chip. It therefore becomes possible to produce new kinds of LSIs with new functions [4–6]. Wafer-to-wafer bonding and die-to-wafer bonding are keys for achieving such 3D LSIs. Various kinds of wafer-to-wafer and die-to-wafer bonding methods have been developed so far [7–16]. We have developed a new 3D bonding technology employing eutectic In–Au bonding and epoxy adhesive injection [17] and fabricated several 3D LSI test chips, such as a 3D image sensor chip, 3D shared memory, 3D artificial retina chip, and 3D microprocessor chip, using this technology [18–21]. In addition, we have succeeded in fabricating high-density In–Au microbumps with a size of $2 \times 2 \mu m^2$ and a pitch of $5 \mu m$ by the modified planarized liftoff method [22].

In this chapter, this new 3D bonding technology employing eutectic In–Au bonding and epoxy adhesive injection is described.

Handbook of Wafer Bonding, First Edition. Edited by Peter Ramm, James Jian-Qiang Lu, Maaike M.V. Taklo.
© 2012 Wiley-VCH Verlag GmbH & Co. KGaA. Published 2012 by Wiley-VCH Verlag GmbH & Co. KGaA.

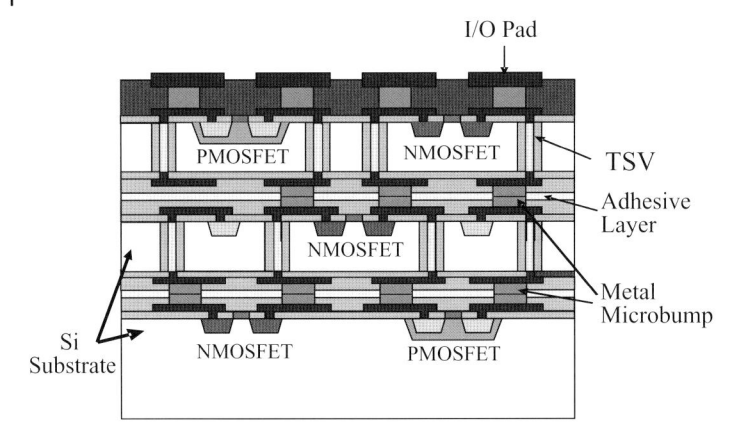

Figure 8.1 Cross-sectional structure of 3D LSI.

8.2
Organic/Metal Hybrid Bonding

The cross-sectional structure of a 3D LSI with through-silicon vias (TSVs) is illustrated in Figure 8.1 [23]. To produce such a 3D LSI, wafer-to-wafer bonding is the key technology. Three kinds of wafer bonding methods have been proposed: (i) adhesive bonding, (ii) direct oxide bonding, and (iii) direct metal bonding [24]. Basically it is necessary to apply a mechanical pressure to wafers and to raise the temperature in these three kinds of bonding methods. Thermal tolerance and shrinkage of adhesive material are issues in the adhesive bonding method. Oxide surfaces have to be atomically flat and special surface treatments are required in direct oxide bonding. In addition, a relatively high temperature is required for bonding. Direct metal bonding is divided into two categories of metal diffusion bonding and metal eutectic bonding. Metal surfaces also have to be atomically flat and special surface treatments are required in metal diffusion bonding such as Cu–Cu direct bonding. The bonding temperature can be reduced in metal eutectic bonding such as Cu–Sn/Cu–Sn bonding. The thermal stability of the intermetallic compound formed after bonding is a very important benefit of this bonding method.

The difficulty of wafer-to-wafer bonding in 3D integration technology is that electrical connections have to be established between the upper wafer and lower wafer even after bonding of the two wafers. Metal microbumps can be used to establish electrical connections between the two wafers. Therefore, bonding methods as shown in Figure 8.2 have been proposed to simultaneously realize wafer-to-wafer bonding (ensuring mechanical connection) and electrical connections between two wafers by metal microbumps. Metal is used for both wafer bonding and microbump bonding in the direct metal bonding shown in Figure 8.2a. An adhesive organic material is used for wafer bonding and metal microbumps are used for electrical connections between the two wafers in the hybrid bonding of Figure 8.2b, whereas direct oxide bonding is employed for wafer bonding in the hybrid bonding of Figure 8.2c.

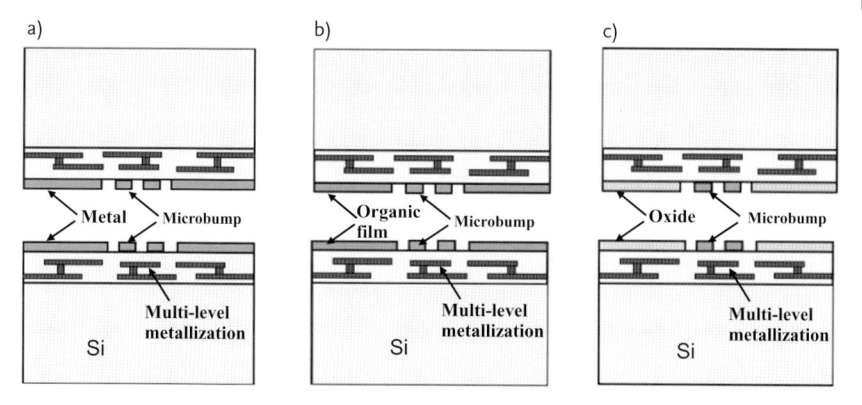

Figure 8.2 Three kinds of wafer bonding methods: (a) direct metal bonding; (b) organic/metal hybrid bonding; (c) oxide/metal hybrid bonding.

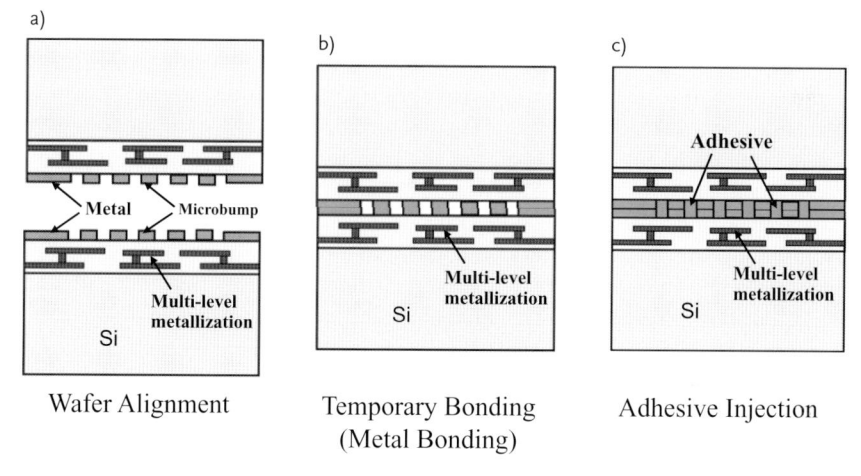

Figure 8.3 New organic/metal hybrid bonding method using direct metal bonding and adhesive injection.

We have developed a 3D integration technology based on organic/metal hybrid wafer bonding. We examined three kinds of methods for organic/metal hybrid bonding. We used a patterned photosensitive polyimide and pre-applied underfill for wafer bonding in the first and second methods, respectively. However, we decided not to employ the second method for wafer bonding because it was difficult to completely push the underfill adhesive out of the narrow gap between two wafers. Therefore, we have employed the second method mainly for die-to-wafer bonding. We developed a new type of organic/metal hybrid bonding as the third method by combining direct metal bonding and adhesive injection in vacuum as shown in Figure 8.3 [17]. In this method, two wafers are temporarily bonded by metal layers and metal microbumps and then a liquid organic adhesive is injected into the gap between the two wafers. Injected adhesive is finally cured to increase

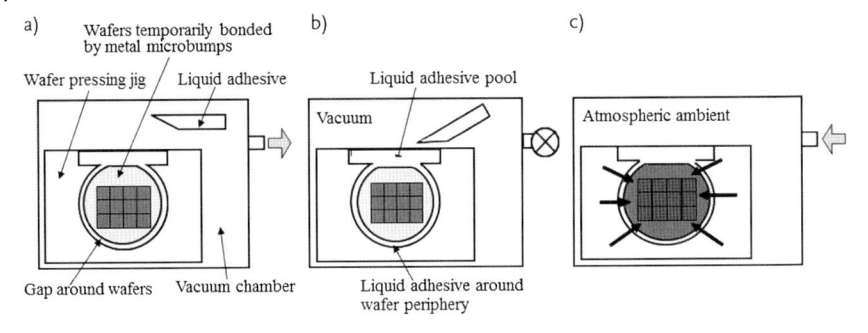

Figure 8.4 Process sequence of adhesive injection method: (a) air evacuated in the gap between wafers with jig; (b) adhesive filled into the gap along the wafer periphery; (c) adhesive injected from the complete wafer periphery by releasing the chamber to atmospheric pressure.

the bonding strength of the two wafers. We have developed special equipment to inject the liquid adhesive in a vacuum ambient as shown in Figure 8.4. The temporarily bonded wafers are installed in a vacuum chamber which is evacuated to approximately 0.1 Pa. A mechanical pressure is applied to the bonded wafers at both sides by a wafer pressing jig. A gap is formed along the wafer periphery in the wafer pressing jig. This gap acts as the path for the liquid adhesive injection and it is connected to a liquid adhesive pool. The air in the gap between the bonded wafers, the liquid adhesive injection path, and the liquid adhesive pool is evacuated by installing the bonded wafers with the wafer pressing jig into the vacuum chamber as shown in Figure 8.4a, and then the liquid adhesive injection path and the liquid adhesive pool are filled with the liquid adhesive as shown in Figure 8.4b. Subsequently, the vacuum chamber is released to atmospheric pressure as shown in Figure 8.4c. As a result, the liquid adhesive is injected from the wafer periphery into the gap between the bonded wafers because of the pressure difference between the inside and outside of the gap. We have mainly used an epoxy resin as the adhesive. It has been confirmed that the gap between bonded wafers with diameters from two to eight inches is successfully filled by the liquid epoxy adhesive when the injection conditions such as the viscosity of the epoxy resin, the gap spacing, and the injection temperature are carefully optimized, although the adhesive injection experiments have been done mainly using two-inch wafers.

8.3
Organic/In–Au Hybrid Bonding

8.3.1
In–Au Phase Diagram and Bonding Principle

We chose indium as a microbump material to lower the bonding temperature. However, indium is easily oxidized, and hence we deposited it in a high-vacuum

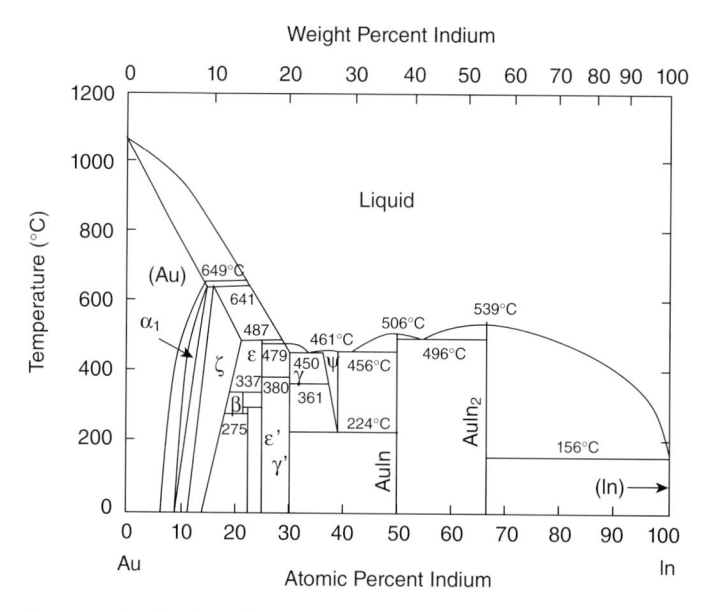

Figure 8.5 In–Au phase diagram.

ambient by an electron beam evaporation method. In addition, a gold cap layer was deposited *in situ* onto the indium surface to protect the indium surface from oxidation. We have to be very careful in controlling the In–Au intermetallic compounds since the In–Au system is represented by a complex phase diagram with 14 equilibrium phases as shown in Figure 8.5 [25]. A major advantage of In–Au microbumps is that we can decrease the bonding temperature. Another advantage is that we can increase the maximum tolerant temperature as the composition changes to a more gold-rich phase during processing as indicated in the phase diagram of Figure 8.5. In addition, the In–Au microbump has a relatively low yield strength, and thus undergoes plastic deformation under a mechanical load, which can help relieve mechanical stresses induced in the bonded wafers. The problems of thermal fatigue and creep movement due to the plastic deformation in the In–Au microbumps can be mitigated by the epoxy adhesive surrounding them.

Figure 8.6 describes the bonding principle of In–Au microbumps [26]. A microbump typically consists of a 3 µm thick indium layer and a 0.3 µm thick gold layer. The $AuIn_2$ interfacial layer is formed between the indium layer and the gold layer right after the In–Au deposition as a result of the reaction of the gold layer with the indium layer as shown in Figure 8.6a. Indium easily diffuses into gold through the grain boundaries of the $AuIn_2$ layer resulting in the formation of a thicker $AuIn_2$ layer which is very stable and not easily oxidized. Upon heating after contacting the upper and lower In–Au microbumps, indium diffuses through the gold layers and another $AuIn_2$ layer is formed at the interface between the upper and lower gold layers as shown in Figure 8.6b. Then, the indium layer in the In–$AuIn_2$

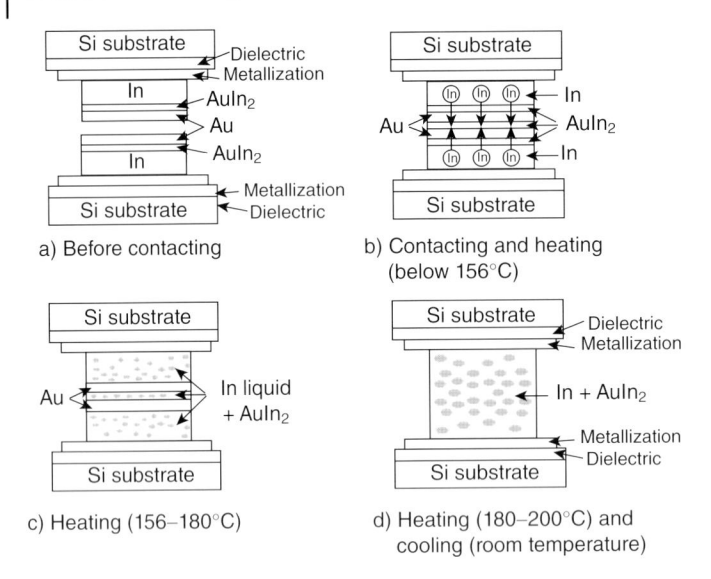

Figure 8.6 Bonding principle of In–Au microbumps.

composite melts at 156 °C. As the temperature increases above 156 °C, the molten indium dissolves the AuIn$_2$ intermetallic layer to form a mixture of indium-rich liquid with AuIn$_2$ grains. This liquid reacts with the remaining gold layer to form more AuIn$_2$ as shown in Figure 8.6c. Consequently, the composition of AuIn$_2$ in the mixture increases. During cooling to room temperature, the solidification of the mixture starts below 156 °C to form indium-rich In–Au alloys.

8.3.2
Formation of In–Au Microbumps by a Planarized Liftoff Method

In–Au microbumps were formed on both surfaces of the upper and lower wafers to be bonded using a new liftoff method called a planarized liftoff method. The fabrication sequence of an In–Au microbump is described in Figure 8.7. Aluminum metallization is formed and a thin tungsten capping layer is deposited on it (Figure 8.7a). Then a polyimide is spin-coated for planarization. After that, a photoresist is coated for liftoff (Figure 8.7b). This photoresist is patterned using an aluminum hard mask (Figure 8.7c). Both photoresist and aluminum hard mask are removed after the In–Au deposition to obtain In–Au microbumps (Figures 8.7d and e). A cross-sectional view of thus fabricated In–Au microbumps with a size of $5 \times 5\,\mu m^2$ and a pitch of $10\,\mu m$ is shown in Figure 8.8. It is clear from the figure that In–Au microbumps of small size and narrow pitch are successfully formed by the planarized liftoff method. Figure 8.9 shows another example of In–Au microbumps formed on TSVs in a CMOS circuit area. Four TSVs with cross-sectional area of $2 \times 12\,\mu m^2$ are electrically connected in parallel by one

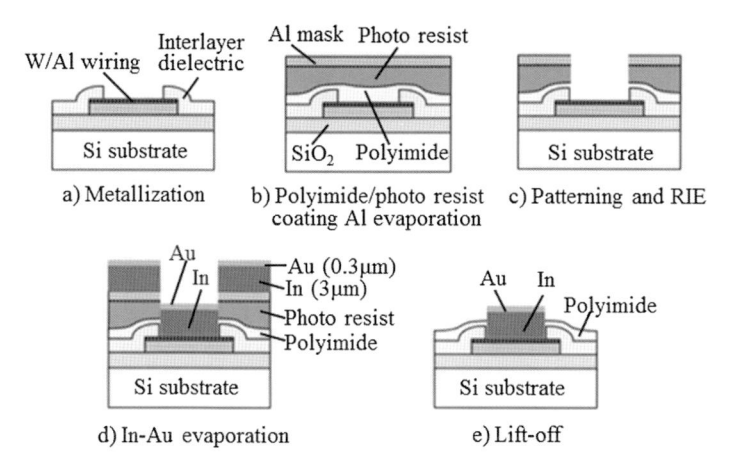

Figure 8.7 Fabrication sequence of In–Au microbump using new planarized liftoff method.

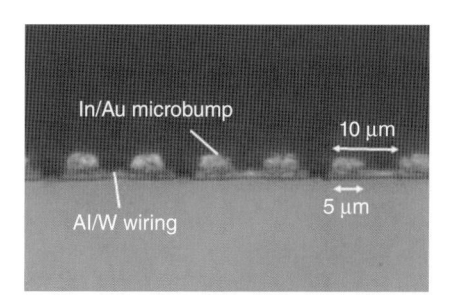

Figure 8.8 Cross-sectional photomicrograph of fabricated In–Au microbumps with a size of 5 × 5 μm² and a pitch of 10 μm.

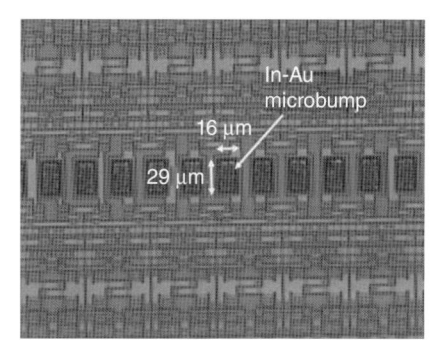

Figure 8.9 In–Au microbumps with a size of 29 × 16 μm² formed on TSVs in a CMOS circuit area. Four TSVs with cross-sectional area of 2 × 12 μm² are electrically connected in parallel by one In–Au microbump.

In–Au microbump with a size of $29 \times 16\,\mu m^2$ to reduce the TSV resistance and to increase the TSV fabrication yield and reliability.

8.3.3
Eutectic In–Au Bonding and Epoxy Adhesive Injection

After the formation of In–Au microbumps by the planarized liftoff method, the upper wafer was aligned and temporarily bonded to the bottom wafer. We have developed a new wafer aligner having an alignment accuracy of $\pm 1\,\mu m$ for 3D integration technology. This wafer aligner can also provide a uniform force and has a function to increase the temperature during temporary wafer bonding to guarantee firm contact between the upper and lower microbumps. The In–Au microbumps and an epoxy adhesive layer are used to bond two wafers. The liquid epoxy adhesive is injected into the gap of approximately $4\,\mu m$ between the two wafers in a vacuum chamber after temporarily bonding the microbumps as previously shown in Figure 8.4. Figure 8.10 shows a cross-sectional photomicrograph of In–Au microbumps after adhesive injection. The size and pitch of the microbumps are $5 \times 5\,\mu m^2$ and $10\,\mu m$, respectively. It is clear in the figure that the upper and lower microbumps are bonded with high alignment accuracy and both microbumps partially melt. It is also obvious in the figure that the epoxy adhesive is completely injected into the narrow gap of $4\,\mu m$ between the upper and lower wafers. No void was found in scanning electron microscopy (SEM) observations. We also confirmed that voids were not present by using test structures in which a quartz glass wafer with microbumps was bonded onto a silicon wafer with microbumps.

The atomic composition of bonded In–Au microbumps was evaluated by electron probe microanalysis (EPMA). Figures 8.11 and 8.12 show the cross-sectional EPMA mapping and depth profile of the In–Au microbumps with size of $5 \times 5\,\mu m^2$ and pitch of $15\,\mu m$ which are bonded at $200\,°C$ for $5\,min$. The spatial resolution and depth resolution in EPMA mapping are approximately 1 and $0.2\,\mu m$, respec-

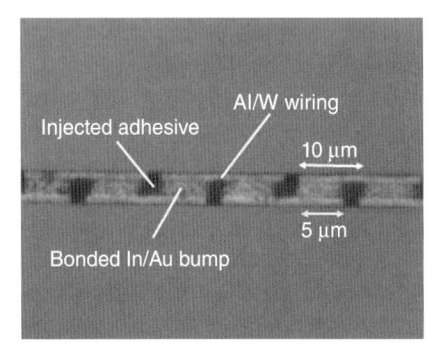

Figure 8.10 Cross-sectional photomicrograph of In–Au microbumps with a size of $5 \times 5\,\mu m^2$ and pitch of $10\,\mu m$ after bonding and adhesive injection.

8.3 Organic/In–Au Hybrid Bonding

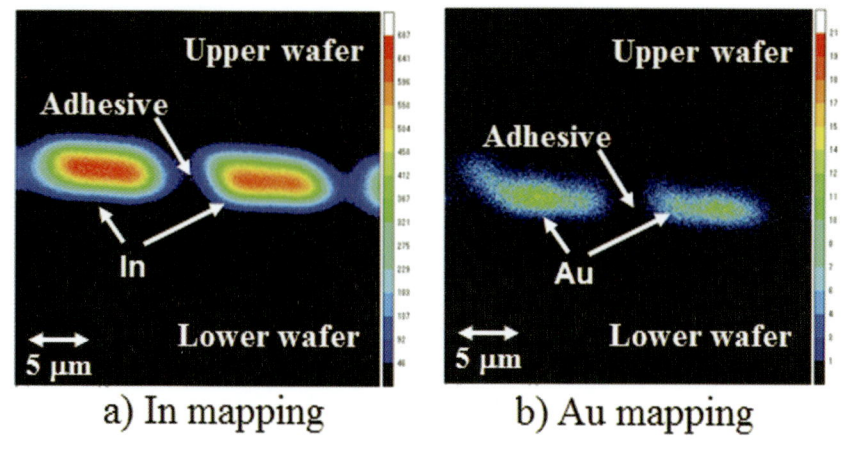

a) In mapping

b) Au mapping

Figure 8.11 Cross-sectional EPMA mapping for indium and gold in In–Au microbumps with a size of 5 × 5 μm² and pitch of 15 μm which were bonded at 200 °C for 5 min.

a) SEM cross-sectional view of In-Au microbump

b) EPMA depth profiles

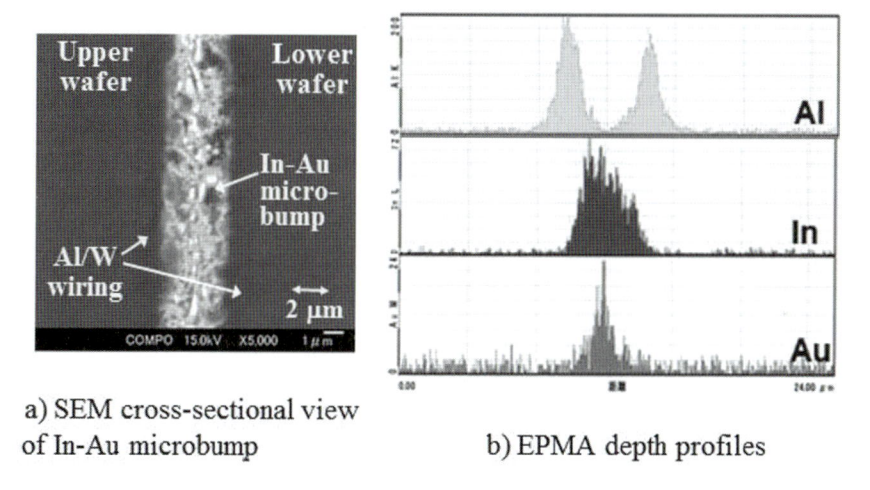

Figure 8.12 EPMA depth profiles for indium and gold in In–Au microbumps with a size of 20 × 20 μm² which were bonded at 200 °C for 5 min.

tively. It is obvious from both figures that indium is distributed over the whole area of bonded In–Au microbumps, whereas gold slightly diffuses into the lower indium layers as expected from the bonding principle of Figure 8.6. Both indium and gold represent the peak intensity at the interface of two microbumps. Figure 8.13 shows the cross-sectional EPMA mapping in the spacing region between the microbumps where the epoxy adhesive is injected. It is clear from the carbon and oxygen atom mapping that the epoxy adhesive is successfully injected into the narrow gap between the two wafers.

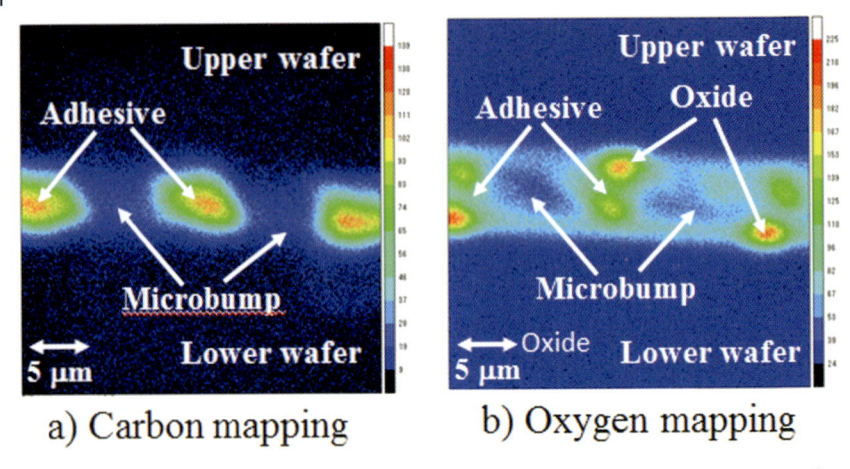

Figure 8.13 Cross-sectional EPMA mapping for carbon and oxygen in the spacing region of In–Au microbumps where epoxy adhesive was injected. In–Au microbumps with a size of 5 × 5 µm² and pitch of 15 µm were bonded at 200 °C for 5 min.

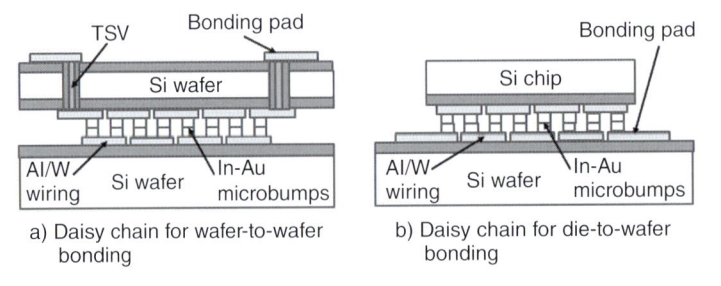

Figure 8.14 Cross-sectional structures of In–Au microbump daisy chains for wafer-to-wafer bonding and die-to-wafer bonding.

8.3.4
Electrical Characteristics of In–Au Microbumps

We designed and fabricated microbump daisy chains to evaluate the microbump resistance after bonding. Microbump daisy chains with two kinds of structures as shown in Figures 8.14a and b were fabricated using wafer-to-wafer bonding and die-to-wafer bonding, respectively. Daisy chains with larger microbumps were fabricated by wafer-to-wafer bonding whereas those with smaller microbumps were fabricated by die-to-wafer bonding. In the structure of Figure 8.14a, the current–voltage characteristics of microbump daisy chains were measured at the bonding pads on the upper wafer which were connected to TSVs. Meanwhile the current–voltage characteristics were measured at the bonding pads on the lower wafer which were not connected to TSVs. Four kinds of microbump patterns were designed for the daisy chains with smaller microbumps as shown in Figure 8.15. Current–voltage characteristics measured in daisy chains with larger and

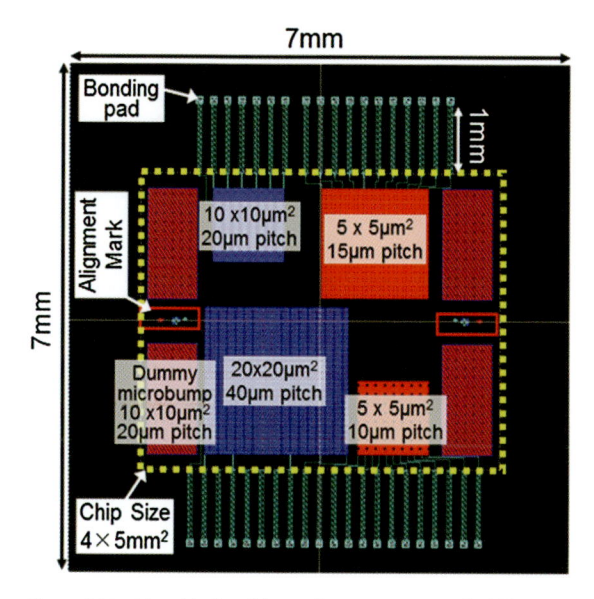

Figure 8.15 Four kinds of daisy chain patterns with different microbump size and pitch.

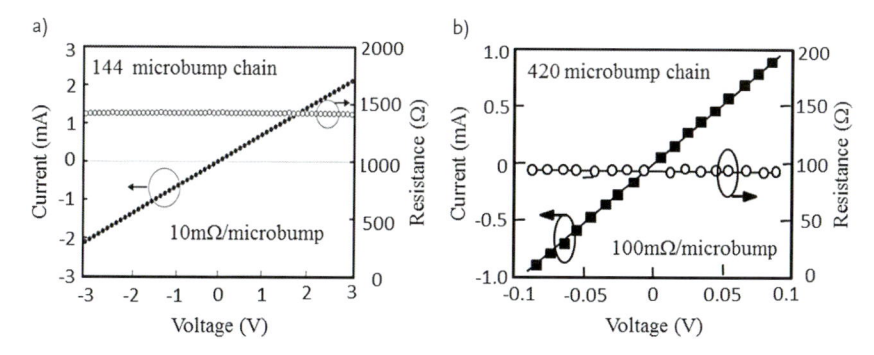

Figure 8.16 Current–voltage characteristics measured in daisy chains with sizes of (a) $29 \times 16\,\mu m^2$ and (b) $10 \times 10\,\mu m^2$.

smaller microbumps are shown in Figures 8.16a and b, respectively. From these characteristics, a very small microbump resistance of approximately $10\,m\Omega$ for each pair of bonded microbumps was obtained for larger microbumps with a size of $29 \times 16\,\mu m^2$, whereas a larger resistance of approximately $100\,m\Omega$ was obtained for smaller microbumps with a size of $10 \times 10\,\mu m^2$.

8.4
Three-Dimensional LSI Test Chips Fabricated by Eutectic In–Au Bonding

We have fabricated 3D LSI test chips using the method of eutectic In–Au wafer bonding and epoxy adhesive injection. The fabrication process flow for 3D LSI test

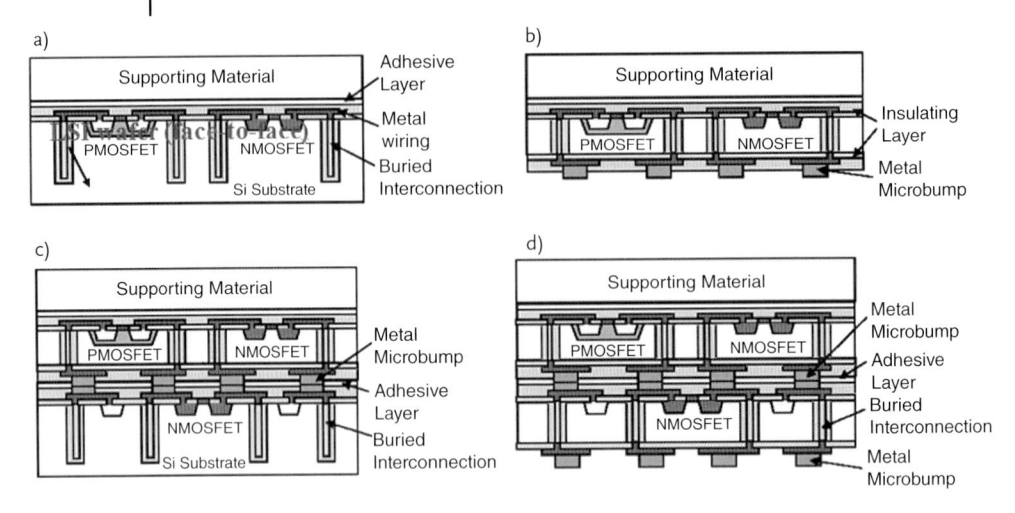

Figure 8.17 Fabrication process flow for 3D LSI.

chips by the front-via method is illustrated in Figure 8.17. First, a thick LSI wafer with TSVs is glued to the supporting material as shown in Figure 8.17a. The LSI wafer glued to the supporting material is thinned from the back surface by mechanical grinding and chemical mechanical polishing to expose the base of TSVs. This is followed by the formation of metal microbumps as shown in Figure 8.17b. The thinned LSI wafer with the supporting material is then bonded to another thick LSI wafer having TSVs as shown in Figure 8.17c. Then the bottom thick LSI wafer is thinned from the back surface and metal microbumps are formed on the base of TSVs By repeating this sequence 3D LSI test chips can be easily fabricated. We can also use an LSI wafer itself as a supporting material. In this case, the first LSI wafer with TSVs and the supporting LSI wafer are bonded face-to-face and the thick supporting LSI wafer remains even after completing the 3D fabrication process. Therefore, a very wide range of thicknesses from several tens of nanometers to several tens of micrometers can be achieved for the thinned wafers bonded to the thick LSI wafer.

Three-dimensional LSIs are suitable for parallel processing and parallel data transferring due to the increased connectivity of their short vertical interconnections compared with conventional LSIs. We can create various kinds of new LSIs with parallel processing and parallel data transferring capabilities by employing 3D stacked structures having many TSVs. So far we have fabricated several 3D LSI test chips, such as a 3D image sensor chip, 3D shared memory, 3D artificial retina chip, and 3D microprocessor chip, using 3D integration technology based on wafer-to-wafer bonding [9–12]. Figure 8.18 shows a configuration of a 3D microprocessor test chip and an SEM cross-sectional view of it. A processor, logic circuits, and cache memory are formed in the first, second, and third layers, respectively, in this 3D microprocessor chip. These three layers are connected by

8.4 Three-Dimensional LSI Test Chips Fabricated by Eutectic In–Au Bonding | 151

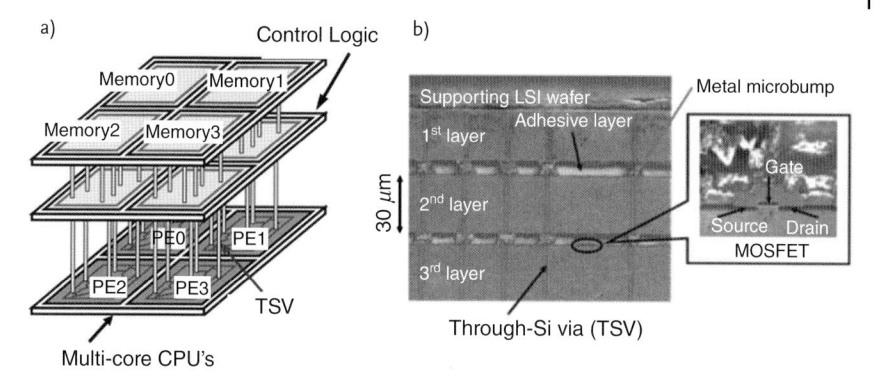

Figure 8.18 (a) Configuration and (b) SEM cross-sectional view of 3D microprocessor test chip fabricated by the eutectic In–Au bonding and epoxy adhesive injection method.

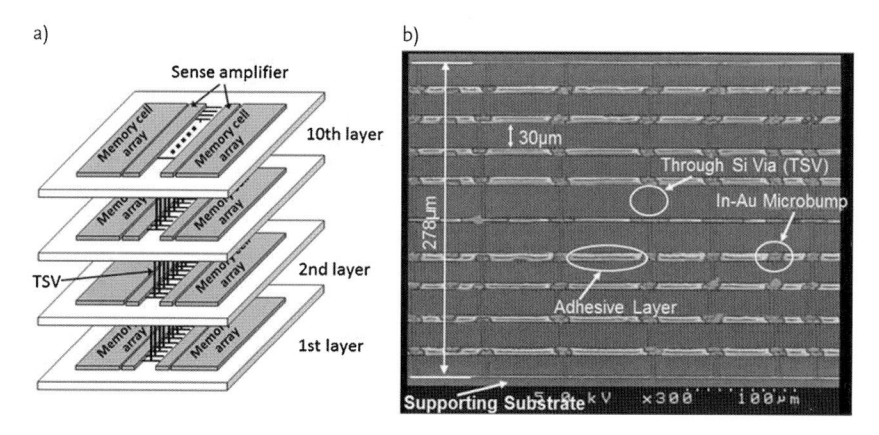

Figure 8.19 (a) Configuration and (b) SEM cross-sectional view of 3D shared memory test chip with ten stacked layers fabricated by the eutectic In–Au bonding and epoxy adhesive injection method.

polysilicon TSVs with a size of $2 \times 12\,\mu m^2$ and In–Au microbumps with a size of $5 \times 15\,\mu m^2$ as shown in Figure 8.18b. The silicon layer thickness is approximately $30\,\mu m$. The epoxy adhesive is injected into the gap of approximately $4\,\mu m$ between the upper and lower wafers. In this test chip, a processor in the first layer operates at a supply voltage of 2.5 V and SRAM cache memory in the third layer operates at 3.3 V. It was confirmed in this test chip that data "1" (V_{in1}) and "0" (V_{in2}) are read from the SRAM cache memory in the third layer and then transferred to the processor in the first layer through the second layer to successfully perform the arithmetic operation. The configuration of the 3D shared memory test chip and an SEM cross-sectional view of it are shown in Figure 8.19. Several blocks of data in a memory layer are simultaneously transferred to another memory layer through a number of TSVs. CPUs are connected to the respective memory layers of this

3D shared memory. Therefore, many CPUs can share identical data without any conflicts after data transfer. The 3D shared memory test chip with ten memory layers was fabricated as shown in Figure 8.19b.

8.5
High-Density and Narrow-Pitch Mircobump Technology

The minimum pitch of TSVs in 3D LSIs will be eventually limited by the keep-out area where the device characteristics are varied by mechanical stresses induced by TSVs. However, the pitch of microbumps can be reduced far more compared to the TSV pitch since it is not limited by such a keep-out area. Microbumps with small size and narrow pitch are indispensable for increasing the data bandwidth and decreasing the thermal resistance between the upper layer and lower layer in 3D LSIs. The effectiveness of high-density microbumps with a small size and narrow pitch increases by combining with high-density redistribution wirings. A typical example of a 3D LSI that requires high-density microbumps with small size and narrow pitch is a 3D stacked image sensor as shown in Figure 8.20, where back-illumination-type photodiodes are stacked on pixel circuits. A large number of photodiodes are directly connected to respective pixel circuits by high-density microbumps of small size and narrow pitch. Only photodiodes are formed on the top layer in this 3D stacked image sensor and hence we can significantly increase the pixel density of the image sensor.

We have developed high-density microbump technology using In–Au microbumps with a size smaller than $2 \times 2 \mu m^2$ and a pitch narrower than $5 \mu m$. To develop such high-density microbump technology, we have proposed a modified planarized liftoff method as shown in Figure 8.21 where clearance grooves are formed around microbumps to accommodate molten indium in microbump bonding. These clearance grooves were formed self-aligned to the microbumps by optimizing In–Au evaporation conditions. Figure 8.22 shows an SEM micrograph

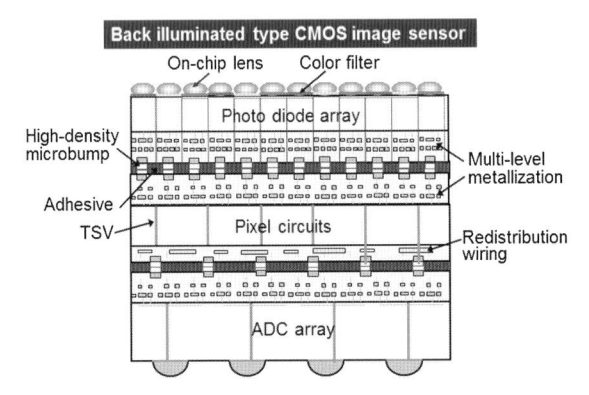

Figure 8.20 Configuration of 3D stacked image sensor with high-density microbumps with small size and narrow pitch.

8.5 High-Density and Narrow-Pitch Mircobump Technology

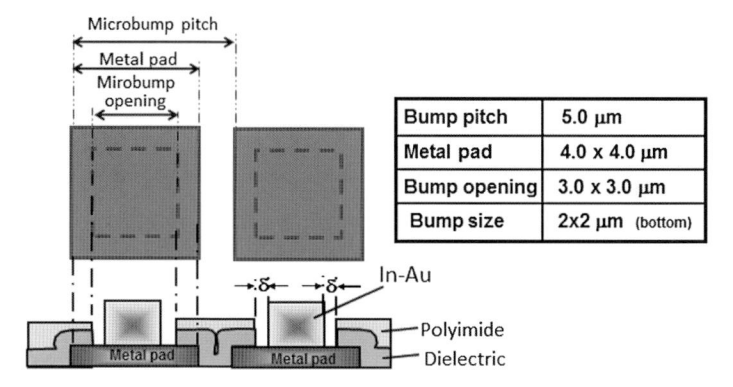

Bump pitch	5.0 μm
Metal pad	4.0 x 4.0 μm
Bump opening	3.0 x 3.0 μm
Bump size	2x2 μm (bottom)

Figure 8.21 Modified planarized liftoff method to fabricate high-density and narrow-pitch mircobumps with clearance groove (δ).

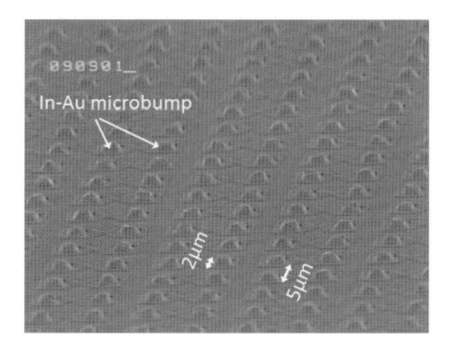

Figure 8.22 SEM micrograph of In–Au microbumps with a size of $2 \times 2\,\mu m^2$ and pitch of $5\,\mu m$ fabricated by the modified planarized liftoff method.

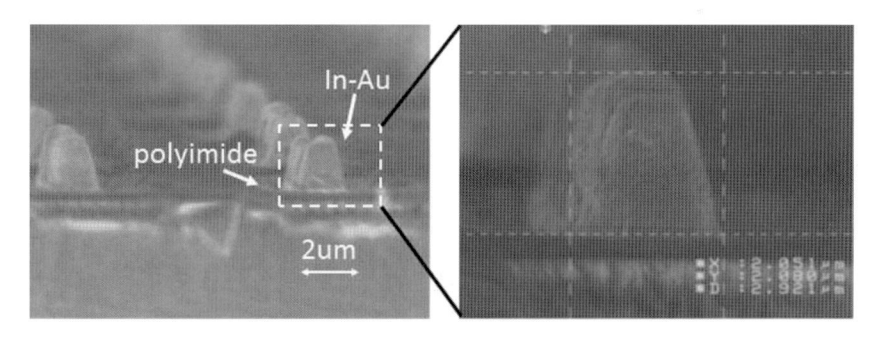

Figure 8.23 SEM cross-sectional view of In–Au microbumps with a size of $2 \times 2\,\mu m^2$ and pitch of $5\,\mu m$ fabricated by the modified planarized liftoff method.

of In–Au microbumps with a size of $2 \times 2\,\mu m^2$ and a pitch of $5\,\mu m$ fabricated by the modified planarized liftoff method. Figure 8.23 shows a magnified view. It is clearly seen in both Figures 8.22 and 8.23 that high-density microbumps with a size of $2 \times 2\,\mu m^2$ and pitch of $5\,\mu m$ were successfully produced using this modified planarized liftoff method. A cross-sectional photomicrograph and an infrared

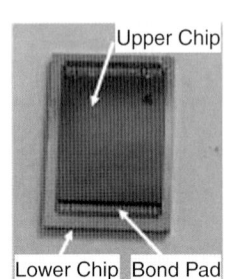

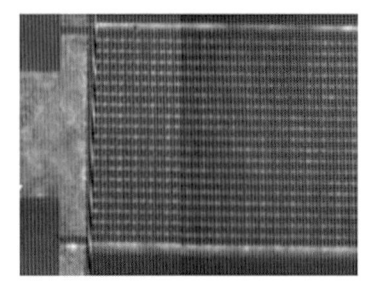

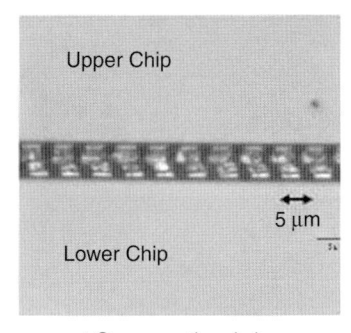

Figure 8.24 (a, c) Photomicrographs and (b) infrared (IR) plan view for a daisy chain with 10 000 pairs of In–Au microbumps after bonding. The size and pitch of the microbumps are $2 \times 2\,\mu m^2$ and $5\,\mu m$, respectively.

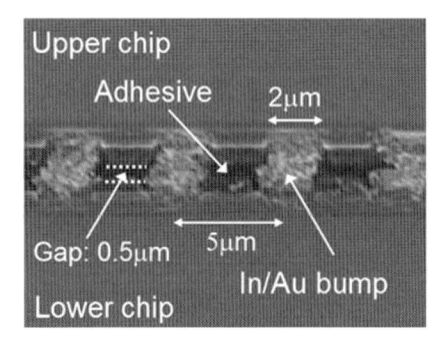

Figure 8.25 SEM cross-sectional view of a daisy chain with 10 000 pairs of In–Au microbumps after bonding. The size and pitch of the microbumps are $2 \times 2\,\mu m^2$ and $5\,\mu m$, respectively.

microscopy plan view for a daisy chain with 10 000 pairs of In–Au microbumps after bonding are shown in Figure 8.24. Figure 8.25 shows a cross-sectional SEM micrograph of it. It is obvious from Figures 8.24 and 8.25 that In–Au microbumps with a size of $2 \times 2\,\mu m^2$ and pitch of $5\,\mu m$ can be successfully bonded with a small misalignment of less than $0.5\,\mu m$, and narrow gaps of approximately $0.5\,\mu m$ between the upper wafer and lower wafer are completely filled with the epoxy adhesive. However, it is very important for achieving a low microbump resistance to optimize the bonding conditions. We obtained a low microbump resistance of $100–120\,m\Omega$ in In–Au microbumps with a size of $2 \times 2\,\mu m^2$ and pitch of $5\,\mu m$ after optimizing the bonding conditions as shown in Figure 8.26. We also succeeded in fabricating further smaller In–Au microbumps with a size of $1.6 \times 1.6\,\mu m^2$ as shown in Figure 8.27. We performed a temperature cycle test on these bonded daisy chains to evaluate the reliability of high-density In–Au microbumps of small size and narrow pitch. The resistance changes of microbumps measured in daisy

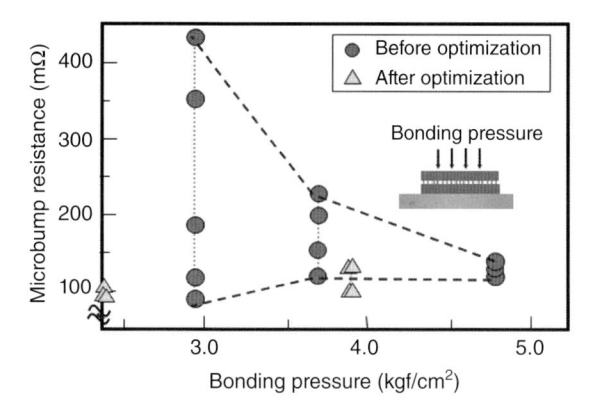

Figure 8.26 Influences of bonding pressure on microbump resistance before and after optimizing the bonding conditions.

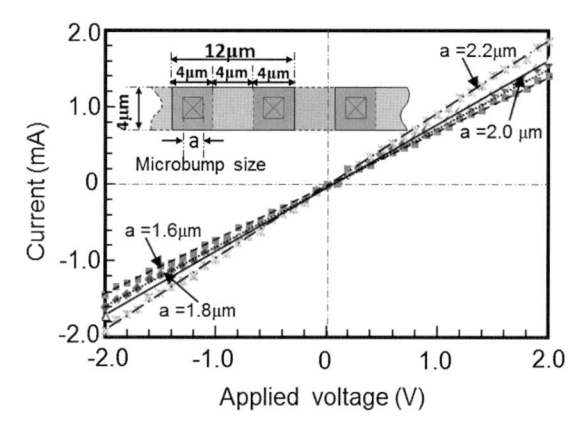

Figure 8.27 Current–voltage characteristics measured in daisy chains with smaller microbumps.

chains with 10 000 pairs of In–Au microbumps are plotted versus the number of temperature cycles in Figure 8.28. The size and pitch are $2 \times 2\,\mu m^2$ and $5\,\mu m$, respectively. The ambient temperature was cyclically varied from −65 to 150 °C with a period of 1 h. A current of $0.5 \times 10^5\,A\,cm^{-2}$ was forced to flow during the test. It is obvious from Figure 8.28 that the microbump resistances hardly change even after more than 500 temperature cycles. Consequently it is confirmed that high-density In–Au microbumps fabricated by modified planarized liftoff and liquid epoxy adhesive injection have a sufficiently high reliability.

A concern in organic/metal hybrid wafer bonding is the mechanical stress induced by the difference in the coefficient of thermal expansion (CTE) between organic material and metal. Organic materials have a larger CTE than metals, and hence more markedly shrink after wafer bonding. Hybrid wafer bonding using

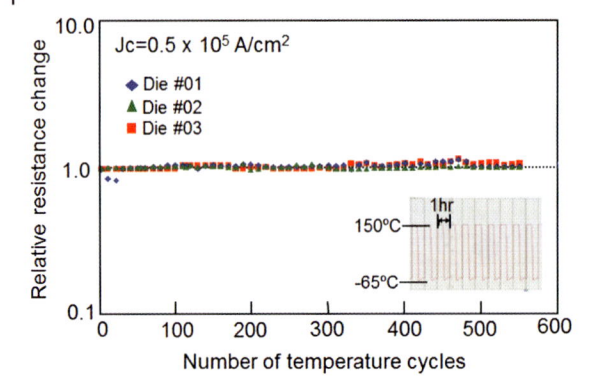

Figure 8.28 Results of temperature cycle testing using daisy chains with 10 000 pairs of In–Au microbumps. The microbump size is $2 \times 2\,\mu m^2$.

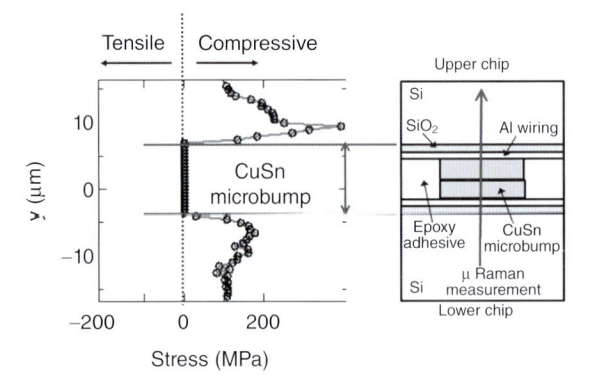

Figure 8.29 Mechanical stress in a silicon substrate induced by Cu–Sn microbumps measured by micro-Raman spectroscopy. The microbump size is $20 \times 20\,\mu m^2$. The thickness of both the SiO_2 film and aluminum wiring is $0.5\,\mu m$.

In–Au microbump bonding and epoxy adhesive injection has the same concern. The epoxy adhesive shrinks more significantly than the In–Au microbumps after wafer bonding. As a result, a large compressive stress is generated in the silicon substrate underneath the microbumps. This compressive stress has an influence on the device characteristics. We evaluated the mechanical stress induced by metal microbumps using micro-Raman spectroscopy as shown in Figure 8.29, where Cu–Sn microbumps were used instead of In–Au microbumps [27]. It is obvious that a large compressive stress is generated in the silicon substrate underneath the Cu–Sn microbumps. This compressive stress will cause serious changes of the device characteristics. We are also evaluating the mechanical stress induced by In–Au microbumps and are to publish the results, including the influences on device characteristics, elsewhere.

8.6 Conclusion

A new 3D integration technology employing eutectic In–Au bonding and epoxy adhesive injection has been developed. We have fabricated several 3D LSI test chips, such as a 3D image sensor chip, 3D shared memory, 3D artificial retina chip, and 3D microprocessor chip, using this technology. We have succeeded in fabricating high-density In–Au microbumps with a size of $2 \times 2\,\mu m^2$ and pitch of $5\,\mu m$ by the modified planarized liftoff method, bonding them with a small misalignment of less than $0.5\,\mu m$ and completely filling the narrow gaps of approximately $0.5\,\mu m$ between the upper wafer and lower wafer with the epoxy adhesive. We obtained a low microbump resistance of $100-120\,m\Omega$ for In–Au microbumps with a size of $2 \times 2\,\mu m^2$ and pitch of $5\,\mu m$ It was confirmed by temperature cycle tests that high-density In–Au microbumps fabricated by modified planarized liftoff and liquid epoxy adhesive injection have a sufficiently high reliability.

Acknowledgment

We would like to thank Prof. T. Tanaka, Assoc. Prof. T. Fukushima, Assoc. Prof. K.-W. Lee, Dr. J. Bea, and Dr. M. Murugesan for their fruitful discussions and collaborations. This work was performed at the Micro/Nano-Machining Research and Education Center (MNC) and the Jun-ichi Nishizawa Research Center at Tohoku University. We would also like to thank the staff of those centers.

References

1 Koyanagi, M. (1989) Roadblocks in achieving three-dimensional LSI. Proceedings of the 8th Symposium on Future Electron Devices, pp. 50–60.

2 Koyanagi, M., Kurino, H., Lee, K.-W., Sakuma, K., and Itani, H. (1998) Future system-on-silicon LSI chips. *IEEE Micro*, **18** (4), 17–22.

3 Davis, J.A., Venkatesan, R., Kaloyeros, A., Beylansky, M., Souri, S.J., Banerjee, K., Saraswat, K.C., Rahman, A., Reif, R., and Meindl, J.D. (2001) Interconnect limits on gigascale integration (GSI) in the 21st century. *Proc. IEEE*, **89** (3), 305–324.

4 Yu, K.-H., Satoh, T., Kawahito, S., and Koyanagi, M. (1996) Real-time microvision system with three-dimensional integration structure. Proceedings of the IEEE International Conference on Multisensor Fusion and Integration for Intelligent Systems, pp. 831–835.

5 Hirano, K., Kawahito, S., Matsumoto, T., Kudoh, Y., Pidin, S., Miyakawa, N., Itani, H., Ichikizaki, T., Tsukamoto, H., and Koyanagi, M. (1996) A new three-dimensional multiport memory for shared memory in high performance parallel processor system. Extended Abstracts of the International Conference on Solid State Devices and Materials, pp. 824–826.

6 Kurino, H., Nakagawa, Y., Nakamura, T., Yamada, Y., Lee, K.-W., and Koyanagi, M. (2001) Biologically inspired vision chip with three dimensional structure. *IEICE Trans. Electron.*, **E84-C** (12), 1717–1722.

7 Takata, H., Nakano, T., Yokoyama, S., Horiuchi, S., Itani, H., Tsukamoto, H., and Koyanagi, M. (1991) A novel fabrication technology for optically interconnected three-dimensional LSI by wafer aligning and bonding technique. Proceedings of the International Semiconductor Device Research Symposium, pp. 327–330.

8 Hayashi, Y., Oyama, K., Takahashi, S., Wada, S., Kajiyana, K., Koh, R., and Kunio, T. (1991) A new three dimensional IC fabrication technology, stacking thin film dual-CMOS layers. Technical Digest of the International Electron Devices Meeting (IEDM), pp. 657–660.

9 Matsumoto, T., Kudoh, Y., Tahara, M., Yu, K.-H., Miyakawa, N., Itani, H., Ichikizaki, T., Fujiwara, A., Tsukamoto, H., and Koyanagi, M. (1995) Three-dimensional integration technology based on wafer bonding technique using micro-bumps. Extended Abstracts of the International Conference on Solid State Devices and Materials, pp. 1073–1074.

10 Fan, A., Rahman, A., and Reif, R. (1999) Copper wafer bonding. *Electrochem. Solid State Lett.*, **2** (10), 534–536.

11 Ramm, P., Bonfert, D., Gieser, H., Haufe, J., Iberl, F., Klumpp, A., Kux, A., and Wieland, R. (2001) Interchip via technology for vertical system integration. Proceedings of the IEEE International Interconnect Technology Conference (IITC), pp. 160–162.

12 Burns, J., McIlrath, L., Keast, C., Lewis, C., Loomis, A., Warner, K., and Wyatt, P. (2001) Three-dimensional integrated circuits for low-power, high-bandwidth systems on a chip. Proceedings of the IEEE International Solid State Circuits Conference, pp. 268–269.

13 Guarini, K.W., Topol, A.W., Ieong, M., Yu, R., Shi, L., Newport, M.R., Frank, D.J., Singh, D.V., Cohen, G.M., Nitta, S.V., Boyd, D.C., O'Neil, P.A., Tempest, S.L., Pogge, H.B., Purushothaman, S., and Haensch, W.E. (2002) Electrical integrity of state-of-the-art 0.13 μm SOI CMOS devices and circuits transferred for three-dimensional (3D) integrated circuit (IC) fabrication. Technical Digest of the International Electron Devices Meeting (IEDM), pp. 943–945.

14 Lu, J.-Q., Jindal, A., Kwon, Y., McMahon, J.J., Rasco, M., Augur, R., Cale, T.S., and Gutmann, R.J. (2003) Evaluation procedures for wafer bonding and thinning of interconnect test structure for 3D ICs. Proceedings of the IEEE International Interconnect Technology Conference (IITC), pp. 74–76.

15 Burns, J.A., Aull, B.F., Chen, C.K., Chen, C.-L., Keast, C.L., Knecht, J.M., Suntharalingam, V., Warner, K., Wyatt, P.W., and Yost, D.-R.W. (2006) A wafer-scale 3-D circuit integration technology. *IEEE Trans. Electron Devices*, **53** (10), 2507–2516.

16 Koyanagi, M., Nakamura, T., Yamada, Y., Kikuchi, H., Fukushima, T., Tanaka, T., and Kurino, H. (2006) Three-dimensional integration technology based on wafer bonding with vertical buried interconnections. *IEEE Trans. Electron Devices*, **53** (11), 2799–2808.

17 Matsumoto, T., Satoh, M., Sakuma, K., Kurino, H., Itani, H., and Koyanagi, M. (1998) New three-dimensional wafer bonding technology using the adhesive injection method. *Jpn. J. Appl. Phys.*, **1** (3B), 1217–1221.

18 Kurino, H., Lee, K.W., Nakamura, T., Sakuma, K., Hashimoto, H., Park, K.T., Miyakawa, N., Shimazutsu, H., Kim, K.Y., Inamura, K., and Koyanagi, M. (1999) Intelligent image sensor chip with three dimensional structure. Technical Digest of the International Electron Devices Meeting (IEDM), pp. 879–882.

19 Lee, K.W., Nakamura, T., Ono, T., Yamada, Y., Mizukusa, T., Hashimoto, H., Park, K.T., Kurino, H., and Koyanagi, M. (2000) Three-dimensional shared memory fabricated using wafer stacking technology. Technical Digest of the International Electron Devices Meeting (IEDM), pp. 165–168.

20 Koyanagi, M., Nakagawa, Y., Lee, K.-W., Nakamura, T., Yamada, Y., Inamura, K., Park, K.-T., and Kurino, H. (2001) Neuromorphic vision chip fabricated using three-dimensional integration technology. Proceedings of the IEEE International Solid State Circuits Conference, pp. 270–271.

21 Ono, T., Mizukusa, T., Nakamura, T., Yamada, Y., Igarashi, Y., Morooka, T., Kurino, H., and Koyanagi, M. (2002) Three-dimensional processor system fabricated by wafer stacking technology. Proceedings of the International Symposium on Low-Power and High-Speed Chips (COOL Chips V), pp. 186–193.

22 Motoyoshi, M., and Koyanagi, M. (2010) Multichip-stack high-functional image sensor, PIXEL 2010 International Workshop.

23 Koyanagi, M., Fukushima, T., and Tanaka, T. (2009) High-density through silicon vias for 3-D LSIs. *Proc. IEEE*, **97** (1), 49–59.

24 Koyanagi, M., Fukushima, T., and Tanaka, T. (2010) Three-dimensional integration technology using through-Si via based on reconfigured wafer-to-wafer bonding. Proceedings of the of Custom Integrated Circuits Conference.

25 Okamoto, H. (2004) Au–In (gold–indium). *J. Phase Equilib. Diff.*, **25** (2), 197–198.

26 So, W.W. and Lee, C.C. (2000) Fluxless process of fabricating In–Au joints on copper substrates. *IEEE Trans. Components Packag. Technol.*, **23** (2), 377–382.

27 Murugesan, M., Ohara, Y., Bea, J.-C., Miyazaki, C., Yamada, F., Kobayashi, H., Lee, K.W., Fukushima, T., Tanaka, T., and Koyanagi, M. (2010) Impact of microbump induced stress in thinned 3D-LSIs after wafer bonding. IEEE 3D System Integration Conference (3D-IC).

9
Thermocompression Cu–Cu Bonding of Blanket and Patterned Wafers

Kuan-Neng Chen and Chuan Seng Tan

9.1
Introduction

This chapter describes Cu–Cu bonding, one key technology for three-dimensional ICs, three-dimensional packaging, and microelectromechanical systems applications. Copper wafer bonding is currently the most promising candidate for wafer-level three-dimensional integration from the application pespective because of the wide use of Cu in semiconductor fabrication and time to market, compared to other bonding materials. Copper can improve the electrical performance, reduce Joule heating, and improve heat transfer because of its high electrical conductivity and thermal conductivity. Therefore, Cu has become a standard interconnect metallization material for advanced electronic applications. In addition, Cu is intuitively to be considered as a candidate for wafer bonding in three-dimensional integration.

Copper wafer bonding was first systematically studied at universities starting in around 2000. Since then, researchers have carried out detailed investigations to understand the mechanism and science of this technology. In addition, Cu bonding parameters of three-dimensional integration for IC applications have been explored to meet corresponding thermal budgets. Currently, both academia and industry are continuing to work on this method for use in real applications. At the same time, how to decrease the bonding temperature and duration but keep a uniform well-bonded quality across the wafer are important topics.

This chapter starts with a classification of two different Cu bonding techniques. The science and technology principles of both techniques are first introduced. Then we focus on a discussion of thermocompression (diffusion) Cu bonding, which is the most popular technique used by industry and academia. The discussion of thermocompression Cu bonding includes blanket Cu bonding and patterned bonding. The bonding mechanism, development, and parameter exploration of reliable bonding processes are presented. Designs of test structures, bonding parameters, and surface preparation to achieve good bonding quality under low bonding temperature are introduced. We then describe the current development and achievements of Cu bonding, including nonblanket bonding, low-temperature

Handbook of Wafer Bonding, First Edition. Edited by Peter Ramm, James Jian-Qiang Lu, Maaike M.V. Taklo.
© 2012 Wiley-VCH Verlag GmbH & Co. KGaA. Published 2012 by Wiley-VCH Verlag GmbH & Co. KGaA.

bonding, and applications. This chapter hopefully can provide a good reference for researchers who are interested in this technology or are applying it in real applications.

9.2
Classification of the Cu Bonding Technique

The surface-activated method and the thermocompression method are currently two approaches to achieve a successful Cu bond. The major difference between these two approaches is the bonding mechanism. The former method can enable two Cu surfaces to be bonded directly at room temperature [1], while the latter uses high temperature and force with simpler design and less cost consideration. These two techniques are described in the following.

9.2.1
Thermocompression Cu Bonding

Thermocompression Cu bonding, also known as diffusion Cu bonding, is another option to achieve good bonding quality in wafer bonding technology. Different from the surface-activated bonding method, thermocompression Cu bonding uses high bonding force to ensure the contact of two Cu wafer/chip surfaces at high temperature to realize the interdiffusion of Cu atoms. The bonding process completes while the diffusion between the two Cu layers saturates – sometimes the original bonding interface disappears. Because of the diffusion bonding mechanism, this method does not require ultraclean surfaces and ultrahigh vacuum chamber conditions.

Since it is simple and less costly, the thermocompression Cu bonding method seems more attractive and has been more developed by industry and academia. One important factor is the bonding temperature. High bonding temperature can boost the diffusion and further improve bonding quality. However, the bonding temperature has to be compatible with the temperature limits of devices or applications in order not to affect device performance. In the description of this chapter we focus on the developments and achievements of the thermocompression Cu bonding technique.

9.2.2
Surface-Activated Cu Bonding

The concept of this method is to achieve bonding energy from the adhesive force between two clean solid surfaces in contact [2, 3]. In order to acquire a well-bonded structure, two Cu surface-activated wafers (chips) have to be kept clean and activated under ultrahigh vacuum conditions (about 10^{-8} torr) before and during the bonding process. By using a dry etching process such as ion beam bombardment or radical irradiation in a certain clean atmosphere, for example in ultrahigh vacuum, the surface contamination and native oxides are removed, so the surfaces can be activated and the bonding process can be completed under ultrahigh vacuum conditions at room temperature.

The bonding structure using this method shows a high bond strength with a void-free interface. This bonding technique can be applied not only to Cu but also to very many other materials because the bonding mechanism is based on contact of activated surfaces. In addition, when bonding two dissimilar materials at room temperature, the thermal mismatch problem is less of a concern. However, the process and equipment of this method are complex, including the requirement of ultrahigh vacuum conditions in the bond chamber and integration of the cleaning equipment, such as an Ar ion beam. These requirements become major concerns when using this technique for mass production.

9.3
Fundamental Properties of Cu Bonding

9.3.1
Morphology and Oxide Examination of Cu Bonded Layer

Figure 9.1 shows a well-bonded Cu–Cu layer. In this image, no original bonding interface can be observed; instead grain structures and twins cross the bonded layer [4]. The two Cu layers were cleaned by HCl and deionized water to remove surface oxide and particles. Then the two layers were bonded at 400 °C for 30 min with a pressure of 4000 mbar and a vacuum of 10^{-3} torr, followed by a nitrogen anneal at 400 °C for 30 min. Appearances of grains and twins indicate two Cu layers have become one bonded layer and the interface is removed [5].

An energy dispersive spectroscopy (EDS) examination shows a small amount and uniform oxygen distribution across the bonded layer, as shown in Table 9.1

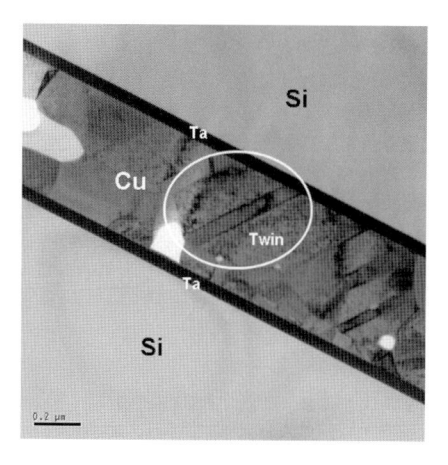

Figure 9.1 Morphology of Cu bonded layers [4]. (Used with permission from TMS/Springer.)

Table 9.1 EDS examination for oxygen in different areas of a bonded layer [5].

Beam size (nm)	Oxygen (wt%)	Location description
500	2.67	Whole bonded layer
25	2.13	Near tantalum layer
5	2.22	Near tantalum layer
25	2.53	Near bonded interface
5	2.78	Near bonded interface
25	2.98	Near tantalum layer
5	2.89	Near tantalum layer

Used with permission from TMS/Springer.

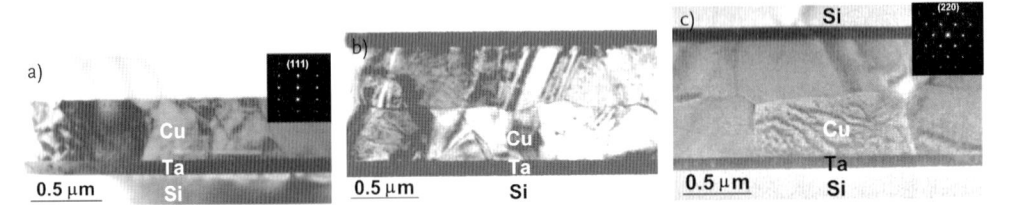

Figure 9.2 TEM images of Cu bonded layers: (a) before bonding; (b) after 400 °C bonding for 30 min; (c) after 400 °C bonding for 30 min followed by nitrogen anneal at 400 °C for 30 min. (Reprinted with permission from [6] Copyright 2002 American Institute of Physics.)

[5]. The detecting ranges are as large as 500 nm or as small as 5 nm. The low quantities of oxygen (<3 wt%), which is below the detection threshold of EDS, indicate that the original oxides on the Cu layer surface may be removed during the HCl pre-clean process, and the remaining oxygen may be considered negligible during bonding. Although the surface oxides on the Cu layer are only partially removed or new oxides form before bonding, during bonding the oxygen atoms can diffuse into the Cu layer easily and then distribute everywhere in the bonded layer for the bonding duration under high temperature [5].

9.3.2
Microstructure Evolution during Cu Bonding

Investigations of the evolution of the bonded layer, including microstructure morphology and grain orientation, can help us to understand the Cu bonding mechanism. Figure 9.2 shows the sequence of changing morphology of the bonded Cu layer, finally reaching a steady state after the anneal process [6]. Before bonding, the Cu grains have an average size of 300 nm, and a major (111) orientation, as shown in Figure 9.2a. After the first 30 min of bonding, in Figure 9.2b, the bonded interface is clearly observed but it is not straight. The grain size distribution ranges

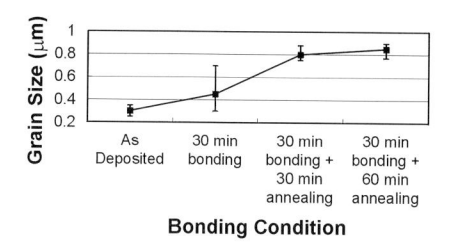

Figure 9.3 Average grain size as a function of different bonding conditions. (Reprinted with permission from [6] Copyright 2002 American Institute of Physics.)

from 300 to 700 nm. Additionally, (111), (200), and (222) grain orientations are observed without a major one. After a further nitrogen anneal for 30 min, a well-developed grain texture with an 800 nm grain size profile and a major (220) orientation are observed, as shown in Figure 9.2c [6].

Based on the results of morphology observation, it can be said that there is strong grain growth during bonding and annealing. After 30 min bonding, the two original Cu layers can still be distinguished. The jagged Cu bonding interface indicates that interdiffusion between the two Cu layers already started during bonding. However, as shown in Figure 9.2b, the bonding duration or supplied energy is not enough to complete grain growth of the bonded layer. Finally, enough energy is provided during the post-bonding anneal to complete grain growth. A stable bonded microstructure is achieved, as shown in Figure 9.2c. The distribution of average grain size under different bonding conditions is shown in Figure 9.3. We observe a large distribution of grain sizes after 30 min bonding due to incomplete grain growth. During the first 30 min of post-bond annealing, a large grain growth phenomenon is observed. But, beyond 30 min of annealing, we do not observe a significant increase of the grain size [6]. In addition, it will be interesting for researchers to study the bond strength if Cu is annealed before bonding.

9.3.3
Orientation Evolution during Cu Bonding

We observe a similar behavior of preferred orientation evolution in the Cu bonded layer as compared to morphology evolution. As shown in Figure 9.4, the preferred orientation changes from (111) to (220) based on X-ray diffraction analysis. Yielding and energy minimization of the whole system lead to this abnormal grain growth [6]. Before bonding, the Cu layer is deposited on a relatively thick Si substrate, so at this stage each Cu bonded layer is in a state of biaxial strain. In addition, the in-plane stress of a grain, the product of biaxial strain and biaxial modulus, is a function of grain orientation factor C_{ijk}. This means the yield stress of the grain also depends on its orientation. Previous research showed that the orientation factor C_{ijk} of (220) has the smallest value of 1.42 while that of (110) has the largest one of 3.46 [7]. Therefore, the yield stress of (110) is much larger than that of (220). If one (111) grain and one (220) grain have the same initial size, the (220) grain

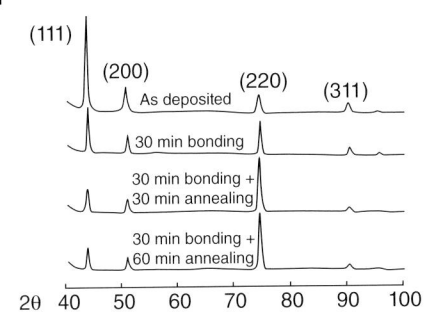

Figure 9.4 X-ray diffraction patterns for different bonding conditions. (Reprinted with permission from [6] Copyright 2002 American Institute of Physics.)

will yield before the (111) grain. Once the (220) grains yield, these grains have an energetic advantage for further growth [7]. Besides, strain energy minimization happens during the yield process [7, 8]. During the further annealing process, the in-plane strain increases again since the bonded wafer is heated again from room temperature to higher temperature. Thus, secondary grain growth occurs with (220) grains yielding. This energetic advantage can minimize the strain energy in the bonded layer. This is the reason why the abnormal (220) grains grow and become the final preferred orientation during bonding [6].

Since Cu with preferred (111) orientation has a longer electromigration lifetime than Cu with preferred (220) orientation, the abnormal (220) grain growth during Cu wafer bonding is not desirable in IC fabrication. Therefore, it is necessary to develop methods to suppress the (111) to (220) transformation while maintaining reasonable bonding strength during Cu wafer bonding.

9.4
Development of Cu Bonding

Bonding quality is strongly related to fabrication and surface preparation, structural design of Cu pads, and bonding parameters. For each factor development, a systematic experimental design needs to be carried out in order to achieve the desired information. The following subsections describe a summary of Cu bonding quality development. Although most results were obtained for the wafer-level scale, the conclusion and guidelines are still valid for die-level Cu bonding technology.

9.4.1
Fabrication and Surface Preparation of Cu Bond Pads

Contamination, native oxide, or other materials make full contact of two Cu surfaces difficult during bonding. Therefore, well-fabricated Cu pads are a first step towards good bonding quality. The Cu pad fabrication process and the surface treatment prior to bonding decide directly the quality of Cu bond pads.

9.4.1.1 Fabrication of Cu Bond Pads

The recesses of surrounding materials, such as oxide, of two Cu damascene bond pads are required in order to make sure the two surfaces of the bond pads can fully contact during bonding. The best bond quality is obtained using a surface treatment: a combination oxide recessing process, including a SiO_2 chemical mechanical polishing (CMP) process, and dilute HF wet etching [9].

9.4.1.2 Surface Cleanliness

When bonding temperature is low or bonding duration is short, the surface cleanliness becomes significant in deciding final bond quality. A wet cleaning, such as HCl solution, plus deionized water rinse usually can effectively remove native oxides and other contaminants prior to the bonding process [5].

9.4.2
Parameters of Cu Bonding

For a Cu bonding process, the parameters include bonding temperature, duration, pressure, chamber ambient condition, and anneal option. Some of these parameters have limits because of device performance and cost considerations. The following are the guidelines for each parameter.

9.4.2.1 Chamber Ambient Condition

Oxide formation and surface contamination, which lead to poor bonding quality, on the Cu surface can be effectively reduced in a high-vacuum environment. The vacuum condition of commercial thermocompression bonders, such as 10^{-4} torr, has been shown to provide good bonding quality [5, 10]. In addition to vacuum, before the bonding process, nitrogen or forming gas purges in the bonder chamber have been shown to be helpful in improving the bonding quality [5].

9.4.2.2 Bond Pressure

Sufficient bond pressure can hold two Cu surfaces close enough for completion of bonding. The maximum bond pressures of current commercial bonders, such as 4000 N for 100 mm wafers and 10 000 N for 200 mm wafers, have already been proved to provide good bond quality [5].

9.4.2.3 Bond Temperature

A higher Cu bond temperature gives better bond quality because of the stronger diffusion activity [11]. However, the bond temperature usually cannot be higher than 400 °C, or the existing devices may be damaged. This parameter is generally the dominant one regarding the final bonding quality.

9.4.2.4 Bond Duration

Although a longer bond duration improves bond quality [11], shorter bonding process times are always preferred in terms of manufacturing cost consideration. For wafer-level Cu bonding, a minimum of at least 30 min is suggested at a bond temperature of 400 °C to achieve good bond quality [5].

9.4.2.5 Bond Profile

For patterned Cu pads on 200 mm wafers, the bond profile for the best bond quality, including strength, alignment, and electrical connectivity, is to apply a small bond pressure (~1000 N) prior to temperature ramping, followed by high bond pressure (~10 000 N). The temperature ramping rate should be kept low; if possible, as low as 6 °C min^{-1} [10].

9.4.2.6 Nitrogen Anneal

When the bonding temperature is above 300 °C, a post-bonding nitrogen anneal has been shown to improve both microstructure and strength of the bonded layer, especially for blanket copper films. During the post-bonding annealing, the bonded layer is provided with enough energy to complete grain growth and thus achieve a stable microstructure [5]. However, the bond strength of low-temperature bonding (below 300 °C) is too low to withstand the thermal stress during post-bond annealing [12].

9.4.3 Structural Design

In a certain available area, the placement of a Cu bond pad means the available area for bonding in the local region, as well as the whole wafer/chip. The placement is related to two structural design factors: the Cu bond pad size (interconnect size) and the pad pattern density (total bond area). Therefore, the structural design of Cu bond pads determines not only the circuit placement but also bond quality.

9.4.3.1 Seal Design

As shown in Figure 9.5, we can provide a seal design to have extra Cu bond area around electrical interconnects, chip edge, and wafer edge. This structure can prevent corrosion and provide extra mechanical support [9].

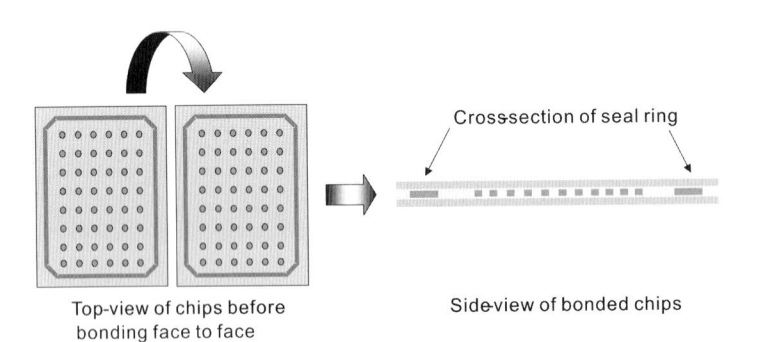

Figure 9.5 Schematic of chip seal design concept. (Reprinted with permission from [9] Copyright 2006 IMIC.)

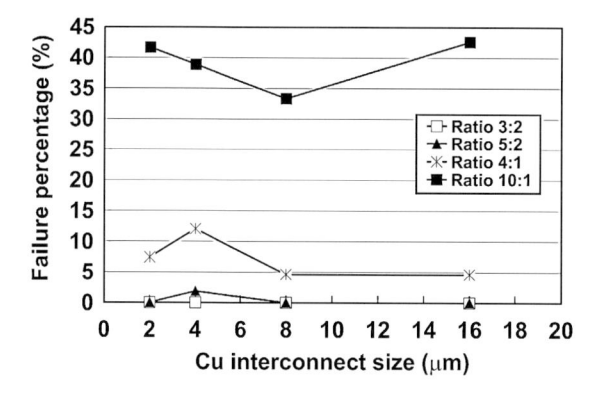

Figure 9.6 Failure percentage results of dicing test for various Cu interconnect pattern geometries. Ratio is taken as that of pitch to diameter. (From [10]. Copyright 2006 IEEE.)

9.4.3.2 Pattern Density

The bonding qualities of several pattern densities from <1% to 35% have been investigated. Results show that higher bond density results in better bond quality. Based on existing results, when pattern density is larger than 13%, the bonded area rarely fails in dicing tests [10].

9.4.3.3 Size of Cu Bond Pads

Several sizes of Cu bond pads, from 2 to 16 μm, have been investigated for the corresponding bonding strength under the same bonding density (total bonding area). Based on experimental results, the bonding quality does not have a strong dependence on bond pad size [10]. Figure 9.6 shows the bond quality for various Cu bond pad structural designs [10].

9.5 Characterization of Cu Bonding Quality

In this section we introduce several approaches to evaluate Cu bonding quality: mechanical tests, image analysis, electrical connectivity, and thermal reliability. Since Cu bonding is a relatively new technology, these characterization results actually help researchers for further study and have already become guidelines for applications.

9.5.1 Mechanical Tests

Mechanical tests are important for understanding the Cu bond strength. The tests can be categorized into quantitative and qualitative approaches, such as the pull test, shear test, dicing test, and tape test.

Among these tests, the dicing test is the easiest and most common approach. When the bonded piece separates due to the sawing stress, it is identified as a "failed piece." For blanket Cu wafers, the optimum Cu bonding strength of pull and shear tests can be as high as 70 MPa and 80 kgw (the unit was from the facility), respectively. As stated previously, from mechanical tests the post-bond thermal anneal improves bond strength when the bonding temperature is above 300 °C [13].

9.5.2
Image Analysis

Although mechanical tests give measurement results of Cu bonding strength, it is still significant to directly observe the morphology of a bonded interface. Electron microscopies, such as transmission electron microscopy (TEM) and scanning electron microscopy (SEM), are methods for examining the morphology of bonded layers. Through TEM observation, we are able to observe the existence of the bonded interface. Besides, when a bonded sample can survive TEM sample preparation, this indicates that the bonded layer has essential bond strength. According to the discussion above, TEM observation results of Cu bonded layers can be categorized into three groups: (i) no interface but grain structure (identified as "grain"), (ii) interface structure (identified as "interface"), and (iii) failure of TEM sample preparation (identified as "TEM failed") [11].

A reference map, including the results from image analysis and dicing test for different bond temperatures and durations, is shown in Figure 9.7 [12]. This map is an important reference for future real applications.

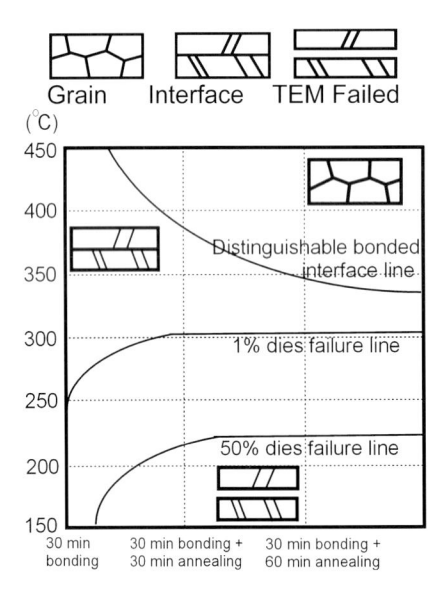

Figure 9.7 Morphology and strength map for Cu wafer bonding [12]. (Reproduced by permission of the Electrochemical Society.)

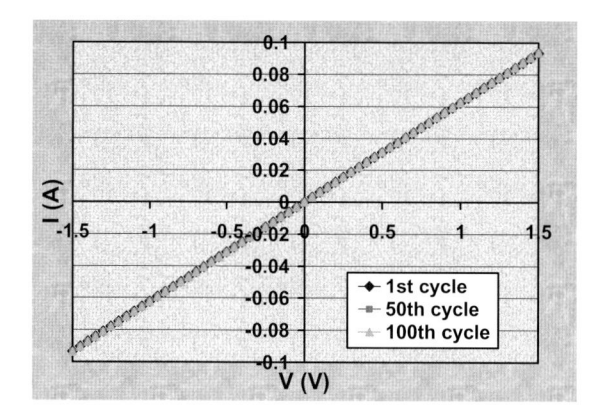

Figure 9.8 Stable current–voltage characteristics of Cu bonded chain after multiple current stressing. (From [10]. Copyright 2006 IEEE.)

9.5.3
Electrical Characterization

Electrical characterization results include specific contact resistance and via chain measurements. The specific contact resistance of bonded Cu interconnects with different sizes is approximately $10^{-8}\,\Omega\,cm^2$ [10, 14]. This value is consistent with measurements taken from similar structures built by depositing and patterning metal lines [15, 16]. In addition, for a well-bonded structure, the chain measurements should show excellent performance and stability of a bonded interface across wafers after multiple stress cycles. Figure 9.8 shows an example [10].

9.5.4
Thermal Reliability

Reliability is definitely an important topic for Cu bonding before its use in real applications. Thermal reliability testing is a typical evaluation method for Cu bonded structures. It has been reported that well-bonded Cu structures across wafers successfully passed thermal reliability tests which subjected the samples to more than 1000 deep thermal cycles between −55 and +125 °C [10]. Further studies are necessary for assessment of the reliability of Cu bonded wafers based on pad size and pitch connection density.

9.6
Alignment Accuracy of Cu–Cu Bonding

Alignment accuracy directly determines the functionality, reliability, and performance of a Cu bonded structure. When designing a structure using Cu bonding

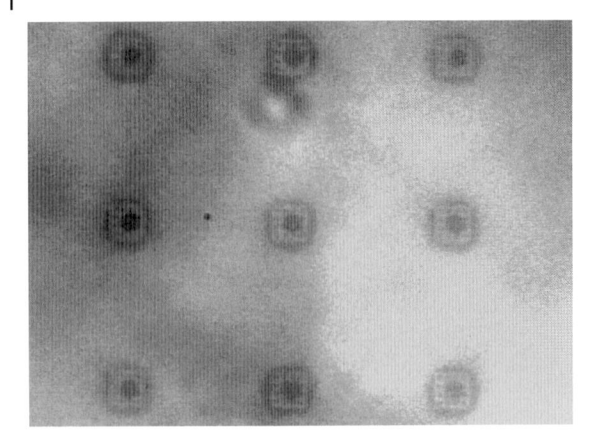

Figure 9.9 Infrared image showing typical misalignment results after bonding [10]. (Copyright 2006 IEEE.)

technology but with a poor alignment accuracy, a more relaxed structure (larger misalignment tolerance) usually needs to be provided.

For the assessment of alignment accuracy, if Cu films are fabricated on transparent wafers, bonded structures can be inspected directly using a microscope. However, infrared inspection is required if both wafers are opaque. Figure 9.9 shows an infrared image of bonded Cu circle to pad structure. By using the best bonding conditions reported in Section 9.5, the misalignment after bonding can be kept within the 0.5 μm range (the alignment resolution of an EVG aligner) [10].

9.7
Reliable Cu Bonding and Multilayer Stacking

Based on previous Cu bonding development, one can achieve a good Cu bond structure which passes the quality characterization requirements with high accuracy alignment. Figure 9.10 shows a high-quality Cu pad–Cu pad bonding result [10].

Multiple Si layer stacking based on Cu bonding, grind back, and etch back has been applied to demonstrate a strong four-layer stack structure, as shown in Figure 9.11 [17]. All bonded Cu layers in this structure became homogeneous layers and did not show original bonding interfaces, as shown in Figure 9.12 [17]. This means that the bonded structure is strong enough to withstand the applied stresses during grind back and etch back. The successful development of multiple Si layer stacking based on Cu bonding means three-dimensional integration using Cu bonding is feasible [17].

The four-layer stacking in Figure 9.12 is a sequential process. It begins with a face-to-face bonding orientation and subsequent face-to-back bonding orientation.

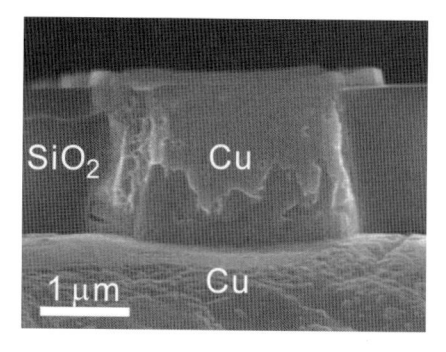

Figure 9.10 SEM image of a 4 μm Cu interconnect bonded to Cu pad showing a high-quality Cu-to-Cu bonding interface [10]. (Copyright 2006 IEEE.)

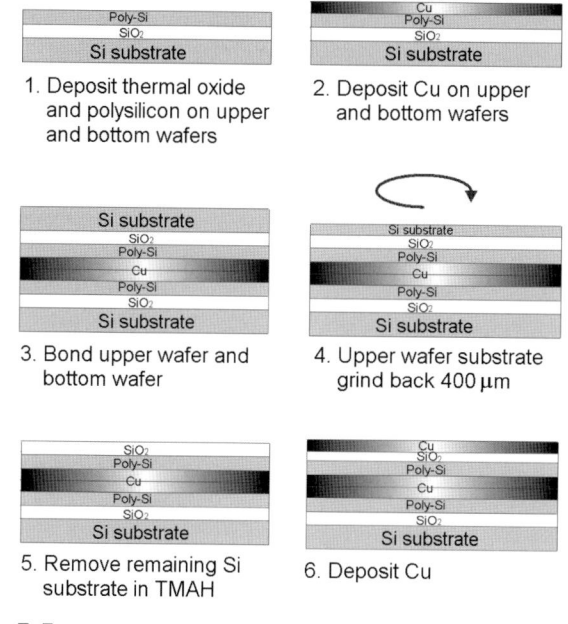

1. Deposit thermal oxide and polysilicon on upper and bottom wafers
2. Deposit Cu on upper and bottom wafers
3. Bond upper wafer and bottom wafer
4. Upper wafer substrate grind back 400 μm
5. Remove remaining Si substrate in TMAH
6. Deposit Cu
7. For next upper wafer bonding, repeat process #3~#6.

Figure 9.11 Schematic of fabrication flow for multiple Si layer stacking based on Cu bonding (TMAH, tetramethylammonium hydroxide). (Reprinted with permission from [17]. Copyright 2005 American Institute of Physics.)

It is also possible to achieve a four-layer stack using a multiplicative approach. One can begin by bonding two wafers face-to-face and perform wafer thinning to achieve a two-layer stack. By bonding two two-layer stacks one can achieve a four-layer stack. Subsequent multiplication is also possible as needed. A process flow for this method is described in [18]. Figure 9.13 shows a four-layer stack achieved using this process flow.

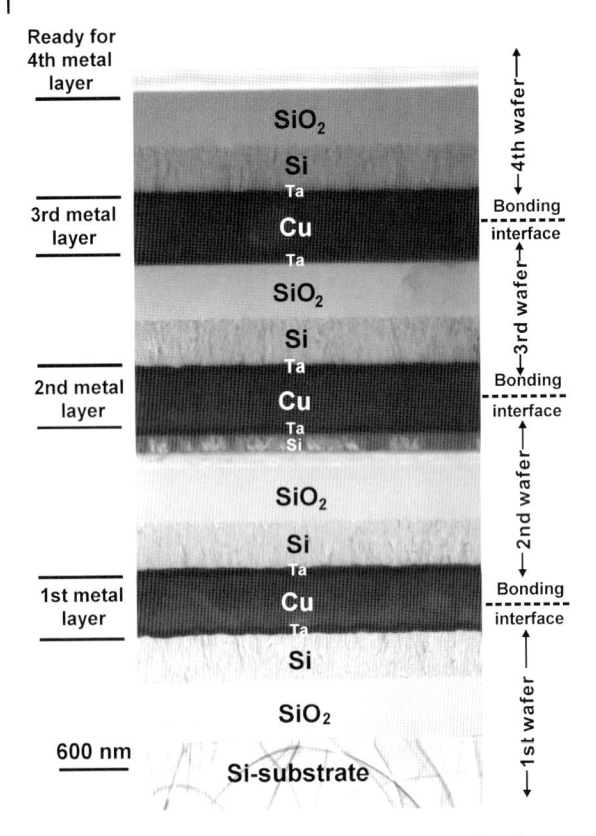

Figure 9.12 Multiple Si layer stacking based on Cu bonding, grind back, and etch back. (Reprinted with permission from [17]. Copyright 2005 American Institute of Physics.)

9.8
Nonblanket Cu–Cu Bonding

Since Cu is a conductive material, a continuous Cu bonding layer between active layers has no practical application. In an actual multilayer three-dimensional IC implementation having Cu as the bonding medium, Cu bonding should be in the form of pad-to-pad or line-to-line bonding with proper electrical isolation. Sections 9.4.3, 9.5, and 9.6 have described the structural design, electrical performance, and alignment accuracy of patterned (nonblanket) Cu–Cu bonding. Figure 9.14 shows a cross section of Cu lines (2–9 µm) that are successfully bonded [19]. The spacing between the bonded lines is 5.3 µm and it is filled with air. Interfacial voids are observed in the bonded lines, which can lead to serious reliability concerns. The bonding process should be optimized to minimize the formation of voids. Another reliability concern is the empty space between the bonded lines that might reduce mechanical support between the active layers. Moisture in the empty space can also potentially corrode the bonded Cu lines.

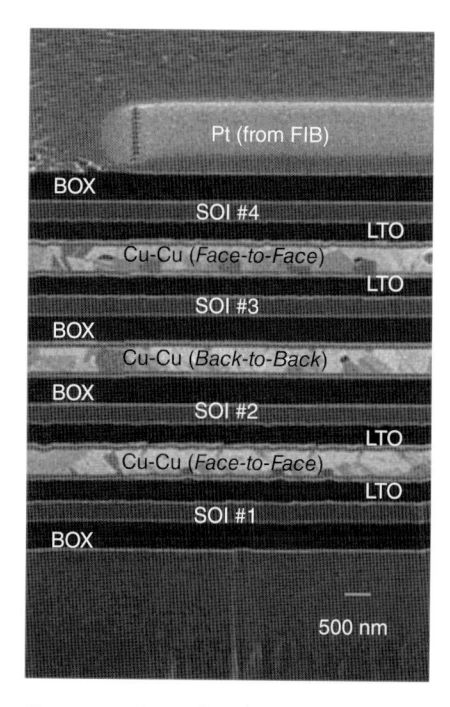

Figure 9.13 Focused ion beam (FIB) SEM image clearly shows a Si four-layer stack achieved by stacking two Si two-layer stacks. This gives a promising path to multilayer and multifunctionality Si stacks [18]. (Reproduced by permission of the Electrochemical Society.)

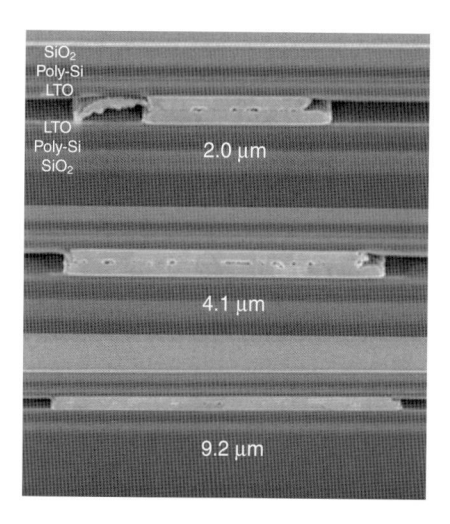

Figure 9.14 Nonblanket bonding of Cu lines. From [19]. (Copyright MRS 2007.)

One solution is to form damascene Cu lines and to perform hybrid bonding of Cu and dielectric. A few examples are the following.

- Jourdain et al. [20] at IMEC have successfully demonstrated the three-dimensional stacking of an extremely thinned IC chip onto a Cu/oxide landing substrate using simultaneous Cu–Cu thermocompression and compliant glue-layer (benzocyclobutene, BCB) bonding. The goal of this intermediate BCB glue layer between the two dies is to improve the mechanical and thermal stability of the bonded stack and to enable separation of die pick-and-place operations from a collective bonding step.
- Lu et al. [21] at RPI have demonstrated another scheme of hybrid bonding using face-to-face bonding of Cu/BCB redistribution layers. The first step is to prepare single-level damascene-patterned structures (Cu and BCB) by CMP in the two Si wafers to be bonded. The second step is to align the two wafers and bond the two aligned wafers.
- The author's (K.-N.C.) previous research group at IBM (now at National Chiao Tung University) demonstrated a 300 mm wafer-level three-dimensional integration process using W through-Si via and hybrid Cu/adhesive wafer bonding [22, 23]. The adhesive used here is polyimide. The bonding approach uses the so-called transfer-join method. The initial reliability evaluations, including interface bonding strength, via chain resistance as a function of deep thermal cycles, ambient permeation oxidation, and temperature and humidity tests, all show favorable results.
- Researchers at Ziptronix have developed a Cu/oxide hybrid bonding technology known as Direct Bond Interconnect (DBI™) [24]. Vertical interconnections in direct oxide bond DBI™ are achieved by preparing a heterogeneous surface of nonconductive oxide and conductive Cu. The surfaces are aligned and placed together to form a bond. The high bond energies possible with the direct oxide bond between the heterogeneous surfaces result in vertical DBI™ electrical interconnections. DBI™ is essentially thermocompression bonding. The exception is that the compression is induced in the Cu structures (due to coefficient of thermal expansion mismatch between Cu and SiO_2) and not externally applied.

9.9
Low-Temperature (<300 °C) Cu–Cu Bonding

Thermocompression bonding of Cu layers is typically performed at a temperature of 300 °C or higher, based on Figure 9.7 [12] and the description in Section 9.4.2. There is a strong motivation to reduce the bonding temperature to even lower values primarily from the point of view of thermal stress induced due to coefficient of thermal expansion mismatch of dissimilar materials in

9.9 Low-Temperature (<300°C) Cu–Cu Bonding

a multilayer stack and temperature swing. A number of approaches have been explored.

- Surface-activated bonding (see Section 9.2.2) [1]. In this method, a low-energy Ar ion beam is used to activate the Cu surface prior to bonding. Contacting two surface-activated wafers enables successful Cu–Cu direct bonding. The bonding process is carried out under ultrahigh vacuum conditions. No thermal annealing is required to increase the bonding strength. Tensile test results show that high bonding strength equivalent to bulk material is achieved at room temperature. In [25], adhesion of Cu–Cu bonds at room temperature in ultrahigh vacuum conditions was measured to be about $3\,J/m^{-2}$ using the atomic force microscope tip pull-off method.

- Copper nanorods [26]. Recent investigation of surface melting characteristics of Cu nanorod arrays shows that the threshold of the morphological changes of the nanorod arrays occurs at a temperature significantly below the bulk Cu melting point. With this unique property of the Cu nanorod arrays, wafer bonding using these arrays as a bonding intermediate layer has been investigated at low temperatures (400°C and lower). Silicon wafers, each with a Cu nanorod array layer, are bonded at 200–400°C. Focused ion beam SEM results show that the Cu nanorod arrays fuse together accompanied by grain growth at a bonding temperature as low as 200°C.

- Solid–liquid interdiffusion (SLID) bonding [27]. This method involves the use of a second solder metal with a low melting temperature such as Sn between two sheets of Cu with high melting temperature. Typically a short reflow step is followed by a longer curing step. The required temperature is often slightly higher than the Sn melting temperature (232°C). The advantages of SLID bonding are that the intermetallic phase is stable up to 600°C and the requirement of contact force is not critical.

- In the DBI™ technology described in [24], a moderate post-oxide bonding anneal may be used to effect the desired bonding between Cu. Due to the difference in coefficient of expansion between the oxide and Cu and the constraint of the Cu by the oxide, Cu is compressed during heating and a metallic bond can be formed.

- Direct Cu–Cu bonding at atmospheric pressure has been investigated by researchers at LETI [28]. By means of CMP, the roughness and hydrophilicity (measure by contact angle) of Cu film are improved from 15 to 0.4 nm and from 50° to 12°. Blanket wafers were successfully bonded at room temperature with an impressive bond strength of $2.8\,J\,m^{-2}$. With a post-bonding annealing at 100°C for 30 min, the bonding strength was improved to $3.2\,J\,m^{-2}$.

- A novel Cu–Cu bonding process has been developed and characterized to create all-Cu chip-to-substrate input/output connections [29]. Electroless Cu plating followed by low-temperature annealing in a nitrogen environment was used to create an all-Cu bond between Cu pillars. The bond strength for the all-Cu structure exceeded 165 MPa after annealing at 180 °C. While this technique has been demonstrated as a packaging solution, it is an attractive low-temperature process for Cu–Cu bonding.

- The author's (C.S.T.) research group at Nanyang Technological University has developed a method of Cu surface passivation using self-assembled monolayers of alkane-thiol. This method has been shown to be effective to protect the Cu surface from particle contamination and to retard surface oxidation. The self-assembled monolayer can be thermally desorbed *in situ* in the bonding chamber rather effectively hence providing clean Cu surfaces for successful low-temperature bonding. Copper wafers bonded at 250 °C present significant reduction in micro-voids and substantial Cu grain growth at the bonding interface [30–32].

9.10
Applications of Cu Wafer Bonding

Copper thermocompression bonding can be applied to fabricate three-dimensional ICs and a number of proposed process flows have been reported [33, 34]. Industries have announced the use of the technology taking advantage of reduced wire length in three-dimensional ICs for both processor-on-processor and memory-on-processor chip designs [35]. These applications can be realized based on Cu bonding with through-Si via technology. Morrow *et al.* [36] at Intel reported the first demonstration of integrating 300 mm wafer stacking via Cu bonding with strained-Si/low-k 65 nm CMOS technology. Sets of active devices such as 65 nm MOSFETs and 4 MB SRAMs were bonded face-to-face using Cu pads. Patti [37] reported Cu bonding technology developed at Tezzaron. A number of three-dimensional devices have been demonstrated using this stacking process, such as a stand-alone memory, CMOS sensor, three-dimensional field-programmable gate array, mixed-signal application-specific IC, and processor/memory stack.

9.11
Summary

In this chapter, a method of wafer bonding using metallic Cu has been described. Bonding is achieved using thermocompression of wafers coated with a Cu layer. Both blanket and patterned wafers have been studied. Fundamental aspects of the bonding mechanism as well as process development of Cu wafer bonding have been examined. Multilayer stacking of Si layers using Cu as the bonding medium has also been presented. A brief summary of methods to achieve low-temperature Cu–Cu bonding is included. Applications of Cu wafer bonding have been described briefly.

References

1 Kim, T.H. *et al.* (2003) Room temperature Cu–Cu direct bonding using surface activated bonding method. *J. Vac. Sci. Technol.*, **A21** (2), 449–453.

2 Shigetou, A. *et al.* (2003) Room temperature ultra-fine pitch and low-profiled Cu electrodes for bumpless interconnect. *Transducers*, 1828–1831.

3 Suga, T. *et al.* (1993) *Ceram. Trans.*, **35**, 323.

4 Chen, K.N. *et al.* (2006) Bonding parameters of blanket copper wafer bonding. *J. Electron. Mater.*, **35** (2), 230–234.

5 Chen, K.N. *et al.* (2001) Microstructure examination of copper wafer bonding. *J. Electron. Mater.*, **30**, 331–335.

6 Chen, K.N. *et al.* (2002) Microstructure evolution and abnormal grain growth during copper wafer bonding. *Appl. Phys. Lett.*, **81** (20), 3774–3776.

7 Thompson, C.V. and Carel, R. (1996) Stress and grain growth in thin films. *J. Mech. Phys. Solids*, **44**, 657–673.

8 Sanchez, J.E. and Artz, E. (1992) Effects of grain orientation on hillock formation and grain growth in aluminum films on silicon substrates. *Scr. Metall. Mater.*, **27**, 285–290.

9 Chen, K.N. *et al.* (2006) Improved manufacturability of Cu bond pads and implementation of seal design in 3D integrated circuits and packages. 23rd International VLSI Multilevel Interconnection (VMIC) Conference, Fremont, CA, 25–28 September 2006.

10 Chen, K.N. *et al.* (2006) Structure design and process control For Cu bonded interconnects in 3D integrated circuits. International Electron Devices Meeting (IEDM), pp. 367–370.

11 Chen, K.N. *et al.* (2003) Temperature and duration effect on microstructure evolution during copper wafer bonding. *J. Electron. Mater.*, **32** (12), 1371–1374.

12 Chen, K.N. *et al.* (2004) Morphology and bond strength of copper wafer bonding. *Electrochem. Solid-State Lett.*, **7** (1), G14–G16.

13 Chen, K.N. *et al.* (2006) Investigations of strength of copper bonded wafers with several quantitative and qualitative tests. *J. Electron. Mater.*, **35** (5), 1082–1086.

14 Chen, K.N. *et al.* (2004) Contact resistance measurement of bonded copper interconnects for three-dimensional integration technology. *IEEE Electron Devices Lett.*, **25** (1), 10–12.

15 Schwartz, G.C. (1998) *Handbook of Semiconductor Interconnection Technology*, Marcel Dekker, New York, p. 187.

16 Rabaey, J. (1996) *Digital Integrated Circuits*, Prentice Hall, Englewood Cliffs, NJ, p. 465.

17 Chen, K.N. *et al.* (2005) Processing development and bonding quality investigations of silicon layer stack using copper wafer bonding. *Appl. Phys. Lett.*, **87** (3), 031909.

18 Tan, C.S. and Reif, L.R. (2005) Multi-layer silicon layer stacking based on copper wafer bonding. *Electrochem. Solid-State Lett.*, **8** (6), G147–G149.

19 Tan, C.S. *et al.* (2007) Silicon layer stacking enabled by wafer bonding. *Mater. Res. Soc. Symp. Proc.*, **970**, 193–204.

20 Jourdain, A. *et al.* (2007) Simultaneous Cu–Cu and compliant dielectric bonding for 3D stacking of ICs. Proceedings of International Interconnect Technology Conference, pp. 207–209.

21 Lu, J.-Q. *et al.* (2008) 3D integration using adhesive, metal, and metal/adhesive as wafer-level bonding interfaces. materials and technologies for 3-D integration. *Mater. Res. Soc. Symp. Proc.*, **1112**, 69–80.

22 Fei, L. *et al.* (2008) A 300-mm wafer-level three-dimensional integration scheme using tungsten through-silicon via and hybrid Cu-adhesive bonding. International Electron Devices Meeting (IEDM), San Francisco CA, 15–17 December 2008.

23 Roy, R.R. *et al.* (2009) Reliability of a 300-mm-compatible 3DI technology based on hybrid Cu-adhesive wafer bonding. Symposium on VLSI Technology and Circuits, Kyoto, Japan, 15–18 June 2009.

24 Enquist, P. (2009) Direct Bond Interconnect (DBI): a multidimensional technology for multi-dimensional ICs. 3D architecture for semiconductor integration and packaging.
25 Tadepalli, R. and Thompson, C.V. (2007) Formation of Cu–Cu interfaces with ideal adhesive strength via room temperature pressure bonding in ultrahigh vacuum. *Appl. Phys. Lett.*, **90** (15), 151919.
26 Wang, P.-I. *et al.* (2007) Low temperature copper-nanorod bonding for 3D integration. *Mater. Res. Soc. Symp. Proc.*, **970**, 225–230.
27 Benkart, P. *et al.* (2005) 3D chip stack technology using through-chip interconnects. *IEEE Des. Test Comput.*, **22** (6), 512.
28 Gueguen, P. *et al.* (2008) Copper direct bonding for 3D integration. IEEE International Interconnect Technology Conference, pp. 61–63.
29 Osborn, T. *et al.* (2008) All-copper chip-to-substrate interconnects. Proceedings of the IEEE Electronic Components and Technology Conference, pp. 67–74.
30 Lim, D.F. *et al.* (2009) Achieving low temperature Cu to Cu diffusion bonding with self assembly monolayer (SAM) passivation. IEEE International Conference on 3D System Integration, 5306545.
31 Lim, D.F. *et al.* (2009) Application of self assembly monolayer (SAM) in Cu–Cu bonding enhancement at low temperature for 3-D integration. Advanced Metallization Conference, Baltimore, MD, 13–15 October 2009.
32 Tan, C.S. *et al.* (2009) Cu–Cu diffusion bonding enhancement at low temperature by surface passivation using self-assembled monolayer of alkane-thiol. *Appl. Phys. Lett.*, **95** (19), 192108.
33 Reif, R. *et al.* (2002) 3-D interconnects using Cu wafer bonding: technology and applications. Advanced Metallization Conference, San Diego, CA, 1–3 October 2002, pp. 37–45.
34 Reif, R. *et al.* (2004) Technology and applications of three-dimensional integration. 206th Electrochemical Society Fall Meeting, pp. 261–276.
35 Morrow, P.R. *et al.* (2006) Design and fabrication of 3D microprocessors. *Mater. Res. Soc. Symp. Proc.*, **970**, 91–102.
36 Morrow, P.R. *et al.* (2006) Three-dimensional wafer stacking via Cu–Cu bonding integrated with 65 nm strained-Si/low-k CMOS technology. *IEEE Electron Device Lett.*, **2** (5), 335–337.
37 Patti, R. (2006) Three-dimensional integrated circuits and the future of system-on-chip designs. *Proc. IEEE*, **94** (6), 1214–1224.

10
Wafer-Level Solid–Liquid Interdiffusion Bonding

Nils Hoivik and Knut Aasmundtveit

10.1
Background

10.1.1
Solid–Liquid Interdiffusion Bonding Process

Solid–liquid interdiffusion (SLID) bonding is a bonding technique based on intermetallic formation that has received much interest in the industrial and scientific community. SLID bonding has several appealing properties, such as [1]:

- high-temperature stability,
- moderate processing temperature in an isothermal process,
- allowing repeated processing without bond melting (e.g., for stacking purposes),
- allowing fine-pitch interconnects,
- thermodynamically stable bonds,
- corrosion-resistant bonds,
- well suited for wafer bonding,
- may use low-cost metallization,
- flux-free processes are possible.

The SLID process makes use of a two-metal system, one metal having a high melting point (M_H) and the other having a low melting point (M_L). The melting points are designated T_H and T_L, respectively, in the following. Typically these metals will be deposited in layers, as shown in the inset in Figure 10.1, for example by electroplating. A metal system for SLID should exhibit thermodynamically stable intermetallic compounds (IMCs) with high melting points, as shown in the schematic phase diagram in Figure 10.1.

Bonding is performed by processing at a temperature somewhat above T_L. M_L melts, and the two metals interdiffuse. In particular, the solid M_H diffuses into the liquid M_L, reacting to form solid IMC. Thus, the bond solidifies at the constant bonding temperature. Note that the interdiffusion process takes place already at lower temperature, although slowly. The kinetics of the process is highly

Handbook of Wafer Bonding, First Edition. Edited by Peter Ramm, James Jian-Qiang Lu, Maaike M.V. Taklo.
© 2012 Wiley-VCH Verlag GmbH & Co. KGaA. Published 2012 by Wiley-VCH Verlag GmbH & Co. KGaA.

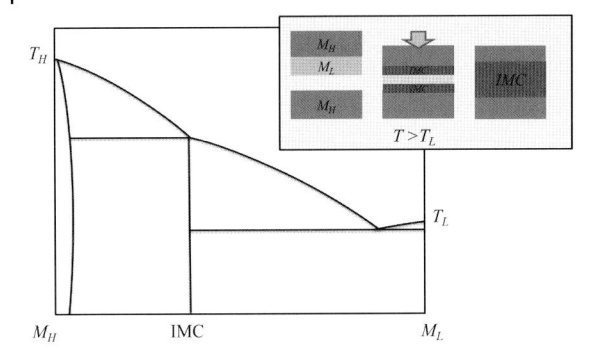

Figure 10.1 Schematic phase diagram, showing a binary metal system with high (T_H) and low melting point (T_L), and with IMC with high melting point. The inset shows a schematic of a typical layer structure for SLID bonding, and the bonding process.

accelerated above T_L. The SLID process should consume all the M_L material, ensuring that the final bond consists only of IMC and M_H, both being high-temperature stable materials, ensuring thermodynamic stability. This requires that the initial metal system be designed such that there is a surplus of M_H in the system. The calculation of the required thickness ratios from the composition of the IMC A_xB_y can be performed as follows:

$$\frac{h_A}{h_B} = \frac{V_A}{V_B} = \frac{x}{y} \frac{M_A}{\rho_A} \frac{\rho_B}{M_B} \qquad (10.1)$$

where h_A and h_B denote the layer thicknesses of the pure metals A and B, V their respective volumes, M their respective molar masses, and ρ their respective densities. A surplus of M_H is important for several reasons:

- to ensure all M_L is converted to IMC (taking process variations and uncertainties into account);
- to prevent diffusing M_L from reaching the under-bump metallization (UBM), potentially creating undesired IMCs and/or giving rise to delamination;
- ensuring a remaining layer of metallic M_H, normally being more ductile than the (often) brittle IMC, for stress handling.

SLID bonding has been investigated by several groups. As a consequence, different terminologies exist. The bonding process is also referred to as transient liquid-phase bonding [2, 3], isothermal solidification [4], or off-eutectic bonding [5]. As they all refer to the same bonding process, the term SLID bonding will be used for all bonding in this context.

10.1.2
SLID Bonding Compared with Soldering

The SLID process shows several similarities with a soldering process: the existence of a liquid metallic phase during the process, and the formation of IMC as the actual bonding material. The same metals are commonly used for both processes: Most solders are based on Sn, with small amounts of alloying metals (such as Cu and/or Ag), and Cu is a commonly used pad material. Similarly, Cu–Sn is the most common metal system for SLID, as discussed below. Processing temperatures are also similar. Another solder system is the Au–Sn solder, with clear similarities to Au–Sn SLID, as will be detailed later in this chapter.

Despite the similarities, SLID and soldering are fundamentally different processes, each with distinct and unique properties. Table 10.1 compares the two bonding techniques. Whereas soldering is a reversible process, where the solder melts and solidifies at the same temperature, SLID bonding is irreversible, resulting in an IMC bond that melts only at temperatures far above the process temperature. A SLID bond thus shares some of the properties of solid-state thermocompression bonding (e.g. Au–Au or Cu–Cu), as well as sharing process similarities with soldering.

The volume of bonding material is far less in a SLID bond than in a solder bond, also having an impact on the various bond properties described in Table 10.1. Note that as solder technology shifts to smaller solder joint volumes for fine-pitch compatibility, the proportion of IMC versus solder in the final joint will increase, as shown in Figure 10.2. Solder technology may thus be expected to approach a SLID process upon future miniaturization [1].

10.1.3
Material Systems for SLID Bonding

As stated above, the SLID bonding process makes use of a high-melting-temperature metal and a low-melting-temperature metal. Candidate high-melting-temperature metals include Cu, Au, Ag, Co, and Ni. Candidate low-melting-temperature metals are Sn, In, and Hg (Hg is not the metal of choice, due to toxicity, but is included here for completeness). Properties for selected combinations of these are listed in Table 10.2.

Table 10.2 lists the melting or phase-transition temperatures for candidate metal systems for SLID bonding. The values are extracted from [8]. Note that the SLID bonding temperatures normally will be somewhat above T_L (about 10–100 °C). For several of the metal systems, there exists a eutectic point with lower melting temperature than T_L. This is not included in the table. Also, several of the metal systems display complex phase diagrams, where details may still be in doubt. The "target IMC" is thus based on an extraction from the phase diagrams that may represent a simplification of a complex system.

It should be mentioned that for most of the systems listed in Table 10.2, the low-temperature metal has a certain solid solubility (typically 5–10 at%) in the high-temperature metal. The solidus temperature of this solid solution is even higher than that of the "target IMC" in Table 10.2 (approaching T_H as the concentration of M_L approaches zero). However, a bond based on such a solid solution would require the M_L/M_H ratio to be significantly smaller than that required for SLID bonding, and will not be covered in this chapter.

Table 10.1 Overview of solder versus SLID properties.

Property	Solder	SLID
Metal system	Excess of low-temperature metal (e.g., Sn)	Excess of high-temperature metal (e.g., Cu)
Temperature profile for bond formation	Heating above melting temperature, followed by cooling	Constant temperature
Effect of repeating the process temperature	Bond melts	No impact
Rework possible?	Yes (but IMC formation also during initial solder process: re-soldering is not 100% reversible)	No (irreversible process)
Self-alignment	Yes	Yes, but limited
Use of flux	Yes, for most solders	Yes/no (dependent on metal system and process)
Pitch	Limited, due to liquid phase (normally >100 μm)	Fine-pitch possible (20 μm demonstrated) [6]
Bond thickness	Typically tens of micrometers	Typically a few micrometers (~10 μm including pads)
Possibility to accommodate different pad heights and irregularities	Good, due to wetting properties of liquid solder	Limited, because of small volumes of liquid phase during process
Importance of IMC	Thin bonding layer between solder and pad	Bonding layer, between two high-temperature metal pads
Temperature stability	Limited (only for applications well below the solder melting temperature)	Excellent
Effect of thermal cycling/thermal aging	Growth of IMC layer: evolution of bond properties over time. Reliability issue	Thermodynamically stable bond. No change in chemical composition
Ductility/stress handling/ability to handle CTE mismatch	Good (for soft solders)	Limited (depending on remaining layers of high-temperature metal)

Various of the metal systems of Table 10.2 have been investigated in early work [4], including studies of the debonding temperature.

Cu–Sn is by far the most studied SLID system. Both metals are of low cost and are readily available. The materials have an extensive heritage in the microsystem/microelectronic sector, as Sn is the basis of all soldering and Cu is a commonly

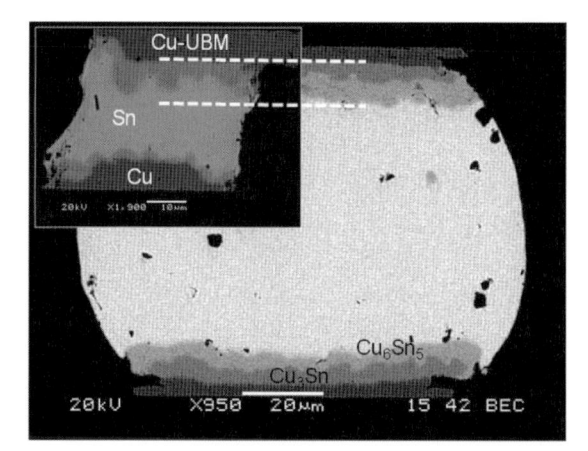

Figure 10.2 SEM image illustrating the difference in dimensions between a conventional and a miniaturized (inset) solder Cu–Sn interconnect [7]. Note that for the conventional solder Cu–Sn interconnect, the majority of the bump remains as pure Sn, while the miniaturized joint (inset) has a higher IMC/Sn ratio, eventually approaching a SLID interconnect (as shown in Figures 10.3 and 10.9).

used substrate and chip pad metallization. The Cu–Sn SLID process can be performed at a moderate temperature (typically 250–300 °C), and results in stable, reliable Cu/Cu$_3$Sn/Cu bonds. The remaining part of this chapter will mainly be devoted to the Cu–Sn SLID process.

Au–Sn is another SLID system that is of particular interest. Although being more costly than Cu–Sn, the use of Au has two important features. First, due to the inertness of Au, flux-free processes are more easily implemented in this system. Second, a SLID bond consisting of Au/IMC/Au layers may take advantage of the ductility of Au for stress handling. Au–Sn SLID bonding is treated in more detail later in this chapter.

Fluxless SLID-like bonding using a Ag–Sn metal system has also been reported [9], with a theoretical thermal stability to 700 °C, and similarly for a Ag–In system [10, 11].

10.1.3.1 IMC Formation Process: Cu–Sn SLID Bonding

Cu–Sn SLID interconnects initially form two distinct phases: η-Cu$_6$Sn$_5$ and ε-Cu$_3$Sn (the phase diagram is shown in Figure 10.3). At a bonding temperature above the melting point of Sn (232 °C), molten Sn dissolves Cu and the IMC η-phase (Cu$_6$Sn$_5$) is formed. This phase actually forms even at room temperature; however, the reaction is greatly accelerated when raising the temperature above the Sn melting point. Since there is excess Cu in the system, the η-phase (Cu$_6$Sn$_5$) will further react with Cu to form the second IMC ε-phase (Cu$_3$Sn). Both phases will increase in thickness as long as there is liquid Sn (and solid Cu) available. When all liquid Sn is consumed, the bond has solidified isothermally to a

Table 10.2 Various metal candidates suitable for SLID bonding.

M_H	M_L	T_H (°C)	T_L (°C)	Target IMC	Temperature stability of target IMC (°C)[a]	Comment
Ag	In	962	157	Ag_2In	315: solid-state PT	
					~600: melting	
				Ag_3In	187: solid-state PT	Nonstoichiometric, uncertain phase
					695: melting	
Ag	Sn	962	232	Ag_3Sn	480: PT ζ + L (partial melting)	
					724: melting	
				ζ (12–23 at% Sn)		
Au	Hg	1064	−39	Au_2Hg	122: PT ζ + L (partial melting)	
				ζ (21–26 at% Hg)	388: melting	
Au	In	1064	157	Au_7In	649: melting	Nonstoichiometric phase
Au	Sn	1064	232	Au_5Sn	190: solid-state PT	Discussed in more detail in the text
					519: melting	
Co	In	1495	157	$CoIn_2$	550: melting	
Co	Sn	1495	232	Co_3Sn_2	500: solid-state PT	
					1170: melting	
Cu	In	1085	157	δ (29–31 at% In)	631: solid-state PT	
					684: melting	
Cu	Sn	1085	232	Cu_3Sn	676: solid-state PT	Discussed in more detail in the text
					~700: Melting	
Ni	In	1455	157	Ni_3In	~850: solid-state PT	
					910: melting	
Ni	Sn	1455	232	Ni_3Sn	~980: solid-state PT	
					~1170: melting	

a) PT, phase transition.

Cu/IMC/Cu structure. The Cu–Cu_6Sn_5 reaction will continue, until also all η-phase (Cu_6Sn_5) is consumed. The left-hand inset in Figure 10.3 shows a cross-sectional image of a bonded sample where both the ε-phase and η-phase can be observed in the bond line. The IMC formation process will terminate upon complete transformation of all IMCs to the stable Cu_3Sn ε-phase. As seen in the phase

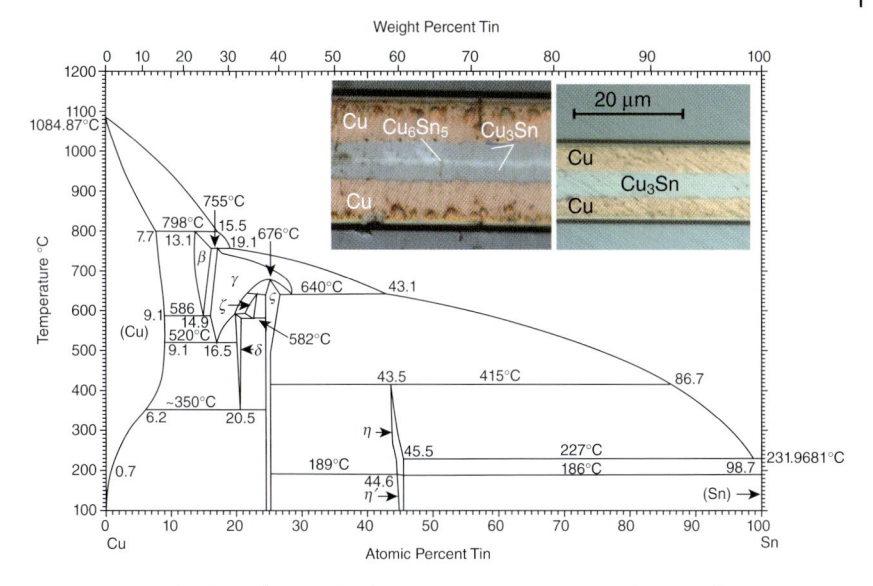

Figure 10.3 Cu–Sn phase diagram [8]. The left-hand inset illustrates a bond line with both the ε-phase and η-phase present. The IMC formation process will eventually terminate upon complete transformation to the stable ε-phase in the bond line, as shown in the right-hand inset.

diagram (Figure 10.3), no thermodynamically stable phases exist for composition between the ε-phase (Cu_3Sn) and the α-phase (Cu with Sn in solid solution), for temperatures up to 350 °C. The $Cu/Cu_3Sn/Cu$ bond line will therefore be thermodynamically stable in this temperature range. Above 350 °C, a further reaction between the layers may lead to the formation of Cu_4Sn, but no liquid phase will occur unless the temperature is increased to about 700 °C.

10.1.3.2 IMC Formation Process: Au–Sn SLID Bonding

Au–Sn solder bonding is a commonly used solder method, and is treated in detail in Chapter 7. Standard Au–Sn solder bonding makes use of the eutectic composition (80 wt% Au), with a melting point at 278 °C. This gives a soldering-type process, with the solidification temperature and the melting/re-melting temperature being equal.

The Au–Sn phase diagram in Figure 10.4 shows the existence of several Au–Sn IMCs. All of these IMCs have a higher melting point than pure Sn, and the δ-phase and ζ/ζ'-phases have melting points significantly higher than the eutectic composition (419 °C and up to 522 °C, respectively). For applications where the bond is to resist higher temperatures, either during later processing or during service, a SLID bond made of one of these IMCs may be appropriate. For a SLID bonding system where there is surplus of pure Au after formation of the IMC bond line, a bond made of the ζ/ζ'-phases will be desired, since further interdiffusion with the Au phase will not result in lower-melting phases. A SLID bond consisting of

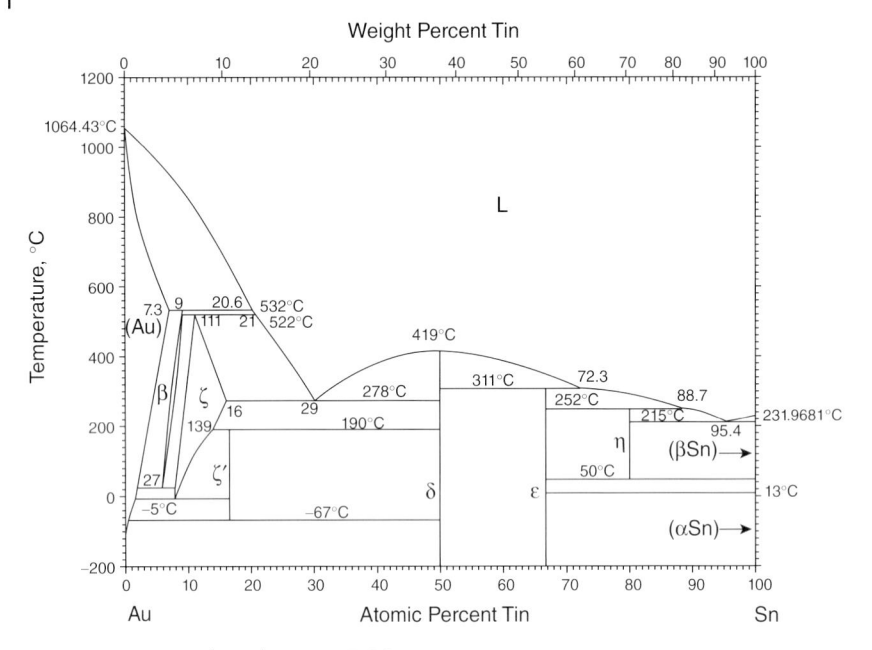

Figure 10.4 Au–Sn phase diagram [12, 13].

δ-phase sandwiched between Au layers, on the other hand, may be susceptible to conversion into a eutectic or near-eutectic structure over time, due to Au–Sn interdiffusion, thus lowering the melting point. A review paper dedicated to Au–Sn SLID is available [14]. Note that the eutectic composition itself is actually composed of the δ-phase and ζ/ζ'-phases, in a ratio corresponding to 80 wt% Au.

An Au–Sn SLID bond may be produced similar to the Cu–Sn SLID bonds discussed earlier: starting with a layered Au–Sn structure, and bonding at temperatures above the melting point of Sn (normally, bonding temperatures above the eutectic temperature will be chosen). The resulting bond line may consist of the various Au–Sn IMCs ($AuSn_4$, $AuSn_2$, $AuSn$, and Au_5Sn). The metal system should be designed with surplus of Au relative to the ζ/ζ'-phases (Au_5Sn), such that the final bond can consist of a Au/ζ/ζ'-phase/Au structure.

The ζ'-phase is the Au-rich IMC phase that is experimentally confirmed to be thermodynamically stable at room temperature, identified as the stoichiometric composition Au_5Sn. At 190 °C, this phase undergoes a solid-state phase transition to the more disordered ζ-phase, which can have a composition ranging from 10 to 16 at% Sn, the actual composition range being temperature dependent. The ζ-phase has a maximum melting temperature (solidus) of 522 °C, at a composition of 11 at% Sn. The ζ-phase is probably stable down to room temperature, for a composition of about 8 at% Sn (not yet experimentally confirmed) [12, 13]. Thus, ζ-phase and ζ'-phase may coexist in thermodynamic equilibrium in the SLID bond, for temperatures up to 190 °C. The layer configuration Au/ζ/ζ'-phase/Au is considered stable. According to the phase diagram, the ζ-phase may react with Au to form the β-phase (traditionally labeled as $Au_{10}Sn$), but no liquid phase will occur unless the temperature is raised to the solidus temperature of the ζ-phase (522 °C at 11 at% Sn composition). A ζ-phase with higher Sn content will have a somewhat lower solidus temperature. However, the Au–Sn interdiffusion is expected to be sufficiently effective at these temperatures for the ζ-phase to increase its Au content to remain solid all the way up to 522 °C, as long as the overall Au content is high enough.

10.2
Cu–Sn SLID Bonding

The benefit of Cu–Sn SLID bonding for interconnects and hermetic sealing, and especially for three-dimensional applications, is quite apparent. With interconnects or bond lines stable for large temperature variations, subsequent bonding steps, processing at elevated temperatures, or thermal annealing, for instance to activate getter materials, will not compromise the initial bonds and integrity of the package. Figure 10.5 shows Cu–Sn SLID bonding where Sn is placed between two Cu pads, a pressure is applied, and the increased temperature will first cause a diffusion process between Sn and Cu to form the η-phase Cu_6Sn_5. Then, given enough time, the final intermetallic compound will be the stable ε-phase Cu_3Sn. Since the metal layers used typically vary between 1 and 8 µm, electroplating is commonly used to deposit both Cu and Sn, as this offers the most economical processing.

The schematic in Figure 10.5 illustrates the bonding process typically applied for die stacking, flip chip (chip to chip, C2C) or chip to wafer (C2W) where the bottom die or wafer is held at an elevated temperature. This does not permit Sn to be deposited on the lower, and heated, dies/wafers since it would otherwise form the intermetallic phases before bonding has taken place.

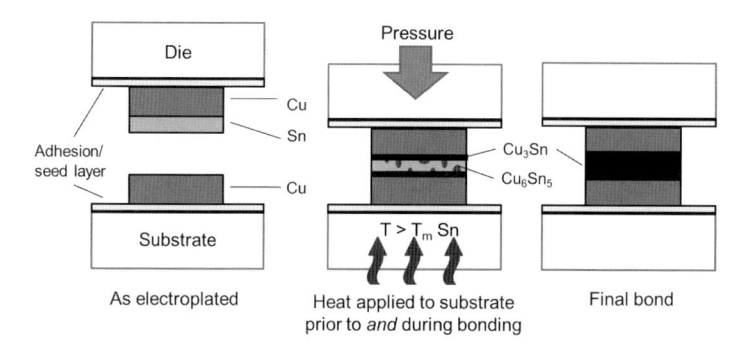

Figure 10.5 Cu–Sn SLID bonding, where typically electroplated Cu and Sn are patterned on respective dies and brought into contact to form the final intermetallic composition Cu_3Sn sandwiched between Cu layers.

Table 10.3 Relevant material properties for Cu, Sn, and the corresponding intermetallic phases present in SLID bonding.

Property/material	M (g)	ρ (g cm^{-3})	M/ρ	E (GPa)	Electrical resistivity ($\mu\Omega$ cm)	Thermal expansion (10^{-6})	Melting temperature (°C)
Cu	63.55	8.92	7.12	110	1.7	17[a]	1084
Sn	118.71	7.28	16.31	41	11.5	23	232
Cu$_3$Sn–ε	309.35	8.9	34.76	133[b] 133[a]	8.93	19	676
Cu$_6$Sn$_5$ (η-phase)	974.83	8.27	117.86	117[b] 113[a]	17.5	16.3	415

a) Ref. [6].
b) Ref. [15].
SOURCE: [1].

10.2.1
Cu–Sn Material Properties and Required Metal Thicknesses

For Cu–Sn SLID bonding, some relevant material properties are presented in Table 10.3.

The minimum Cu/Sn thickness ratio for obtaining a steady-state Cu$_3$Sn SLID bond, meaning that the diffusion process has come to completion, can be calculated from Eq. (10.1) and the properties given in Table 10.3. Thus, the required layer thickness ratio for Cu to Sn is ≥ 1.3. This means that for a Sn layer of 2.5 μm, a Cu layer of a minimum thickness of 3.25 μm is required. This amount of Cu could theoretically be the sum of the amount of Cu on both sides of the joint; however, the diffusion process may not be fully symmetric due to different temperatures of the top and bottom wafer, and because interdiffusion will occur faster where the layers are deposited on top of each other than at the bonding interface. Thus, to ensure a sufficient amount of Cu is present in the joint, if one side as a source should be somewhat inhibited due to oxides or similar, a Cu thickness of 5 μm on each bonding partner, with a total thickness of 3 μm for Sn, has been found to work well.

In order to prevent squeeze-out of molten Sn during wafer-level bonding, one may be tempted to increase the Cu thickness and leave the Sn layer relatively thin; or to use thicker Sn as a compliant layer to take up any nonplanarity across the Cu features. However, there is a tradeoff often overlooked which relates to the relationship between the critical interlayer thickness and pore-free bonds. Bosco and Zok [16] studied the critical Sn thickness based upon the fact that during heating, and depending on the heating rate, the interlayer material (Sn) reacts with the base material (Cu) through interdiffusion forming Cu$_6$Sn$_5$. Thus, for a good

10.2 Cu–Sn SLID Bonding

Scalloped Interface

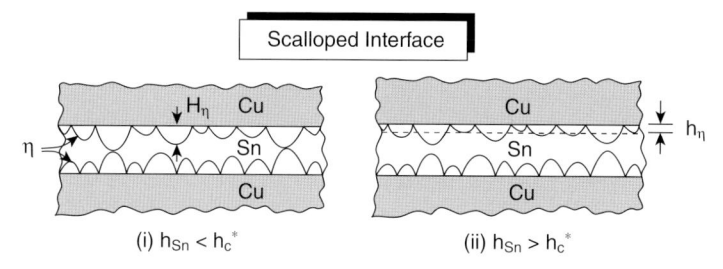

(i) $h_{Sn} < h_c^*$ (ii) $h_{Sn} > h_c^*$

Figure 10.6 Schematics of the two scenarios for the formation of Cu_6Sn_5 scallops and the related critical thickness relationship. (Reprinted from [16] with permission from Elsevier.)

and solid bond to form, the thickness of Sn must exceed the amount of intermetallic material formed *before* reaching the melting point of Sn. In other words, if all available Sn has been consumed and transformed to Cu_6Sn_5, no liquid Sn will be present when reaching the bonding temperature leading to voids or pores in the bond line. Figure 10.6 illustrates how the formation of the scalloped Cu_6Sn_5 intermetallic may contact from either side of the Cu before melting begins.

The critical thickness, h_C^*, can be expressed from a mass-balance relationship:

$$h_C^* = h_C \Omega = h_\eta C_\eta \left(\frac{\rho_\eta}{\rho_{Sn}} \right) \Omega \tag{10.2}$$

where $2h_C$ is the total critical thickness of the η-phase when the melting point of Sn is reached, C_η is the mass fraction of Sn in the η-phase, and ρ_η and ρ_{Sn} are the mass densities of η-phase and Sn, respectively. Ω is a geometrical factor which takes into account the height, h_η, of the scalloped Cu_6Sn_5 grains, which again is a function of the heating rate used. In general, Ω will be larger than 1 unless a complete planar interface is formed, and will depend on both the heating rate and the metallic grain structure of Cu since grain boundary diffusion is the rate-controlling mechanism for these thin films (explained further in the following section). This geometrical factor can be calculated based upon direct measurements of the IMC growth. If the Sn thickness (h_{Sn}) is larger than the critical thickness, h_C, there will be enough Sn present in the bond line, and thus it is recommended to use a thickness only slightly above the critical thickness.

10.2.2
Bonding Processes

Measurements of the phase growth have been extensively researched, and a good estimate for time-to-completion for a SLID bond is shown in Figure 10.7. The reaction process between Cu and Sn is diffusion controlled; the first reaction to take place is the fast Cu–Sn → Cu_6Sn_5 which is a convection-assisted reaction via the grain boundaries into the liquid Sn [17]. This is also the reaction in which the scallops are formed. Several researchers have reported a clear dependency on the relative size of these scallops as a function of heating rate [16, 18, 19]. However,

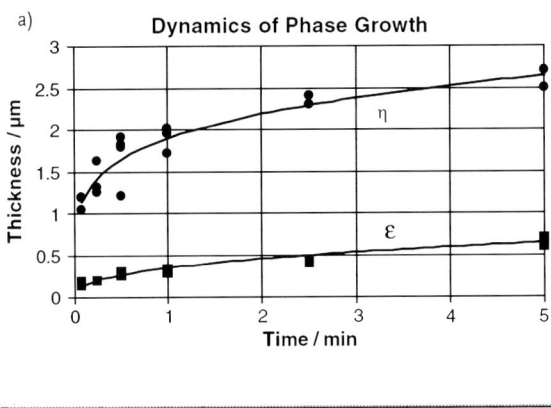

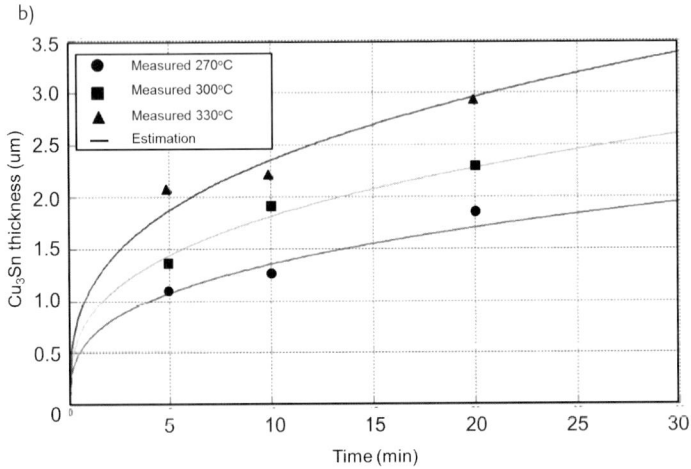

Figure 10.7 (a) Phase growth velocities of Cu–Sn intermetallic (based on sufficient supply of Cu and Sn). (Reprinted from [17] with permission from Elsevier.) (b) measured Cu_3Sn thickness for films annealed at 270, 300, and 330 °C.

there appears to be some inconsistency between the reported measurements since the size of the scallops will depend on the grain size of the Cu layer. For very pure and fine-grained Cu films, the phase growth velocity and hence diffusion rate will be greater than for Cu films with large grains. The second reaction process in which the η-phase is converted to ε-phase requires solid-state diffusion through the Cu_3Sn layer, and thus will progressively be slowed down by the growing thickness of the ε-phase, which is shown in Figure 10.7a [17]. Whereas the solidification of a 3 μm thick Sn layer can be achieved in some 10 s, the complete transformation to the stable Cu_3Sn needs more than 10 min, depending on temperature, as shown in Figure 10.7b.

From a manufacturing point of view the complete process time, and especially time spent in the wafer bonder, should be as short as possible. Fortunately, the bonding process can be arranged such that, once a complete intermetallic bond line has formed, the final annealing step to convert all available Cu_6Sn_5 to Cu_3Sn

Table 10.4 Overview of bonding parameters used in Cu–Sn SLID bonding.

Bond	Metal thickness (μm)	Bond pressure or force	Chamber pressure/ atmosphere	Bond temperature (°C)	Time (min)	Source
C2W	Cu: 5 Sn: 3	Dies held in place by organic polymer	Vacuum oven	300	ND	[17, 20]
C2C	Cu: 5 SnAg: 1.5	49 N	N_2 gas flow	240–400	1	[6]
C2C	Cu: 7 Sn: 2, 3.7, and 6	1.3–66.7 N	N_2 gas flow	275	1	[21]
C2W	Cu: 4 Sn: 2–3	200–400 kPa	Formic vapor	300	10	[22]
C2W	Cu: ND Sn: ND	5 MPa	ND	250	ND	[23]
W2W	Cu: 6 Sn: 4	0.02 MPa	5×10^{-4} mbar	350	10	[19]
W2W	Cu: 5 Sn: 3	3–20 MPa	Vacuum	300	10	[24]
W2W	Cu: ND Sn: 3.5	5.5 MPa	ND	ND	20	[25]
W2W	Cu: 5 Sn: 1.5	ND	Vacuum	280	15	[26]

C2C, chip to chip; C2W, chip to wafer; W2W, wafer to wafer, ND, not disclosed.

can be done in a batch oven. Furthermore, significant reductions in both the critical interlayer thickness and the bonding time can be achieved through an increase in the heating rate en route to the bonding temperature [16]. Table 10.4 presents some processing parameters successfully used in Cu–Sn SLID bonding.

From Table 10.4, it is apparent that the bonding pressure used varies significantly. It is natural to try to use a low pressure in order to prevent the liquid Sn from flowing out uncontrollably from the bond region when reaching the melting temperature of Sn. An elegant process technique where no force is applied was demonstrated by Huebner et al. for C2W bonding. The soldering process does not use any solid or liquid flux during the pick-and-place routine, but instead a low-melting-temperature organic (Bibenzyle, with a melting temperature of 55 °C) is applied, which vaporizes completely during the heating cycle. This organic is inert to the metal surfaces and does not affect the bonding process [17, 20]. The great benefit of this process is that the lack of a bonding pressure allows for the bonded dies to achieve a small degree of self-alignment. Bonded dies that initially were misaligned by ±10 μm reliably self-aligned to within ±1 μm [17].

Using little or no bonding force requires very good uniformity of the Cu and Sn pads to be bonded. Thus, most processes use some level of pressure to ensure contact between the metals. In particular for high-density interconnect arrays, there is an upper limit to how much pressure can be applied from a bonding tool, and there are several studies in the literature of yield and Sn flow-out versus bonding pressure [1, 21]. In general, with a thicker Sn layer (3–6 μm) and lower pressures (to prevent flow-out of liquid Sn), the bond line will comprise both intermetallic phases, unless the time at elevated temperature is prolonged.

As an alternative to using a continuous bond force, a more detailed bonding profile can take advantage of the ductile Sn layers to conform better to each other, given any nonuniformity across the wafer, compared to the significantly harder Cu surface, and the interdiffusion process between Cu and Sn would occur with a slow temperature ramp past the melting point of Sn. Combining this with reducing the bond force at 240 °C, a little above the melting temperature for Sn (232 °C) to account for any thermal gradients between the wafers, the reduction in bonding pressure can reduce squeeze-out of any molten Sn.

Figure 10.8 shows a schematic of such a bonding profile where a greater temperature ramp rate to 150 °C (7 °C min^{-1}) is followed by a smaller temperature ramp rate (3 °C min^{-1}) until reaching the melting temperature of Sn and the final soak temperature. This was done so as to reduce the amount of Sn left at the interface when reaching 240 °C, and to allow for the diffusion process to occur in the bond line early in the process. The wafers were brought into contact with a bond force of 7 kN (20 MPa) after the temperature had reached 150 °C. Following making of

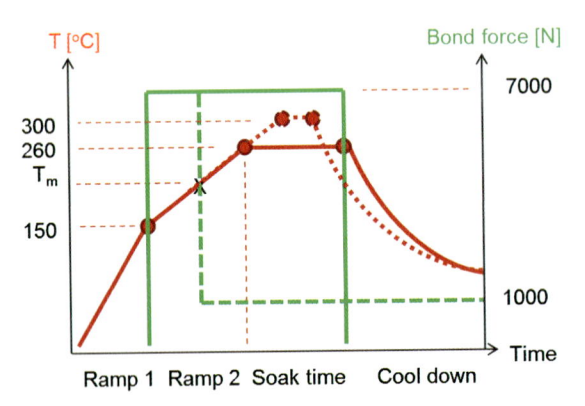

Figure 10.8 Schematic of a bonding profile. The bonding force is applied at 150 °C and either kept constant (7 kN/20 MPa) throughout or reduced at a temperature near the melting point of Sn (240 °C) and kept at a lower value (1 kN/3 MPa).

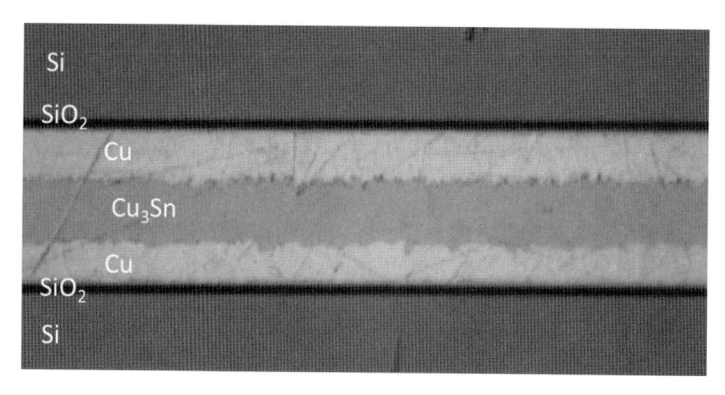

Figure 10.9 Cross-sectional image of a bond line processed with a two-step temperature profile. All available Sn has been converted to Cu_3Sn, as verified by SEM-EDX analysis.

contact, the force was either reduced to 1 kN (3 MPa) at a temperature right above the melting point of Sn (240 °C), or maintained high throughout the soak time.

Using this two-step temperature profile, combined with a soak time of at least 10 min at 260 °C, the bond line will be fully converted to IMC. Figure 10.9 shows a photograph of a Cu–Sn bond using a soak temperature of 300 °C for 10 min. From energy dispersive X-ray (EDX) analysis of the cross section, it is determined that all available Sn has been converted to Cu_3Sn (measured 24.1 at% Sn, versus stoichiometric 25 at% Sn). The intermetallic phase is surrounded by pure Cu on either side.

From bonding experiments performed, it has been observed that a continuous high contact force in general leads to a higher bonding yield. In contrast to what has been presented in earlier studies using a low, even zero, bonding force, there appear to be only small amounts of squeezed-out Sn – even with a high pressure of 20 MPa applied. No large areas of pure Sn outside the bond frames could be observed, and any excess typically measured about 10–15 µm. It is believed that this is a result of the wafer bonding surfaces being brought into contact at a low temperature (150 °C) and kept in contact while the temperature is increased at a low rate (3 °C min^{-1}) past the melting point of Sn. Thus, the interdiffusion process will be near symmetric and lead to a good IMC formation in the bond line.

10.2.3
Pretreatment Requirements for SLID Bonding

One of the key requirements for successful SLID bonding is to remove, or convert, the oxides that are formed on the bonding surface. Both Cu and Sn oxidize, but Cu oxidation is known to be faster than Sn oxidation. Thin SnO layers can be tolerated and incorporated into the molten Sn when bonding. Therefore, the most critical oxide to remove is that of Cu prior to, and during, bonding, which can be accomplished by:

- bonding in hydrogen environment,
- exposing Cu to formic acid vapor to reduce Cu oxide back to Cu,
- use of an acid (HCl) to etch the surface to remove Cu oxide.

These, and other methods described in [27], have also been used to attempt to convert SnO to Sn, with formic acid vapor having a good effect in reducing the stable oxides [28]. For Cu the reduction processes are as follows:

$$CuO(s) + H_2(g) \times Cu(s) + H_2O(g)$$

$$CuO(s) + HCOOH(g) \rightarrow Cu(s) + CO_2(g) + H_2O(g)$$

Table 10.5 lists some of the demonstrated methods for pretreatment of Cu surfaces prior to Cu–Sn SLID wafer-level bonding. Use of formic acid vapor usually requires temperatures above 100 °C to be effective. Note that both etching and formic acid vapor treatment are followed by a rinse step in deionized water to remove any residual compounds. To avoid a rinse step, a plasma process may be used; however, one must ensure that the temperature is kept low so as not to unintentionally melt the Sn layer. A high bonding yield is also ensured by bonding in an inert atmosphere, use of a reactive gas, or bonding in vacuum, to avoid any re-oxidation of the metal surfaces at elevated temperatures.

10.2.4
Fluxless Bonding

A bond process that can be achieved without use of fluxes is called a fluxless or a flux-free process. This has been successfully achieved by using a passivation layer to prevent an oxide from forming on the bonding surface, or to modify the bonding surface to have a higher tolerance to oxidation. A few selected examples are described here.

10.2.4.1 Use of Passivation Layer
Lee and Chen demonstrated fluxless oxidation-free Cu–Sn bonding in which they deposited both a thin Cu layer on a thick Sn layer and a Au layer on a Cu layer,

Table 10.5 Various pretreatments used to ensure a clean Cu surface before bonding.

Pretreatment method	Temperature	Rinse	Dry	Source
HCOOH vapor	100 °C	Megasonic deionized water	Spin dry	[29]
5% HCl	Room temperature	Dionized water	N_2 dry	[30]
HCOOH vapor	150 °C	–	–	[22]
Ar plasma, 500 W, 170 s	–	ND	ND	[19]

ND, not disclosed.

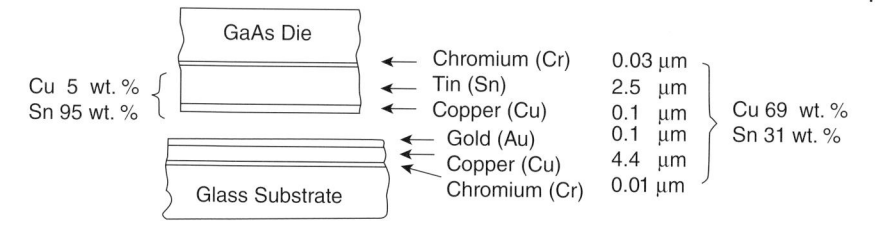

Figure 10.10 Schematic of layered bonding surfaces in fluxless Cu–Sn bonding with both Cu_6Sn_5 and Au to reduce the oxidation of Cu. (Reprinted from [31] with permission from Elsevier.)

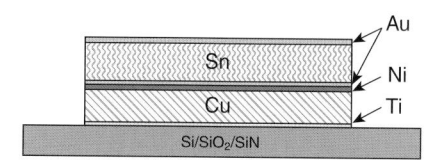

Figure 10.11 Schematic layer structure of Cu–Sn bonding using a Au layer to protect the Sn surface from oxidizing [25].

as shown in Figure 10.10. The thin Cu layer will initially react with Sn and form the η-phase. This outer layer of Cu_6Sn_5 then blocked any Sn oxides from forming. On the Cu side, the thin layer of Au, which is noble, effectively prevents any oxidation of the Cu. During bonding, the thin Au layer is then dissolved into the Sn to form a small amount of $AuSn_2$, the effects of which on the resulting joint can be neglected [31].

Yu and Thew demonstrated fluxless bonding, although in N_2 atmosphere, using a symmetric bonding scheme where a layer of Ni–Au between the Cu and Sn is used to slow down the diffusion rate between Cu and Sn. A thin layer of Au is deposited on top of Sn to prevent any oxidation (Figure 10.11). After fabrication, a standard O_2 plasma de-scum process was carried out to clean the surfaces. The bonding process was done using a pressure of 5.5 MPa and a temperature of 280 °C for 20 min [25].

Liu et al. demonstrated a version of the aforementioned method in which a thin layer of Sn is added to the Cu surface. This makes it possible to perform the bonding in air, without any flux agent, and without introducing any other materials (such as Au) to the bonding interface. As mentioned earlier, the anticorrosion behavior of Cu–Sn alloy (heterogeneous film comprised of Cu_3Sn, Cu_6Sn_5, and β-Sn) has been reported to be better than that of Cu [32]. The bottom dies have an additional 200 nm Sn electroplated on top of 5 μm Cu, and are batch annealed in vacuum (5 mbar) at 250 °C for 5 min to fully convert the Sn to Cu_3Sn (Figure 10.12). Flip chip bonding was carried out in ambient air at 260 °C with a pressure of 10 MPa [33].

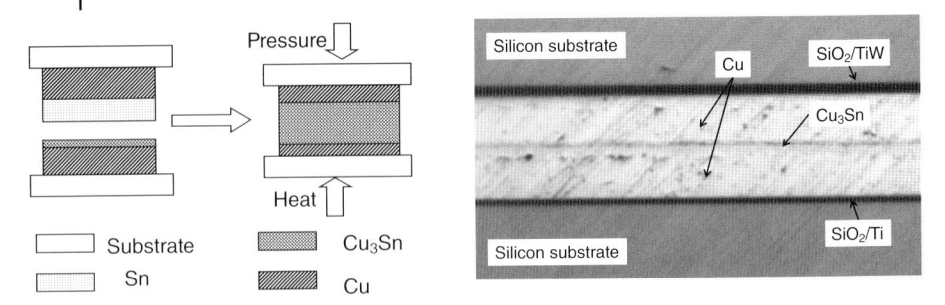

Figure 10.12 Fluxless bonding with a thin Cu_3Sn layer as an oxidation barrier (not drawn to scale) [33].

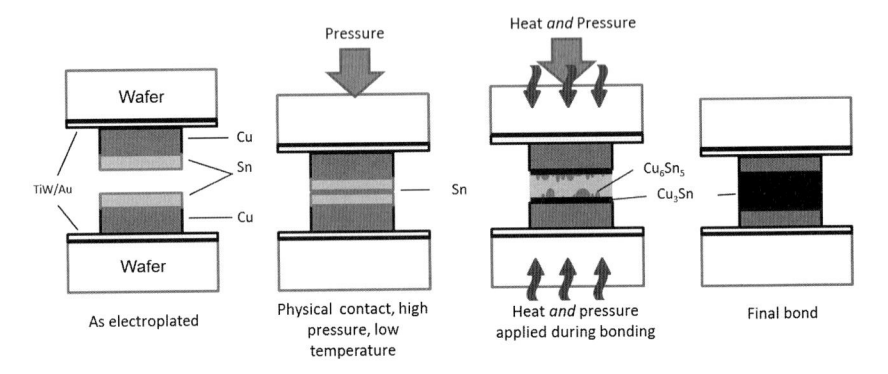

Figure 10.13 Schematic of fluxless symmetrical bonding of Cu–Sn to Cu–Sn for wafer-level applications.

10.2.4.2 Cu–Sn to Cu–Sn Symmetric Bonding

Cu–Sn SLID bonding can also be performed without the use of any pretreatment of the bonding surface or use of flux agent. This may be of significant importance for wafers with integrated and released sensitive micro- and nanosystems where a fluxless wafer-level hermetic packaging solution is required. Since Sn oxidizes significantly less than Cu, and if sufficiently thin, any Sn oxide can be absorbed into the bond line. Thus, the wafers can be electroplated as usual, then placed in a wafer bonder and brought into contact at a low temperature (150°C). Furthermore, having Sn on both sides of the bonding surface further relaxes the uniformity requirement across the wafers surface. Since Sn is a ductile and soft material, the Sn surface will deform and break any Sn oxide when the wafers are brought into contact using a high pressure, thus creating a good bonding interface.

Figure 10.13 illustrates the concept of using symmetric metal thicknesses for fluxless wafer-level bonding, where an electroplated 5 μm thick Cu layer is capped with a 1.5 μm thick Sn layer. The resulting bond line is shown in the cross-sectional image in Figure 10.14.

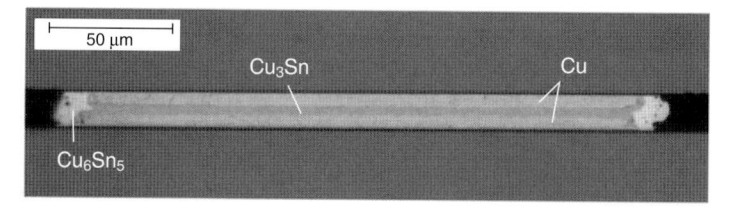

Figure 10.14 Cross-sectional image of a symmetrically bonded Cu/Cu$_3$Sn/Cu bond frame. Outside the edges of the frame Cu$_6$Sn$_5$ remains. With further annealing, this will be converted to Cu$_3$Sn [24].

This process was also applied to wafers stored for one month and bonded using no flux or pretreatment, proving that Sn, with a small amount of oxide on the surface, combined with a high bonding force does indeed permit fluxless wafer-level Cu–Sn SLID bonding [24]. This was further demonstrated in [26] in which identical sets of wafers with symmetric Cu–Sn layers were bonded with, and without, the use of forming gas (5% H$_2$/95% N$_2$). No differences in bond integrity were observed.

10.3
Au–Sn SLID Bonding

10.3.1
Au–Sn Material Properties and Required Metal Thicknesses

Some properties of relevant Au–Sn compounds are listed in Table 10.6.

As discussed above, a high-temperature stable Au–Sn SLID bond should consist of the ζ/ζ' (Au$_5$Sn) phase. The minimum Au/Sn thickness ratio for obtaining a steady-state Au$_5$Sn SLID bond can be calculated from Eq. (10.1).

Thus, the required layer thickness ratio for Au to Sn is ≥3.1. This means that for a Sn layer of 2 µm, a Au layer of a minimum thickness of 6.2 µm is required (being the sum of the Au supplied from both sides). Significantly thicker Au layers should be used to ensure that Au layers are intact at both sides of the IMC bond line, and to allow for transformation to the high-temperature ζ-phase, being somewhat richer in Au.

10.3.2
Bonding Processes

Our group has studied Au–Sn SLID bonding, with a special emphasis on verifying the high-temperature stability expected from the above discussion based on the phase diagram [36, 37]. Au and Sn layers were electroplated, in addition to depositing a thin Au layer on top of Sn to ensure a flux-free bonding process [38]. The metal thicknesses are designed such that the overall Sn content

Table 10.6 Overview of the thermomechanical properties of selected Au–Sn phases [8, 34, 35].

Phase	T_m (°C)	Density, ρ (kg m^{-3})	CTE (ppm K^{-1})	E (GPa)
Sn	232	7 280	24	41
δ-AuSn	419	11 700	14	70–87
Eutectic AuSn	278	14 700	16	69–74
ζ'-Au$_5$Sn	190	16 300	18	62–76
ζ-AuSn 0.18–0.10 at%	519	–	20	58
β -Au$_{10}$Sn	532	–	–	88
Au	1064	19 320	14.4	77.2

SOURCE: www.matweb.com.

is 11 at% [36] (13 at% Sn in [37]), thus ensuring Au surplus relative to the Au$_5$Sn phase, and also Au surplus versus the ζ-phase with a high solidus point, as discussed above.

Bonding was performed at 300, 315, and 350 °C, for different bonding times in the range 2–20 min. The high-temperature stability was tested by applying a shear force while heating the samples to 400 °C, a total of 25 samples being tested (including samples bonded with an alternative process where a eutectic Au–Sn preform (80 wt% Au) was sandwiched between Au layers, the amount of Au and Sn being the same as for the samples with electroplated layers). No delamination or movement of the uppermost chip was found for all but two of the samples (these two delaminating at 350–375 °C, believed to be a result of process variations in a process that is not yet fully optimized). Note that this high-temperature stability applies even for samples bonded at 300 °C, being below the melting temperature of the Sn-rich IMC AuSn$_2$. If such an IMC were present in the bond line, this might have affected the high-temperature stability.

Cross sections of bonded samples showed indeed a layer structure similar to Cu–Sn SLID bonds, with a uniform and homogeneous IMC phase sandwiched between Au layers (Figure 10.15). The different bonding temperatures, bonding times, and the two different processes all gave similar results. The bond line IMC was found by EDX analysis to contain about 10 at% Sn. The precision of EDX analysis at these relatively low Sn concentrations is limited, and the IMC was concluded to be Au$_5$Sn (17 at% Sn), since the confirmed room temperature stable compounds are α (Au with maximum 3 at% Sn), Au$_5$Sn (17 at% Sn), AuSn (50 at% Sn), and AuSn$_2$ (67 at% Sn). Also, the amount of Sn in the original metal system was consistent with the observed Au/IMC/Au structure, with the IMC being Au$_5$Sn. Note that EDX analysis does not have sufficient resolution to distinguish the ζ'-phase (Au$_5$Sn) from the ζ- or β-phase (Au$_{10}$Sn) (that might be present, possibly in a metastable state). X-ray diffraction studies to address this question are in progress.

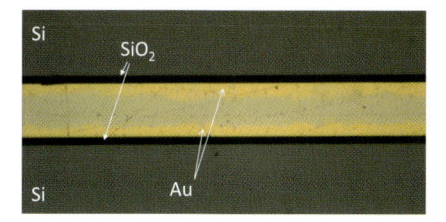

Figure 10.15 Cross-sectional optical micrograph of bonded sample (bonding at 350 °C for 2 min [37]), where the bond line was identified as Au/Au$_5$Sn/Au. (Reproduced by permission of IMAPS-Europe.)

10.4 Application of SLID Bonding

10.4.1 Cu–Sn Bonding

Cu–Sn SLID bonding for the electronics industry was first developed focusing on high-density interconnects for stacked three-dimensional IC systems, initially developed by IZM Fraunhofer in Germany. In particular, as the I/O density increases, the interconnect pitch shrinks, and there was a desire to replace solder soft bumps with interconnects of Cu–Sn IMC. A significant amount of characterization was demonstrated by Infineon in Germany led by Huebner et al. and a very good overview of this work and related applications can be found in [1].

Both Intel and Samsung have demonstrated prototype IC stacks using Cu–Sn SLID bonding [39]. However, lately Cu–Sn SLID bonding has also shown great promise for hermetic sealing for microelectromechanical systems (MEMS) devices, which is discussed in Section 10.4.1.2.

10.4.1.1 Through-Si Vias and Interconnects Using Cu–Sn SLID

Three-dimensional stacking, and especially in conjunction with vertical system integration technology, with no need for additional process steps at the stack level was early identified as one of the main applications for Cu–Sn SLID bonding. The compatibility with standard BEOL layer Cu interconnects and Cu through-Si vias (TSVs) make it very appealing to include a Sn layer for bonding these stacks.

One of the earlier demonstrations of this application was performed by Ramm et al. The interchip via (ICV)-SLID concept is based on bonding of top chips to a bottom wafer (C2W) by very thin Cu–Sn pads (thickness approximately 8 µm) which provide both the electrical and the mechanical interconnect, as shown in Figure 10.16 [40]. Note that the TSVs in this case are not Cu, but W. The stack was bonded at a temperature of 300 °C, using bonding pressure, and complete transformation of Sn to Cu$_3$Sn was obtained.

Tanida et al. demonstrated a slight variation of Cu–Sn SLID bonding in which a small amount of Ag was incorporated into the Sn layer, as shown in Figure 10.17

Figure 10.16 ICV-SLID technology: focused ion beam image of a three-dimensional integrated test structure, showing a cross section of a 10 μm thin chip with W-filled ICVs connected to the bottom device wafer by a SLID metal system (Cu, Cu$_3$Sn ε-phase, Cu) [40].

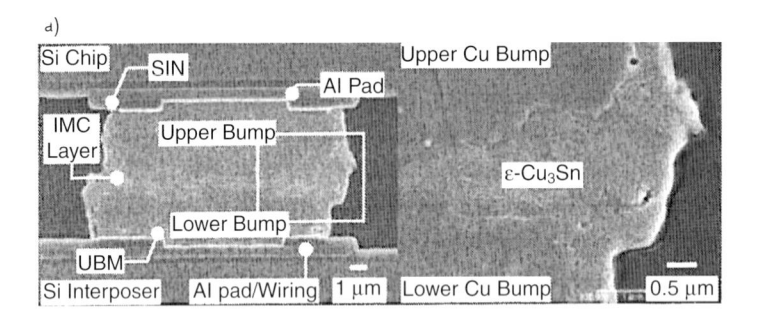

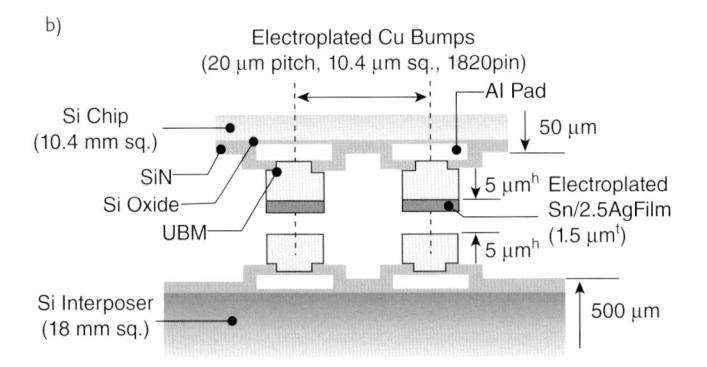

Figure 10.17 (a) SEM cross-sectional image of an interconnect bonded at 350 °C; (b) schematic of the micro-Cu bumps [6]. (Reproduced by permission of APEX/JJAP.)

[6]. This approach has also been incorporated by NEC, Oki, and Toshiba in Japan [39]. The interconnects had a pitch of 20 μm and targeted to be directly combined with Cu-based TSVs. Bonding (C2C) was performed using N_2 gas around the process area to prevent oxidation of the Cu and a force of 49 N for 60 s.

10.4.1.2 Hermetic Sealing at Wafer Level

MEMS and nanoelectromechanical systems (NEMS) devices often require vacuum conditions inside the package to operate at optimum performance, requiring a hermetic package. MEMS/NEMS devices are very fragile and sensitive after they have been released, thus no post-processing may be permitted after the micro- and nanodevices have been released and left free-standing on the wafer. Flux-free bonding, as discussed in Section 10.2.4, is crucial for these hermetic packages. Fluxless SLID bonding has been demonstrated using symmetric Cu–Sn bond frames [24, 26], or using a thin layer of Au deposited on Cu and Sn in the case of nonsymmetric Cu–Sn bonding [25]. Both of these approaches eliminate the requirement to include expensive modification of the bonding tools to use flux gases in the chamber, and also speed up the bonding process.

Yu et al. bonded square Cu–Sn bond frames with a width of 300 μm and length of 11 mm on 8-inch wafers [25] (Figure 10.18). The metal layers and bonding parameters are described in Section 10.2.4.1. The hermeticity was evaluated using He leak testing. This method is suitable since a cavity in the package is included to increase the volume to $0.02 \, cm^3$. The recorded leak rate is lower than $5 \times 10^{-8} \, atm \, cm^3 \, s^{-1}$, which satisfies the general MIL-STD 883E.

Yuhan and Le have successfully demonstrated a package design which incorporates both Cu vias and cavity hermetic sealing using Cu–Sn SLID bonding [19]. The bond frame is 200 μm wide, and encloses a $520 \times 540 \, \mu m^2$ square frame (Figure 10.19a). The bonding pressure is relatively low (200 N), to prevent squeeze-out of Sn, and the bonding temperature is 350 °C for 10 min. It is interesting to note that even with this high bonding temperature, the final bond line is not homogeneous Cu_3Sn (Figure 10.19b). The hermeticity was also verified using He leak detection and gross leak testing, although with a very small cavity ($0.001 \, cm^3$).

Figure 10.18 (a) Scanning acoustic image of a 300 μm wide Cu–Sn bond frame around a cavity; (b) cross-section of the bond line [25].

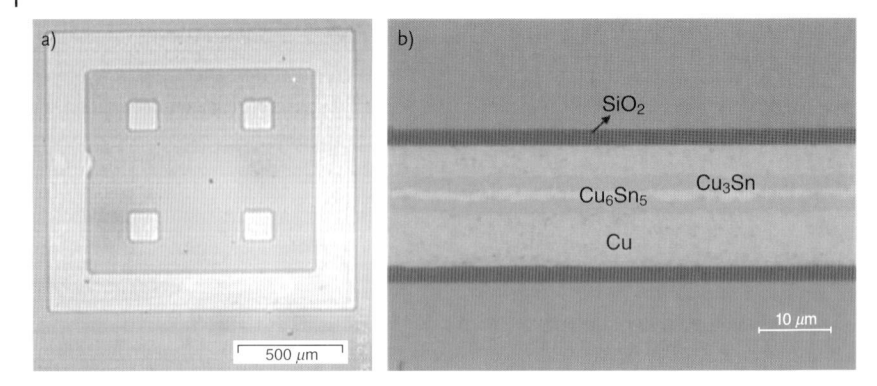

Figure 10.19 (a) Infrared image of a square bond frame; b) cross-sectional image of the bond line which consists of Cu_3Sn and Cu_6Sn_5 [19].

A recent example using a fluxless Cu–Sn bonding process for encapsulation of sensitive pixels on a bolometer was demonstrated by Lapadatu et al. (Figure 10.20). For bolometers to operate at optimal performance, the vacuum level in the package must be below 0.1 mbar. This can be obtained using a getter inside the cavity. These getter materials are activated by a thermal annealing step, typically at 350 °C. This restricts the vacuum encapsulation process: (i) the bonding temperature must be low enough to be compatible with the CMOS wafer as well as the pixel materials and structure, while at the same time the joint must resist the temperature of the subsequent getter activation; (ii) the bonding frames must withstand the etching of the sacrificial layer for the release of the pixels; and (iii) due to the presence of the released pixels, no wet chemical treatment of that wafer is allowed prior to bonding. As described earlier, Cu–Sn SLID bonding fulfills these requirements related to both material compatibility and temperature, and by using Sn to protect the Cu layer from oxidizing the process does not require pre-bonding cleaning of the metal surfaces [24, 26]. Furthermore, an innovative design for the bond frame was applied in which the frame on the top wafer (the cap) consists of a solid metal line, and the frame on the bottom wafer (device wafer) comprises several parallel rails. The empty spaces between the rails are intended for the confinement of the potential excess Sn flow when the temperature increases above the Sn melting point.

10.4.2
Au–Sn Bonding

Au–Sn SLID bonding has the advantage over Cu–Sn SLID bonding in that the remaining Au layer is softer than Cu, and can thus act as a stress-absorbing layer. Au–Sn SLID bonding can therefore be an alternative for bonding of dissimilar materials, with different coefficient of thermal expansion (CTE). SLID bonding in general is particularly useful for high-temperature applications, such applications

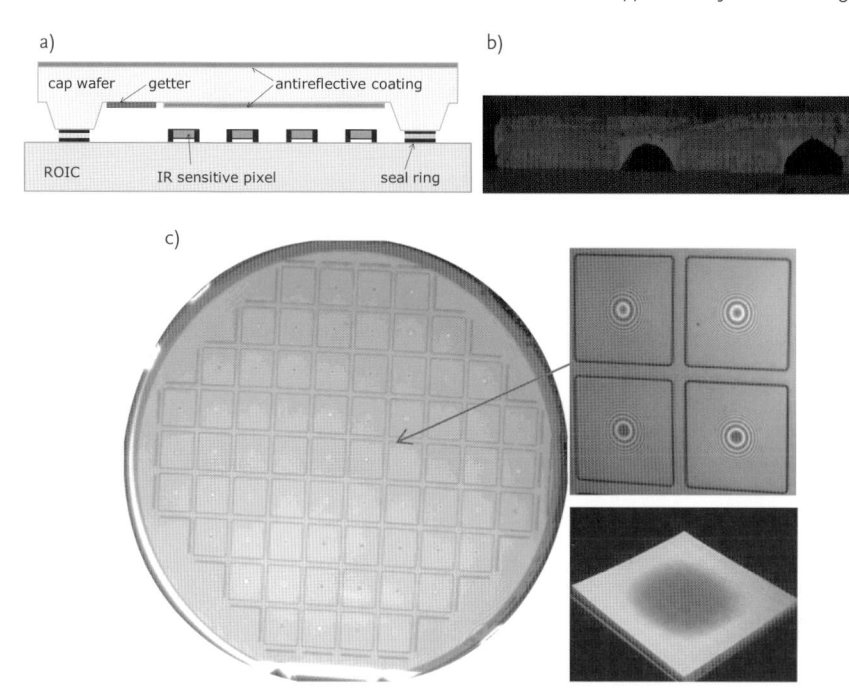

Figure 10.20 (a) Bolometer pixels encapsulated by Cu–Sn wafer-level bonding. (b) Cross-sectional photograph of a bond frame which combines divided and continuous sections. (c) Infrared image of 150 mm wafer pair containing Cu–Sn frames bonded in vacuum. The cap deflection under atmospheric pressure demonstrates that the pressure in the bonded cavities is low. Interferometer measurements (example shown at lower right) confirm the deflection [26].

often using SiC or other high-temperature components. Bonding of SiC components to substrates is therefore a particular case where Au–Sn SLID bonding is of interest [5, 14, 41, 42].

Johnson and coworkers have studied SLID-like Au–Sn bonding [5, 41, 42], focusing on die attach for SiC chips to substrates (AlN and direct bonded copper (DBC)), thus performing the bonding on a system with CTE mismatch. Their metallization structure deviates from the "typical" SLID structure in that it is unsymmetric (different thickness of Au on substrate and chip), such that there is not necessarily an Au buffer layer on each side of the IMC after bonding. They use several different, rather complex metal systems. The actual bonding metal systems make use of preforms: either eutectic preforms (80 wt% Au) sandwiched between Au layers [5, 41], or "off-eutectic preforms" (100 μm thick Au foil with 4 μm plated Sn on both sides) [42]. The resulting microstructure, as investigated by cross-sectional microscopy and EDX analysis, is complex. The lack of Au buffer layer on both sides of the IMC apparently leads to Sn reactions both with the UBM layers and with the Cu metallization (for DBC substrate).

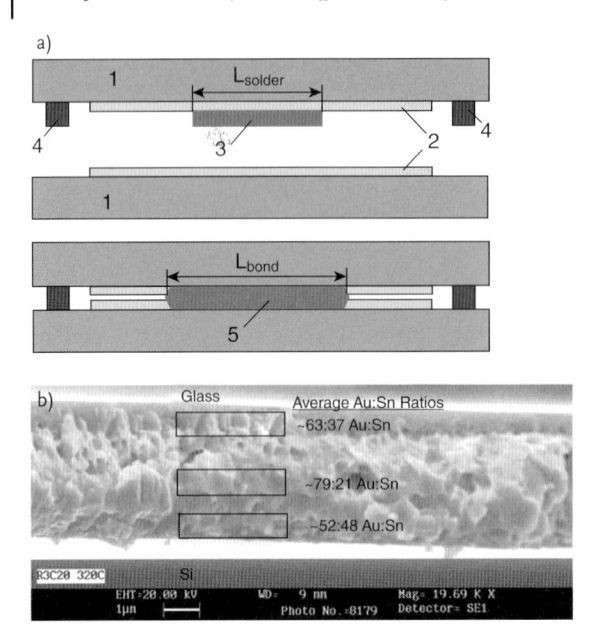

Figure 10.21 (a) Schematic of electroplated solder and Au layer. The stand-offs (numbered "4") control the final thickness of the bond line. (b) Au–Sn compositions measured at different locations of the bond line [43].

To the best of our knowledge, true Au–Sn SLID bonding has not yet been studied at the wafer level. However, the sub-processes involved (photolithography, electroplating, and bonding) are well suited for wafer-level implementation. Since Au–Sn solder processes are readily performed at the wafer level (Chapter 7), Au–Sn SLID bonding has a similar potential of wafer-level bonding implementation.

Belov *et al.* demonstrated wafer-level Au–Sn bonding, using thin bond lines (2–3 µm) for sealing and interconnections [43]. In order to ensure appropriate spacing, stand-offs were used on the wafers which defined the bond line thickness, and reduced any unwanted squeeze-out of liquid Sn, as shown in Figure 10.21a. The metal layers were deposited by electroplating, and bonding performed at 300–350 °C. They were able to produce strong bonds and sealed cavities proven to be liquid-proof. Their Au–Sn metal system contained 65% Au/35% Sn, hence being on the Sn-rich side of the eutecticum (Figure 10.21b). Their processes and layer thicknesses are very similar to those used for SLID bonding, except for the Au/Sn ratio. If true high-temperature stability is desired, the metal system should be designed to have a larger Au/Sn ratio, as discussed in Section 10.3.1.

10.5
Integrity of SLID Bonding

Compared to eutectic soldering, interconnects with pure intermetallic will be stronger, but also more brittle. Reliability investigations on high-temperature storage, temperature cycling, and high-humidity testing, in addition to electromigration tests, have been carried out by several researchers. Further details can be found in [1]. The electrical performance and mechanical strength both for interconnects and bond frames are of importance. Void formation in the bond line can be detrimental to mechanical integrity and reliability.

10.5.1
Electrical Reliability and Electromigration Testing

Huebner et al. tested the electrical reliability using chains of 500 standard Cu–Sn SLID bonded interconnects. This was done to resemble a product with high I/O and interconnect density. Bonded interconnect measuring $15 \times 15\,\mu m$ had a contact resistance of $7.5\,m\Omega$ [20]. After high-temperature storage and temperature cycling, the integrity of the interconnects was maintained, and the measured resistance values only dropped by about 2% [17].

Earlier electromigration tests of Cu–Sn IMC joints at 150 °C and $0.6\,mA\,\mu m^{-2}$ have already shown an increased electromigration performance compared to traditional solder flip chip interconnections [44]. While at 150 °C electromigration damage is triggered in solder joints for current densities as low as $0.06\,mA\,\mu m^{-2}$, the IMC bumps stressed with a 10 times higher current density do not show any failures, or even degradation, after more than 1000 h [7]. Chao et al. both simulated and experimentally verified that electromigration stressing enhanced the IMC growth in joints that had both Cu_3Sn and Cu_6Sn_5 present. The conversion to Cu_3Sn followed a reaction-controlled mechanism compared to the diffusion-controlled mechanism under thermal aging [45].

10.5.2
Mechanical Strength of SLID Bonds

For a homogeneous bond line, Cu_3Sn is stronger than pure Sn and cracks usually propagate between Cu and Cu_3Sn, or at the interface between Cu_3Sn and Cu_6Sn_5 if both phases are present [18]. With a different modulus for Cu_6Sn_5 and Cu_3Sn (Table 10.3) the strength of joints will therefore be a function of the intermetallic phases present in the bond line. Bond lines that comprise a homogeneous Cu_3Sn layer will thus have higher shear strength than bond lines with $Cu_3Sn/Cu_6Sn_5/Cu_3Sn$ structures [21]. Tanida et al. also reported that the tensile strength increases with increasing bonding temperature, reaching a maximum at 350 °C, at which point the bond line is homogeneous Cu_3Sn [6]. Dimcic et al. observed that the formation of cracks is very sensitive to the cooling rate after aging (thermal anneal-

Table 10.7 Measured mechanical strength of Cu–Sn SLID bonds.

Bonding parameters		Average shear strength (MPa)	Maximum/minimum shear strength (MPa)[a]	Source
°C	MPa			
350	200 N	19.5	13.4 to 25.3	[19]
250	96	60	ND	[46]
280	5.5	75.2	ND	[25]
275	25.3	9	3.6 to 16.9	[21]
–[b]	–	19	ND	[20]
260	20	39.8	16 to 80	[24]
350	1.25	20	10 to 27	[47]

a) ND, not disclosed.
b) 5 mm × 5 mm chip area of 10 μm contacts with 20 μm pitch; Sn thickness: 3 μm.

ing). The tendency for crack formation is increased with higher cooling rate and higher thermal stress. This offers the possibility of prolonging the lifecycle of the device in which these interconnects are implanted by controlling their heating and cooling rate during operation [23]. Table 10.7 presents a collection of measured mechanical strengths of Cu–Sn SLID bonds reported in the literature. Note the large spread in values, which may be due to sample structure and preparation, as well as IMC composition in the bond lines.

For Au–Sn SLID bonds, very high shear strength (>70 MPa) has been reported by different researchers [14, 42].

10.5.2.1 Temperature Cycling/Annealing Tests

For most Cu–Sn bonded dies and bond frames which are thermally annealed or cycled, the mechanical integrity is maintained, and the spread in the measured strength is reduced. In cases where the SLID bond is not completed (incomplete conversion to Cu_3Sn), a thermal process will further drive the diffusion process to create homogeneous bonds with only Cu_3Sn and Cu present, for which the shear strength will increase.

Table 10.8 presents some annealing and temperature cycle tests found in the literature.

Johnson and coworkers [5, 41, 42] performed thermal storage tests for SiC chips bonded, using the Au–Sn SLID method, to DBC and AlN substrates. The bonds survive this thermal storage without significant deterioration of the bond strength. Bonds surviving severe thermal cycling tests (35 to 500 °C, 1000 cycles) were also demonstrated [42], even when bonding materials with different CTE (SiC and AlN). The CTEs of the two materials are quite closely matched (4.0 and 4.5 ppm °C^{-1}, respectively), but are not equivalent. Whereas CTE mismatch is considered a major challenge for SLID bonding, due to the brittle nature of IMCs, the remaining Au layer in a Au–Sn SLID bond is believed to contribute to the ductility for stress absorption.

Table 10.8 Annealing and temperature cycle tests performed on Cu–Sn SLID bonds.

Test performed			Comments	
Temperature storage	175 and 200 °C	1000 hours	No significant change in measured resistance; 2% reduction indicates reaction of IMC	[17]
Temperature cycling	−65 to 150 °C	1000 cycles	Morphology remained unchanged; >2.5% change in resistance	[17]
Temperature cycling	−40 to 150 °C	1000 cycles	Small reduction in shear strength (39.8 to 33.3 MPa)	[24]
Annealing	350 and 400 °C	30 minutes	No loss of vacuum; integrity is intact	[26]
Temperature cycling	−40 to 125 °C	3900 cycles	No open failures detected	[7]

An ongoing study [14] has confirmed that Au–Sn SLID bonding of SiC chips to substrates with CTE mismatch can indeed survive thermal cycling tests.

10.5.2.2 Void Formation

Voids are often found in the region of Cu_3Sn and near the interface between Cu_3Sn and Cu. If voids are present in the region of brittle material, cracking and fractures may occur. This is also the reason why most of the fractures observed are located between the Cu and IMC layer. The exact origin of these voids remains a topic for discussion, and a plethora of articles can be found discussing the source, but it is generally agreed that they are a combination of Kirkendall voids and impurities left in the metals present before bonding. The impurities may come from organic residual material, especially for electroplated metals, but also from trapped oxides.

The Kirkendall voids arise from an imbalance in flux of Cu atoms which have to diffuse through the Cu_3Sn layer to react with the Cu_6Sn_5 layer, and the corresponding reaction and flux of Sn atoms between Sn and Cu_6Sn_5 and through the Cu_3Sn to the Cu interface. This interfacial reaction is increased at elevated temperature, which results in vacancies in the Cu and Cu_3Sn layer, which can be observed as voids. The number and size of the voids have been observed to increase with increasing time and annealing temperatures [48]. To suppress the formation of the microvoids, a Cu pad with small grains is preferred to disperse the vacancies into the grain boundaries [49].

Figure 10.22 shows an SEM cross section of a Cu–Sn SLID joint in which the small Kirkendall voids can be seen near the Cu–Cu_3Sn interface, in addition to entrapped impurities within the IMC phases [23].

Furthermore, it is believed that impurities from residual cleaning agents may increase the amount of Kirkendall voids in a joint. Tests carried out by Dimcic *et al.*, in which deposited Cu/Sn/Cu films were compared to bonded interconnects,

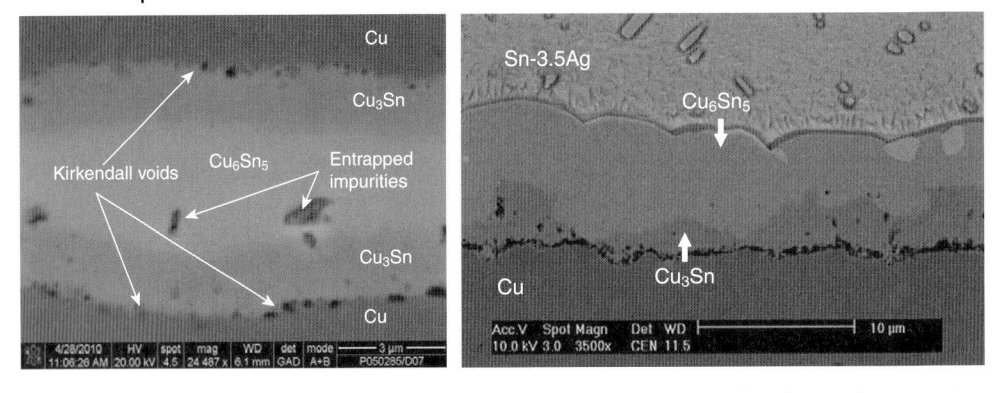

Figure 10.22 SEM cross-sectional images showing small Kirkendall voids near the Cu$_3$Sn–Cu interface [23]. (Reprinted from [50] with permission from Elsevier.)

reveal that the voids appear more randomly spread out for annealed films compared to voids merging together for the bonded interconnects, and do not form cracks [23]. This is supported by the work of Yu and Kim in which they compared the amount of Kirkendall voids present on pure Cu foils and electroplated films. For joints created using pure Cu foil, no voids were present, whereas for joints made from electroplated Cu films voids occurred at the Cu–Cu$_3$Sn interface. A clear relationship was found between the amount of S (from the electroplating bath) in the Cu film and the void nucleation rate [50].

Labie *et al.* studied the effect of these voids on the electrical performance. Small cracks appear between the particles that are entrapped at the bonding interface, but despite the large area of defects and particles, this does not result in a resistance increase of the *in situ* monitored daisy chains. Furthermore, after temperature cycling and high-temperature storage, the Cu–Cu$_3$Sn interface is almost completely covered with voids, as shown in Figure 10.23. Nevertheless, this degradation does not seem to affect the monitored resistance since most of the resistance arises from the Cu interconnect daisy chain tracks and the IMC bumps only contribute a small part (a few milliohms) to the total resistance (about 150 Ω) [7]. It is interesting to point out that although voiding is detrimental to the reliability and stability of Cu–Sn SLID interconnects, the overall performance of these bonds is surprisingly high. Ongoing research efforts are targeted towards reducing the amount of voiding.

10.6
Summary

SLID bonding is a bonding technique based on intermetallic formation that has attracted much interest in the industrial and scientific community due to the high-

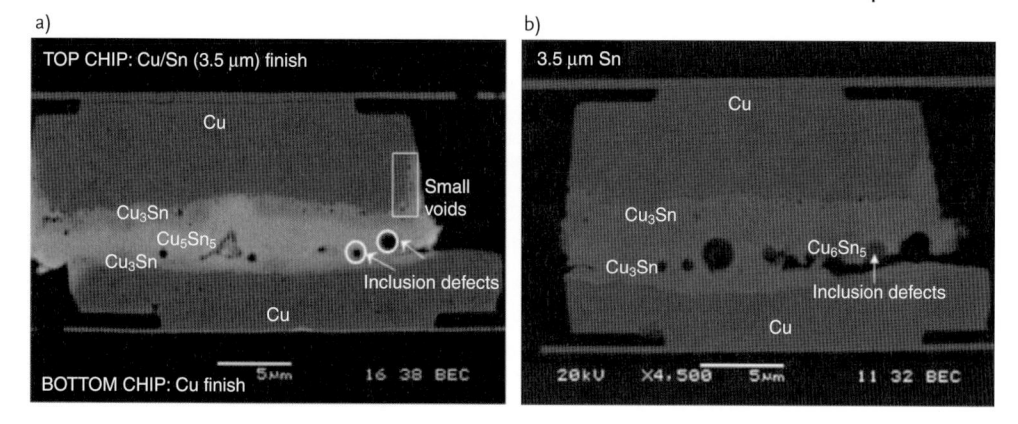

Figure 10.23 SEM cross-sectional image of Cu–Sn SLID joint (a) after bonding and (b) after thermal cycling for 1800 cycles between −40 and 125 °C [7]. The Cu_3Sn layer has grown at the expense of Cu_6Sn, and an increased amount of small voids at the Cu_3Sn–Cu interface is visible after thermal cycling.

temperature stability and possibility for repeated processing without melting of bonds. Cu–Sn is by far the most studied SLID system. Both metals are of low cost and are readily available. The materials have extensive heritage in the microsystem/microelectronic sector, as Sn is the basis of all soldering and Cu is a commonly used substrate and chip pad metallization in addition to BEOL and TSV interconnections. Au–Sn SLID bonding has also been demonstrated, and is of particular interest for CTE-mismatched systems and because a fluxless process is easily implemented.

Since the first demonstrations of Cu–Sn SLID bonding as an excellent interconnection method for three-dimensional and die-stacking applications, it has over recent years been demonstrated for wafer-level bonding with TSVs together with hermetic sealing and packaging at the wafer level. The bonding process can be fluxless which makes it very suitable for protection and packaging of sensitive MEMS and similar devices.

Compared with eutectic soldering, SLID interconnects with pure intermetallic will be stronger, but also more brittle. Reliability investigations such as shear strength after high-temperature storage and temperature cycling, in addition to electrical tests, have been – and continue to be – research topics. The electrical performance and mechanical strength both for interconnects and bond frames demonstrate very high yield and excellent performance. However, in particular for Cu–Sn SLID bonds, voids forming at the interface of Cu and IMC are detrimental to mechanical integrity and overall reliability. Ongoing research efforts are targeted towards reducing the amount of, as well as understanding the mechanism behind, void formation.

References

1 Munding, A. *et al.* (2008) Cu/Sn solid–liquid interdiffusion bonding, in *Wafer Level 3-D ICs Process Technology* (eds C.S. Tan, R.J. Gutmann, and L.R. Reif), Springer Science, New York, pp. 131–170.

2 Iino, Y. (1991) Partial transient liquid-phase metals layer technique of ceramic–metal bonding. *J. Mater. Sci. Lett.*, **10**, 104–106.

3 Duvall, D.S. *et al.* (1974) TLP bonding: a new method for joining heat resistant alloys. *Weld. J.*, **53**, 203–214.

4 Schmid-Fetzer, R.L. (1996) Fundamentals of bonding by isothermal solidification for high temperature semiconductor applications, in *Design Fundamentals of High Temperature Composites, Intermetallics, and Metal-Ceramic Systems* (eds R.Y. Lin, Y.A. Chang, R.G. Reddy, and C.T. Liu), TMS, Warrendale, PA, pp. 75–98.

5 Johnson, R.W. *et al.* (2007) Power device packaging technologies for extreme environments. *IEEE Trans. Electron. Packaging Manuf.*, **30**, 182–193.

6 Tanida, K. *et al.* (2004) Micro Cu bump interconnection on 3D chip stacking technology. *Jpn. J. Appl. Phys.*, **43**, 2264–2270.

7 Labie, R. *et al.* (2010) Reliability testing of Cu–Sn intermetallic micro-bump interconnections for 3D-device stacking. 3rd Electronic System-Integration Technology Conference (ESTC).

8 Massalski, T.B. (1990) *Binary Alloy Phase Diagrams*, ASM International, Materials Park, OH.

9 Kim, J.S. *et al.* (2007) Fluxless bonding of silicon to Ag-cladded copper using Sn-based alloys. *Mater. Sci. Eng. A*, **458**, 116–122.

10 Kim, J.S. *et al.* (2008) Very high-temperature joints between Si and Ag–copper substrate made at low temperature using InAg system. *IEEE Trans. Components Packag. Technol.*, **31**, 782–789.

11 Kim, J.S. *et al.* (2007) Fluxless bonding of Si chips to Ag-copper using electroplated indium and silver structures. 12th International Symposium on Advanced Packaging Materials: Processes, Properties, and Interfaces (APM 2007), pp. 199–204.

12 Okamoto, H. (2007) Au–Sn (gold-tin). *J. Phase Equilib. Diffus.*, **28**, 490.

13 Liu, H.S. *et al.* (2003) Thermodynamic modeling of the Au–In–Sn system. *J. Electron. Mater.*, **32**, 1290–1296.

14 Tollefsen, T.A. *et al.* (2011) Au-Sn SLID bonding – Properties and Possibilities. *Metall. Mater. Trans. B*, submitted.

15 Yang, P. *et al.* (2008) Nanoindentation identifications of mechanical properties of Cu_6Sn_5, Cu_3Sn, and Ni_3Sn_4 intermetallic compounds derived by diffusion couples. *Mater. Sci. Eng. A*, **485**, 305–310.

16 Bosco, N. and Zok, F.W. (2004) Critical interlayer thickness for transient liquid phase bonding in the Cu–Sn system. *Acta Mater.*, **52**, 2965–2972.

17 Huebner, H. *et al.* (2006) Microcontacts with sub-30 μm pitch for 3D chip-on-chip integration. *Microelectron. Eng.*, **83**, 2155–2162.

18 Bartels, F. *et al.* (1994) Intermetallic phase formation in thin solid-liquid diffusion couples. *J. Electron. Mater.*, **23**, 787–790.

19 Yuhan, C. and Le, L. (2009) Wafer level hermetic packaging based on Cu–Sn isothermal solidification technology. *J. Semiconductors*, **30**, 086001.

20 Huebner, O.E.H. *et al.* (2002) Face-to-face chip integration with full metal interface. Presented at the Advanced Metallization Conference (AMC 2002).

21 Huffman, A. *et al.* (2007) Effects of assembly process parameters on the structure and thermal stability of Sn-capped Cu bump bonds. Presented at the Electronic Components and Technology Conference.

22 Benkart, P. *et al.* (2007) Three-dimensional integration scheme with a thermal budget below 300 °C. *Sens. Actuators A*, **139**, 350–355.

23 Dimcic, B. *et al.* (2010) Influence of the processing method on the amount and

development of voids in miniaturized interconnections. 3rd Electronic System-Integration Technology Conference (ESTC).

24 Hoivik, N. *et al.* (2010) Fluxless wafer-level Cu–Sn bonding for micro- and nanosystems packaging. 3rd Electronic System-Integration Technology Conference (ESTC).

25 Yu, D.Q. and Thew, M.L. (2010) Newly developed low cost, reliable wafer level hermetic sealing using Cu/Sn system. 3rd Electronic System-Integration Technology Conference (ESTC).

26 Lapadatu, A. *et al.* (2010) Cu–Sn wafer level bonding for vacuum encapsulation of microbolometer focal plane arrays. *Electrochem. Soc. Trans.*, **33**, 73–82.

27 Lee, C.C. *et al.* (2009) Advanced bonding/joining techniques, in *Materials for Advanced Packaging* (eds D. Lu and C.P. Wong), Springer Science, New York, pp. 51–76.

28 Munding, A. (2007) Interconnect technology for three-dimensional chip integration. Doctoral thesis, Fakultät für Ingenieurwissenschaften und Informatik, Universität Ulm.

29 Farrens, S. (2008) Precision wafer-to-wafer packaging using eutectic metal bonding. Presented at the SMTA Pan Pacific Symposium, Hawaii.

30 Chanchani, R. (2009) 3D Integration technologies – An overview, in *Materials for Advanced Packaging* (eds D. Lu and C.P. Wong), Springer Science, New York.

31 Lee, C.C. and Chen, Y.-C. (1996) High temperature tin–copper joints produced at low process temperature for stress reduction. *Thin Solid Films*, **286**, 213–218.

32 Subramanian, B. *et al.* (2006) Structural, microstructural and corrosion properties of brush plated copper–tin alloy coatings. *Surf. Coat. Technol.*, **201**, 1145–1151.

33 Liu, H. *et al.* (2010) Intermetallic Cu_3Sn as oxidation barrier for fluxless Cu–Sn bonding. Proceedings of the 60th Electronic Components and Technology Conference (ECTC), pp. 853–857.

34 Yost, F.G. *et al.* (1990) Thermal-expansion and elastic properties of high gold–tin alloys. *Metall. Trans. A*, **21**, 1885–1889.

35 Chromik, R.R. *et al.* (2005) Mechanical properties of intermetallic compounds in the Au–Sn system. *J. Mater. Res.*, **20**, 2161–2172.

36 Aasmundtveit, K.E. *et al.* (2010) Au–Sn fluxless SLID bonding: effect of bonding temperature for stability at high temperature, above 400 °C. 3rd Electronic System-Integration Technology Conference (ESTC).

37 Aasmundtveit, K.E. *et al.* (2009) Au–Sn SLID bonding: fluxless bonding with high temperature stability to above 350 °C. European Microelectronics and Packaging Conference (EMPC 2009), pp. 1–6.

38 Wang, K. *et al.* (2008) Surface evolution and bonding properties of electroplated Au/Sn/Au. 2nd Electronics System-Integration Technology Conference (ESTC 2008), pp. 1131–1134.

39 Lu, D. and Wong, C.P. (eds) (2009) *Materials for Advanced Packaging*, Springer Science, New York.

40 Ramm, P. *et al.* (2003) 3D system integration technologies. *Mater. Res. Soc. Symp. Proc.*, **766**, E5.6.1–E5.6.12.

41 Johnson, R.W. and Williams, J. (2005) SiC power device packaging technologies for 300 to 350 °C applications. *Mater. Sci. Forum*, **483–485**, 785–790.

42 Zheng, P. *et al.* (2008) Metallurgy for SiC die attach for operation at 500 °C. Proceedings of IMAPS International Conference on High Temperature Electronics (HiTEC 2010), Albuquerque, NM, pp. 8–17.

43 Belov, N. *et al.* (2009) Thin-layer Au–Sn solder bonding process for wafer-level packaging, electrical interconnections and MEMS applications. IEEE International Interconnect Technology Conference (IITC 2009), pp. 128–130.

44 Labie, R. *et al.* (2008) Resistance to electromigration of purely intermetallic micro-bump interconnections for 3D-device stacking. International Interconnect Technology Conference (IITC 2008), pp. 19–21.

45 Chao, B. *et al.* (2006) Electromigration enhanced intermetallic growth and void formation in Pb-free solder joints. *J. Appl. Phys.*, **100**, 084909.

46 Lannon, J. *et al.* (2009) High density Cu–Cu interconnect bonding for 3-D integration. 59th Electronic Components

and Technology Conference (ECTC 2009), pp. 355–359.

47 Du, M. *et al.*(2005) Exploration of a new wafer-level hermetic sealing method by Cu/Sn isothermal solidification technique for MEMS/NEMS devices. *Proc. SPIE*, **5650**, 332–336.

48 Tang, W.-M. *et al.* (2010) Solid state interfacial reactions in electrodeposited Cu/Sn couples. *Trans. Nonferrous Met. Soc. China*, **20**, 90–96.

49 Peng, W. *et al.* (2007) Effect of thermal aging on the interfacial structure of SnAgCu solder joints on Cu. *Microelectron. Reliab.*, **47**, 2161–2168.

50 Yu, J. and Kim, J. (2008) Effects of residual S on Kirkendall void formation at Cu/Sn–3.5Ag solder joints. *Acta Mater.*, **56**, 5514–5523.

D. Hybrid Metal/Dielectric Bonding

11
Hybrid Metal/Polymer Wafer Bonding Platform
Jian-Qiang Lu, J. Jay McMahon, and Ronald J. Gutmann

11.1
Introduction

Three-dimensional (3D) hyper-integration[1)] is an emerging technology that can form highly integrated systems by vertically stacking and connecting together various materials, technologies, and functional components [1–4]. The potential benefits of 3D integration vary for each 3D platform; they include multifunctionality, small form factor, increased speed and data bandwidth, reduced power, reduced component packaging, increased yield and reliability, flexible heterogeneous integration, and reduced overall costs. Figure 11.1 shows schematic representations of major wafer-level 3D integration platforms pursued currently [1].

There are four key enabling technologies for 3D integration [1]: wafer alignment, wafer bonding, wafer thinning, and interstrata interconnection. The wafer alignment technology enables accurate alignment of the circuits or devices on two or more wafers. The wafer bonding technology enables two or more processed wafers to be thermomechanically bonded together. The wafer thinning technology enables the circuits or devices on two or more wafers to be closely connected by short interstrata interconnections. The interstrata interconnection technology (i.e., intertier, interchip, or interwafer interconnections as also called in the literature) enables electrical, thermal, and optical interstrata interconnects or fluidic channels between the circuits or devices on the stacked wafers. The interstrata electrical interconnection passing through thinned Si substrate is called a through-Si via, or, in general, a through-strata via (TSV). The interstrata electrical interconnections at the bonding interface could be called "bonded interstrata via" (BISV). Among the four key enabling technologies, wafer bonding technology enables not only the stacking of circuits or devices on two or more wafers, but also BISV and wafer thinning, thus a massive number of short TSVs.

As can be seen from Figure 11.1, currently there are four major bonding and interstrata interconnection approaches, as highlighted in Figures 11.1b–e [5–29]:

---

1) The term "hyper-integration" means to integrate various materials, processing technologies, and functions beyond ultralarge-scale integration (ULSI) or gigascale integration.

Handbook of Wafer Bonding, First Edition. Edited by Peter Ramm, James Jian-Qiang Lu, Maaike M.V. Taklo.
© 2012 Wiley-VCH Verlag GmbH & Co. KGaA. Published 2012 by Wiley-VCH Verlag GmbH & Co. KGaA.

11 Hybrid Metal/Polymer Wafer Bonding Platform

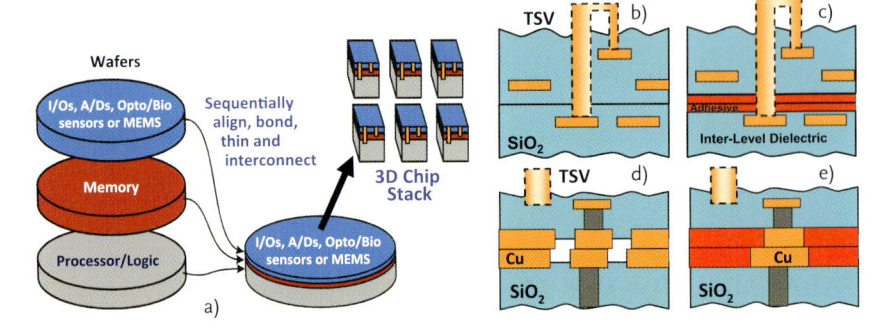

Figure 11.1 Schematics of a major wafer-level 3D integration platform with (a) four key 3D enabling processes, and (b–e) four major wafer bonding approaches. (Figure adapted from [1]. © 2009 IEEE.)

- oxide-to-oxide bonding (Figure 11.1a) [28, 29],
- adhesive (polymer) bonding (Figure 11.1b) [5–13],
- Cu-to-Cu bonding (Figure 11.1d) [14–19],
- bonding of hybrid metal/polymer redistribution layer (Figure 11.1e) [20–27].

Since the key advantage of 3D integration is to vertically integrate multiple circuits, devices, and/or systems on separately processed wafers, the following conditions to bond two wafers are desired for any 3D approaches, if not strictly required:

- compatible with back-end-of-the-line (BEOL) IC process, that is, with a bonding temperature at, or lower than, 400 °C;
- high thermal and mechanical stability of the bonding interface over the ranges of BEOL and packaging processing conditions;
- no outgassing during bonding to avoid void formation;
- seamless bonding interface with high bond strength to prevent delamination.

To satisfy the last condition, strong chemical bonds must be formed between the bonding surfaces over the entire wafer pair during the bonding process. This further requires (i) topography dictated by characteristics of the materials at the bonding interface (e.g., atomically flat bonding surfaces for oxide bonding) and/or (ii) diffusion of a massive number of atoms/molecules across the bonding interface during bonding or post-bond anneal.

Various wafer bonding approaches are described in this book. The hybrid wafer bonding approach using damascene-patterned metal/dielectric layers [30–32] involves combining metal bonding for direct electrical interstrata interconnections and dielectric (oxide or polymer) bonding for mechanical bonding strength, thus facilitating electrical and mechanical bonds between a pair of wafers. Specific details of a hybrid metal/polymer wafer bonding platform using damascene-patterned Cu/benzocyclobutene (BCB) layers as the wafer bonding intermediate layers are discussed in this chapter. Details of metal/oxide bonding [31, 32] are described in the following chapters. Since the Cu and BCB bonding are relevant to this chapter, their advances and advantages are summarized in the following.

For polymer bonding, particularly with BCB as the bonding polymer, as described in Chapter 3, major research advances include [5–13, 33, 34]:

- free of voids, defects, and structural damage;
- sufficiently high bond strength with a critical adhesion energy of the order of $30 \, \text{J m}^{-2}$;
- no degradation of electrical characteristics on wafers with Cu interconnect test structures and wafers with CMOS Si-on-insulator Cu/low-k devices and circuits after multiple bonding and thinning processes;
- significantly reduced wafer edge chipping during wafer thinning [33] and comparable chip sawing results to two-dimensional IC wafers [13];
- no degradation in bond strength after conventional die packaging reliability tests [13].

The key advantages of this approach include:

- the ability of the polymer to accommodate wafer surface nonplanarity (e.g., due to the last BEOL metallization) and particulates at the bonding interfaces;
- edge chipping protection during wafer thinning [33] and dicing [13];
- no handling wafers are required, that is, thinned Si is not transferred as in some other wafer-level 3D approaches;
- stacks of three or more wafers can be fabricated without changing the processing approach since the wafer bond with fully cured BCB sandwiched between two Si wafers is thermally stable up to 400 °C [7, 34].

For metal bonding, particularly with Cu as the bonding metal, as described in previous chapters, the key advances include that [14–19]:

- the Cu bond can be formed at a temperature of 350 °C or lower;
- the Cu grain passes though the bonding interface;
- the specific contact resistance of the bonded Cu is as low as $10^{-8} \, \Omega \, \text{cm}^2$.

The key advantages of this approach are:

- interstrata electrical vias, or BISVs, can be formed during the bonding process;
- density of BISVs (not TSVs) can be very high because they are very short; the via density is largely limited by the alignment accuracy.

11.2
Three-Dimensional Platform Using Hybrid Cu/BCB Bonding

The 3D technology platform using hybrid metal/polymer bonding [20–25, 35] employs wafer bonding of damascene-patterned metal/polymer redistribution

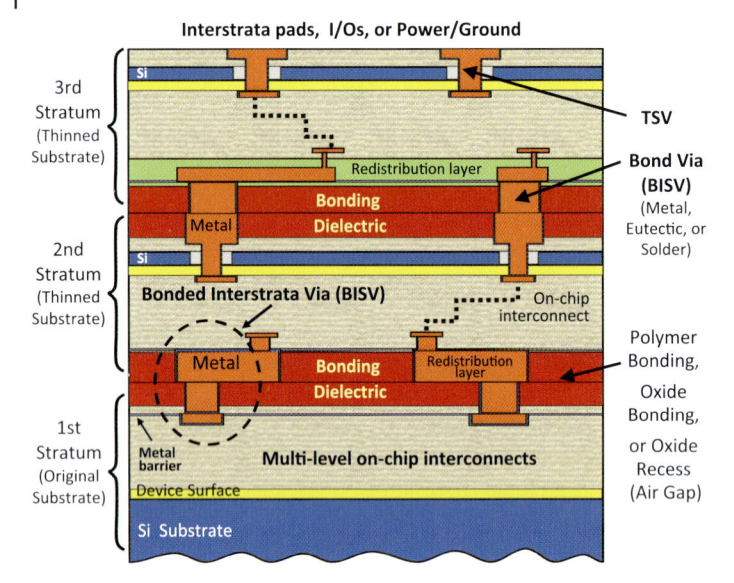

Figure 11.2 Schematic of a 3D platform using hybrid metal/dielectric bonding. For hybrid Cu/BCB redistribution layer bonding: Cu–Cu bonds provide interstrata electrical interconnects (i.e., BISVs) and BCB–BCB polymer bonds provide mechanical wafer attachment. Two redistribution layer options are shown on the second and third wafers, respectively [22].

layers on two wafers, thus providing interstrata electrical interconnects (i.e., BISVs) and polymer bonding of two wafers in one unit processing step. A conceptual schematic of this approach is illustrated in Figure 11.2 and key points are listed as follows:

- Cu/Ta and BCB are selected as the metal/liner and polymer for feasibility demonstration of this 3D platform.

- The bottom wafer (i.e., the first stratum) has a full wafer thickness as a base wafer to mechanically support the 3D fabrication processes of multiple strata of circuits/devices, which are fabricated separately on different wafers with their optimized materials and fabrication processes. Multicore processor chips with multilevel interconnects could be fabricated on this base wafer, serving as the central processing unit (CPU) for the whole 3D system. A heat sink could be attached to the backside of the processor chip, perhaps after thinning to about 200 μm for improved heat sinking as done with conventional two-dimensional ICs.

- A damascene-patterned Cu/BCB layer using chemical-mechanical planarization (CMP) technology is formed over the uppermost metal layer of the bottom wafer.

- A damascene-patterned Cu/BCB redistribution layer is formed over the uppermost metal layer of a second wafer. Memory chips with L3 cache and/or advanced DRAMs could be fabricated on this wafer, providing the memory needed for the CPU on the base chip.

- This second wafer is then flipped, aligned, and bonded to the patterned Cu/BCB layer on the base wafer. Note that the patterned Cu/BCB layer on the base wafer can also be a Cu/BCB redistribution layer if needed (not shown in Figure 11.2).

- The substrate of the face-down bonded second wafer is then thinned by mechanical grinding, CMP, and wet-etching.

- The process can be extended to multiple wafer stacks by etching through the thinned second wafer of the bonded pair to create another damascene-patterned layer, which mates with a third wafer. Note that the patterned Cu/BCB layer on the thinned second wafer substrate can also be a Cu/BCB redistribution layer if needed (not shown in Figure 11.2).

- An extra Cu/oxide (or Cu/BCB, or other metal/dielectric) redistribution layer, for example, that over the uppermost metal layer of the third wafer as shown in Figure 11.2, can be added prior to the patterning process of any Cu/BCB bonding layer, thus simplifying the patterning process of Cu/BCB bonding layer because only Cu bonding posts (vias) are needed. This approach also offers a simple bonding scheme, that is, with minimum misalignment one is always bonding Cu posts to Cu posts and BCB field to BCB fields, avoiding undesirable contact (i.e., bonding) of long Cu lines with BCB fields (see the bonding layers between second and third wafers as shown Figure 11.2). Moreover, this approach provides much more redistribution capability than that combining Cu bonding vias with the redistribution layer (e.g., the Cu/BCB redistribution layer on front side of the second wafer as shown in Figure 11.2).

This 3D platform using hybrid metal/polymer wafer bonding would provide:

- both electrical and mechanical interstrata connections/bonds (combining advantages of both BCB/BCB and Cu/Cu bonding);

- thermal management options: a Cu/BCB "redistribution layer" could serve as a good thermal conductor and/or spreader (with large percentage of Cu area), or as a poor thermal conductor (with large percentage of BCB area), based on local design considerations;

- high interstrata interconnectivity while allowing large alignment tolerance by eliminating direct bonding between deep interstrata vias (i.e., TSVs) and bonding pads;

- a "redistribution layer" as interstrata interconnect routing for wafers, on which the interstrata interconnect pads are not matched, which further reduces the process flow and is compatible with wafer-level packaging technologies;

- potential edge chipping protection during wafer thinning and dicing.

This platform is attractive for applications of monolithic wafer-level 3D integration (e.g., 3D interconnects, 3D ICs, wireless, smart imagers, etc.) as well as wafer-level packaging, passives, microelectromechanical systems (MEMS), optical MEMS, bio-MEMS, and sensors. In particular, we believe that the platform is attractive as an intermediate between wafer-level 3D ICs and more conventional wafer-level 3D packaging, where the fully functional individual wafers are equivalent to hard IP cores [36].

Clearly, the bonding process for such a hybrid metal/dielectric technology platform would be more challenging than simple Cu, oxide, or BCB bonding, as a variety of surfaces are exposed including the dielectrics, diffusion barriers, and electrical conductors. Ideally, all should be capable of being bonded to one another without interfering with the electrical characteristics of the Cu-to-Cu interconnection. Surface preparation techniques for improving adhesion of BCB to Si and silicon nitride as well as Cu have been discussed in the literature, but not with respect to a wafer bonding application. Further, wafer bonding of soft-baked BCB has been well documented for 3D applications [5–13, 33], as has damascene patterning of Cu in fully cured BCB [37, 38]. Based on what is known about Cu and BCB processing, a baseline process for a Cu/BCB hybrid bonding platform is developed and its details are discussed in the following section.

11.3
Baseline Bonding Process for Hybrid Cu/BCB Bonding Platform

The two main challenges for any hybrid metal/dielectric bonding are (i) selection of metals and dielectrics for damascene patterning and (ii) wafer-level feature-scale CMP to ensure bonding integrity and low resistance of electrical BISVs. For hybrid metal/polymer bonding, a Cu/BCB system is selected to demonstrate this bonding approach, where BCB is partially cured to enable Cu/BCB CMP [21, 23] because a partially cured BCB layer offers the best compromise between patterning capability and bond quality, as confirmed in the next section.

Figure 11.3 illustrates the key steps of the Cu/BCB wafer bonding process. Table 11.1 gives the typical 3D integration processing flow with hybrid wafer bonding using damascene-patterned metal/polymer redistribution layers. This wafer bonding process flow is a combination of BCB bonding and Cu bonding processes.

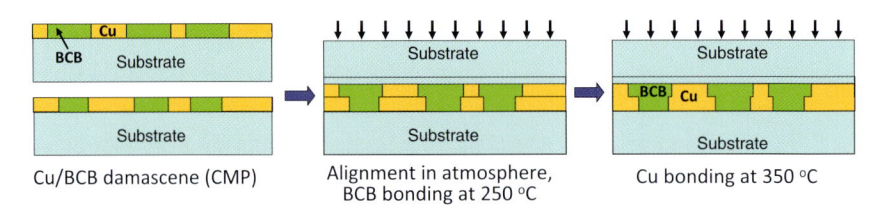

Figure 11.3 Simplified set of wafer bonding steps using damascene-patterned Cu/BCB surfaces.

Table 11.1 Typical 3D integration processing steps with Cu/BCB wafer bonding.

No.	Process step	Purpose of the process step
1	Wafer pre-cleaning and drying of the wafers	Remove particles, contaminations, and moisture from the wafer surfaces
2	Polymer application on wafers	Apply BCB uniformly (<2% nonuniformity)
	• Treating the wafer surfaces with an adhesion promoter (AP3000)	• Adhesion promoter can enhance the adhesion between the wafer surfaces and the polymer adhesive
	• BCB spin-coating and soft-bake at 170 °C for 1–5 min with N_2 flow	• Soft-bake removes solvents and volatile substances and hardens the surface to allow surface contact during wafer alignment
	• Partial cure of BCB at 250 °C for 60 s or more	• Partial cure results in >55% BCB crosslinking, enabling Cu/BCB CMP process [23]
3	Damascene-patterned metal/polymer bonding interface	Form metal bonding pads, posts, and "dummy" bonding surfaces
	• Cu/BCB damascene (CMP) and post-CMP brush cleaning	• CMP and post-CMP brush cleaning to form flat, clean surface across the wafers
4	Wafer alignment	Wafers are aligned with pre-patterned alignment marks, allowing TSV formation at the desired locations
5	Wafer bonding	Bond two wafers seamlessly
	• Vacuum pump to 0.1 mtorr	• Vacuum to prevent oxidation of metal and adhesives
	• Temperature ramp-up to 250 °C	• Allow BCB polymerization (crosslinking or cure)
	• Apply bond pressure of >0.3 MPa	• Force intimate contact of bonding surfaces
	• Hold for <60 min for bonding	• Cure the BCB to almost 100%
	• Temperature ramp-up to 350 °C	• Allow Cu interdiffusion and grain growth across the bonding interfaces, and eliminate small voids at the bonding interface. Optional anneal allows shorter bonding time in the bonding chamber
	• Hold for <60 min for bonding	
	• Release bond pressure, cool wafer, and unload (optional further anneal)	
6	Wafer thinning	Allow formation of high-density TSVs with micrometer diameter and relative low height-to-diameter aspect ratio (~5:1)
	• Top wafer back-grinding	
	• CMP and tetramethylammonium hydroxide etch	
7	TSV formation (similar to Cu damascene process)	Connect the ICs in the stack to next stratum and/or bring out I/Os from the stack
8	Repeat steps 1–7 for more strata	Form multiple strata of 3D ICs

The major differences are (i) partially cured BCB is used instead of soft-baked BCB to enable Cu/BCB damascene and (ii) BCB is still bonded at relatively low temperature of 250 °C as for soft-baked BCB bonding, followed by Cu bonding at elevated temperature of 350 °C. The detailed wafer bonding results and evaluation for this bonding approach are discussed in the following section.

11.4
Evaluation of Cu/BCB Hybrid Bonding Processing Issues

To evaluate hybrid metal/polymer wafer bonding, a via-chain structure is designed and fabricated following the baseline process flow as described in Table 11.1. The damascene-patterned Cu/BCB layer is fabricated as follows [21, 25]. BCB is spun onto oxidized 200 mm Si wafers using the manufacturer's specifications. These films are soft-baked at 170 °C for three minutes in nitrogen, followed by a partial curing process at 250 °C in nitrogen until the films are more than 55% crosslinked. These films are then photolithograhically patterned and etched using inductively coupled plasma etching. The etch process consumes photoresist at about the same rate as BCB, so the photoresist mask is fully etched away during this process. Following etching, the patterned BCB is filled with a sputtered Ta liner and sputtered Cu material for interconnects. Damascene patterning is accomplished using an IPEC 372M polisher with IC1400 k-groove pads and commercially available slurries for Cu and barrier removal. Both the Cu removal phase (i.e., the first-stage CMP) and the barrier removal phase (i.e., the second-stage CMP) consist of an abrasive slurry and an oxidizer. The Cu removal slurry mixture is specifically formulated to be selective over Ta, while the barrier removal slurry mixture is specifically formulated to be selective over Cu. After CMP, the damascene-patterned films are brush-cleaned in deionized water before aligning and bonding. After the baseline step 5 as shown in Table 11.1, bonded pairs are thinned in a three-step process: grinding and polishing, followed by a third step of wet processing that completely removes the top Si substrate. Optical microscopy and focused ion beam scanning electron microscopy (FIB-SEM) inspections of the bonded interfaces follow. Electrical measurements are conducted to characterize the contact resistance of the bonded Cu vias (i.e., BISVs).

11.4.1
CMP and Bonding of Partially Cured BCB

BCB films cured with different temperature and time cycles are evaluated for the baseline second-stage CMP process. BCB films on SiO_2/Si wafer are baked at 190, 220, and 250 °C to a crosslink percentage of 45 to 95%. Controllable CMP removal rate and uniformity can be obtained for BCB films that are cured at a temperature of 220 °C or higher to a crosslink percentage of 50% or higher [21, 23].

The four-point bending technique [9] is used to quantify the critical adhesion energy (i.e., bonding strength) of partially cured BCB bonded Si wafers under a

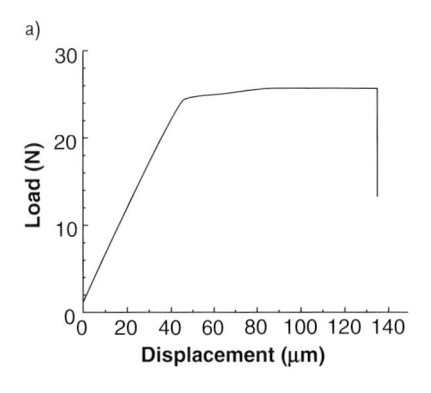

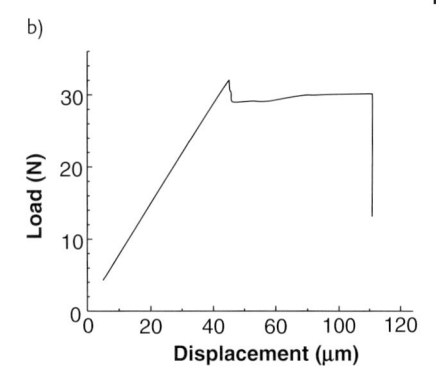

Figure 11.4 Load–deflection curves of four-point bending measurements for two specimens bonded with blanket partially cured BCB. (a) Typical result for wafers bonded using 50% crosslinked BCB followed by bonding at 250°C for 60 min. (b) Typical result for wafers bonded using this process, plus additional 60 min bonding at 350°C. Adhesion energy values are similar to those for soft-baked BCB [9]. (Adapted from [24]. © 2007 MRS.)

variety of surface preparation conditions [24]. Figure 11.4 shows typical load–deflection curves of four-point bending measurements for two specimens bonded with blanket partially cured BCB [24]. A load plateau region indicates quasistatic delamination of the weakest interface and allows extraction of a critical adhesion energy. For samples prepared under conditions such as partial cure, CMP/brush clean, and/or sulfuric acid dip, the typical delamination occurs at the BCB-to-SiO_2 interface (using DOW AP3000 as an adhesion promoter) and critical adhesion energy values are similar to that of soft baked BCB at about $30\,J\,m^{-2}$ [9, 24].

The high bonding strength and controllable CMP process confirm that partially cured BCB would offer the best compromise between patterning capability and bond quality for hybrid metal/polymer bonding using damascene-patterned Cu/BCB layers. More evidence can be seen from the bonding results as discussed in the following sections.

11.4.2
Cu/BCB CMP Surface Profile

In general, the CMP process is well developed for Cu/polymer damascene patterning, such as Cu/SiLK [39, 40] and Cu/BCB [37, 38]. However, since the surface topography of damascene-patterned Cu conductors with Ta liners and partially cured BCB dielectric plays an important role in the bondability of patterned wafers, the step heights of various features in patterned Cu/BCB wafers are inspected. Figure 11.5 shows a feature-level nonplanarity for a centrally located structure versus a structure near the edge of a 200 mm wafer [23]. The Cu features are slightly raised with respect to the partially cured BCB surface. Step heights over patterned features between Cu and BCB vary from 60 to 120 nm over the radius

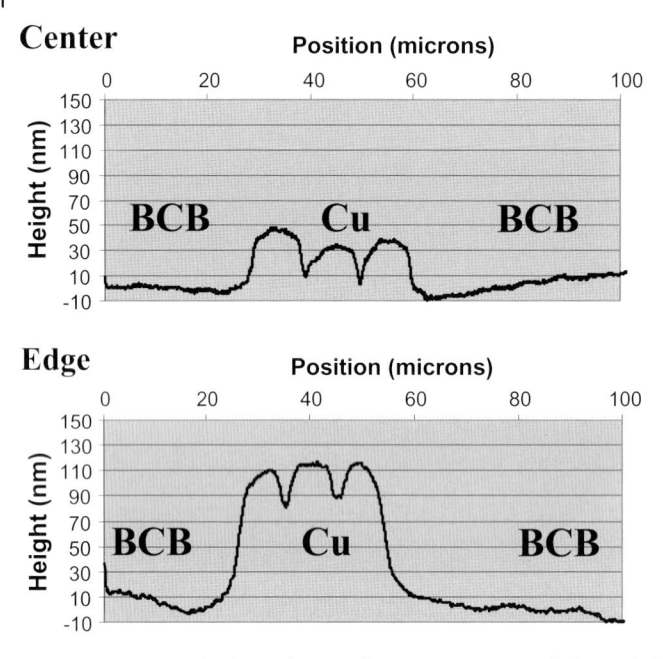

Figure 11.5 Feature-level nonplanarity for structure in centrally located die versus structure in die near the edge of 200 mm wafer [23]. (© 2005 MRS.)

of a 200 mm wafer pair, with some well-bonded areas and some poorly bonded areas. This implies that good bonding can be obtained when step heights of features with a pitch of about 10 μm are as large as 60 nm.

Further work evaluated the effect of wafer-scale topography by employing a nonideal die layout [41]. Dies were placed with closely spaced rows, but widely spaced columns. As a result, the BCB field was high relative to the Cu-patterned regions. This nonplanarity was approximately 500 nm over a distance of several dies. Although this large nonplanarity can easily be prevented by closely spaced Cu/BCB patterns, even these nonplanarities may not be a barrier to good bondability.

11.4.3
Hybrid Cu/BCB Bonding Interfaces

Observation of Cu–Cu, BCB–BCB, and Cu–Ta–BCB bonding interfaces is accomplished using two methods: (i) by milling away material using a dual-beam FIB-SEM instrument and (ii) optical observation through a transparent film or substrate. Cross-sectional FIB-SEM is useful for inspecting bonding interfaces, such as grain structure in post-bonded Cu–Cu and Cu–Ta–BCB interfaces as well as interfacial voids. Interstrata via-chain structure is optically visible after the damascene-patterned Cu/BCB bonding and wafer thinning steps are completed if the entire top substrate is removed during thinning. Optical observation of BCB–BCB defects is possible in areas where BCB–BCB topography is not accommodated.

Figure 11.6 shows several FIB-SEM cross sections of an interstrata via-chain structure fabricated by damascene patterning Cu/Ta into 50% crosslinked BCB on two wafers and bonding the pair using the baseline bonding process. Figure 11.6a presents a selected long bonding interface of the via-chain structure, showing Cu–Cu bonding, BCB–BCB bonding, and voids at BCB–BCB bonding interfaces. Figures 11.6b, c, and f show details of three Cu–BCB bonding interfaces. Figures 11.6d and e show more details from Figures 11.6c and f, respectively. Note that the BCB–BCB bonding is performed at 250 °C; reliable Cu–Cu bonding would not occur at this temperature until the temperature is ramped up to 350 °C as previously published results [17] indicate that a temperature greater than 300 °C is required to achieve lower than 1% die failure during dicing of Cu–Cu bonded structures, and that an interface-free bond is achievable when bonding and annealing temperatures are at or greater than 350 °C for an hour or more.

All the FIB-SEM cross sections show (i) excellent Cu–Cu bonding as Cu grains pass across the original Cu–Cu bonding interface and (ii) seamless BCB–BCB bonding. All the FIB-SEM cross sections also show that Cu voids tend to form at the Cu/Ta line edge (see Figure 11.6d and region B of Figure 11.6e). These Cu voids are known to be often formed after the CMP process, so-called "Cu line edge trenching" [42]; they could also be formed due to thermal stress during the thermal processing, such as post-CMP anneal or high-temperature bonding [24, 43]. This Cu void formation should not be an intrinsic issue for Cu/BCB bonding because it could be controlled by a combination of improved Cu filling, prior-CMP anneal, and CMP process.

Even more interesting results about these Cu voids at the Cu/Ta line edge are that:

- almost all the Cu voids at the Cu/Ta line edge on one wafer are filled by Cu diffusion from another wafer when Cu voids oppose the Cu surface (marked as region A in Figure 11.6e), indicating Cu fusion bonding;
- almost all the Cu voids at the Cu–Ta–BCB interface on one wafer are not filled by BCB from another wafer when Cu voids oppose the BCB surface (marked as region B in Figure 11.6e), indicating no BCB reflow, thus minimizing the possibility of Cu bond contamination by BCB.

The Cu-to-Cu fusion bonding and no BCB reflow are clearly confirmed in all cases.

From Figure 11.6b, nanovoids are observed typically at Cu grain boundaries. These voids may be formed during Cu grain growth and could be enhanced or created by the ion and/or electron bombardment from FIB-SEM because the Cu grains are weakly contacted at the grain boundaries. These nanovoids could be prevented by improved Cu filling, prior-CMP anneal, CMP process, and post-bonding anneal.

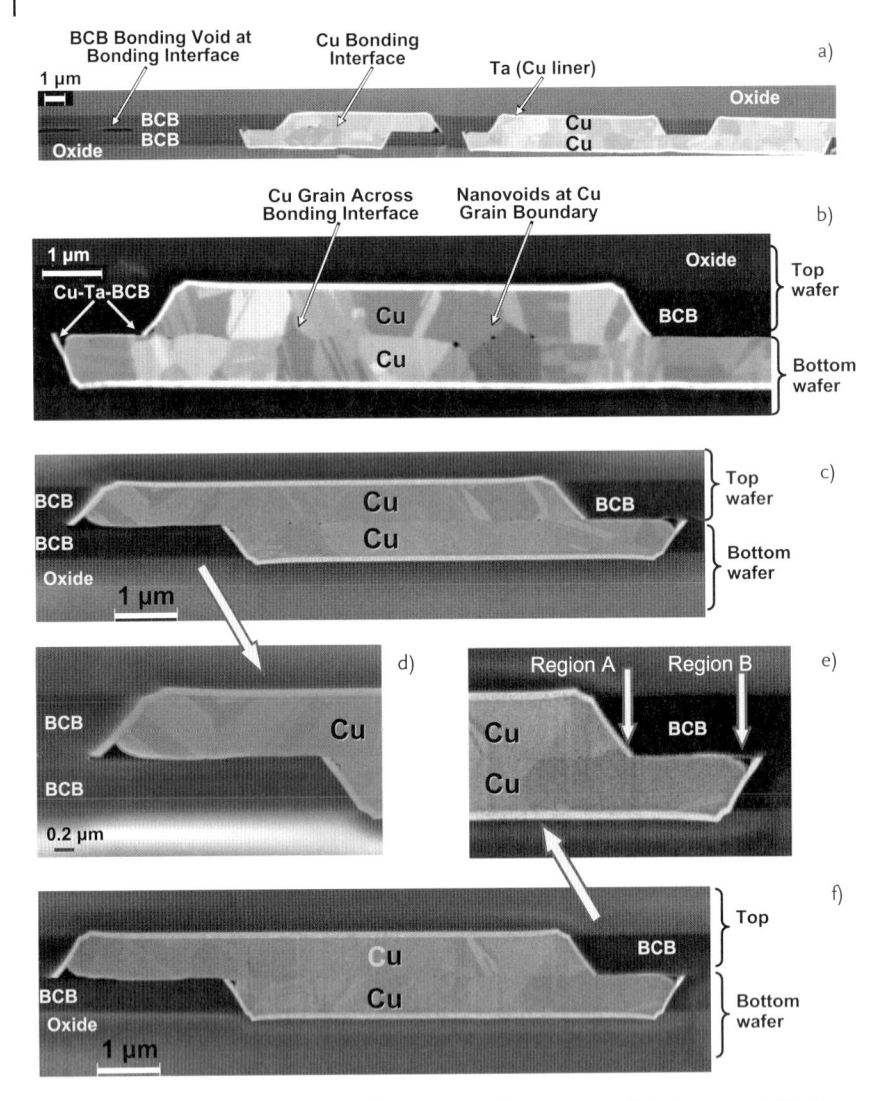

Figure 11.6 FIB-SEM cross sections of an interstrata via-chain structure using hybrid Cu/BCB bonding, showing: Cu–Cu bonding with Cu grains across the bonding interface; seamless BCB–BCB bonding; filled Cu voids on one wafer by Cu from another wafer when the Cu voids oppose the Cu surface as marked as region A in (e), indicating Cu fusion bonding; Cu voids (at Cu–Ta–BCB interface on one wafer) that are not filled by BCB when Cu voids oppose the BCB surface as marked as region B in (e), indicating no BCB reflow, thus minimizing possible Cu bond contamination by BCB; nanovoids at the Cu grain boundaries; and BCB bonding voids several micrometers away from the Cu BISV structure. See text for detailed description and discussion.

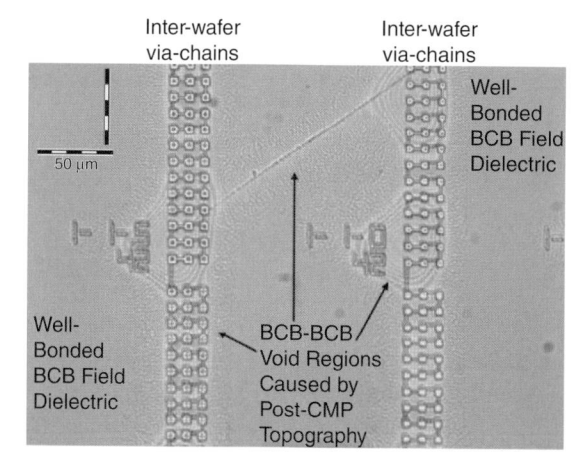

Figure 11.7 Optical micrograph of an interstrata via-chain structure using Cu/BCB bonding, showing an area where post-CMP topography is too large to allow well-bonded BCB–BCB interfaces in all areas. Micrograph with the scratch was chosen as an indication that BCB–BCB defect density can be controlled without Cu/Ta in the structure [25]. (© 2008 IEEE.)

BCB bonding voids are also observed several micrometers away from the Cu BISV structure as shown in Figure 11.6a. Voiding can also be observed in plan view via optical microscopy. Figure 11.7 shows defects in a region where post-CMP topography is too large to be accommodated. Voids are visible at the BCB–BCB interfaces near the Cu structure where post-CMP topography is expected due to dielectric loss [23, 44]. An area with a visible scratch is also shown in Figure 11.7 to illustrate that voiding can be produced in areas where no Cu structure exists. This suggests that the observed BCB–BCB defects can be caused by BCB topography alone.

The Cu nanovoids and BCB bonding voids are concerns, but both can be addressed in further development. In particular, optimized processes for Cu/BCB damascene patterning and bonding are expected to solve these problems. However, the good Cu/BCB bonding achieved with a relatively high Cu/BCB surface nonplanarity as shown in Figure 11.5 and discussed in Section 11.4.2 suggests that the partially cured BCB can accommodate certain Cu/BCB surface nonplanarity. Experiments have been designed and conducted to investigate the BCB accommodation mechanisms, as described in the following section.

11.4.4
Topography Accommodation Capability of Partially Cured BCB

Bonding experiments [25] utilizing structured Si wafers are depicted schematically in Figure 11.8. This procedure has a simplified material set in comparison to the full integration with via chains; however, it is useful in demonstrating how

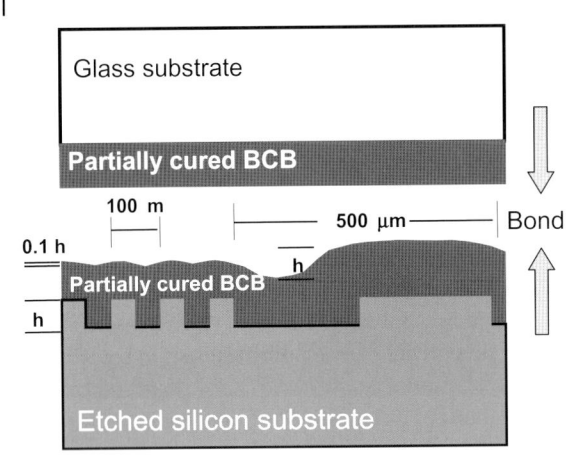

11 Hybrid Metal/Polymer Wafer Bonding Platform

Figure 11.8 Schematic cross section of bonded wafers used in the characterization of the topography accommodation capability of partially cured BCB. A blanket film of partially cured BCB on a glass wafer is bonded to a blanket film of partially cured BCB that has been coated over structured Si [25]. (© 2008 IEEE.)

partially cured BCB films behave under various bonding conditions, with varying surface step heights and feature sizes. Bare Si wafers are photolithographically patterned with lines of pitch varying between 1 and 500 μm, and etched in a SF_6 plasma. This process produces steps from 0.12 to 1.5 μm deep in the Si. These structured surfaces are then coated with BCB, which is partially cured at 250 °C to a degree of crosslinking between 50 and 90%. The degree of planarization (DoP) for the partially cured BCB film is expected to be about 90% for small features, based on previously published data [45].

Figure 11.9 compares two surface profile measurements of a structured Si wafer. These profiles represent the surface before and after BCB coating with a nominal BCB thickness of 1 μm, and the height shift is equal to the BCB film thickness. Note that the etched Si step height of 500 nm is translated to the same topography in the 250 μm wide features, that is, no planarization, but is reduced to about 50 nm in the 50 μm wide features, that is, a DoP of about 90%. These BCB-coated, structured Si surfaces are bonded against a flat BCB-coated glass wafer to examine the topography accommodation capability of the BCB. The glass wafer has a coefficient of thermal expansion close to that of the Si wafer. The bonding recipe used is identical to that described in Section 11.3. After bonding, optical inspection of the interface is possible because the top substrate (glass) is transparent.

The resulting observations include a difference in voiding presentations for differing crosslink percentages: BCB films crosslinked to a degree of 50 or 60% present voids that are typical of a solid/solid bond as well as "hazy" or "dendritic" type voids. Figure 11.10 shows optical micrographs of bonding of partially cured BCB over structured Si with 1.5 μm trenches. The high regions of the Si trenches are well bonded for 50–90% crosslinked BCB. Hazy defects are observed at the

Profiles for 50 and 250 μm Features after Si Etch and BCB Coating

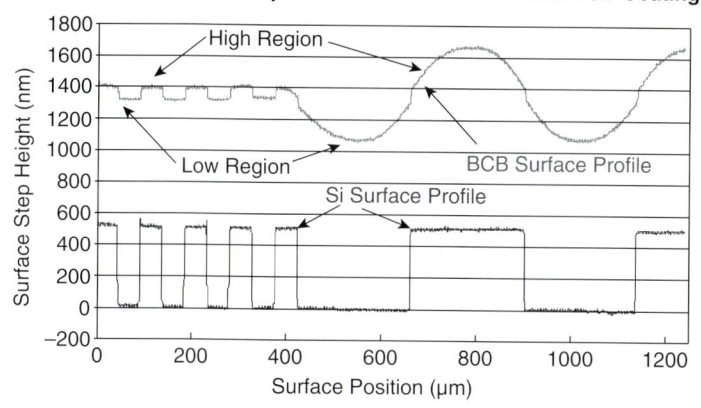

Figure 11.9 Profilometry measurement of the DoP for partially cured BCB spun over structured Si (compare to bottom wafer in Figure 11.8). Post-etch profile has lowest features at zero height; post-BCB profile has lowest features at about 1100 nm height (for a nominal BCB thickness of 1 μm). Features less than 100 μm are typically about 90% planarized [25]. (© 2008 IEEE.)

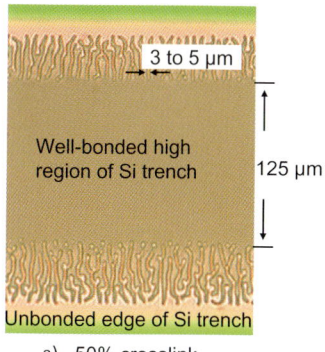

a) 50% crosslink

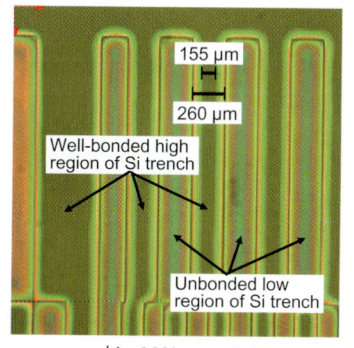

b) 90% crosslink

Figure 11.10 Optical images of bonding of BCB over structured Si with trenches etched to a depth of 1.5 μm. (a) Well-bonded high region of Si trench with voids encroaching on the edge of Si trench for a 50% crosslinked BCB. (b) Well-bonded high region of Si trench with well-defined edges and no bonding (void) at the low region of Si trench [25]. (© 2008 IEEE.)

edge of Si trenches due to "crosslink-percentage-controlled" voids as shown in Figure 11.10a. These small crosslink-percentage-controlled voids are consistently observed at the edge of features with steps that are too large to be completely accommodated in cases of 50% crosslinked BCB. Once the crosslinking percentage is raised to 70% or greater, this type of void is not observed; voids exhibit smooth edges after bonding as shown in Figure 11.10b for the case of 90%

crosslinked BCB. The BCB at low regions of the Si trench is not bonded due to the large feature size of the trench that BCB cannot accommodate.

Table 11.2 shows the progression of voiding types as observed in the structured bonding experiments as surface topography is increased [25]. Since both crosslink-percentage-dependent voids and feature-size-dependent voids are dependent on the step height, the initial Si step height at which voids disappear is of interest. For the smallest topography, all steps can be accommodated, resulting in a void-free structure as shown in the top image of Table 11.2, that is, both high and low regions of the BCB surface are well bonded. For the 50% crosslinked case, the BCB-coated structures first begin to exhibit accommodation with an initial Si step

Table 11.2 Progression of void classifications as the magnitude of topography increases. In the top row, all structure is bonded; in the middle row, the topography dictates the bonded area; on the bottom row, no area is bonded. The intermediate rows show where "hazy" or "dendritic" type voids appear when 50 and 60% crosslinked BCB is used in the structured bonding experiments.

Example	High region	Low region	Nomenclature
	Defect-free	Defect-free	Step accommodated
	Defect-free	Hazy dendritic	Step influenced
	Defect-free	Void	Patterned
	Hazy dendritic	Void	Externally influenced
	Void	Void	Externally dominated

Adapted from [25]. © 2008 IEEE.

height of 1 μm, but do not show consistent accommodation until the Si step is decreased to 500 nm. At that point, BCB surface topography of about 50 nm exists in the small features, and both high and low regions of the BCB surface are bonded. For cases where the level of crosslinking is increased to 70%, accommodation does not begin until the Si step is at 500 nm, and significant accommodation is not observed until this step is decreased to 120 nm. Thus, for 70% crosslinked BCB, the accommodation point is when the BCB surface topography of micrometer-scale features measures about 12 nm or smaller, that is, about 10% of the step height of 120 nm.

As the Si trench step height is increased in structured bonding experiments utilizing 50–60% crosslinked BCB, "hazy" or "dendritic" type voids begin to appear in the bottom of the patterned features. Once the topography dominates the capability for accommodation, voids follow the structure and create a patterned bonding layer. If structure external to that topography (such as a particle) provides influence, then it is possible to observe the "hazy" or "dendritic" type voids at the top of the features as well. Once that external structure dominates the capability for the partially cured BCB to bond, the interface will be completely unbonded. Not only is it noteworthy that this additional defect mechanism is observed in the 50 and 60% crosslinked BCB films, but also that the magnitude of topography accommodation (as defined by the number of dies exhibiting complete accommodation) is observed to be about four times greater in comparison to films with 70–90% crosslinking.

Defects at the partially cured BCB–BCB interface are found to be controllable based on the magnitude of the post-CMP topography. Since the magnitude of topography acceptance is found to be nearly constant for wafers constructed in 50 and 60% crosslinked BCB, at a level that is about four times greater than that found for 70–90% crosslinked BCB, the mechanism is attributed to deformation of the partially cured BCB film. The sharp difference in topography accommodation for the partially cured films used in this study coincides with the structural change between the sol–gel/rubber region and the gel–glass region of BCB according to its time–temperature transformation curve [46].

11.4.5
Electrical Characterization of Hybrid Cu/BCB Bonding

Electrical characterization of hybrid Cu/BCB bonding is conducted using four-probe measurements of specific contact resistance of the via-chain structure as illustrated in Figure 11.11 [24]. The cross-sectional area of the contacts is measured by optical microscopy and assumed to be uniformly bonded over areas that exhibit continuous chains. Specific contact resistance of the order of $10^{-7}\,\Omega\,cm^2$ is higher than reported values for bonded evaporated Cu [47] and for surface-activated bonding [48]. Contacts in the surface-activated bonding method are very clean, as *in situ* cleaning is done in an ultrahigh vacuum system just prior to alignment and bonding of the wafers. Contacts in the evaporated Cu approach have been reported to have oxygen content of the order of a few percent [49], but after annealing at

Chain Size (mm)	Number of Contacts	Resistance (Ω)	Contact Area ($cm^2 \times 10^{-8}$)	Specific Contact Resistance ($10^{-7} \Omega\text{-}cm^2$)
4	6	20	5	1.7
8	2	0.35	63	1.1
8	8	1.2	63	0.9
8	32	14	63	2.7

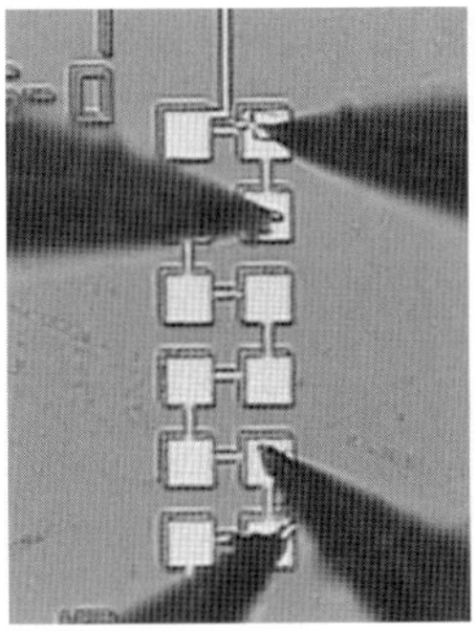

Figure 11.11 (a) Electrical testing of 12 μm pads (8 μm via interconnects). (b) Summary of several tested structures [24]. In all cases the bond process included bonding at 250 °C for 60 min followed by bonding at 350 °C for 60 min. (© 2007 MRS.)

400 °C this oxygen is diffused throughout the film. It is likely that the contact resistance reported in [47, 49] is lower than that for the hybrid metal/polymer bonding approach because of the difference in annealing temperatures and, particularly, the absence of post-CMP cleaning (as needed with Cu damascene patterning).

11.5
Summary and Conclusions

This chapter describes a hybrid metal/polymer bonding 3D platform with damascenepatterned Cu/BCB. Many aspects of the processing and characterization can be applied to other hybrid bonding platforms. The hybrid metal/polymer bonding platform combines the advantages of metal/metal bonding (for direct electrical interstrata interconnections) and polymer bonding (for robust thermo-mechanical wafer bonding strength as well as surface topography accommodation). The redistribution layers (Cu/BCB or other metal/dielectric combinations) at the bonding interface provide design tradeoffs between thermal management, BISV routing, and alignment relaxation.

Partially cured BCB offers the best compromise between patterning capability and bond quality, and thus enables the hybrid Cu/BCB wafer bonding 3D platform. Controllable CMP process can be achieved when BCB is cured to a degree of crosslinking of more than 50%; void-free bonding with high bond strength (i.e., about 30 J m^{-2}, similar to that of soft-baked BCB bonding) is possible with partially cured BCB up to a degree of crosslinking of 90% under various surface processing conditions.

A structured bonding procedure that allows isolation of the partially cured BCB topography accommodation capability demonstrates that voids begin to appear for different magnitudes of topography based on the pitch of the features and the crosslinking of the partially cured BCB. BCB with a degree of crosslinking of 70–90% is found to accommodate BCB surface topography depth of about 12 nm for features of about 1 μm with a nominal BCB thickness of 1 μm. Topography accommodation is observed to increase by a factor of about 4 when the BCB crosslinking is reduced to 50%.

The processing flow of the hybrid metal/polymer bonding is a combination of polymer-to-polymer bonding and metal-to-metal bonding, without introducing extra difficult processing steps. A via-chain structure is designed, fabricated, and characterized to demonstrate this 3D platform. Cu–Cu bonding with Cu grain across the bonding interface and seamless BCB–BCB bonding are demonstrated. A contact resistance of the order of $10^{-7}\,\Omega\,cm^2$ is obtained. Although BCB bonding voids, nanovoids at Cu grain boundaries, and voids at Cu–Ta–BCB interface (i.e., Cu/Ta line edge) are found and the contact resistance is still high, these problems could be solved by process optimization of Cu/BCB damascene patterning and bonding.

These promising results demonstrate the feasibility of this 3D technology platform using hybrid metal/polymer bonding. This platform is attractive for all applications of monolithic wafer-level 3D integration and wafer-level 3D packaging. In particular, we believe that the platform is attractive as an intermediate between wafer-level 3D ICs and more wafer-level 3D packaging, where the fully functional individual wafers are equivalent to hard IP cores.

Acknowledgments

This research has been partially supported by the Interconnect Focus Center, sponsored by MARCO, DARPA, and NYSTAR. The Interconnect Focus Center is one of five research centers funded under the Focus Center Research Program, a DARPA and Semiconductor Research Corporation.

References

1 Lu, J.-Q. (2009) 3D hyper-integration and packaging technologies for micro-nanosystems. *Proc. IEEE*, **97** (1), 18–30.
2 Garrou, P., Ramm, P., and Bower, C. (eds) (2008) *Handbook of 3D Integration: Technology and Applications of 3D Integrated Circuits*, Wiley-VCH Verlag GmbH, Weinheim.
3 Lu, J.-Q., Rose, K., and Vitkavage, S. (2007) 3D integration: why, what, who, when? *Future Fab Int.*, **23**, 25–27.
4 Meindl, J.D., Davis, J.A., Zarkesh-Ha, P., Patel, C.S., Martin, K.P., and Kohl, P.A. (2002) Interconnect opportunities for gigascale research. *IBM J. Res. Dev.*, **46** (2/3), 245–263.
5 Lu, J.-Q., Cale, T.S., and Gutmann, R.J. (2008) 3D integration based upon dielectric adhesive bonding, in *Wafer Level 3-D ICS Process Technology* (eds C.S. Tan, R.J. Gutmann, and R. Reif), Springer, pp. 219–256.
6 Niklaus, F., Stemme, G., Lu, J.-Q., and Gutmann, R. (2006) Adhesive wafer bonding. *J. Appl. Phys.*, **99** (3), 031101.
7 Lu, J.-Q., Kwon, Y., Kraft, R.P., Gutmann, R.J., McDonald, J.F., and Cale, T.S. (2001) Stacked chip-to-chip

interconnections using wafer bonding technology with dielectric bonding glues. IEEE International Interconnect Technology Conference (IITC 2001), 4–6 June 2001, pp. 219–221.

8 Lu, J.-Q., Cale, T.S., and Gutmann, R.J. (2005) Wafer-level three-dimensional hyper-integration technology using dielectric adhesive wafer bonding, in *Materials for Information Technology: Devices, Interconnects and Packaging* (eds E. Zschech, C. Whelan, and T. Mikolajick), Springer-Verlag, Berlin, pp. 386–397.

9 Kwon, Y., Seok, J., Lu, J.-Q., Cale, T.S., and Gutmann, R.J. (2006) Critical adhesion energy of benzocyclobutene (BCB)-bonded wafers. *J. Electrochem. Soc.*, **153** (4), G347–G352.

10 Lu, J.-Q., Lee, K.W., Kwon, Y., Rajagopalan, G., McMahon, J., Altemus, B., Gupta, M., Eisenbraun, E., Xu, B., Jindal, A., Kraft, R.P., McDonald, J.F., Castracane, J., Cale, T.S., Kaloyeros, A., and Gutmann, R.J. (2003) Processing of inter-wafer vertical interconnects in 3D ICs. *MRS Proc.*, **V18**, 45–51.

11 Lu, J.-Q., Jindal, A., Kwon, Y., McMahon, J.J., Rasco, M., Augur, R., Cale, T.S., and Gutmann, R.J. (2003) Evaluation procedures for wafer bonding and thinning of interconnect test structures for 3D ICs. IEEE International Interconnect Technology Conference (IITC 2003), June 2003, pp. 74–76.

12 Gutmann, R.J., Lu, J.-Q., Pozder, S., Kwon, Y., Menke, D., Jindal, A., Celik, M., Rasco, M., McMahon, J.J., Yu, K., and Cale, T.S. (2003) A wafer-level 3D IC technology platform. Advanced Metallization Conference (AMC 2003), pp. 19–26.

13 Pozder, S., Lu, J.-Q., Kwon, Y., Zollner, S., Yu, J., McMahon, J.J., Cale, T.S., Yu, K., and Gutmann, R.J. (2004) Back-end compatibility of bonding and thinning processes for a wafer-level 3D interconnect technology platform. IEEE International Interconnect Technology Conference (IITC 2004), June 2004, pp. 102–104.

14 Chen, K.-N., Lee, S.H., Andry, P.S., Tsang, C.K., Topol, A.W., Lin, Y.-M., Lu, J.-Q., Young, A.M., Ieong, M., and Haensch, W. (2006) Structure design and process control for cu bonded interconnects in 3D integrated circuits. Technical Digest of the IEEE International Electron Devices Meeting (IEDM 2006), December 2006, pp. 367–370.

15 Chen, K.N., Tsang, C.K., Topol, A.W., Lee, S.H., Furman, B.K., Rath, D.L., J.-Q. Lu, Young A.M., Purushothaman, S., and Haensch, W. (2006) Improved manufacturability of cu bond pads and implementation of seal design in 3D integrated circuits and packages. 23rd International VLSI Multilevel Interconnection (VMIC) Conference, September 2006, pp. 195–202, IMIC.

16 Chen, K.N., Fan, A., Tan, C.S., and Reif, R. (2002) Microstructure evolution and abnormal grain growth during copper wafer bonding. *Appl. Phys. Lett.*, **81** (20), 3774–3776.

17 Chen, K.N., Tan, C.S., Fan, A., and Reif, R. (2004) Morphology and bond strength of copper wafer bonding. *Electrochem. Solid-State Lett.*, **7** (1), G14–G16.

18 Morrow, P., Park, C.-M., Ramanathan, S., Kobrinsky, M.J., and Harmes, M. (2006) Three-dimensional wafer stacking via Cu–Cu bonding integrated with 65-nm strained-Si/low-k CMOS technology. *IEEE Electron Dev. Lett.*, **27** (5), 335–337.

19 Patti, R. (2006) Three-dimensional integrated circuits and the future of system-on-chip designs. *Proc. IEEE*, **94** (6), 1214–1222.

20 Lu, J.-Q., Jay McMahon, J., and Gutmann, R.J. (2007) Wafer bonding of damascene-patterned metal/adhesive redistribution layers. US Patent Application 20070207592, 6 September 2007.

21 McMahon, J.J., Lu, J.-Q., and Gutmann, R.J. (2005) Wafer bonding of damascene-patterned metal/adhesive redistribution layers for via-first 3D interconnect. 55_{th} IEEE Electronic Components and Technology Conference (ECTC 2005), pp. 331–336.

22 Lu, J.-Q., McMahon, J.J., and Gutmann, R.J. (2006) Via-first inter-wafer vertical interconnects utilizing wafer-bonding of damascene-patterned metal/adhesive

redistribution layers. 3D Packaging Workshop at IMAPS Device Packaging Conference, Scottsdale, AZ, 20–23 March 2006, paper WP64.
23. McMahon, J.J., Niklaus, F., Kumar, R.J., Yu, J., Lu, J.-Q., and Gutmann, R.J. (2005) CMP compatibility of partially cured benzocyclobutene (BCB) for a via-first 3D IC process. *MRS Proc.*, **867**, W4.4.1–W4.4.6.
24. Gutmann, R.J., McMahon, J.J., and Lu, J.-Q. (2007) Damascene patterned metal/adhesive redistribution layers. *MRS Proc.*, **970**, 205–214.
25. McMahon, J.J., Chan, E., Lee, S.H., Gutmann, R.J., and Lu, J.-Q. (2008) Bonding interfaces in wafer-level metal/adhesive bonded 3D integration. 58th Electronic Components and Technology Conference (ECTC 2008), May 2008, pp. 871–878.
26. Swinnen, B., Jourdain, A., De Moor, P., and Beyne, E. (2008) Direct hybrid bonding, in *Wafer Level 3-D ICs Process Technology* (eds C.S. Tan, R.J. Gutmann, and R. Reif), Springer Science, New York, pp. 257–267.
27. McMahon, J.J., Gutmann, R.J., Smith, G., and Lu, J.-Q. (2008) A wafer-level 3D integration technology platform using damascene patterned metal/adhesive hybrid bonding. Proceedings of 25th International VLSI Multilevel Interconnection (VMIC) Conference (MIC 2008), Fremont, CA, 28–30 October 2008, pp. 337–342.
28. Guarini, K.W., Topol, A.W., Ieong, M., Yu, R., Shi, L., Newport, M.R., Frank, D.J., Singh, D.V., Cohen, G.M., Nitta, S.V., Boyd, D.C., O'Neil, P.A., Tempest, S.L., Pogge, H.B., Purushothaman, S., and Haensch, W.E. (2002) Electrical integrity of state-of-the-art 0.13 mm SOI CMOS devices and circuits transferred for three-dimensional (3D) integrated circuit (IC) fabrication. Technical Digest of the IEEE International Electron Devices Meeting (2002 IEDM), pp. 943–945.
29. Burns, J.A., Aull, B.F., Chen, C.K., Chen, C.-L., Keast, C.L., Knecht, J.M., Suntharalingam, V., Warner, K., Wyatt, P.W., and Yost, D.-R.W. (2006) A wafer-scale 3-D circuit integration technology. *IEEE Trans. Electron Dev.*, **53** (10), 2507–2516, October 2006.
30. Lu, J.-Q., McMahon1, J.J., and Gutmann, R.J. (2008) 3D integration using adhesive, metal, and metal/adhesive as wafer-level bonding interfaces. *MRS Symp. Proc.*, **1112**, 69–80.
31. DiCioccio, L., Gueguen, P., Gergaud, P., Zussy, M., Lafond, D., Gonchond, J.P., Rivoire, M., Scevola, D., and Clavelier, L. (2008) 3D vertical interconnects by copper direct bonding. *MRS Symp. Proc.*, **1112**, 81–89.
32. Enquist, P., Fountain, G., Petteway, C., Hollingsworth, A., and Grady, H. (2009) Low cost of ownership scalable copper direct bond interconnect 3D IC technology for three dimensional integrated circuit applications. Proceedings of the IEEE International 3D System Integration Conference (3D IC), San Francisco, CA, 28–30 September 2009.
33. Lu, J.-Q., Kwon, Y., Jindal, A., McMahon, J.J., Cale, T.S., and Gutmann, R.J. (2003) Dielectric glue wafer bonding and bonded wafer thinning for wafer-level 3D integration, in *Semiconductor Wafer Bonding VII: Science, Technology, and Applications*, (eds F.S. Bengtsson, H. Baumgart, C.E. Hunt, and T. Suga), Electrochemical Society, Pennington, NJ, pp. 76–86.
34. Kwon, Y., Seok, J., Lu, J.-Q., Cale, T.S., and Gutmann, R.J. (2005) Thermal cycling effects on critical adhesion energy and residual stress in benzocyclobutene (BCB)-bonded wafers. *J. Electrochem. Soc.*, **152** (4), G286-G294.
35. McMahon, J.J. (2008) Damascene patterned metal/adhesive wafer bonding for three-dimensional integration. PhD thesis, Rensselaer PolytechnicInstitute.
36. Gutmann, R.J., Lu, J.-Q., Kraft, R.P., Belemjian, P.M., Erdogan, O., Barrett, J., and McDonald, J.F. (2001) IP core-based design, high-speed processor design and multiplexing LAN architectures enabled by 3D wafer bonding technologies. Proceedings of DesignCon 2001: Wireless and Optical Broadband Design Conference, Santa Clara, CA, 29 January–1 February 2001, paper WB-15.

37 Price, D.T., Gutmann, R.J., and Murarka, S.P. (1997) Damascene copper interconnects with polymer ILDs. *Thin Solid Films*, **308–309**, 523–528.

38 Borst, C.L., Thakurta, D.G., Gill, W.N., and Gutmann, R.J. (2002) Chemical-mechanical planarization of low-k polymers for advanced IC structures, *Trans. ASME J. Electron. Packag.*, **124** (4), 362–366.

39 Goldblatt, R.D., Agarwala, B., Anand, M.B., Barth, E.P., Biery, G.A., Chen, Z.G., Cohen, S., Connolly, J.B., Cowley, A., Dalton, T., Das, S.K., Davis, C.R., Deutsch, A., DeWan, C., Edelstein, D.C., Emmi, P.A., Faltermeier, C.G., Fitzsimmons, J.A., Hedrick, J., Heidenreich, J.E., Hu, C.K., Hummel, J.P., Jones, P., Kaltalioglu, E., Kastenmeier, B.E., Krishnan, M., Landers, W.F., Liniger, E., Liu, J., Lustig, N.E., Malhotra, S., Manger, D.K., McGahay, V., Mih, R., Nye, H.A., Purushothaman, S., Rathore, H.A., Seo, S.C., Shaw, T.M., Simon, A.H., Spooner, T.A., Stetter, M., Wachnik, R.A., and Ryan, J.G. (2000) A high performance 0.13 μm copper BEOL technology with low-k dielectric. Proceedings of the IEEE International Interconnect Technology Conference (IITC), June 2000, pp. 261–263.

40 Kuchenmeister, F., Stavreva, Z., Schubert, U., Richter, K., Wenzel, C., and Simmonds, M. (1998) A comparative CMP study of BCB and SiLK for copper damascene technologies. *MRS Conf. Proc.*, **14**, 237–248.

41 Gutmann, R.J., McMahon, J., and Lu, J.-Q. (2005) Global planarization requirements for wafer-level three-dimensional (3D) ICs. Proceedings of Surface Engineering and Coatings, Session on Chemical Mechanical Polishing (CMP), World Tribology Congress III, Washington, DC, 12–16 September 2005.

42 Lefevre, P. (2008) Defects observed on the wafer after the CMP process, in *Microelectronic Applications of Chemical Mechanical Planarization* (ed. L. Yuzhuo), John Wiley & Sons, Inc., New York, pp. 511–561.

43 Tan, C., Reif, R., Theodore, N., and Pozder, S. (2005) Observation of interfacial void formation in bonded copper layers. *Appl. Phys. Lett.*, **87**, 201909.

44 McMahon, J.J., Lu, J.-Q., and Gutmann, R.J. (2008) Three Dimensional (3D) integration, in *Microelectronic Applications of Chemical Mechanical Planarization* (ed. Y. Li), John Wiley and Sons, pp. 431–465.

45 Stokich, T., Fulks, C., Bernius, M., Burdeaux, D., Garrou, P., and Heistand, R.H. (1993) Planarization with Cyclotene™ 3022 (BCB) polymer coatings. *MRS Symp. Proc.*, **308**, 517–526.

46 Dibbs, M.G., Townsend, P.H., Stokich, T.M., Huber, B.S., Mohler, C.E., Heistand, R.H., Garrou, P.E., Adema, G.M., Berry, M.J., and Turlik, I. (1992) Cure management of benzocyclobutene dielectric for electronic applications. SAMPE Conference, pp. 1–10.

47 Chen, K.N., Fan, A., Tan, C.S., and Reif, R. (2004) Contact resistance measurement of bonded copper interconnects for three-dimensional integration technology. *IEEE Electron Dev. Lett.*, **25** (1), 10–12.

48 Shigetou, A., Itoh, T., Matsuo, M., Hayasaka, N., Okumura, K., and Suga, T. (2006) Bumpless interconnect through ultrafine Cu electrodes by means of surface-activated bonding method. *IEEE Trans. Adv. Packag.*, **29** (2), 218–226.

49 Chen, K.N., Tan, C.S., and Reif, R. (2005) Abnormal contact resistance reduction of bonded copper interconnects in three-dimensional integration during current stressing. *Appl. Phys. Lett.*, **86**, 011903.

12
Cu/SiO$_2$ Hybrid Bonding

Léa Di Cioccio

12.1
Introduction

Bonding of metal surfaces is extensively used for microelectromechanical system sealing, power devices, heat dissipation or three-dimensional interconnections. For these applications, techniques such as thermocompression, with or without eutectic alloys or adhesives layers, bumps with low-temperature solders, or direct bonding have been implemented [1–4].

Moreover, for more Moore and more than Moore applications, low-temperature bonding and metal bonding are becoming the main drivers of the latest developments. As copper is the main metal used for CMOS interconnects, a high-density copper interconnection between layer structures is expected to be important for future three-dimensional integration of discrete electronic devices fabricated on the basis of various technology/design concepts.

To address that, copper/oxide surface direct bonding presents many attractive advantages: compatibility with front-end or back-end requirements for a sequential approach, low-temperature process, through-interconnect layer compatibility, high accuracy of alignment during bonding, as the surfaces are bonded at room temperature no underfill is needed, and a very high interconnect density is possible.

Direct wafer bonding refers to a process by which two mirror-polished wafers are put into contact and held together at room temperature by adhesive forces, without any additional intermediate materials [5]. Direct wafer bonding can be achieved in clean rooms at room temperature with ambient air, without any pressure and without any adhesive materials. However, high bond strength and a void-free bonding interface are required in order to provide reliable and high-performance devices.

Copper/oxide surface direct bonding is a technology that enables three-dimensional metal interconnects at the same time as a bond between two piled

Handbook of Wafer Bonding, First Edition. Edited by Peter Ramm, James Jian-Qiang Lu, Maaike M.V. Taklo.
© 2012 Wiley-VCH Verlag GmbH & Co. KGaA. Published 2012 by Wiley-VCH Verlag GmbH & Co. KGaA.

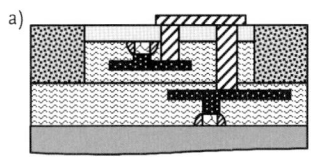

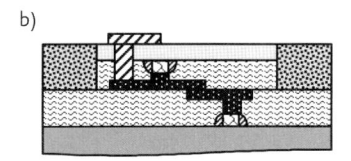

Figure 12.1 (a) Die-to-wafer integration scheme with oxide bonding. This bonding technology needs two through-silicon vias. (b) Die-to-wafer integration scheme with patterned copper bonding.

up two-dimensional processed strata. Otherwise, with an insulating layer at the bonding interface, two through-silicon vias (TSVs) of different depths are needed to connect the upper and the lower layer. A single TSV technology is feasible because of localized electrical conductive pads at the bonding interface obtained with copper/oxide surface direct bonding (Figure 12.1).

Direct copper bonding experiments have led to 8 μm pitch interconnections of thin copper electrodes at around 350 °C using oxide bonding technologies for local copper thermocompression bonding [6]. The diffusion of copper atoms across a thick adsorbate (mainly oxide) layer was obtained after long-time annealing in order to obtain sufficient bonding strength and electrical conductivity. Bumpless copper electrodes with a pitch of 6 μm fabricated with the applied damascene process were also demonstrated. The surfaces were bonded at room temperature under vacuum after Ar sputtering of the surfaces in order to remove the native copper oxide layer. The use of high-vacuum conditions increased the bonding process complexity, such as difficulties in sample handling and alignment procedure, as well as leading to low manufacturing throughput and high cost. More recently this process was modified allowing bonding at atmospheric pressure in an O_2 atmosphere after the same surface preparation under vacuum [7].

The best process has to be easily integrated into the dual damascene back-end-of-the-line (BEOL) process, using standard BEOL process steps. So we aimed to develop a process based on surface preparation using chemical mechanical polishing (CMP), with bonding at room temperature in ambient air without added bonding pressure during the bonding. The bonding annealing temperature should be in the range allowed by the BEOL requirements, that is, under 400 °C. This process is described below.

More recently, we developed a Cu/Cu direct bonding process on either plain or patterned surfaces at room temperature in ambient air without bonding pressure [8–12]. Bonding of 5 and 20 μm copper pads with a pitch of 10 and 40 μm, respectively, was performed with wafer-to-wafer alignment of 0.6 μm (mean value) and a standard deviation of $\sigma = 0.4$ μm. A specific contact resistivity of $1\,\Omega\mu m^2$ was obtained from a Kelvin structure [13–16] and was reduced to $47\,m\Omega\mu m^2$ in this work.

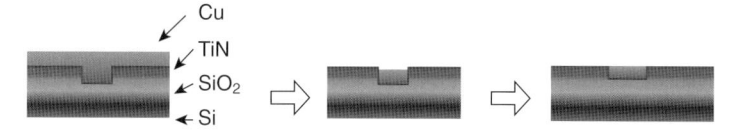

Figure 12.2 Damascene CMP process with dishing controlled on copper pads.

12.2
Blanket Cu/SiO$_2$ Direct Bonding Principle

12.2.1
Chemical Mechanical Polishing Parameters

Damascene copper CMP is widely used in production lines as an efficient back-end process for copper removal down to the oxide. The standard damascene CMP process is the process most used to planarize metal levels in microelectronics. However, to avoid leakage, the process usually induces a dishing of the copper pads on patterned surfaces. This surface topology leads to important inconveniences: copper pad dishing and oxide erosion. These are critical issues if one wants a direct copper contact between two face-to-face copper pads. At CEA-Leti, an optimized CMP-based process was implemented on patterned surfaces to reach a high surface planarization level (Figure 12.2). The CMP optimization (pad/slurry matching) target is to minimize copper dishing and oxide erosion with respect to the lay-out, and a standard wet cleaning is applied to remove the slurry residues and the remaining particles.

Wafers with 5 μm copper pads and a copper density of 20% were designed to achieve direct Cu/SiO$_2$ bonding. After CMP-based preparation, a dishing of less than 10 nm was obtained on 5 μm width pads over all of the wafer.

After cleaning, wafers were introduced separately in an EVG SmartView™ alignment tool. Alignment was optically achieved using specific alignment marks. The bonding wave was then initiated with a tip at the center of the wafers. Bonding occurs at room temperature, with ambient air and atmospheric pressure.

An optimized CMP process has been developed to control copper dishing and oxide erosion. The key point is to obtain a perfectly flat surface to enable electrical contact on each copper pad. Copper oxidation during the surface preparation is controlled (Figures 12.3 and 12.4).

For oxide bonding using this CMP process, a bonding a bonding energy of 1.4 J m^{-2} was measured after a 400 °C annealing step for 30 min, which is compatible with further post-processes such as thinning. Figure 12.5 shows that when using this process, the sealing of the different interfaces is perfect.

Since infrared radiation does not propagate through the copper materials, scanning acoustic microscopy (SAM) has to be used to characterize patterned bonding quality. SAM has the advantage of being a nondestructive technique at the wafer

240 | *12 Cu/SiO₂ Hybrid Bonding*

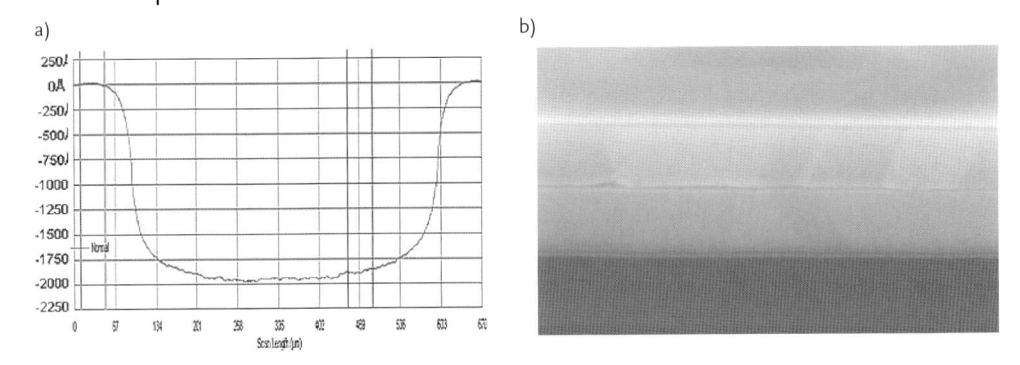

Figure 12.3 (a) Dishing characterized by a profilometer scan of a 500 μm copper pad after a standard CMP process. (b) Cu/Cu SEM cross section of two 20 μm copper pads bonded with dishing after a 200 °C annealing step for 30 min.

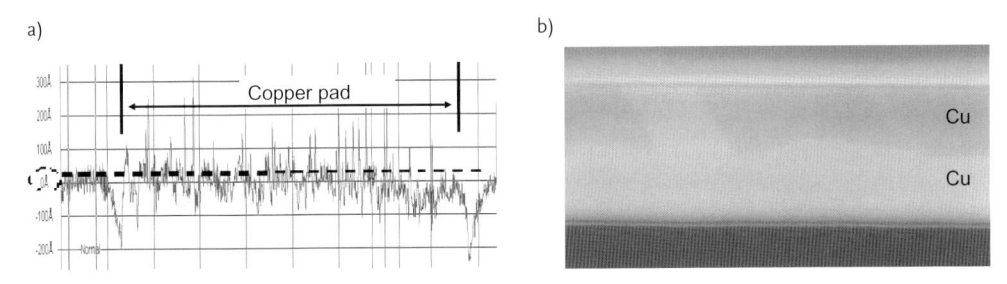

Figure 12.4 (a) Dishing characterized by a profilometer scan of a 500 μm copper pad after an optimized CMP process. The maximum roughness is less than 10 nm.

(b) Cu/Cu SEM cross section of two 20 μm wide and 0.5 μm high copper pads bonded without dishing after a 200 °C annealing step for 30 min.

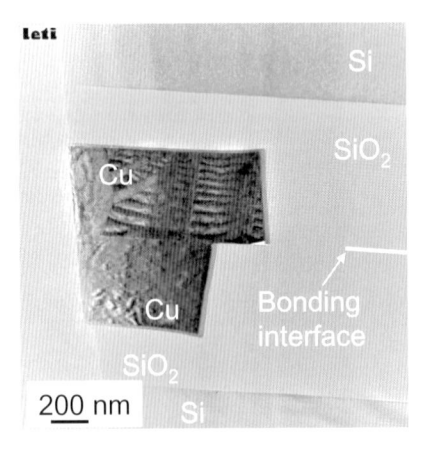

Figure 12.5 TEM image of bonded copper pad. It can be seen that the sealing of the different interfaces is perfect.

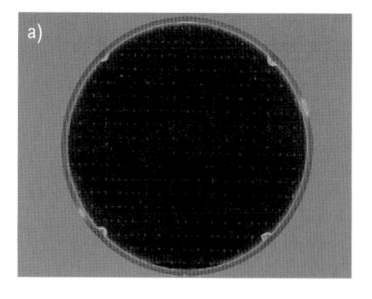

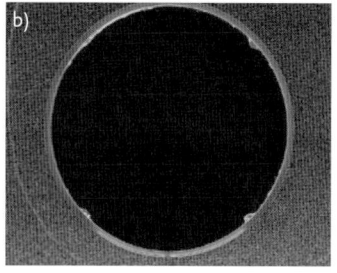

Figure 12.6 SAM images of 200 mm direct Cu/SiO$_2$ bonded pair (a) without post-bonding anneal, and (b) after a post-bonding anneal for 2 h.

Table 12.1 Bonding strength ($G = \Gamma_0 + \Gamma_p$) of a bonding pair as a function of post-bond annealing temperature. The duration for the annealing step was 2 h.

Annealing temperature (°C)	Bonding strength (J m^{-2})
200	1.14
400	6.6

scale. Without post-bond annealing, the whole wafer is bonded; no echo is detected from the bonding interface, as shown in Figure 12.6a. The four marks present at the edge of the bonding pair are induced by the bonding tool spacers. After post-bond annealing at up to 400 °C for 2 h, no bonding degradation is seen (Figure 12.6b). This high bonding quality is reached because of a perfect homogeneity between the center and the edge of the wafer after the CMP process.

After bonding of 5 μm patterned copper pads, three kinds of contacts are obtained at the bonding interface: SiO$_2$/SiO$_2$, SiO$_2$/Cu (if misalignment occurs), and Cu/Cu. Since SiO$_2$/Cu does not allow interdiffusion, its bonding strength is lower than those of SiO$_2$/SiO$_2$ and Cu/Cu. For this reason, the SiO$_2$/Cu contact density at the bonding interface has to be minimized to obtain higher bonding strength.

Bonding strength can be determined using the double cantilever beam (DCB) technique [17]. This technique consists of measuring the debonding length induced by the insertion of a blade at the bonding interface. It is important to note that the interface fracture energy for ductile materials has two major contributions: the chemical work of adhesion Γ_0 and the work of plastic deformation Γ_p. When measurements are done with mode I methods, such as the DCB technique, there is a minimum plasticity contribution [18]. After a post-bond annealing step at 200 °C for 2 h, the bonding strength obtained is 1.14 J m^{-2}. When a 400 °C post-bond annealing step for 2 h is performed on the bonding pair, the bonding strength reaches 6.6 J m^{-2} (Table 12.1). This high bonding strength (three times higher than that for blanket SiO$_2$/SiO$_2$ bonding; see Figure 12.12) is induced by direct

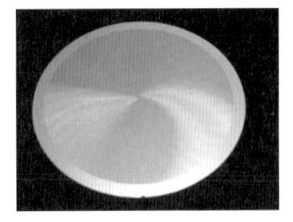

Figure 12.7 Bonded pair (200 mm) with 200 °C post-bonding anneal. The top silicon was thinned down to 5 µm with successive coarse and fine grinding.

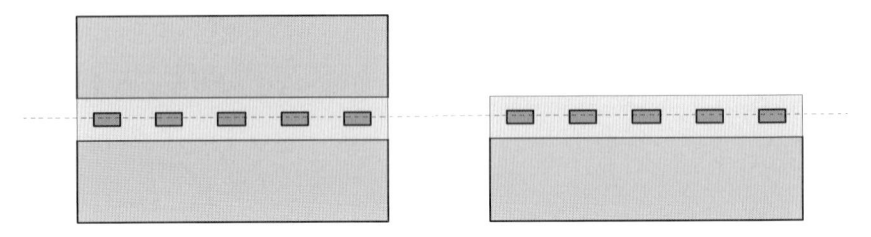

Figure 12.8 Two wafers with mixed copper oxide surfaces bonded face-to-face in an alignment tool. Then, one wafer was ground and its top silicon was removed down to the oxide layer.

copper to copper contact zones. Since interdiffusion and copper grain growth can happen in these zones, the bonding strength becomes very important (see Section 12.3).

With a bonding strength of $1.14\,J\,m^{-2}$ (200 °C post-bond anneal), the bonding pair is able to sustain post-processes such as silicon thinning. We have demonstrated the possibility of thinning the top silicon wafer down to 5 µm. The thinning process was performed by successive coarse and fine grinding. With 5 mm edge trimming after grinding, the 5 µm thin silicon substrate was free of delamination (Figure 12.7). A silicon CMP step can then be performed on the top silicon to remove post-grinding residual strain. The thinning process is very important for post-processing such as the formation of TSVs for three-dimensional technology.

Achieving Cu/SiO_2 bonding with this designed surface leads to a bond pitch of less than 10 µm. To allow optical observation of the pad bonding, after thinning down the top silicon wafer to 50 µm, the residual silicon was chemically removed with a tetramethylammonium hydroxide (TMAH) bath (Figure 12.8). Figure 12.9 presents a top view of Cu/SiO_2 bonding with a 10 µm pitch. Since misalignment is less than 1 µm, the bottom pads are barely visible.

A focused ion beam (FIB) was used for cross-sectional secondary electron microscopy (SEM) analyses. The dark area observed above the copper pad in Figure 12.10 is an electron charging effect induced by the presence of silicon oxide at the top of the copper pads. At the bonding interface, after a 300 °C post-bond annealing step for 2 h, small voids are observed at the Cu/Cu bonding interface. Regarding

12.2 Blanket Cu/SiO₂ Direct Bonding Principle | 243

Figure 12.9 top view of a 10 µm pitch direct Cu/SiO$_2$ bonding after total top silicon removal.

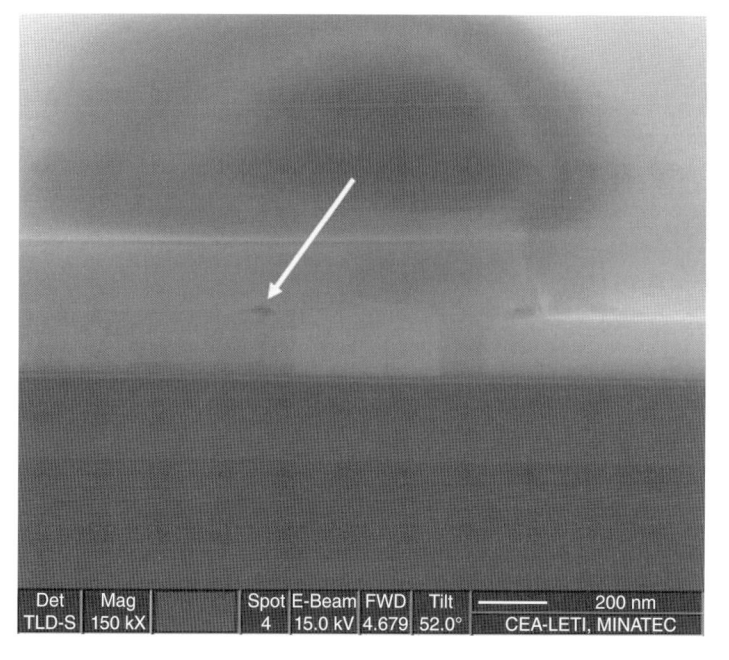

Figure 12.10 FIB-SEM cross section of direct Cu/SiO$_2$ bonding with 10 µm pitch. The Cu/Cu interface is sealed and reveals copper grain growth from one layer to the other. Tiny voids (about 20 nm in diameters) are observed at the bonding interface.

the Cu/Cu bonding interface, one can see a zigzag-shaped interface due to growth of copper grains from one layer to the other. The voids might arise from residual surface roughness prior to the bonding or vacancies in the copper layers which would migrate at the bonding interface.

12.2.2
Bonding Quality and Alignment

Bonding of patterned surfaces can be achieved at atmospheric pressure in an EVG SmartView alignment tool. The alignments marks used are standard crosses in a

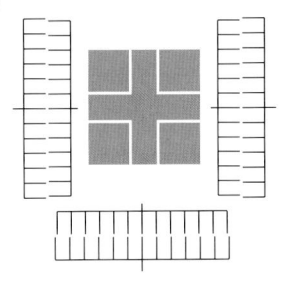

Figure 12.11 Alignment marks after bonding of patterned wafers.

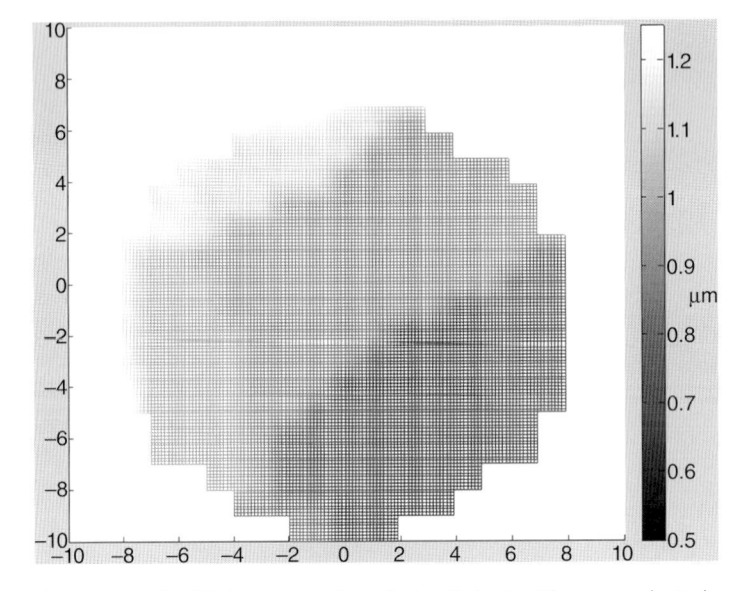

Figure 12.12 Graphical representation of a bonded pair with measured misalignment. The maximum measured misalignment is 1.2 μm all over the bonded pair.

box with verniers allowing measurement accuracy of 0.2 μm or less depending on the specific mark design. Alignment measurements are done with an infrared microscope (Figure 12.11).

Since alignment marks are present in every die of the wafer, it is possible to create a mapping of the misalignment of the bonded pair. Figure 12.12 shows a graphical representation of this misalignment for the whole of a bonded pair. The mean value and standard deviation can be determined using Eqs. (12.1) and (12.2), respectively:

$$\bar{x} = \frac{1}{n}\sum_{i=1}^{n} x_i \qquad (12.1)$$

Table 12.2 Evolution of the mean misalignment and its standard deviation for a Cu/SiO$_2$ bonded pair after a grinding process. An evolution of 0.1 μm is meaningless as it is the accuracy of the measurement.

Infrared alignment measures after bonding	x (μm)	σ (μm)
Cu/SiO$_2$ bonding as bonded	0.6	0.4
Cu/SiO$_2$ bonding after grinding down to 50 μm	0.7	0.3
Cu/SiO$_2$ bonding after grinding down to 15 μm	0.7	0.3

$$\sigma = \sqrt{\frac{1}{n-1}\sum_{i=1}^{n}(x_i - \bar{x})^2} \tag{12.2}$$

No misalignment above 1.2 μm was recorded and the mean value was 0.6 μm with σ = 0.4 μm. These results confirm that since the process is done at room temperature no very large misalignment is induced during the bonding.

The same measurements were done on a bonded pair after a 400 °C annealing for 2 h, and also for two subsequent thinning processes to 50 and 15 μm. The measured misalignments were stable as is evident from Table 12.2.

12.3
Blanket Copper Direct Bonding Principle

Blanket copper-to-copper direct bonding can be achieved at room temperature, with ambient air and atmospheric pressure. Evolution of the bonding strength ($G = \Gamma_0 + \Gamma_p$) is characterized by two different techniques: the DCB method and four-point bending. With four-point bending, the angle of mixed modes is higher than with the DCB method. The measured bonding strength with ductile materials such as copper is then expected to be higher with four-point bending than with the DCB method due to an increase of the plastic contribution Γ_p. Figure 12.13 presents the evolution of the bonding strength as a function of 30 min post-bond annealing temperature. At room temperature and without post-bond annealing, bonding strength values of 2.5 and 6.2 J m^{-2} are measured with DCB and four-point bending, respectively. With a 100 °C annealing step for 30 min, the bonding strength increases and reaches 3 and 12 J m^{-2} with DCB and four-point bending, respectively. These bonding strength values are high compared to those of thermal oxide. Above 200 °C, for blanket copper bonding, the initiated fracture propagates directly through the silicon bulk. The copper-to-copper bonding strength becomes so high that the sample breaks, making the measurements of bonding strength impossible. As previously mentioned in this chapter the same phenomenon was observed for a patterned surface.

Transmission electron microscopy (TEM) observations of a room temperature Cu/Cu bonding interface reveals a very sharp interface as shown in Figure 12.14.

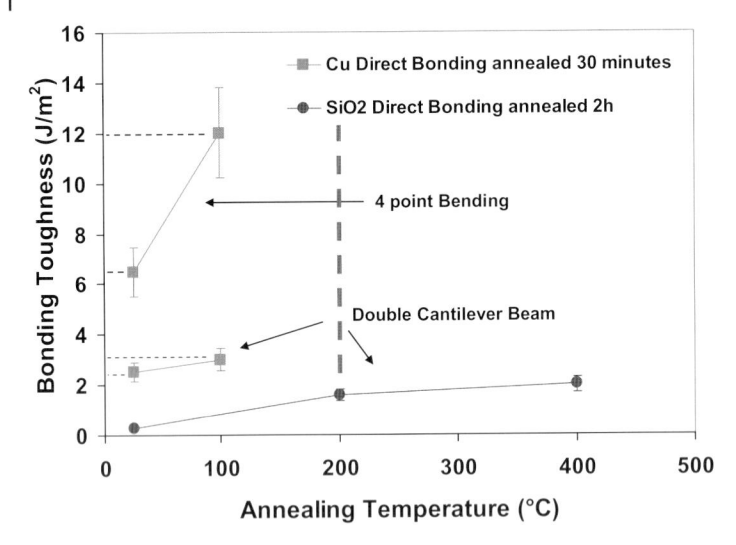

Figure 12.13 Bonding strength ($G = \Gamma_0 + \Gamma_p$) of a bonding pair as a function of the post-bond annealing temperature.

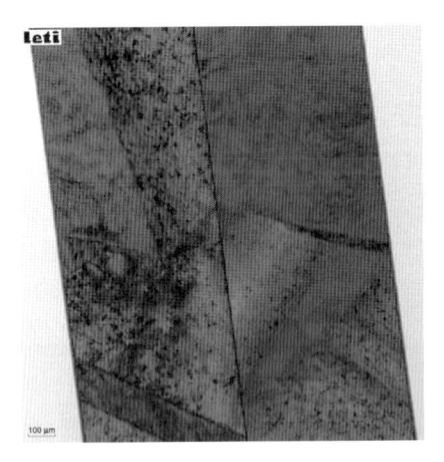

Figure 12.14 TEM cross-sectional image of direct copper bonding without post-bonding anneal. The copper layer thickness is 500 μm.

This is perfectly in accordance with the copper direct bonding process since very flat and low-roughness surfaces are mandatory.

When higher magnification TEM is used to characterize the bonding interface, one can see an interfacial layer with a width of about 4 nm (Figure 12.15). Since the bonding process is perfectly symmetric, a sharp interface is expected rather than a boundary layer. This layer is crystalline with a different orientation from the copper. Since the bonding is done without additive material, at the first stages of bonding, rough surfaces contact theory has to be considered.

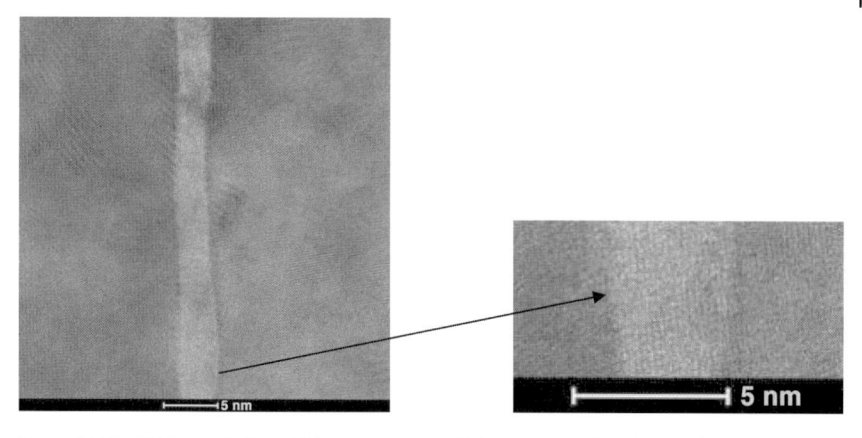

Figure 12.15 High-resolution TEM cross section of direct copper bonding without post-bonding anneal. A 4 nm thick crystalline layer is present at the bonding interface.

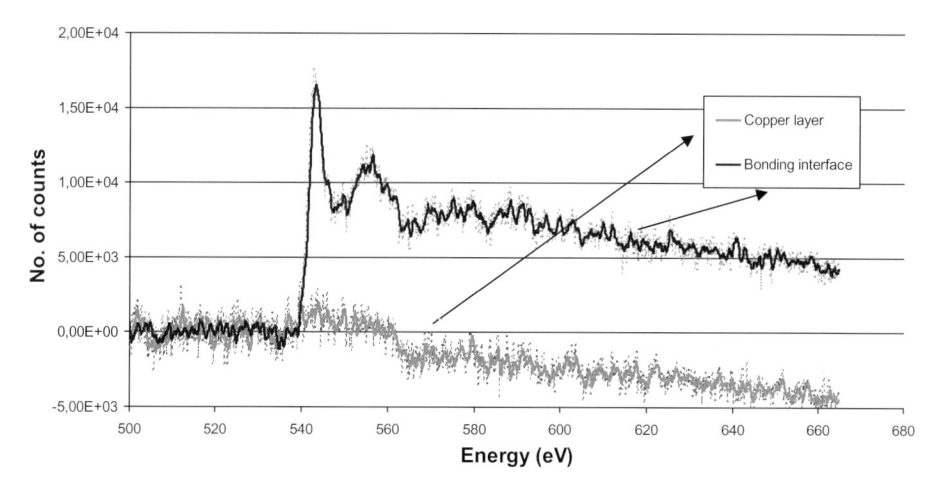

Figure 12.16 Electron energy loss spectra at a bonding interface compared to a copper layer. A peak of oxygen is detected when the analysis is carried out on the bonding interface.

Identification of elements present at the bonding interface was carried out with electron energy loss spectroscopy. Both the copper layer and the interfacial layer were analyzed with this technique. Results are presented in Figure 12.16. A peak is detected at 543 eV for the interfacial layer. This peak indicates the presence of oxygen within the interfacial layer. Since copper oxide has crystalline properties [19], this interfacial layer can be identified as a copper oxide layer.

The high bonding strength achieved at room temperature without additional annealing can be explained by the presence of a homogeneous copper oxide. When a crack is initiated, it has to propagate through a crystalline structure instead of a

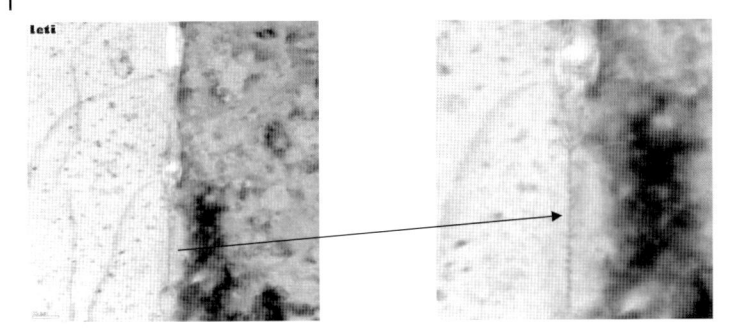

Figure 12.17 High-magnification TEM cross section of direct copper bonding after 30 min post-bond annealing at 200 °C. Dislocations are observed at the bonding interface, meaning grain boundary creation and interface sealing.

single interface. This propagation path through the crystalline layer needs more energy and so the measured bonding strength is increased.

Above 200 °C post-bond annealing, measurement of the bonding strength of copper direct bonding is impossible. To understand the impact of this thermal treatment on the bonding interface and interfacial layer, TEM was first used to visualize the situation. Figure 12.17 shows TEM images of the bonding interface after a 200 °C annealing step. The initial crystalline interfacial layer evolves along the bonding interface and is no longer continuous. Intimate contact between the two copper layers is still possible. From the high-magnification cross-sectional TEM image, one can see dislocations at this new bonding interface. These dislocations are the signature of grain boundary creation between the two layers and the interface sealing.

In situ thermal evolution X-ray reflectivity (XRR) characterization of the bonding interface was performed at the European Synchrotron Radiation Facility with 27 keV X-rays. This technique has already been used for detailed understanding of silicon direct bonding [20]. Thermal cycles from room temperature up to 400 °C were used for this experiment. XRR measurements were done every 25 °C increment, once the temperature stabilized. Figure 12.18 presents the superposition of the fits of the bonding interface electron density for room temperature and 175, 200, and 225 °C. At room temperature, the decreased electron density of the bonding interface is due to copper oxide, residual roughness, or buried nanovoids. Around 200 °C, the electron density of the bonding interface increases. At this temperature the bonding interface is becoming closed, that is to say, some parts of the interface are creating true copper-to-copper contact and making the electron density of the bonding interface closer to that of copper.

The sealing of the bonding interface at 200 °C may explain the impossibility of characterizing the bonding strength. At this temperature, the two copper layers seem to have created an intimate bond from one layer to the other. That is why during DCB or flexion tests it is easier for the initiated fracture to propagate through the silicon bulk than through a copper grain boundary.

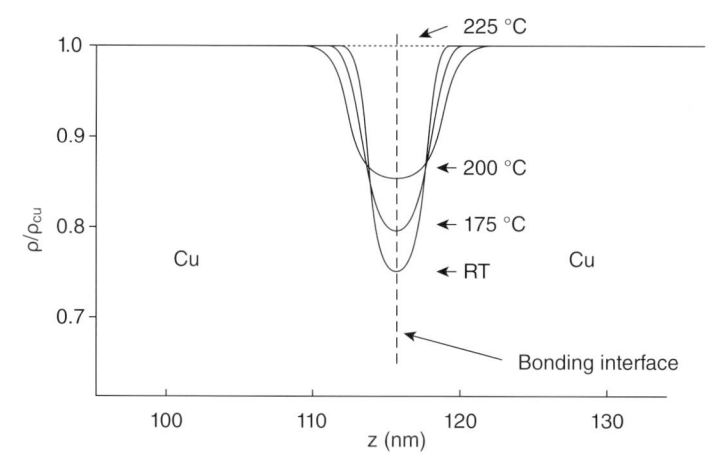

Figure 12.18 XRR fit of electron density close to a bonding interface. At 225 °C, the bonding interface is becoming sealed. (RT, room temperature.)

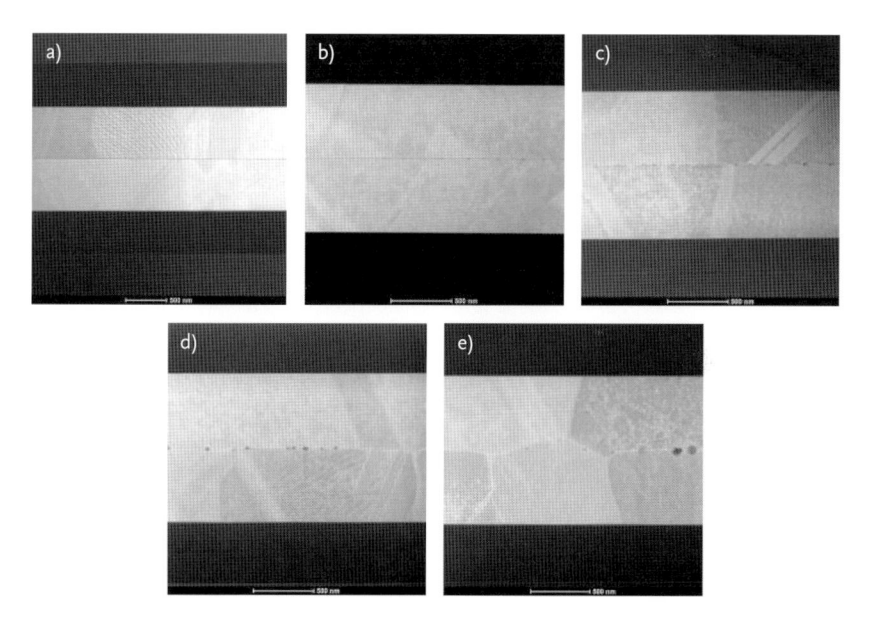

Figure 12.19 TEM cross sections of direct copper bonding with successive 30 min post-bond annealing: (a) room temperature; (b) 100 °C; (c) 200 °C; (d) 300 °C; (e) 400 °C.

Macroscopic evolution of direct copper bonding with temperature has been studied with low-magnification TEM. Figure 12.19 presents copper bonding cross sections at room temperature and after successive 100, 200, 300, and 400 °C annealing steps with an annealing time of 30 min for each step. For each temperature, a sample was taken from the same bonded pair prior to subsequent thermal

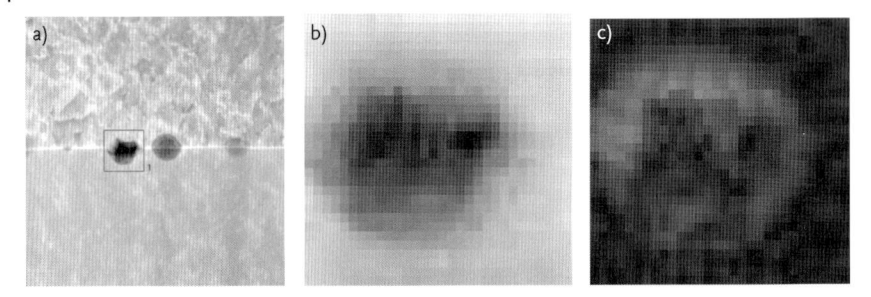

Figure 12.20 (a) Selected zone for EDX analyses. (b) Copper mapping and (c) oxygen mapping.

treatment. Imaging was performed using dark field, which makes less dense constituents appear darker. Since the density of copper oxide is less than that of copper, the interfacial layer appears darker in dark-field TEM cross sections. Even after a 100 °C annealing step for 30 min, the oxide at the bonding interface starts to flow and agglomerate. After a 300 °C post-bond anneal, copper grains start to grow from one layer to the other. The bonding interface assumes a zigzag shape. During the annealing steps, the copper oxide diffuses and agglomerates along the bonding interface, and then an intimate copper-to-copper bond is created to allow macroscopic copper grain growth.

With post-bond annealing, 20 nm nodules appear at the bonding interface. Characterization of these nodules was carried out with electron dispersive X-ray (EDX) analysis. EDX mapping of a selected zone (Figure 12.20a) was performed. Copper (Figure 12.20b) and oxide (Figure 12.20c) constituents were followed over all of the selected zone. The brighter a pixel, the greater the amount of element is present. For copper mapping, a hole is present at the nodule center with no copper inside it. Looking at the oxygen mapping, a crown of oxide around the nodule is apparent. Nodules seem to be voids covered with copper oxide.

However, after thermal treatments above 200 °C, the bonding interface becomes a grain boundary with intimate copper-to-copper contact. This true copper-to-copper bond enables electrons to pass from one copper layer to the other and to reinforce the bonding.

These observations confirm that the sealing of the interface is driven by two physical mechanisms. The first and main one is the direct bonding adhesion of surfaces at room temperature without any pressure or compression. The second mechanism is a modified diffusion bonding.

Diffusion bonding is an extensively studied mechanism in sintering of metals [21] and consists of a plastic deformation of the asperities of both metal films. Such deformation can induce elliptical voids along the bonding interface. In diffusion bonding the strengthening of the bonding interface is obtained by diffusion at the interface, bulk diffusion, grain growth, and power-law creep and is conditioned by temperature, pressure, and initial surface conditions [22].

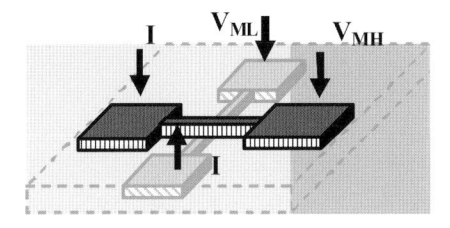

Figure 12.21 Schematic of a Kelvin cross structure.

In our case, we have a modified diffusion bonding since no pressure is applied during bonding and annealing, and the process is driven mainly by temperature, metal oxide thermodynamic instability, and the very small initial roughness.

For patterned surfaces, the oxide zones are bonded at room temperature and the copper might be slightly dished and still unbonded. The mechanism discussed above is, however, still possible. The thermal expansion of the copper pads in the vertical direction (horizontal expansion is impeded by the silicon wafer), greater than that of the oxide, allows the aligned copper pad surfaces to come into contact and bond because of the diffusion bonding.

12.4
Electrical Characterization

Determination of contact resistance is commonly carried out with Kelvin structures. Figure 12.21 shows a schematic representation of such a Kelvin cross where two of the tips force the current to pass through the bonding contact and the other two are used to measure the voltage from both parts of the contact. This technique is widely used because it is a free access resistance method. One branch of the Kelvin cross is designed on the top wafer, while the second is on the bottom wafer. With this configuration, the misalignment will not affect the measured resistance.

To a first approximation, the contact resistance R_C can be directly determined using the relation

$$R_C = \frac{V_{MH} - V_{ML}}{I} \quad (12.3)$$

where V_{MH} and V_{ML} are the electrical potential of the top and bottom layer, respectively. One can then determine the specific contact resistivity:

$$\rho_C = R_C \times A_C \quad (12.4)$$

where ρ_C is the specific contact resistivity and A_C the contact area.

In order to obtain contact resistance at a Cu/Cu direct bonding interface, 200 mm silicon wafers were used with 800 nm silicon oxide deposited. Kelvin branches (500 nm) were dry etched over the wafer and 20 nm Ti/TiN (chemical

Table 12.3 Repartition of different contacts at the bonding interface for patterned Kelvin structures.

Contact type	% of the bonding interface
SiO_2/SiO_2	60
Cu/SiO_2	39.9
Cu/Cu	0.14

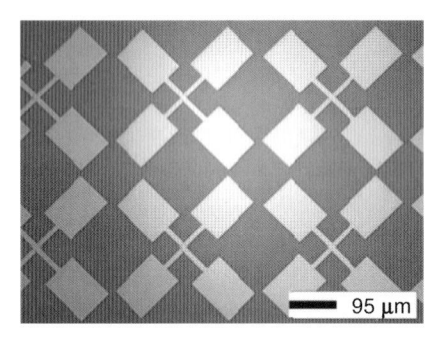

Figure 12.22 Optical microscopy top view of a bonded pair after top wafer substrate removal.

vapor deposition) followed by 1 μm copper (physical vapor + electrochemical deposition) were deposited. Surface preparation was performed by CMP-based activation as described in Section 12.2. Kelvin branches have been designed to obtain a copper surface density of 20%. After bonding, three kinds of contacts are obtained at the bonding interface (Table 12.3).

After room temperature bonding and post-bond annealing, grinding plus TMAH were used to entirely etch the top silicon substrate, as previously explained. A dry silicon oxide etching step is finally used to expose the direct contacts between measurement probes and $75 \times 95\,\mu m^2$ copper pads (Figure 12.22).

Measured resistances after 2 h thermal treatment at 200 and 400 °C are presented in Figures 12.23 and 12.24, respectively. Since the Kelvin structure appears over all of the wafer, multiple measurements were possible.

Voltage–current, $V(I)$, curves were obtained with a HP4155 analyzer. The current was varied from −100 up to +100 mA with a step of 1 mA. More than 20 Kelvin structures were characterized over the whole of the wafer.

For all measurements, the obtained curve revealed an ohmic behavior since the voltage is proportional to the current and the curve passes through the origin of the axes. No hysteresis issues were recorded for these electrical measurements. Forward current was exactly the same as reverse current. The resistance R_C was determined by a linear fit of the $V(I)$ curve. For all measurements, the obtained

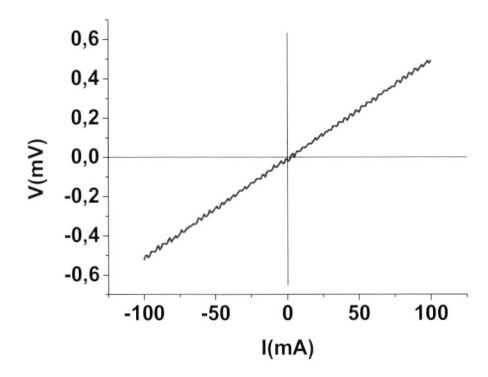

Figure 12.23 Single $V(I)$ curve of wafer-to-wafer $10 \times 10\,\mu m^2$ Cu/Cu contact after post-bonding anneal at 200 °C for 2 h. R_C and ρ_C are equal to $5\,m\Omega$ and $0.5\,\Omega\,\mu m^2$, respectively.

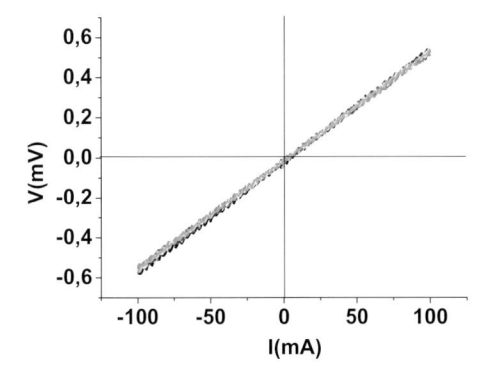

Figure 12.24 Twenty $V(I)$ curves of wafer-to-wafer $10 \times 10\,\mu m^2$ Cu/Cu contact after post-bonding anneal at 400 °C for 2 h. R_C and ρ_C are equal to $5\,m\Omega$ and $0.5\,\Omega\,\mu m^2$, respectively.

contact resistance was $5\,m\Omega$ with $\sigma = 0.1\,m\Omega$. Good electrical property homogeneity was achieved over the whole of the bonded pair with direct Cu/SiO_2 bonding after post-bond annealing.

When the same experiment is carried out on Kelvin structures with different contact areas, after an annealing step at 400 °C for 2 h, the measured resistance is always in the region of $5\,m\Omega$. Table 12.4 summarizes the resistance for contact areas from 3×3 up to $25 \times 25\,\mu m^2$.

Since the contact resistance has to be inversely proportional to the bonding area, second-order effects affect the contact resistance determined using Eq. (12.3). Indeed, with such low resistances, Eq. (12.3) is not accurate enough to describe the electrical behavior of the bonding interface. Furthermore, with the Kelvin crosses used for this study, current is forced to pass across a 90° angle. Figure 12.25 shows a simulation of equipotentials all along the current path. A drop of

12 Cu/SiO₂ Hybrid Bonding

Table 12.4 Evolution of measured contact resistance and contact resistivity as a function of contact area.

Contact area (μm^2)	Mean R_C (mΩ)	Mean ρ_C ($\Omega \mu m^2$)
3 × 3	5.28	0.047
5 × 5	5.18	0.129
7 × 7	5.20	0.255
10 × 10	5.05	0.505
25 × 25	5.40	3.375

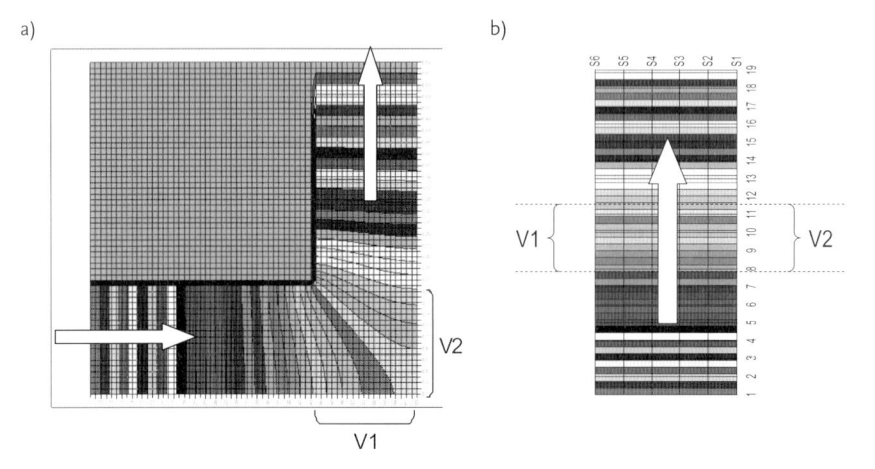

Figure 12.25 Comparison of equipotential simulation all along (a) angular and (b) straight current path during Kelvin measurement. A drop of 2% of the total voltage occurs between each different scale of gray.

2% of the total voltage occurs between each different gray color. For an angular current path, a parasitic voltage drop affects the measured resistance around the contact area.

When the contact resistance becomes negligible compared to the parasitic resistance $(V_1 - V_2)/I$, the measured resistance remains constant. The measured resistance R_M can then be expressed by

$$R_M = R_C + \frac{V_1 - V_2}{I} \tag{12.5}$$

Taking into account this parasitic effect, specific contact resistance cannot be directly determined using Eq. (12.3). When contact resistances are in the region of a few milliohms, Kelvin structures with angular current flow are not accurate enough to obtain a good characterization of the electrical properties of the bonding interface.

12.5
Die-to-Wafer Bonding

The process used for wafer-to-wafer bonding can be adapted to die-to-wafer bonding. After a surface-activation process, dies can be directly bonded onto a host substrate with an SET FC 150 alignment tool. SAM characterization of the bonding interface after a 400 °C post-bonding anneal for 2 h reveals only few unbounded zones (Figure 12.26). Defects at the die corner are due to handling issues. A misalignment was found of 0.5 μm after infrared characterization.

Using the same protocol with die-to-wafer bonding, the top silicon die substrate was removed by successive grinding, TMAH treatment, and silicon oxide dry etching. Electrical characterization was performed on a 400 °C/2 h annealed Kelvin cross. The very first $I(V)$ curve shows perfect ohmic behavior (Figure 12.27).

The measured resistance was 12 mΩ for a $10 \times 10\,\mu m^2$ contact area. The increase of contact resistance compared to wafer-to-wafer bonding can be explained by the additional steps required for dies or a slight copper oxidation. Ways to improve this electrical behavior are under investigation. However, the Cu/Cu contact resistance obtained remains negligible compare to that of TSV resistance used for three-dimensional IC schemes.

Figure 12.26 SAM image of Cu/SiO_2 die-to-wafer bonding after a 400 °C post-bonding anneal for 2 h.

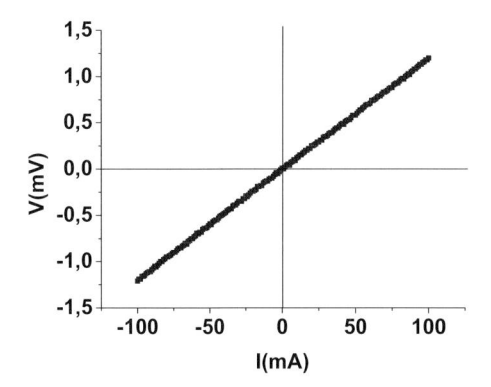

Figure 12.27 $V(I)$ curve of die-to-wafer $10 \times 10\,\mu m^2$ Cu/Cu contact after a post-bonding anneal at 400 °C for 2 h. R_C and ρ_C were determined as 12 mΩ and $1.2\,\Omega\,\mu m^2$, respectively.

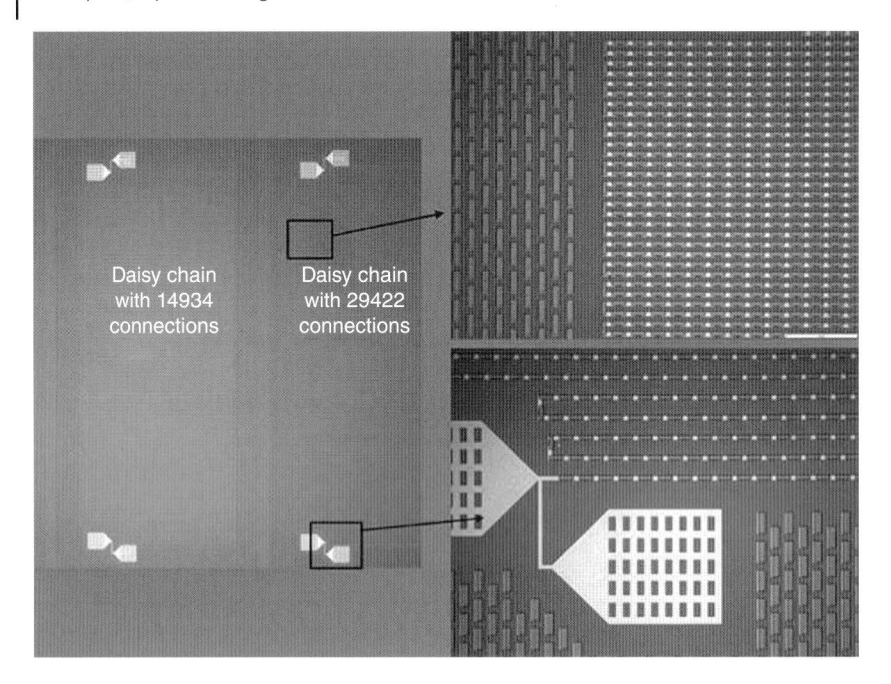

Figure 12.28 Optical observation of daisy chains after the complete removal of top silicon.

12.5.1
Daisy Chain Structures [23]

Daisy chains with up to 29 422 connections have been designed to limit the angular current flow parasitical resistance effect and to demonstrate the reliability copper direct bonding electrical contact. Copper lines were 3 µm wide, 500 nm thick, and 10 µm long. The pitch was 14 µm and contact area was $3 \times 3\,\mu m^2$ (Figure 12.28). After annealing the bonded pair at 400 °C for 2 h, the $V(I)$ curve obtained was perfectly ohmic across the 29 422 connections. The total resistance was measured at about 2.34 kΩ. That is to say, the resistance for each "Cu contact + line" element was 79.5 mΩ.

A schematic cross section of the daisy chain is presented in Figure 12.29. A copper film resistivity ρ of 2.1×10^{-2} Ωµm was measured on specific electrical characterization of the structure. The experimentally measured resistance of 79.5 mΩ for "Cu contact + line" can be separated into three different structures (Figure 12.29). Theoretical values for R_1 and R_3 can be obtained using

$$R = \frac{\rho L}{S} \tag{12.6}$$

Using $\rho = 2.1 \times 10^{-2}$ Ωµm, $L = 2\,\mu m$, and $S = 0.5 \times 3\,\mu m^2$, one can obtain a value for R_1 and R_3 of about 28 mΩ. If R_2 were induced only by copper material, the ideal

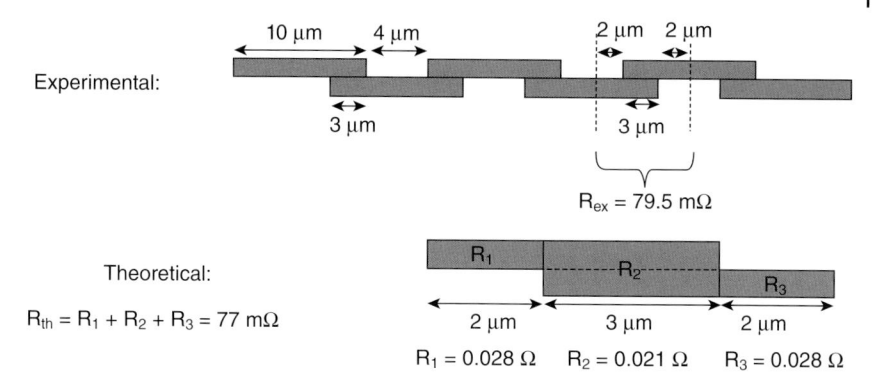

Figure 12.29 Experimental and theoretical measurements for the resistance per node on daisy chain with $3 \times 3\,\mu m^2$ contact area.

resistance would be $R_{2th} = 21\,m\Omega$. Since the resistance obtained for "Cu contact + line" is $79.5\,m\Omega$, the resistance $R_{2real} = R - R_1 - R_3 = 23\,m\Omega$. That is to say, for a daisy chain with 29 422 connections, after a 400 °C annealing step, the impact of the copper-to-copper bonding interface for $3 \times 3\,\mu m^2$ contact areas is $R_{2real} - R_{2th} = 2.5\,m\Omega$.

This difference is induced by copper line resistance variations (affecting R_1, R_2, and R_3), misalignment error during bonding (affecting mainly the contact area and so R_2), and the real bonding interface resistance R_C. To a first approximation we assume that this difference is the maximum value of R_C, though one can estimate the specific contact resistance ρ_C using Eq. (12.4). The specific contact resistance ρ_C is then about $22.5\,m\Omega\,\mu m^2$.

12.6 Conclusion

Cu/SiO$_2$ bonding has been demonstrated for wafer-to-wafer bonding as well as die-to-wafer bonding. In both cases, high bonding quality and alignment are evident from SAM and infrared analyses. A very low contact resistance is achieved with multiple contact areas, and the contact resistivity is reduced down to $\rho_C = 47\,m\Omega\,\mu m^2$ for Kelvin structures and down to $\rho_C = 22.5\,m\Omega\,\mu m^2$ for daisy chains with 29 422 connections. With post-bond annealing, the bonding interface acts as a copper grain boundary.

Acknowledgment

This work was supported by SOITEC and STMicroelectronics.

References

1 Chen, K.N., Fan, A., Tan, C.S., and Reif, R. (2004) Contact resistance measurement of bonded copper interconnects for three-dimensional integration technology. *IEEE Electron Devices Lett.*, **25** (1), 10–12.

2 Labie, R., Ruythooren, W., Baert, K. *et al.* (2008) Resistance to electromigration of purely intermetallic micro-bump interconnections for 3D-device stacking. Proceedings of the International Interconnect Technology Conference (IITC 2008), pp. 19–21.

3 McMahon, J.J., Lu, J.-Q., and Gutmann, R.J. (2005) Wafer bonding of damascene-patterned metal/adhesive redistribution layers for via-first three-dimensional (3D) interconnect. Proceedings of the Electronic Components and Technology Conference, pp. 331–336.

4 Temple, D., Malta, D., Lannon, J.M. *et al.* (2008) Bonding for 3-D integration of heterogeneous technologies and materials. *ECS Trans.*, **16**, 3–13.

5 Ventosa, C., Rieutord, F., Libralesso, L. *et al.* (2008) Hydrophilic low-temperature direct wafer bonding. *J. Appl. Phys.*, **104**, 123524.

6 Enquist, P., Fountain, G., Petteway, C. *et al.* (2009) Low cost of ownership scalable copper direct bond interconnect 3D IC technology for three dimensional integrated circuit applications. IEEE International Conference on 3D System Integration.

7 Shigetou, A. and Suga, T. (2009) Modified diffusion bond process for chemical mechanical polishing (CMP)-Cu at 150 °C in ambient air. Proceedings of the Electronic Components and Technology Conference.

8 Cioccio, L.D., Gueguen, P., Grossi, F. *et al.* (2008) 3D technologies at CEA-Leti MINATEC. Proceedings of the International Symposium on Microelectronics (IMAPS 2008).

9 Gueguen, P., Di Cioccio, L., and Rivoire, M. (2008) Copper direct bonding for 3D integration. Proceedings of the International Interconnect Technology Conference (IITC 2008), pp. 61–63.

10 Leduc, P., Di Cioccio, L., Charlet, B. *et al.* (2008) Enabling technologies for 3D chip stacking. International Symposium on VLSI Technology, Systems and Applications, pp. 76–78.

11 Sillon, N., Astier, A., Boutry, H. *et al.* (2008) Enabling technologies for 3D integration: from packaging miniaturization to advanced stacked ICs. Proceeding of the IEEE International Electron Devices Meeting, pp. 1–4.

12 Gueguen, P., Di Cioccio, L., Gergaud, P. *et al.* (2009) Copper direct bonding characterization and its interests for 3D integration. *J. Electrochem. Soc.*, **156** (10), H772–H776.

13 Gueguen, P., Di Cioccio, L., Gonchond, J.P. *et al.* (2009) 3D vertical interconnects by copper direct bonding. *MRS Symp. Proc.*, **1112**, 81–90.

14 Gueguen, P., Ventosa, C., Di Cioccio, L. *et al.* (2010) Physics of direct bonding: applications to 3D heterogeneous or monolithic integration. *J. Microelectron. Eng.*, **87**, 477–484.

15 Di Cioccio, L., Gueguen, P., Signamarcheix, T., *et al.* (2009) Enabling 3D interconnects with metal direct bonding. Proceedings of the International Interconnect Technology Conference (IITC 2009), pp. 152–154.

16 Di Cioccio, L., Gueguen, P., Taibi, R. *et al.* (2009) An innovative die to wafer 3D integration scheme: die to wafer oxide or copper direct bonding with planarised oxide inter-die filling. IEEE International Conference on 3D System Integration.

17 Maszara, W.P., Goetz, G., Caviglia, A. *et al.* (1988) Bonding of silicon wafers for silicon-on-insulator. *J. Appl. Phys.*, **64** (10), 4943–4950.

18 Tadepalli, R., Turner, K.T., and Thompson, C.V. (2008) Mixed-mode interface toughness of wafer-level Cu–Cu bonds using asymmetric chevron test. *J. Mech. Phys. Solids*, **56**, 707–718.

19 Lawless, K.R. (1974) The oxidation of metals. *Rep. Prog. Phys.*, **37**, 231–316.

20 Rieutord, F., Eymery, J., Fournel, F. et al. (2001) High-energy x-ray reflectivity of buried interfaces created by wafer bonding. *Phys. Rev. B*, **67**, 125408.
21 Munir, A. (1979) Analytical treatment of the role of surface oxide layers in the sintering of metals. *J. Mater. Sci.*, **14**, 2733–2740.
22 Diest, K., Archer, M.J., Dionne, J.A. et al. (2008) Silver diffusion bonding and layer transfer of lithium niobate to silicon. *Appl. Phys. Lett.*, **93**, 092906.
23 Taibi, R., Di Cioccio, L., Chappaz, C. et al. (2010) Full characterization of Cu/Cu direct bonding for 3D integration. Proceedings of the Electronic Components and Technology Conference.

13
Metal/Silicon Oxide Hybrid Bonding
Paul Enquist

13.1
Introduction

This chapter describes a metal/silicon oxide hybrid bonding technology that is an ideal solution for the fabrication of three-dimensional (3D) integrated circuits (ICs). This bonding technology is part of a broader class of hybrid bonding technologies that are distinct from conventional bonding technologies in that a heterogeneous bond comprising more than one type of bond component is utilized as opposed to a homogeneous bond that comprises a single type of bond component. The heterogeneous nature of the hybrid bond generally offers advantages over a homogeneous bond. For example, a homogeneous flip-chip bond utilizing solder bumps can result in yield and reliability limitations due to the lack of bonding between the reflowed bumps. Adding a heterogeneous underfill bond component between the reflowed bumps significantly improves the reliability of the solder interconnects at the expense of increased stress in the silicon die [1].

While the flip-chip/underfill hybrid bond is an excellent solution for most 3D packaging applications, the inherent nonplanarity of this bond limits the interconnect pitch and makes it unsuitable for 3D integration applications that require a bond with a micrometer scalable interconnect. Two alternative planar hybrid technologies have been proposed and developed to meet the needs of these 3D IC applications. One of these is a metal/adhesive hybrid technology using copper/benzocyclobutene (BCB) that is the subject of Chapter 11 [2]. The other is a metal/non-adhesive hybrid technology that includes a variety of metal/silicon oxide combinations that is the primary subject of this chapter. The relationship between these hybrid bonding technologies is summarized in Figure 13.1.

13.2
Metal/Non-adhesive Hybrid Bonding – Metal DBI®

The metal/non-adhesive hybrid bonding technology is a class of direct bonding that has been developed at Ziptronix, Inc. and trademarked as Direct

Handbook of Wafer Bonding, First Edition. Edited by Peter Ramm, James Jian-Qiang Lu,
Maaike M.V. Taklo.
© 2012 Wiley-VCH Verlag GmbH & Co. KGaA. Published 2012 by Wiley-VCH Verlag GmbH & Co. KGaA.

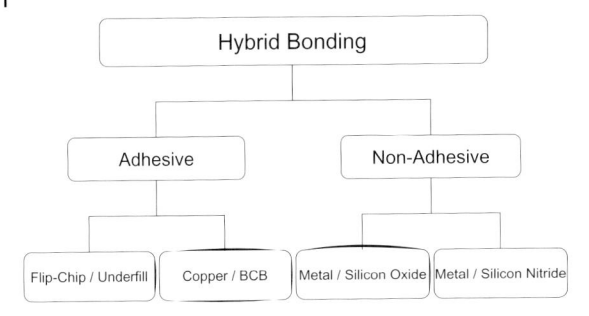

```
                    Hybrid Bonding
                   /              \
            Adhesive            Non-Adhesive
           /        \           /            \
  Flip-Chip/   Copper/BCB   Metal/Silicon   Metal/Silicon
  Underfill                 Oxide           Nitride
```

Figure 13.1 Various hybrid bonding technologies.

Bond Interconnect (DBI®) (e.g., [3–7]). DBI® is covered by a number of issued patents and patent applications (e.g., [8, 9]).

A key advantage of the metal/non-adhesive or metal/direct hybrid bonding technology compared to adhesive-based hybrid bond technologies is the breadth of post-bond fabrication processes possible. Adhesive-based hybrid bond technologies have a bond temperature characteristic of the adhesive and are typically limited after bonding to the glass transition temperature which can be close to the bond temperature. In contrast, the non-adhesive or direct hybrid bond technology is capable of achieving high bond energy at low temperatures and is capable of withstanding temperatures after bonding well in excess of that which would damage the materials that are hybrid bonded, for example ICs.

Another advantage of non-adhesive or direct hybrid bonding technology is compatibility with established CMOS wafer foundry volume manufacturing materials and processes. The materials and processes used to implement this hybrid bonding technology can leverage those currently used in volume CMOS wafer manufacture, lowering the barrier for adoption and facilitating process integration of hybrid bonding with CMOS back-end-of-line (BEOL) for lowest cost-of-ownership (CoO) 3D IC manufacture.

13.3
Metal/Silicon Oxide DBI®

The majority of direct hybrid bonding development has focused on the use of metal/silicon oxide hybrid bonding. Many different configurations have been evaluated including various combinations of hybrid bond surface fabrication processes, surface topographies, types of metal, hybrid bond alignment, and hybrid surface contact techniques. These configurations are discussed below in detail.

13.3.1
Metal/Silicon Oxide DBI® Surface Fabrication

The hybrid metal/silicon oxide bonding surface can be fabricated with a number of different process flows. The ability to implement this direct hybrid bonding with a variety of process flows facilitates adoption of the technology in a range of different environments and an assortment of different metals. Damascene, planarized stud, and nonplanar process flows suitable for aluminum, copper, gold, and nickel metals are described below.

13.3.1.1 Damascene Meta/Silicon Oxide DBI® Process Flow

The damascene hybrid bonding process flow is similar to that used in volume CMOS copper BEOL wafer fabrication. A single or dual damascene can be implemented depending on the desired surface patterning as described below. This process flow typically includes deposition of a silicon oxide layer, followed by via etch (single damascene) or via and trench etch (dual damascene) through the silicon oxide layer to a last metal layer. The desired hybrid bond metal can then be deposited in via or via and trench, for example by physical vapor deposition or electroplating. This metal deposition may be preceded by a conductive barrier liner deposition to impede diffusion of the metal into the surrounding silicon oxide and improve reliability. The deposited metal is then polished with chemomechanical polishing (CMP) to the desired surface planarity as discussed below. This process can scale to a submicrometer pitch as demonstrated by interlevel vias in CMOS copper BEOL. This process flow has been proven with both aluminum and copper hybrid bond metals as described in more detail below.

13.3.1.2 Planarized Stud Metal/Silicon Oxide DBI® Process Flow

The planarized stud hybrid bonding process flow differs from the damascene hybrid bonding process flow in that the hybrid bond metal is formed prior to instead of after the hybrid bond silicon oxide. The hybrid bond metal is first deposited on a last metal or via layer, for example using electroplating on a seed layer with a photoresist mask on a tungsten plug CMP surface on a CMOS aluminum BEOL wafer. The seed layer is then etched using the hybrid bond metal or photoresist as a mask, followed by silicon oxide deposition and CMP to the desired surface planarity as discussed below. This process can also scale to a submicrometer pitch. This process flow has been proven using nickel hybrid bond metal as described in more detail below.

13.3.1.3 Nonplanar Metal/Silicon Oxide DBI® Process Flow

The nonplanar hybrid bonding process flow is suitable for hybrid bond metals that are not well suited for CMP in either the damascene or planarized stud process. This process flow can begin with either the damascene or planarized stud process flow wherein the hybrid bond metal is replaced with a metal that is suitable for CMP. The hybrid bond metal can then be selectively deposited or deposited and patterned on the planar damascene or planarized stud hybrid surface with electrical contact to the CMP-suitable metal. The thickness of the metal is typically kept within 100 nm to minimize the nonplanarity of the hybrid surface and allow high bond energy as described below. This process flow has been proven using gold hybrid bond metal as described in more detail below.

13.3.2
Metal/Silicon Oxide DBI® Surface Patterning

The hybrid bond surface is patterned to define regions of metal and silicon oxide according to the process flow used to define the surface. These regions are designed to mate with another hybrid bond surface after these surfaces are aligned and brought into contact. The surface can be predominantly metal, silicon oxide, or comparable amounts of metal and silicon oxide portions. The geometries defining the metal or silicon oxide portions can be similar to those used in CMOS BEOL wafer fabrication, for example (sub)micrometer-scale interlevel vias or intralevel routing features. Pairs of surfaces consisting of vias and/or routing features with one being a mirror image of the other may be patterned for subsequent flipping of one surface, alignment, and contact to the other surface as described below. Surfaces that are not mirror image pairs may also be patterned for subsequent contact, for example bonding via to routing or crossover routing features.

13.3.3
Metal/Silicon Oxide DBI® Surface Topography

There are basically three DBI® surface topography configurations that can result from the DBI® surface fabrication process flows described above. These are DBI® metal planar with the adjacent silicon oxide, or above or below the adjacent silicon oxide. A planar metal/silicon oxide DBI® surface topography is one where the metal and silicon oxide relative heights are within 1 nm of each other. The metal and silicon oxide relative heights of the nonplanar metal/silicon DBI® surface topographies are typically within about 10 nm. It is also possible to accommodate a larger variation of greater than 100 nm, for example if a larger interconnection pitch is acceptable [3].

13.3.4
Metal/Silicon Oxide DBI® Surface Roughness

The silicon oxide component of the DBI® surface is typically less than 0.5 nm root mean square (RMS) to facilitate high direct bond energy between silicon oxide components of the hybrid bond. A similar low RMS for the metal component is conducive for a direct metal bond and further increase in bond energy. However, high hybrid bond energies and reliable 3D interconnections can be achieved with higher metal component RMS less than about 10 nm as described below. These roughness values can be readily achieved with commercial CMP tools using the appropriate combination of pads and slurries for the type of metal and silicon oxide, surface patterning, and desired surface topography.

13.3.5
Metal/Silicon Oxide DBI® Surface Activation and Termination

After the appropriate metal/silicon oxide DBI® surface topography and roughness have been met, the surface can be activated and terminated to enable high bond energy at low temperature after alignment and placement of two DBI® surfaces into contact in an ambient manufacturing environment. Effective activation can be achieved with a slight etch that removes contaminants, breaks surface atomic bonds, and maintains or does not significantly increase surface roughness. Terminating this surface allows the surface to be handled in an ambient manufacturing environment before it is aligned and placed into contact with another DBI® surface to form a DBI® bonded pair. The termination may also facilitate a chemical reaction between surfaces placed together that enables high bond energy at low temperature, for example below 200°C. Different types of terminations, for example based on nitrogen and hydroxyl, are possible. Higher bond energies may be obtained if both surfaces are activated and terminated prior to contact. High bond energies may also be obtained if only the silicon oxide is activated and terminated, although both metal and silicon oxide components of the hybrid surface may be exposed to the activation and termination process. The activation and termination can be accomplished with a number of different methods. For example, a plasma process can be used to both activate and terminate or a wet process can be used after a plasma activation process to terminate. These processes can be implemented with a discrete tool set or a cluster tool that includes aligning and placing DBI® bond surfaces together (www.suss.com; www.evgroup.com). The capability of high bond energy at low temperature of the technology enables a low CoO manufacturing solution by eliminating the need for a bond chamber that increases tool cost and lowers tool throughput. Surface activation and termination methods, devices that can be made with these methods, and tools that can be used to implement these methods are described more fully in a number of issued patents and patent applications (e.g., [10–16]).

13.3.6
Metal/Silicon Oxide DBI® Alignment and Hybrid Surface Contact

A number of options are available for the alignment and placement of two DBI® surfaces into contact. The preferred implementation can depend on a variety of requirements including alignment accuracy, surface form factor, and throughput. For example, a pick-and-place tool (www.set-sas.fr) may be preferred for a die-to-wafer application with a less than $10\,mm^2$ die size requiring more than 100 placements per wafer and $10\,\mu m$ alignment accuracy. Alternatively, a high-accuracy wafer alignment and placement tool (www.suss.com; www.evgroup.com) may be

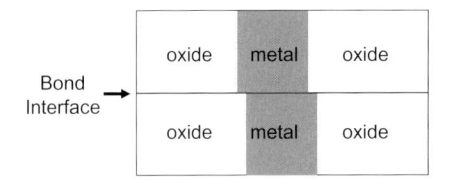

Figure 13.2 Generic metal/silicon oxide DBI® hybrid bond cross section showing silicon oxide-to-silicon oxide, metal-to-metal, and silicon oxide-to-metal bond components. The silicon oxide-to-metal bond component(s) may result from misalignment of mirror image patterned wafers.

required for a 200 or 300 mm wafer-to-wafer application requiring submicrometer alignment accuracy.

Different types of bonds are formed when two metal/silicon oxide hybrid surfaces are bonded. Various aspects of how these bonds are formed depend on the properties of the silicon oxide and metal and surface topography. Details of these bonds and how they are formed in the context of bonding hybrid patterned wafers with the metal surface level with, above, or below the silicon oxide surface are described below.

13.3.6.1 DBI® Bond Components

There are generally three types of bonds when two hybrid metal/silicon oxide surfaces have been aligned and placed together. These are silicon oxide-to-silicon oxide, metal-to-metal, and silicon oxide-to-metal bonds, as shown in Figure 13.2. These bonds are typically direct bonds without any intervening adhesive material and the metal-to-metal or silicon oxide-to-metal direct bond may be to a native oxide on the metal. The metal-to-metal direct bond can make an electrical interconnection as a result of direct contact or after a temperature cycle as described below. If desired, fusing, melting, or reflow of the metallic bond may also occur after a temperature cycle depending on the metal and temperature cycle used. Some of these features are reviewed below in the context of bonding mirror image patterned wafers with the metal surface height equal to, above, or below the silicon oxide height. The silicon oxide-to-metal bond component may be intentional, for example when bonding a via pattern to a routing pattern or bonding dissimilar routing patterns, or unintentional, for example from misalignment when bonding mirror image via or routing patterns to each other.

The bonding of these various metal/silicon oxide hybrid surfaces results in a wide variety of relative areas of these three different types of bonds. This range is generally limited by the design rules associated with the patterning and alignment accuracy of the surface contact tool and not the DBI® technology, assuming the appropriate combination of surface roughness, termination, topography, metal, and silicon oxide is used. For example, when bonding a closest pack density of mirror image vias to form a high-density 3D interconnect, the silicon oxide-to-silicon oxide and metal-to-metal bond can be about 80 and 20%, respectively, of the hybrid surface if the alignment accuracy is good enough for the misaligned metal-to-silicon oxide area to be negligible. Conversely, the metal-to-metal bond component can be larger than either of the silicon oxide-to-silicon oxide or metal-to-silicon oxide components if a network of power or ground planes are bonded together. The metal-to-silicon oxide component can also be larger than either of the silicon oxide-to-silicon oxide or metal-to-metal components if the misalignment is large or different routing layers are bonded together. The capability of the DBI® technology to accommodate these different types of bonds as part of the hybrid bond enables high bond energy regardless of the hybrid patterning or misalignment and facilitates submicrometer scaling of a 3D interconnect pitch.

13.3.6.2 Metal Surface Planar with Silicon Oxide Surface

Planarity of the metal and silicon oxide components on the hybrid surfaces allows these components to come in direct contact with each other to form direct silicon oxide-to-silicon oxide, metal-to-metal, or silicon oxide-to-metal bonds according to the respective patterns on these surfaces. This allows for a contiguous bond across the entire hybrid bond interface that affords high bond energy and isolated electrical interconnections. Both high bond energy and isolated electrical interconnections can be achieved by simply placing together suitably prepared surfaces with the requisite alignment in an ambient manufacturing environment, for example if the silicon oxide surface is appropriately activated and terminated and the metal does not have a native oxide of an extent that could preclude electrical interconnections. Planarity between the metal and silicon oxide components further enables these electrical interconnections with an interconnection pitch to be realized that is limited only by the patterning of these components and the accuracy with which the hybrid surfaces can be aligned and placed.

It is also possible that electrical interconnections are not formed upon initial placement of the hybrid surfaces together. This can happen, for example, if there is a native oxide of an extent on the metal surface that inhibits the electrical interconnection. The extent of this native oxide can depend on a number of factors including the type of metal and surface preparation or cleaning prior to bonding. Electrical interconnections may then be formed after placement, for example with a temperature cycle that creates pressure at the metal-to-metal bond from high silicon oxide-to-silicon oxide bond strength and a higher coefficient of thermal expansion (CTE) of the metal than the silicon oxide. The temperature cycle resulting in electrical interconnections depends on a number of factors including the type of metal, extent of native oxide, and silicon oxide-to-silicon oxide temperature-dependent bond strength.

13.3.6.3 Metal Surface Above Silicon Oxide Surface

High bond energy and isolated electrical interconnections can also be achieved with hybrid surfaces where the metal component is above the silicon oxide surface. The bond energy of this configuration after placing hybrid surfaces together is not limited to the metal-to-metal bonding if the material on which the hybrid patterns are built (e.g., CMOS wafers) has sufficiently low Young's modulus to allow the silicon oxide-to-silicon oxide or metal-to-silicon oxide components to come into contact. The Young's modulus required depends on a number of factors including the hybrid surface patterning, interconnection pitch, silicon oxide-to-silicon oxide and metal-to-silicon oxide bond energy, and extent of the metal above the silicon oxide surface. For example, 20 nm of extent may allow a 20 μm pitch, while 100 nm of extent may only allow a 1 mm pitch when DBI® bonding gold/silicon oxide hybrid surfaces [3]. A native oxide on the metal may be present to an extent to further limit the interconnection pitch or metal extent above the silicon oxide by requiring additional silicon oxide-to-silicon oxide or metal-to-silicon oxide bond area to provide adequate bond energy to compress the metal-to-metal during a post-contact heat treatment.

13.3.6.4 Metal Surface Below Silicon Oxide Surface

High bond energy and isolated electrical interconnections can also be achieved with hybrid surfaces where the metal component is below the silicon oxide surface. Electrical connections with this configuration are not made after hybrid surface contact unless the Young's modulus of the material on which the hybrid patterns are built is sufficiently low to allow the metal-to-metal components to come into contact. Electrical interconnections with this configuration can also be made with a post-contact temperature cycle. A larger temperature excursion than that of the other configurations may be required to close the gap formed by thermal expansion between two metal surfaces recessed below silicon oxide surfaces. Alternatively, this gap may be closed by reflow, for example if a suitable solder-based or alloyed metal is used.

This configuration may provide some advantages in volume manufacturability. For example, the surface topography of some combinations of metal/silicon oxide may be easier to control within a specification if a metal recess is allowed to be within a certain range as opposed to a more precise restriction of less than 1 nm recess. This control can be achieved with a small recess less than about 5–10 nm whose gap can be closed with a minor temperature cycle after surface contact. For example, a 1 μm tall copper metal with a recess or CMP dishing maintained within 2 nm below a silicon oxide surface can be expected to have the resulting metal-to-metal gap within 4 nm after silicon oxide-to-silicon oxide contact at ambient to be closed at about 100 °C above ambient. The use of nickel instead of copper is expected to require about 150 °C to close this gap due to the smaller CTE of nickel compared to copper.

13.3.7
Metal Parameters Relevant to DBI® Surface Fabrication and Electrical Interconnection

Metal/silicon oxide hybrid bonding is generically applicable to a wide variety of types of metals. The implementation details for a particular type of metal are typically different for a different type of metal. However, the four properties of metals summarized below provide some basic guidelines on how to prepare a hybrid surface for bonding and how to achieve isolated electrical interconnections with a scalable pitch within a low thermal budget after bonding.

13.3.7.1 Polishing Rate

The polishing rate of a metal is the rate at which metal is removed from a metal/silicon oxide hybrid surface. It is dependent on a number of factors including the metal hardness, malleability, ductility, surface patterning, and polishing conditions including pad, slurry, downforce, and rotation. A metal polishing rate similar to, lower than, or higher than the silicon oxide polishing rate will tend to yield a DBI® bond surface with metal that is planar, higher than, or lower than, respectively, the silicon oxide. Appropriate selection of polishing conditions will generally result in an appropriate surface topography for DBI® bonding.

13.3.7.2 Native Oxide Stability

Metals that have a stable oxide, like nickel and aluminum, are more likely to have this oxide inhibit electrical interconnections upon initial metal-to-metal contact. An electrically conducting bond may be achieved by removing this oxide prior to bonding or reduction of the oxide interfacial layer after metal-to-metal contact. Removal of the oxide prior to bonding may allow electrical interconnection upon direct metal contact, but may also require handling, alignment, and placement in a non-oxidizing environment to prevent re-oxidation of the metal surface. Reduction of the oxide interfacial layer after metal-to-metal contact eliminates this environmental restriction and can be accomplished with a post-contact process step, for example a temperature cycle that sufficiently compresses the metal-to-metal bond with a combination of high silicon oxide-to-silicon oxide bond energy and metal-to-silicon oxide CTE mismatch. Note that this process step has a very low CoO because the compression is provided intrinsically by the silicon oxide direct bond eliminating the need for external compression tooling and enabling a batch process. It is also possible for this process step to be eliminated completely if this temperature cycle can be provided by a downstream process step.

13.3.7.3 Coefficient of Thermal Expansion

The CTE of a metal can vary by a factor of about two depending on the type of metal. This is about 25–50 times or 12–25 ppm °C^{-1} greater than that of the silicon oxide. This significant difference in CTE results in considerable compressive stress at the metal-to-metal interface and tensile stress at the silicon oxide-to-silicon oxide interface at low temperatures above the hybrid surface fabrication and bonding ambient. This compressive stress can exceed the yield strength of the metal and facilitate formation of a metal-to-metal direct metallic bond with reliable electrical performance.

13.3.7.4 Yield Strength

The metal yield strength can vary by two or three orders of magnitude depending on the metals selected and their deposition methods. Metals with low yield strength will typically require less compressive stress to form electrical interconnections than metals with high yield strength and comparable native oxide. This lower stress may or may not occur at lower temperatures depending on the relative metal CTE and topography.

13.3.8
DBI® Metal/Silicon Oxide State of the Art

DBI® is a generic hybrid bonding technology that includes a wide range of metal/silicon oxide combinations. This section reviews four of these combinations as evaluated by a number of independent organizations and summarizes the state-of-the-art capability with regard to interconnect pitch, thermal budget, resistivity, yield, and reliability.

13.3.8.1 Aluminum/Silicon Oxide DBI® Hybrid Bonding

Aluminum/silicon oxide DBI® hybrid bonding has been reported by Edinburgh University [17]. A damascene process was used to build the hybrid surface with 20–40 nm of aluminum CMP dishing. Bonded hybrid wafers were heated to 435 °C to increase the bond strength. About 20 nm of the 40–80 nm metal-to-metal gap after placing hybrid surfaces into contact was closed during the post-contact temperature cycle. The residual gap was closed by stress relief-induced hillock formation resulting in a metal-to-metal bond and electrical interconnections. A minimum metal-to-metal contact area of 10 μm by 10 μm was needed to obtain electrical connections. Contact resistivity was over $1\,k\Omega\,\mu m^2$ and the yield was 20%.

13.3.8.2 Copper/Silicon Oxide DBI® Hybrid Bonding

Copper/silicon oxide DBI® hybrid bonding has been reported by Ziptronix, Inc. and CEA-Leti [7, 18]. Damascene processes were used to build the hybrid surfaces with a copper CMP planar or dished topography of about 2 nm. The Ziptronix process included a TiW via and trench conductive diffusion barrier liner for improved reliability. CEA-Leti reported single connections with a contact resistivity of $0.98\,\Omega\,\mu m^2$ and a 200 °C heat treatment after hybrid surface contact. Ziptronix reported fully functional serial daisy chains with 72 500 elements on a 25 μm pitch with contact resistivity less than $0.45\,\Omega\,\mu m^2$ after a 125 °C heat treatment following hybrid surface contact. These parts exceed the JEDEC temperature cycling and HAST reliability requirement. Ziptronix has also reported scaling of the direct copper DBI® process to smaller pitch and higher yield with fully functional 463 000 element daisy chains on a 10 μm pitch while maintaining contact resistivity less than $0.45\,\Omega\,\mu m^2$ using an alignment/placement pick-and-place tool with ±1 μm over 3σ accuracy. This process is based on existing CMOS industry standard volume production copper BEOL tools and processes.

13.3.8.3 Gold/Silicon Oxide DBI® Hybrid Bonding

Gold/silicon oxide DBI® hybrid bonding has been reported by Ziptronix, Inc. [3]. A nonplanar process was used to build the hybrid surface using 20–100 nm of gold on top of a planar surface. Direct metal-to-metal bonding at room temperature was reported for 5 mm diameter gold pads on a 6 mm pitch with an unbonded annulus of less than 20 μm around 20 nm pads.

13.3.8.4 Nickel/Silicon Oxide DBI® Hybrid Bonding

Nickel/silicon oxide DBI® hybrid bonding has been reported by Ziptronix, Inc. (e.g., [4–6]). A planarized stud process was used to build the hybrid surface with a nickel CMP planar or dished topography of about 2 nm. Ziptronix reported fully functional serial daisy chains with 1 000 000 elements on an 8 μm pitch with contact resistivity less than $0.5\,\Omega\,\mu m^2$ with 300 °C heat treatment subsequent to hybrid surface contact. These parts exceeded ten times the JEDEC temperature cycling and three times the JEDEC HAST reliability requirement. Ziptronix has also reported scaling of the direct nickel DBI® process to 1.5 μm pitch with sub-micrometer pitch capability subject to adequate wafer alignment and placement accuracy. This process has also been used to three-dimensionally integrate industry standard CMOS aluminum and copper BEOL interconnect wafers. Wafer start options for this process include a planarized tungsten plug CMP aluminum BEOL surface and a planarized passivated copper BEOL surface. A lower CoO alternative for implementing nickel direct DBI® process in volume production is replacement of the tungsten plug or copper-filled via with nickel at the end of the BEOL.

13.4
Metal/Silicon Nitride DBI®

Another type of direct hybrid bonding is metal/silicon nitride hybrid bonding. This is of particular interest for applications in which reliability issues result from diffusion of the metal into the surrounding silicon oxide that can be resolved by the use of silicon nitride. An example of a metal for which this can be a concern is copper, as illustrated by the use of silicon nitride capping layers in copper dual damascene CMOS BEOL fabrication.

Figure 13.3 provides an example of a copper/silicon nitride hybrid bonding structure with intrinsically high reliability resulting from complete enclosure of copper with either a conducting barrier liner or insulating silicon nitride barrier. The preparation of this surface can be carried out with the damascene process flow described above with a capping layer of silicon nitride added on top of the silicon oxide prior to trench and/or via etch. The silicon nitride is thick enough to provide sufficient residual silicon nitride diffusion barrier thickness after the DBI® CMP and depends on CMP details, typically 0.1 to 0.5 μm.

This copper/silicon nitride hybrid bond surface is similar to a typical CMOS copper BEOL surface after deposition of silicon nitride on the copper CMP surface. The primary difference is exposed copper on the hybrid bond surface that will be subsequently electrically interconnected to another hybrid surface, for example copper/silicon nitride, as shown in Figure 13.4. Note that in the case of copper/silicon nitride to copper/silicon nitride hybrid bonding, there will generally be

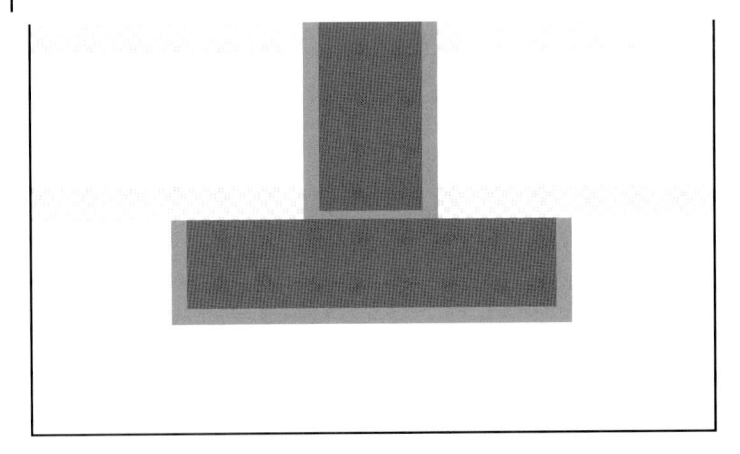

Figure 13.3 Hybrid copper/silicon nitride surface before hybrid bonding.

Figure 13.4 Bonded hybrid copper/silicon nitride hybrid surfaces.

some copper that is not bonded to copper due to misalignment resulting from imperfect alignment and placement of the two hybrid bonding surfaces together. The surface silicon nitride effectively caps this misaligned copper allowing a fully encapsulated, reliable 3D IC interconnect.

13.5 Metal/Silicon Oxide DBI® Hybrid Bonding Applications

Metal/silicon oxide DBI® hybrid bonding can be used in a wide variety of 3D packaging and integration applications. This section focuses on those 3D applications that require a process capability beyond that available from technology currently qualified for volume production that can be accommodated by DBI® and are thus likely to drive the adoption of this nascent hybrid technology.

Key distinguishing advantages of the DBI® technology compared to other bonding technologies are the ability to create large arrays of high-density 3D interconnections with high bond energy at low thermal budget with minimal applied external stress. Applications for which these advantages are required include pixelated 3D ICs, 3D heterogeneous integration, and CMOS (ultra) low-k 3D integration. The suitability of DBI® for these applications is described in more detail below.

13.5.1 Pixelated 3D ICs

A pixelated 3D IC is a type of 3D IC containing a periodic repetition of standard cells or pixels. Minimizing the pixel size with CMOS scaling reduces die size and cost assuming availability of a likewise scalable 3D interconnect bond technology. The capability of DBI® to support submicrometer 3D interconnect scalability is very well suited for this type of application.

An example of a pixelated 3D IC application is 3D backside illuminated (BSI) CMOS image sensors (CIS). Conventional two-dimensional (2D) BSI-CIS technology, consisting of bonding a CIS to a handle wafer and removing the CIS substrate to allow imaging from the backside, is experiencing rapid adoption due to improved pixel scaling, resolution, quantum efficiency, sensitivity, crosstalk, fill factor, chief ray angle, and module thickness. A 3D BSI-CIS differs from a conventional 2D BSI-CIS in that the handle wafer is a CMOS wafer that contains circuitry otherwise exterior to the pixels, enabling a 50% form factor reduction when stacked underneath the pixels. This standard CMOS handle wafer without the PMOS detectors is considerably less expensive than a conventional CIS CMOS wafer. The combination of reduced CMOS cost and reduced die size enables a 30% cost reduction in a smaller package for a 3D BSI-CSI part compared to a 2D BSI-CSI part with comparable resolution. Alternatively, a 3D BSI-CSI part can have double the pixels or resolution of a 2D BSI-CSI part in the same size package (R. Guidash, personal communication).

Advanced 3D BSI-CIS designs will require at least one 3D interconnection per pixel requiring 3D interconnect that can scale with the pixel pitch. The migration of frontside illuminated (FSI) CIS pixel pitch from the 1.6 to 1.4 μm node and the fundamental capability of BSI to scale better than FSI is expected to result in a required 3D BSI-CIS pixel pitch of 1 μm and below. This submicrometer pitch scalability is within DBI® capability.

The suitability of DBI® for this application has been investigated by Kodak (unpublished data). A 1.5 megapixel, 1.25 μm pixel pitch, 3D BSI-CIS was designed and constituent 110 nm aluminum BEOL CMOS sensor and circuit wafers built through tungsten plug CMP, as described above, by a commercial foundry. The wafers were then hybrid bonded using a direct nickel DBI® planarized stud process and the sensor wafer was thinned to 10 μm using back-grinding and silicon CMP by Ziptronix. Bonded wafers were annealed at about 400 °C to improve dark current. Low sensor wafer distortion was enabled by high DBI® bond energy at low temperature that facilitated wafer-scale color filter array fabrication. 3D BSI-CIS parts were then singulated, packaged, and tested yielding the results described below.

A scanning electron micrograph cross section of a single DBI® 3D interconnection at the hybrid interface within the 3D BSI-CIS part is shown in Figure 13.5. This micrograph shows the nickel DBI® interconnection of metal 4 in the sensor wafer to metal 4 in the circuit wafer using a nickel diameter and thickness of about 3 and 0.5 μm, respectively.

The quantum efficiency of the 1.5 megapixel, 1.25 μm pixel pitch 3D BSI-CSI parts built with nickel DBI® and bulk silicon CMOS detector wafers were measured after application of an antireflection coating. Ideal blue response was obtained as shown in Figure 13.6. Lag, linearity, and charge capacity were as good as FSI-CSI parts. Operability of the 1.5 megapixels was 100%. An unfocused image obtained from the prototype parts is shown in Figure 13.7.

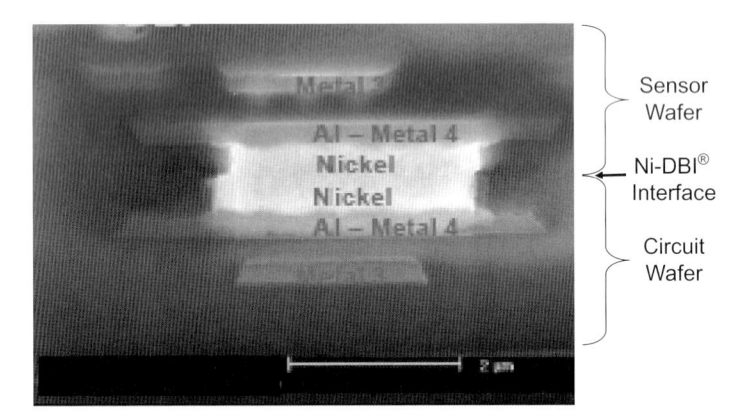

Figure 13.5 Scanning electron micrograph cross section of 3 μm diameter nickel DBI® interconnecting CMOS sensor and circuit wafers in prototype 1.5 megapixel, 1.25 μm pitch 3D BSI-CIS.

13.5 Metal/Silicon Oxide DBI® Hybrid Bonding Applications

minf F0.4 urn pird GT-Variable Share

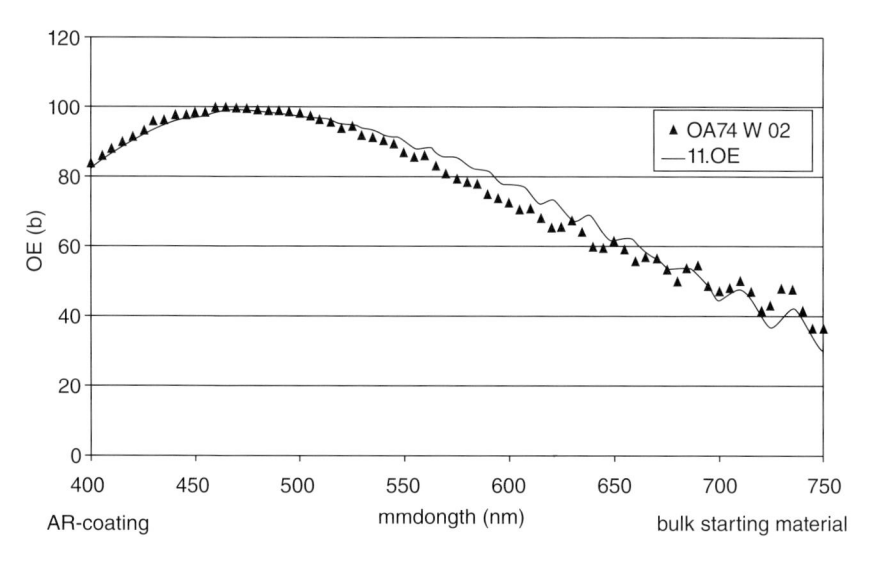

Figure 13.6 Quantum efficiency (QE) versus wavelength for antireflective (AR)-coated 3D BIS-CIS built with a bulk silicon CMOS detector wafer and nickel DBI®.

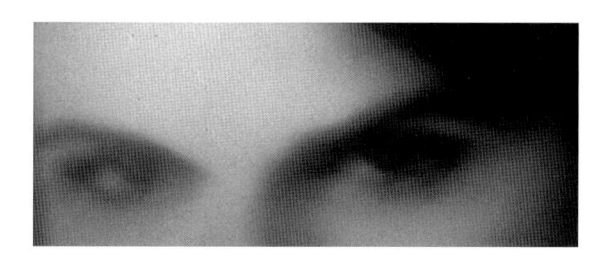

Figure 13.7 Unfocused image from 1.5 megapixel, 1.25 µm pixel pitch, 3D BIS-CIS built with nickel DBI® hybrid bonding.

Scaling the format to 8 megapixels and the pixel pitch and DBI® interconnect to 1.1 µm pitch is projected to enable a high-resolution 8 megapixel CIS in an industry-standard SMIA-65 package for mobile imaging applications including cell phones. The excellent performance and yield of these prototype 3D BSI-CIS parts indicate that DBI® is a suitable candidate technology for next-generation BSI-CISs.

13.5.2
Three-Dimensional Heterogeneous Integration

Three-dimensional heterogeneous integration (HI) involves bonding of materials like III/V semiconductors that can have significantly lower thermal budgets than

CMOS. These materials also have significantly different CTE that can further limit the thermal budget of a bond or post-bond fabrication process. The precise temperature limit depends on the combination of materials integrated, so the 3D bonding and interconnect technology with the lowest thermal budget will generally be able to accommodate the widest variety of HI. Applications enabled by HI include improved spectral response image sensors by combining non-silicon detector material with CMOS readout integrated circuits and high-resolution pixilated displays by combining nitride-based visible emitting material with CMOS driver circuits.

The fundamental high bond energy at low temperature capability of DBI(R) is ideal for HI. For example, the 125 °C result reported by Ziptronix for direct copper DBI® with 2 nm dishing is well within the thermal budget requirements of most HI combinations. Reduction of this thermal budget to room temperature is possible with the appropriate surface roughness, planarity, and activation and termination as described above.

13.5.3
CMOS (Ultra) Low-k 3D Integration

Advanced CMOS nodes require (ultra) low-dielectric materials to minimize interconnect delays. These materials are very brittle and can be easily damaged due to their poor mechanical properties. A 3D integration technology that applies minimum stress is thus expected to be compatible with advanced CMOS nodes. The capability of DBI® to achieve high bond energy at low temperature by simply placing hybrid surfaces together without any externally applied pressure indicates that this is a suitable technology for this application.

The suitability of DBI® for CMOS low-k 3D integration was evaluated by Ziptronix with a direct nickel DBI® process using 300 mm, 65 nm, 12-level metal, copper low-k FPGA wafers and a silicon test wafer. After DBI® bonding, the silicon test wafer was thinned to about 25 μm and silicon vias were cut to expose FPGA pads for testing. A scanning electron microscopy cross section showing DBI® integration of the 65 nm, copper low-k FPGA wafer to the silicon test wafer is shown in Figure 13.8. Comparison of on-wafer FPGA tests before and after DBI® bonding, test wafer thinning, and FPGA pad cut indicated no degradation in FPGA performance due to the DBI® and post-bond processes. This initial DBI® low-k result indicates compatibility of DBI® with CMOS low-k 3D integration.

13.6
Summary

Metal/silicon oxide hybrid bonding enables direct bonding between both metal and silicon oxide components with high bond energy and low thermal budget required for 3D IC applications. The technology is robust in that it can be implemented with industry-standard CMOS wafer fabrication tools and processes and

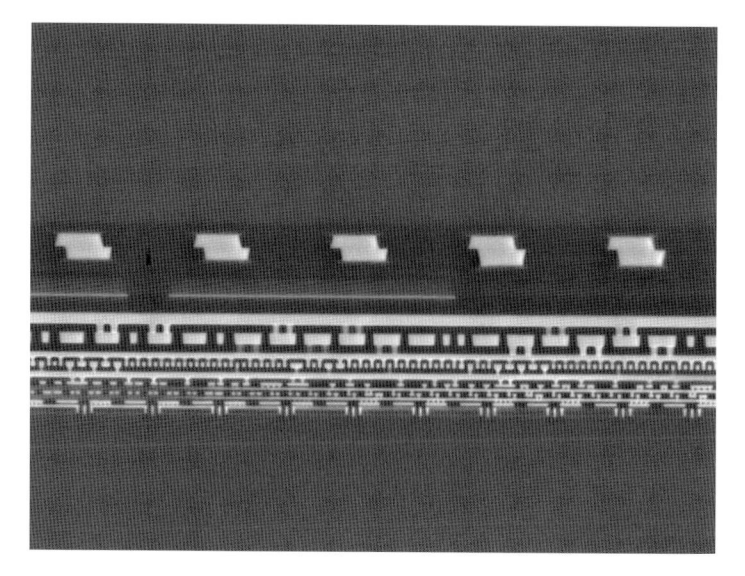

Figure 13.8 Scanning electron micrograph cross section of 10 μm pitch nickel DBI® 3D interconnecting 65 nm, 12-level metal, copper/low-k FPGA CMOS wafer-to-silicon test wafer.

a variety of metals, process flows, surface topographies, surface patterning, and bond configurations. High bond energy with submicrometer 3D interconnect scalability at low or room temperature without externally applied pressure designate this technology as a preferred platform solution for emerging 3D IC applications including pixelated ICs, HI, and copper/low-k CMOS.

References

1 Palaniappan, P., Baldwin, D.F., Selman, P.J., Wu, J., and Wong, C.P. (1999) Correlation of flip chip underfill process parameters and material properties with in-process stress generation. *IEEE Trans. Electron. Packag. Manuf.*, **22** (1), 53–62.

2 Gutmann, R.J., McMahon, J.J., and Lu, J.Q. (2007) Damascene-patterned metal-adhesive (Cu-BCB) redistribution layers. *MRS Symp. Proc.*, **970**, 205–214.

3 Tong, Q.Y. (2006) Room temperature metal direct bonding. *Appl. Phys. Lett.*, **89**, 182101.

4 Enquist, P. (2007) High density Direct Bond Interconnect (DBI™) technology for three-dimensional integrated circuit applications. *MRS Symp. Proc.*, **970**, 13–24.

5 Enquist, P. (2009) Scalability and low cost of ownership advantages of Direct Bond Interconnect (DBI®) as drivers for volume commercialization of 3D integration architectures and applications. *MRS Symp. Proc.*, **1112**, 33–42.

6 Enquist, P. (2008) Direct bonding processes for 3-D integration, in *Handbook of 3D Integration*, vol. 2, (eds P. Garrou, C. Bower, and P. Ramm), Wiley-VCH Verlag GmbH, Weinheim, p. 487.

7 Enquist, P., Fountain, G., Petteway, C., Hollingsworth, A., and Grady, H. (2009) Low cost of ownership scalable copper direct bond interconnect 3D IC technology for three dimensional integrated circuit applications. IEEE International Conference on 3D System Integration, 28–30 September 2009.

8 Tong, Q.Y., Enquist, P.M., and Rose, A.S. (2005) Method for room temperature metal direct bonding. US Patent 6,962,835, 8 November 2005.

9 Tong, Q.-Y., Fountain, G., and Enquist, P. (2009) Method for low temperature bonding and bonded structure. US Patent 7,602,070, 13 October 2009.

10 Tong, Q.-Y., Fountain, G., and Enquist, P. (2005) Method for low temperature bonding and bonded structure. US Patent 6,902,987, 7 June 2005.

11 Tong, Q.-Y., Fountain, G., and Enquist, P. (2006) Method for low temperature bonding and bonded structure. US Patent 7,041,078, 9 May 2006.

12 Tong, Q.-Y. (2006) Method of room temperature covalent bonding. US Patent 7,109,092, 19 September 2006.

13 Tong, Q.-Y., Fountain, G., and Enquist, P. (2008) Method for low temperature bonding and bonded structure. US Patent 7,335,572, 26 February 2008.

14 Tong, Q.-Y. (2008) Method of room temperature covalent bonding. US Patent 7,335,996, 26 February 2008.

15 Tong, Q.-Y., Fountain, G., and Enquist, P. (2008) Method for low temperature bonding and bonded structure. US Patent 7,387,994, 17 June 2008.

16 Tong, Q.-Y., Fountain, G., and Enquist, P. (2009) Method for low temperature bonding and bonded structure. US Patent 7,553,744, 30 June 2009.

17 Lin, H., Stevenson, J.T.M., Gundlach, A.M., Dunare, C.C., and Walton, A.J. (2008) Direct Al–Al contact using low temperature wafer bonding for integrating MEMS and CMOS devices. *Microelectron. Eng.*, **85**, 1059–1061.

18 Gueguen, P., Cioccia, L.D., Gonchond, J.P., Gergaud, P., Rivoire, M., Scevola, D., Zussy, M., Lafond, D., and Clavelier, L. (2009) 3D vertical interconnects by copper direct bonding. *MRS Symp. Proc.*, **1112**, 81–90.

Part Two
Applications

14
Microelectromechanical Systems

Maaike M.V. Taklo

14.1
Introduction

Wafer bonding is a central process step within microsystem technology. Several microelectromechanical systems (MEMS) that are available on the market today would not have been successful without the use of wafer bonding technology. Three of the wafer bonding techniques described earlier in this book have been especially central for MEMS in the last three decades: field-assisted (anodic) bonding, glass frit bonding, and direct wafer bonding. However, presently there seems to be a trend in which adhesive bonding and metal bonding are gaining ground, though within separate application areas. Adhesive bonding can be a low-cost alternative for products with relaxed requirements regarding hermeticity and dimensional control. The use of adhesives has proven to be useful for instance within the field of bio-MEMS, and for micro total analysis systems in particular. Furthermore inkjet heads are commonly polymer bonded. The seal for these systems must normally be leak-tight for fluids but can accept a certain leak of gas, as opposed to devices like pressure sensors, accelerometers, and gyros where hermeticity is of primary concern. Metal bonding has lately been mentioned to have favorable properties concerning hermeticity. Bond frames where glass is replaced by metal are expected to allow a reduction in width from 150–300 μm down to about 30 μm and still ensure a satisfactory lifetime for a device [1]. Typically, a reduction in the footprint of a device corresponds directly to a gain in profit, and thus a reduced bond frame width is most often targeted. On the other hand, re-qualification of products already manufactured using glass-based bonding can be expensive.

From observations of existing and planned devices for the market, there still seems to be a large variety in the selection of applied wafer bonding technologies. The wide range of both applications and participants in the field of MEMS, each participant with its own intellectual property, tool park, and history, indicates that a large range of technologies can be expected to be applied also in future products. Development of a wafer bonding technology for a specific device or as part of a

---

Handbook of Wafer Bonding, First Edition. Edited by Peter Ramm, James Jian-Qiang Lu, Maaike M.V. Taklo.
© 2012 Wiley-VCH Verlag GmbH & Co. KGaA. Published 2012 by Wiley-VCH Verlag GmbH & Co. KGaA.

company's technology platform can take years, be extremely costly, and should not be underestimated.

This chapter looks into how wafer bonding can be used for device encapsulation, but also how wafer bonding can be used to build up advanced three-dimensional MEMS devices. Examples of devices where wafer bonding is an essential and enabling process step are presented together with their various requirements for the bonding process. Finally, we look into some concerns that are important for three commonly used wafer bonding techniques (fusion bonding, anodic bonding, and eutectic AuSn bonding) if they are to be applied for mechanically structured wafers.

14.2
Wafer Bonding for Encapsulation of MEMS

14.2.1
Protection during Wafer Dicing

The most obvious application of wafer bonding is probably its use as a zero-level packaging step for devices before dicing. Most MEMS devices need some kind of encapsulation in a final package for mechanical protection or to have a certain atmosphere in which they are operated. Caps can naturally be mounted one by one onto singulated MEMS devices during first-level packaging, but substantial cost reduction can normally be achieved for every process step that can be transferred from chip to wafer level.

It should be mentioned that several opportunities exist for both dicing and packaging of MEMS devices having delicate and/or released structures. If the structures cannot survive the water jet used for regular dicing (one may experience breakage of, for example, beams and membranes, and stiction of released structures) the devices can be diced with a laser saw. However, the cost efficiency of this kind of process and its applicability to all kinds of wafer materials and thicknesses are somewhat limited. Another alternative is to postpone the release process itself until after dicing at chip level, but this again has cost implications. It is commonly accepted that the most cost-efficient solution is rather to perform wafer-level bonding before regular dicing. A watertight seal is then obviously required for the wafer bonding process.

14.2.2
Routing of Electrical Signal Lines

The challenge for most MEMS devices that are to be encapsulated at the wafer level is how to combine transfer of electrical signals with a sealed bond frame. If through-silicon vias (TSVs) are realized in the MEMS wafer itself or in the cap wafer, signal transfer is quite easily combined with a wide range of wafer bonding methods since the topography in the bond frame region will not depend on the electrical routing. Early examples of such design solutions were presented by Esashi *et al.* [2], where signal lines were routed through the cap wafer. However, since the manufacturing of TSVs so far has been considered as relatively expensive and complicated, it has been more common to route signal lines in horizontal electrical feedthroughs crossing the bond frame. Critical issues have been the influence on the topography of the routing across the bond frame and the ability to electrically insulate the routing when crossing the bond frame. One solution is to use for instance adhesive bonding or glass frit bonding, since both techniques can tolerate some topography in the bond areas and the bond frame material is itself electrically insulating. Only the glass frit solution can lead to a hermetic seal, but an adhesive like benzocyclobutene (BCB) can at least lead to a watertight seal which may be good enough to withstand wafer dicing. Glass frit bonding in combination with horizontal electrical feedthroughs has been used [3]. BCB has been used in several microfluidic devices where hermeticity is normally not required [4]. When horizontal feedthroughs are applied, the cap wafer must be removed in the regions of the device where the electrical pads are to be connected, typically by wire bonding, to the remaining system. So-called bridge dicing can be applied to achieve access to the pads. An example of how this can be realized is shown in Figure 14.1 for an accelerometer bonded with BCB [5]. A cavity is defined in the cap wafer above the pads, and the depth of this cavity is made large enough to allow the dicing blade to stop within this cavity.

A solution patented by Sensonor [6] is to use what they call buried conductors to route electrical signals below an anodically bonded frame. The electrical signal is transferred in highly doped pn-insulated silicon regions when crossing the bond frame. The benefit of the buried conductors is that the surface topography in the bond frame region is reduced to a level that does not hinder a hermetic seal by anodic bonding. The relatively high resistance of such conductors can be a limitation for some radio-frequency MEMS devices, but buried conductors are highly applicable for devices like pressure sensors, accelerometers, or gyros.

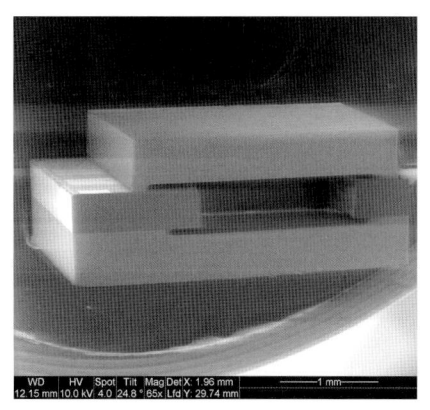

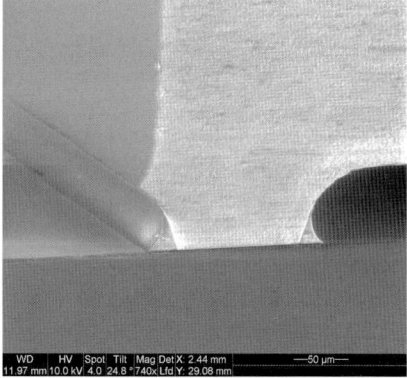

Figure 14.1 An accelerometer with horizontal feedthroughs where the electrical pads are revealed by bridge-dicing after wafer-level encapsulation by BCB bonding [5].

As solutions for vertical feedthroughs are now offered commercially (for instance TSVs in MEMS wafers [7], silicon vias in glass cap wafers [8], or tungsten vias in glass cap wafers [9]), we can probably expect to see devices designed with TSVs or other vertical feedthroughs more frequently in the near future. The cost of including TSVs is expected to be decisive for most companies when choosing their technology platform. Two obvious benefits of TSVs related to wafer bonding are the removal of the challenge of having routing crossing the bond frame and the removal of the need to reveal the bond pads after bonding. Another benefit of introducing TSVs is that the footprint of the device can be reduced since the bond pad area can be cut off. The combination of wafer bonding with TSVs is essential to achieve truly wafer-level packaged devices.

14.3
Wafer Bonding to Build Advanced MEMS Structures

14.3.1
Stacking of Several Wafers

By using bonding in the manufacturing of MEMS, real three-dimensional devices can be fabricated. Buried channels, cavities of various sizes, and movable parts can be combined to build advanced miniaturized sensors and actuators. A differential pressure sensor is shown in Figure 14.2 as an example of a device built up by fusion bonding of a silicon triple stack [10]. Bare silicon surfaces were bonded to thermally oxidized surfaces. The oxide layers were included as spacers in the device.

Direct wafer bonding, and fusion bonding in particular, has proven to be a versatile tool for building up complicated MEMS devices since extended postprocessing of such wafer laminates is feasible. As long as the wafer bonding is well performed, the laminate can be handled afterwards as a regular, but thicker wafer by most automatic laboratory tools. Bonded wafers are accepted by several

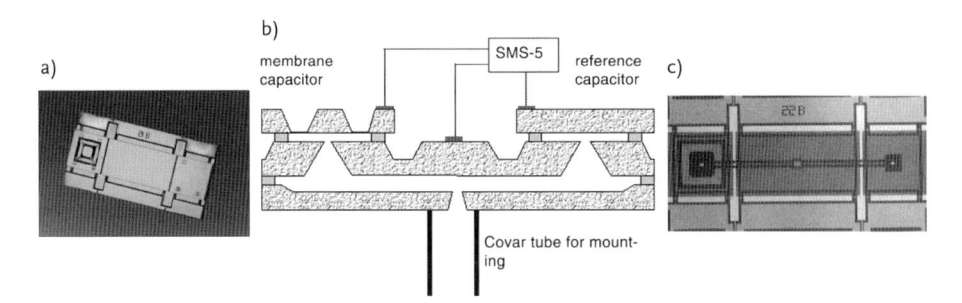

Figure 14.2 A differential pressure sensor built up by fusion bonding of a triple stack of silicon wafers [10]. (a) An image of the sensor is shown together with (b) a sketch of the cross section and (c) an infrared image of the device.

commercial actors providing process services such as ion implantation and epitaxial deposition. If several wafers are to be stacked, one must keep in mind that the stack becomes gradually stiffer for every wafer that is added, resulting in reduced adaptability to a subsequent wafer. The yield of each bonding step must be high if the overall yield of a multiple stack is to become reasonable. Despite such challenges, successful examples of multiple stacks of bonded wafers for MEMS have been presented, for instance by Mehra *et al.* [11]. That group has presented a range of combustion engines throughout the last few years manufactured by fusion bonding of up to five silicon wafers.

14.3.2
Post-processing of Bonded Wafers

When bonded wafer pairs, also called laminates, are to be post-processed, it is important to focus not only on the quality of the complete bonded laminate but also on the quality of the wafer bonding locally. A range of problematic handling issues will arise if the laminate edges or surfaces do not fall within certain standards, especially if the post-processing is mainly to be done by automatic handling. Alignment of the wafers during bonding is a major concern for overall size control. If the wafers are supposed to be patterned at the time of bonding, it is important that the individual patterns are well centered on the wafers and well aligned to wafer flats if such are present. Otherwise the final laminate will be larger than a standard wafer and the combined wafer flat may not be well enough defined for certain automatic flat finders.

If silicon-on-insulator (SOI) wafers are used in combination with wafer bonding and etch-back to realize certain structures, the shape of the wafer edges is most critical. Unless these edges are given special attention, they may end up like uneven razor blades that efficiently cut up materials like Teflon cassettes. Residues of either the laminate itself or cut up material may contaminate, for instance, rinse baths, etch baths, and spin rinse dryers. Several wafer bonding processes are prone to result in imperfect bonding at the wafer edges. Imperfectly bonded wafer edges may be slightly yield-reducing when bonding is used for wafer-level encapsulation, but these edges may be serious yield reducers when bonding is combined with post-processing. If layers on the laminate edges flake off during post-processing, this may be extremely harmful to both the laminate itself and the tools involved. Harmful edges may be etched away in separate process steps, or may sometimes be completely avoided if various combinations of edge protections (resist/mechanical) are used.

The laminate surface topography must be well controlled. Voids are well known as indicators of imperfectly bonded wafers. Several bonding processes result in residual gas either as a product of the bonding process or as a result of improperly cleaned or not well enough dried wafer surfaces [12]. Desorbed gas molecules tend to gather in air gaps if such are available or create voids locally if energetically favorable. Such voids may contain an overpressure large enough to cause the laminate surface to bulge upwards. These voids may be harmless if limited in size and numbers, but become problematic if they are too large. Outward bulging areas are for instance problematic during photo processes if minimum alignment and exposure gaps are required because the bulging areas can collide in the photo mask. Bulging surfaces may crack if placed upside down on chucks, and they may burst if post-processed at an elevated temperature. Bonded wafers should therefore typically be characterized with regard to voids before post-processing and a completely void-free bonding process should be targeted.

14.4
Examples of MEMS and Their Requirements for the Bonding Process

To give an idea of the notably large variety of bonding methods that have been used for various MEMS devices over the last few years, some examples are presented in Table 14.1. The large range of MEMS devices made from SOI, and often

Table 14.1 Examples of some MEMS devices and the applied bonding technology. Requirements for the bonding are indicated. For some devices wafer bonding is used to fabricate (build up) the device, whereas in other examples wafer bonding is used primarily for encapsulation.

Example of device	Bonding method	Vacuum requirement	Bond strength requirement	Ref.
Pressure sensor	Anodic	Hermetic	High	[13]
Accelerometer	Al–Al	Hermetic, leak rates quantified	High, shear tests performed	[14]
Microfluidic chip	Adhesive, SU-8	Nonhermetic, liquid-tight	~20 MPa expected	[15]
Microphone	Anodic or eutectic suggested	Hermetic		[16]
Interlayer cooling in three-dimensional chip stack	AuSn	Leak-tight	7 bar overpressure	[17]
Microactuator for HDD R/W heads	Adhesive, SU-8	Nonhermetic	Up to 20.6 MPa expected	[18]
Micromirror	Adhesive, PMGI	Nonhermetic	High expected	[19]
Micromotor	Fusion	Hermetic		[11]
Liquid lens	Dry-film photoresist	Nonhermetic		[20]
Gyroscope	Solder	Hermetic		[21]

bonded SOI, wafers are not included, but an overview of devices based on SOI technology can be found in Usenko and Carr [22].

The choices that have been made with regard to bonding technology are probably a mixture of technologies that were already available in the laboratories where the device was developed and technologies specially developed for the device given certain specifications. The most important specifications are vacuum level required and bond strength required. According to [23] there is nothing called 100% hermetic. The permeability rate of a material is the rate at which gas atoms diffuse through a material. By definition materials with less than 1 day of sealing capacity ($10^{-14}\,\mathrm{g\,cm^{-1}\,s^{-1}\,torr^{-1}}$) are considered nonhermetic. An indication of the required vacuum level for encapsulation of certain applications can be found in [24]:

- accelerometers: 300–700 mbar
- infrared sensor arrays: $<10^{-2}$ mbar
- pressure sensors: 1–10 mbar
- resonators/electromechanical filters: 10^{-1} mbar
- bolometers/radio-frequency switches: $<10^{-4}$ mbar.

The acceptable leak rate must then be seen in accordance with the required lifetime of the device. The bond strength required will also vary with application. A bond strength of 4–5 MPa is the minimum acceptable for most wafer-level bonded devices since weaker interfaces will normally not survive regular wafer dicing. Higher bond strength is needed for bond seals that will experience differential pressures. A wide range of bond strength evaluation methods have been presented in the literature. A review paper on the topic has been published [25].

14.5
Integration of Some Common Wafer Bonding Processes

A wafer bonding process may show promising features such as having a high bond strength or extreme hermeticity, but the process is not highly valuable unless it can be integrated with the process flow for the complete MEMS device. Typical challenging factors of bonding processes are high bonding temperatures, high bonding pressures, demand for perfectly smooth and flat wafers, and patterning of bonding materials. We will look into some details related to combining three commonly used wafer bonding techniques (fusion bonding, anodic bonding, and eutectic AuSn bonding) with process sequences where mechanical structuring of the wafers to be bonded is included.

14.5.1
Fusion Bonding of Patterned Wafers

When fusion bonding is to be applied for a MEMS device, either to cap a device or to build up a certain structure in a device, the wafers will have to satisfy some rather extreme demands. The surface roughness of both wafers is recommended to be <0.5 nm (root mean square) and the bow or warp of, for instance, a 100 mm wafer of 525 μm thickness should be less than about 25 μm. If possible, at least one of the surfaces should be a bare silicon surface, and the other surface should preferably be a thermally oxidized silicon surface or another bare silicon wafer. The motivation for having an oxide layer on one side of the stack is the ability of this layer to absorb gas products from the bonding process during annealing [26]. The reason for preferring a bare silicon surface on the other side of the stack is that the bonding process can be regarded as an oxidation process needing fresh silicon.

The cleanliness and the appropriate surface preparation of the wafers to be bonded are of utmost importance. The surfaces are most often rendered as hydrophilic (terminated by Si–OH groups) as possible, though also bonding of extremely hydrophobic surfaces (terminated by Si–HF or Si–H groups) may be applied. Fusion bonding of hydrophilic surfaces will result in a layer of oxide several nanometers thick at the bond interface, as also described in Chapter 5. This oxide layer can be omitted by bonding hydrophobic surfaces, but hydrophobic bonding is normally considered as more demanding due to an especially low pre-bonding strength. It is normally only applied if the electrically insulating effect of the oxide growth on the bond interface cannot be accepted from a design point of view.

An unprocessed or only thermally oxidized silicon surface satisfies the requirements for fusion bonding. However, if the oxide or the silicon needs to be removed or patterned before bonding, one must be careful with the surface treatment. Removal of the oxide by dry etching is, for example, not recommended as the bombardment of the ions during etching may increase the surface roughness of the underlying silicon despite a high selectivity in the etch recipe. Wet removal of oxide layers is rather recommended. Dry removal of a resist layer from a surface that is to be bonded is, on the other hand, normally allowable as long as the stripping is done thoroughly and in a Faraday cage to avoid ion bombardment. It is critical that all hydrocarbons are removed as these are efficient hinders for good bonding. Wet and dry removal of the resist can preferentially be combined. Reticulation of resist, typically on wafer edges, is a known problem especially during deep reactive ion etching (DRIE). Reticulated resist should be avoided as it otherwise may be very hard to avoid residues completely. A non-optimized DRIE process may result in the formation of damaged side walls of structures and even in black silicon (an unwanted side effect of reactive ion etching resulting in very low reflectivity due to "grass" of silicon) in certain regions of a wafer. Such etch damages may also result in residues on the wafer surfaces that are hard to remove. Tiny parts of the silicon may come loose during rinsing, etc., and later stick onto the surfaces.

When designing for fusion bonding, one must keep in mind that the motivation for the wafers to stick together after pre-bonding is that the total energy of the system can be reduced. Wafer curvatures will fight against the pre-bonding whereas all areas that can be bonded will assist in the energy reduction. For that reason it should be obvious that to optimize the bond area at the wafer level is beneficial, but that the closer one gets to a situation with perfectly flat wafers with flawless surfaces, this demand will be more relaxed. A definite minimum required total bond area for a given wafer size is therefore hard to give, but the demand for a large bond width is in general not very strong. Well-defined bond widths of 30–50 μm of limited extension can become well bonded as long as the total bond area at the wafer level is sufficiently large.

Cleaning and surface modification (for either hydrophilic or hydrophobic bonding) must be done within some hours before bonding if achieved by wet baths. Wafer handling should be reduced to a minimum after wafer cleaning. It can be a good solution to sort all wafers to be bonded in their cassettes before cleaning and to avoid any handling with tweezers after cleaning. If tweezers are required, a vacuum tool should at least be used so that the surfaces to be bonded are not touched at all. Various wet baths have been suggested to render the surfaces hydrophilic, but the easiest available solution is simply to use an RCA1 clean (a mixture of NH_4OH/H_2O_2). A piranha clean (a mixture of H_2SO_4/H_2O_2) can be used in addition before hydrophilization if a more efficient cleaning of the surfaces seems to be required. If delicate MEMS structures are present that will not survive any wet processing, solutions with for instance oxygen plasmas can be used instead to achieve the correct surface conditions.

Pre-bonding, typically performed at 50–90 °C in a vacuum better than about 5×10^{-3} mbar, should preferentially not result in stressed delicate structures like bulging membranes or bended beams. Plastic deformation may occur during the subsequent high-temperature annealing caused by slip in the silicon crystal lattice. Slip lines may for instance appear if undesirable bubbles or voids develop during annealing as shown in Figure 14.3. In this example an oxide layer was not well enough removed before bonding and the oxide residues resulted in voids and related slip lines.

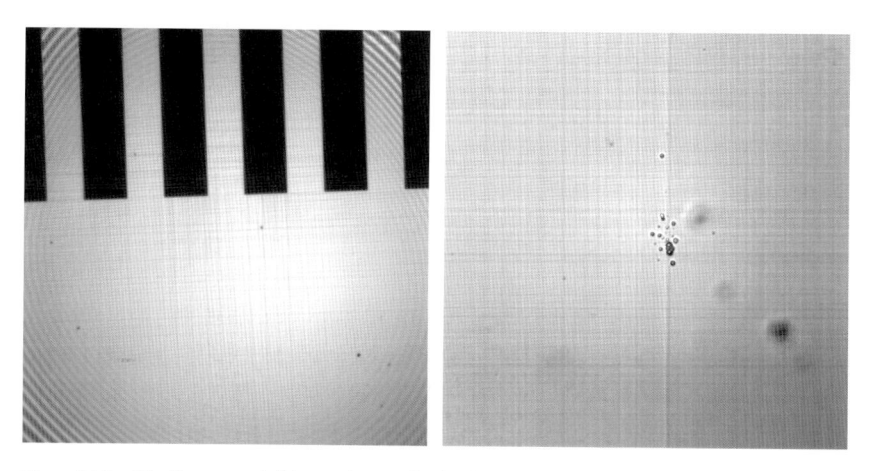

Figure 14.3 Slip lines are visible in silicon after high-temperature annealing of fusion bonding. The silicon was stressed during annealing because oxide residues resulted in the evolution of voids. (Light microscope images by SINTEF.)

If a delicate membrane is needed in a device, it is a good solution to bond together, for example, a bulk wafer with a cavity and an SOI wafer. The handle layer of the SOI wafer can be removed after performing the complete fusion bonding process and slip lines should be avoided.

If two structured silicon wafers for some reason turn out to be impossible to bond, some criteria could be checked:

- inspect the surface cleanliness (with optical microscopy, scanning electron microscopy (SEM/energy dispersive X-ray spectroscopy (EDX));
- measure the wafer bows;
- measure the surface roughness (atomic force microscopy is required);
- try to bond each of the surfaces individually to a perfectly clean test wafer;
- consider including a chemical mechanical polishing (CMP) step;
- measure the water contact angle to ensure that one has the desired surface (hydrophilic/hydrophobic).

If the wafer surfaces still do not bond, an alternative may be to deposit oxide layers on the surfaces and perform a CMP step where the oxide is partly removed (e.g., deposit 2 µm of oxide by plasma-enhanced chemical vapor deposition and remove about 1 µm). Plasma-activated bonding, which is another version of direct wafer bonding, has proven to be a valuable process for wafers with deposited oxide layers that have been subjected to CMP. This process is described in Chapter 6.

14.5.2
Anodic Bonding of Patterned Wafers

Anodic bonding is normally considered as less problematic to perform on structured wafers than fusion bonding, especially because the demands for an extremely low surface roughness and limited wafer bow or warp are less stringent. However, as mentioned in Chapter 4, the surfaces should be rather near perfect even for anodic bonding, especially if a hermetic seal is required.

When anodic bonding is planned for a device, there are certain issues that must be checked in the design. The design of the wafers must ensure that the electric field required for the process can be formed at the interfaces to be bonded. Electrical contact to the silicon wafer is needed, and thus the backside of the wafer should not have a very thick dielectric layer. However, since bonding is performed at an elevated temperature and with voltages up to 2000 V, thinner layers of, for instance, oxide can be tolerated. The acceptable thickness of the backside dielectric layer will depend on its electrical quality since the bonding process must rely on breaking through it with the available voltage. Oxide layers may also be present anywhere inside the silicon wafer, as is the case for SOI wafers with buried oxide (BOX) layers. The presence of an oxide layer directly on the bonding surface must be considered somewhat differently. This oxide layer will not act only as a temporary hindrance for the bonding that can be broken through, but rather as a permanent contributor to a reduced voltage drop across the bonded interface in the same manner as the initial air gap between the wafers. A piezoresistive silicon/glass

Figure 14.4 A piezoresistive silicon/glass cantilever die fabricated using anodic bonding. Both a backside and a buried oxide layer were present during bonding. The element is seen from the top and from the bottom side. (Photo: SINTEF.)

cantilever die that was anodically bonded at SINTEF despite the presence of several dielectric layers is shown in Figure 14.4. There was a thermally grown oxide layer (370 nm) on both the front and backside of the silicon wafer, and a BOX layer of 500 nm in addition. A voltage of 1000 V was applied at 400 °C to enable bonding of this wafer to a Pyrex 7740 glass wafer. Even though p–n junctions can be insulating at room temperature, such junctions are not a concern during anodic bonding. At the bonding temperature, p and n regions behave almost the same electrically.

If dielectric layers thicker than those that are possible to break through during bonding are present in the silicon wafer, one may get around this by contacting the silicon wafer from the front with a so-called side pin rather than from the back with a plate electrode. A small portion of the glass wafer must be removed to allow a side pin to contact the silicon wafer, and the silicon wafer must be designed in such a way that the electric voltage is distributed well from the contact point to all the regions of the silicon wafer that is to be anodically bonded. An example of a device that was fabricated at SINTEF by using a side pin and anodic bonding is shown in Figure 14.5. Both nitride and oxide layers were present in the design which made it hard to perform the bonding using the standard configuration with electrical contact from the backside of the silicon wafer by a plate electrode.

Another concern relevant for ensuring the required electric field is to avoid short circuiting somewhere in the glass. This is a topic of current interest as compound wafers become more common. Compound wafers of silicon and glass or tungsten and glass are commercially available [8, 9]. One must ensure that the electronically conductive parts of the compound wafer are not in direct contact with the electrode on the glass wafer (which is a so-called ion conductor). This can be avoided for instance by carefully designing the electrode or by adding another glass wafer or

Figure 14.5 An optical microphone fabricated using anodic bonding. A side pin was used during anodic bonding because thick dielectric layers were present in the design that could not be broken through with available parameter settings of the bonder. (Photo: SINTEF.)

layer in the system that can be removed again after bonding. The former solution may be an option for a large-volume product, whereas the latter is at least the preferred choice for prototyping. Conductive layers like metals may be included on the bonding side of the glass wafer, for example as electrodes within the device or as optical elements. One must check that these metal elements do not short circuit the system. A special precaution must be taken if thin aluminum layers are supposed to be present on the bond interface of glass wafers. When a voltage is applied, a bonding process will start at the interface between the metal and the glass. Since the bonding process includes oxidation, as described in Chapter 4, there is a risk that the aluminum layer be completely "oxidized away" and become a part of the glass wafer. A large portion of Pyrex 7740, for example, is actually alumina. The effect was demonstrated by for instance Veenstra et al. [27].

When anodic bonding is used for patterned silicon and glass wafers, one should adjust the bonding parameters to the areas to be bonded. The amount of charge moved during a bonding sequence of a specific wafer design can be calculated by integrating the current moved as a function of time. The density of charge can be a parameter that one strives to keep constant as an indicator of bonding quality. However, this value is only an indicator and not a guarantee of good anodic bonding quality. If a rather low current limitation (e.g., 5–10 mA) is set in the bonder, the time can be a parameter that is easily adjusted to achieve the required charge density. The current as a function of time will then be a simple step function for several minutes.

In the same manner as fusion-bonded laminates, anodically bonded laminates can be further processed. However, since the sodium contamination level of wafers bonded to Pyrex 7740, for example, is not acceptable for a silicon technology line, separate tooling must typically be used. If the laminates are to be post-processed, the wafers must be well aligned during bonding to ensure a final size acceptable for the tools. Post-processing at temperatures up to 400–500 °C can be performed, but it must be kept in mind that glass is actually a liquid and is not in an energetically stable state. Stress may thus be introduced or relaxed during post-processing at elevated temperatures. Glasses are complex structures where both anneal temperatures and cooling procedures may play a role, but a role that is hard to predict.

14.5.3
Eutectic Bonding of Patterned Wafers: AuSn

Bonding with metal frames seems to becoming gradually more popular, for many reasons. The expected leak-tightness of a well-bonded metal frame is better than for most glass seals. This indicates that narrower bond frames can be used and the overall footprint of a device, and with it its cost, is minimized. A rather unpredictable yield has been a recognized issue for various eutectic and thermocompression bonding processes on the wafer scale for years. However, both metal deposition tools and bonding tools have lately improved and various metal bonding technologies are presently frequently recommended on the wafer scale even when a high yield is of primary concern [28]. Metal deposition can now be made with improved thickness uniformity across large wafers. In addition, critical bonding parameters like chuck temperature, chamber pressure, and especially tool pressure distribution can be better controlled than before.

Eutectic bonding using AuSn solder without the need of fluxes has been offered commercially for encapsulation (www.hymite.com) and is for instance well suited for optical components. If used for mounting, AuSn offers a low thermal resistance coupling to the substrate, which for instance is important for laser diodes where the junction temperature must be kept under tight control. The required bonding temperature is relatively high (about 310 °C), but still CMOS compatible. AuSn is chemically quite inert which can be an advantage for MEMS devices like microfluidic systems. The ability of the material to adapt to underlying topography is rather good since the liquid phase of the material system is normally reached for at least a part of the volume of the bonding system. Wafer bonding with AuSn is described in detail in Chapter 7, but we will here elaborate on some issues relevant for process integration of AuSn bonding for structured MEMS devices in particular.

The AuSn solder can either be plated or sputter-deposited. Additional metal layers underneath the solder take the role of adhesion, barrier, and/or seed layers. Materials like titanium, TiW, or tantalum are often used to ensure adhesion through covalent bonds to silicon or silicon oxide. If electrical conduction through the bonding pad is needed, as for instance when combined with TSVs, an electrically conductive layer consisting of aluminum, copper, or gold is deposited on top of the adhesion layer. Next, a layer of platinum, nickel, or chromium can be deposited as a barrier and wetting layer. This top layer can prevent the formation of intermetallic phases between the AuSn solder and the adhesion or conduction layer. Intermetallic phases may lead to de-wetting of the solder or brittle, and therefore unreliable, bonds. The top layer can be omitted, but one should then make sure that there is a layer of pure gold left on both sides of the final bonded interface instead. Layers of unreacted gold can also prevent de-wetting. The adhesion, barrier, and/or seed layers can be deposited by sputtering or evaporation since only some hundreds of nanometers are required. In the case of nickel and chromium, an additional Au layer of about 50 nm thick can favorably be deposited in the same sputter run to prevent the formation of oxide layers. Oxide layers are problematic in the case of later electroplating. The choice of materials will depend on the specific application in addition to availability. Gold and copper, for example, are metals with a high electrical conductivity that may favorably be used in the design also for electrical routing, whereas platinum, nickel, or chromium can be used if only a mechanical connection is needed. Satisfactory adhesion of the layers can be a challenge, especially if deposited onto an oxide layer. Fine tuning of sputtering parameters and layer thickness control are required to ensure a rather stress-free system. Sometimes a compromise must be made between optimal adhesion, wetting, and minimum stress. By comparing for example Ti/Pt, Ti/Ni, and Ti/Cr [29] for flip-chip bonding, it was found that chromium was the best compromise:

- platinum dissolved quickly in gold resulting in a shift in the phase composition causing an increased liquidus and therefore bonding temperature;
- nickel could have a high intrinsic stress after deposition and be prone to delamination; special process control during sputtering (low pressure, bias sputtering) was needed;
- chromium has a low solubility in AuSn, but exhibited the lowest wettability of these metals.

A consequence of the complexity of the layers is that short-loop experiments are normally recommended before any bond trials are made where the layered system is exposed to relevant temperature treatments. Thermal cycling followed by tape tests, scratch tests or wafer dicing, and SEM/EDX analysis can reveal weak interfaces within the systems.

When the layers are to be deposited onto structured wafers, further challenges occur. Patterned thin underlying layers or shallow wet etched structures are normally not a concern as long as the topography is within several micrometers. However, if angles of etched walls are larger than 90°, which can be the case for dry-etched structures, the deposited layers may not be continuous across the wafer, and electrical contact to regions within etched regions may be lost. This is naturally only a problem if the deposited material is to function as a seed layer during electroplating of AuSn, and not if the solder will be sputtered. Typically, however, areas that are plated for the purpose of bonding are positioned either on the wafer surface or inside only shallowly etched recesses, and thus any potential lack of contact to a seed layer inside deeply etched cavities is normally not a concern.

The most common solution for the deposition of the AuSn material has been electroplating, where the material is plated inside a mold of resist. But as mentioned, at least at chip level, bonding of sputtered layers of eutectic or near-eutectic

AuSn layers has also been demonstrated [17]. When the composition is richer in tin, the system is less sensitive to phase changes as can be seen from the phase diagram for AuSn presented for instance in [30]. Liftoff was used to pattern the 4 μm thick AuSn layer. The same patterning method can probably be extended to slightly thicker layers as long as the sputtering parameters can be properly adjusted. The thermal exposure of the underlying thick photoresist system must be minimized to prevent decomposition. When plating is applied, gold and tin are normally plated as separate layers on one wafer and gold on the other wafer [31]. The minimum thickness of each plated layer is normally in the range of 1–3 μm due to limited thickness control of thinner layers. The total thickness of the layers has been recommended to be at least about 10 μm before bonding to ensure a thick enough liquid phase between the high-melting intermetallic phases formed within 1–3 μm at the bonding interface. The final bond frame thickness, the final standoff height, may differ substantially from the initial total layer thickness depending on bonding parameters like temperature, tool pressure, tool pressure distribution, and bonding time. A rather large and unpredictable standoff height with a variation in thickness of the order of (at least) a micrometer is obviously a design restriction for MEMS devices. Capacitive devices are examples where smaller and better controlled standoff heights are normally required. Thinner layers are as mentioned achievable by using sputtered layers, but thinner layers may result in reduced adaptability to underlying topography. For the most demanding MEMS devices, only the introduction of additional structures like recesses or bumpers are expected to give an acceptably well-controlled standoff height. An example of an AuSn bond frame positioned within a recess is shown in Figure 14.6. The

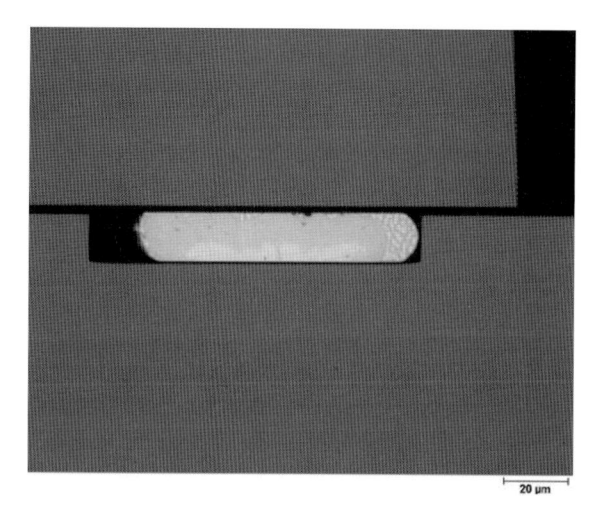

Figure 14.6 An example of an AuSn bond frame positioned inside a shallow recess on the bottom wafer to ensure a well-controlled standoff height. A deep cavity was dry etched in the upper wafer as seen to the right in the image. (Light microscope image by SINTEF.)

drawback with having to include additional structuring and complexity is that this normally adds cost and reduces yield.

The most important issue for process integration of eutectic bonding in MEMS devices is probably how to combine deposition of the required metal layers with mechanical structuring of the wafers. Plating may be performed before mechanical structuring for instance if the wafers are etched from the opposite side of where the plated layers are deposited. Special protective masks, such as ProTEK® [32], or single-side wafer holders, such as those available from SPS (www.SPS-Europe.com), may be used to protect the AuSn even during wet etching. A possible disadvantage with the use of coatings is the need of pre- or soft baking at temperatures where the AuSn starts to interdiffuse and oxidize. Some compromises may be made between optimal coating conditions and optimal conditions for the AuSn layer. Release of structures by removal of sacrificial layers of for instance oxide using HF vapor may also be done after plating. However, short-loop experiments should then be performed to ensure compatibility between all metal layers in the system and the HF vapor process that is to be used. It is extremely important to make sure that the HF vapor process is completely dry (anhydrous).

Another alternative is to perform all mechanical structuring first and metal deposition for bonding afterwards. This is a possibility if the photoprocessing required can be performed on the structured surface despite the topography present and if the structures themselves will survive mechanically the processes needed for the deposition. Wet- or dry-etched cavities and trenches are examples of structures that mostly can be defined before metal deposition. However, special attention is needed if the structures are etched deeply into the bulk of the wafers. If the AuSn material is to be plated, a thick layer of photoresist (about 10 µm) is patterned on top of the seed layer as a mold for the plating. The resist coating can be a challenge for heavily structured wafers. Spin coating can be used as long as the wafer topography is not too large, but spray coating or coating with electrode-positable resist is recommended when etched depths exceed several micrometers or whenever etched angles are large. Thickness uniformity is especially challenging after hard bake of most resists in regions with convex edges. A resist frequently used for the purpose of AuSn plating of structured wafers, Eagle 2100 ED from Shipley, can preferentially be baked in a special atmosphere to ensure thickness uniformity [33].

The electroplating deposition rate inside the openings of the resist is roughly proportional to the local current density. The size of the openings in the mask design should therefore be studied well to ensure a good thickness uniformity of the plated regions. Simulations can be made to optimize designs for a uniform current distribution and advanced plating tools can ensure a better uniformity than simple laboratory equipment. It is important not to overlook special features like alignment marks and test structures in the design. These are often small structures where the current density can become large, and if precautions are not made the structures will stick up like towers on the wafers and hinder the wafer-level bonding. Solutions can be to avoid such structures, to manually cover such structures with resist before plating, or to position such structures in recesses so that they will not be the highest structures on wafer level in the end.

When MEMS wafers with patterned AuSn structures are to be bonded, some special care must be taken: The alignment accuracy may have to be improved from a typical range of ±5–10 µm to ±1 µm if the bond frame width is reduced as suggested in [1]. The bonding tool pressure should not be too high since structured wafers are more prone to fracture than regular integrated circuit wafers (however, eutectic bonding is normally not expected to require especially high bonding tool pressures, as opposed for instance to thermocompression bonding). MEMS wafers are preferentially bonded using a fluxless AuSn bonding process. If a certain atmosphere is required in sealed cavities after bonding, desorption of gases from for instance sputtered layers must be taken into consideration (a getter material can be required if a vacuum level below about 1 mbar is needed). Otherwise, MEMS wafers can be bonded with AuSn bonding processes developed for flat wafers.

14.6
Summary

The role of wafer bonding for MEMS has been and will be significant without doubt. The need for integration and miniaturization of devices will push the bonding technology forward. As we have seen in this chapter, process integration of wafer bonding for structured wafers is somewhat more demanding than for flat integrated circuit wafers, but normally fully achievable. The choice of bonding method for a given application will depend on a combination of bonding requirements and availability of relevant processes and know-how. Different bonding methods are still expected to be found for future MEMS devices.

References

1 Martin, J. (2009) Commercial MEMS case studies: the impact of materials, processes and designs. *MRS Symp. Proc.*, **1139**, 3–12.

2 Esashi, M., Ura, N., and Matsumoto, Y. (1992) Anodic bonding for integrated capacitive sensors. Proceedings of Micro Electro Mechanical Systems (MEMS '92), Travemünde, 4–7 February 1992, pp. 43–48.

3 Notzold, K., Dresbach, C., Graf, J., and Bottge, B. (2009) Temperature dependent fracture toughness of glass frit bonding layers. Symposium on Design, Test, Integration & Packaging of MEMS/MOEMS, 1–3 April 2009, pp. 378–383.

4 Niklaus, F., Andersson, H., Enoksson, P., and Stemme, G. (2001) Low temperature full wafer adhesive bonding of structured wafers. *Sens. Actuators A*, **92** (1–3), 235–241.

5 Taklo, M.M.V., Bakke, T., Vogl, A., Wang, D.T., Niklaus, F., and Balgård, L. (2008) Vibration sensor for wireless condition monitoring. Proceedings of the Pan Pacific Microelectronics Symposium, pp. 305–310.

6 Jakobsen, H. and Kvisterøy, T. (1996) Sealed cavity arrangement method. US

Patent 5,591,679, filed 30 January 1996 and issued 7 January 1997.
7 Silex Microsystems (2009) *News and Press*, http://www.silexmicrosystems.com/News_Press.asp?active_page_id=176 (accessed 24 November 2009).
8 Plan Optik AG (2009) *Products*, http://www.planoptik.com/produkte/produkte_en.html (accessed 26 August 2011).
9 Schott UK (2010) *Schott HermeS® Substrate*, http://www.schott.com/epackaging/english/auto/others/hermes.html (accessed 26 August 2011).
10 Visser, M.M., Plaza, J.A., Wang, D.T., Moe, S.T., Schjølberg, K., and Hanneborg, A.B. (1999) Pressure sensor for harsh environments realized by triple stack fusion bonding, *ECS Proc.*, **99-35**, 362–367.
11 Mehra, A., Ayon, A.A., Waitz, I.A., and Schmidt, M.A. (1999) Microfabrication of high-temperature silicon devices using wafer bonding and deep reactive ion etching. *J. Microelectromech. Syst.*, **8**, 152–160.
12 Mack, S., Baumann, H., Gösele, U., Werner, H., and Schlögl, R. (1997) Analysis of bonding-related gas enclosure in micromachined cavities sealed by silicon wafer bonding. *J. Electrochem. Soc.*, **144**, 1106–1111.
13 Kvisterøy, T. (2005) From research to industry. Proceedings of MicroMechanics Europe (MME '05), Göteborg, Sweden, 4–6 September 2005.
14 Yun, C.H., Martin, J.R., Tarvin, E.B., and Winbigler, J.T. (2008) Al to Al wafer bonding for MEMS encapsulation and 3-D interconnect. 21st IEEE International Conference on Micro Electro Mechanical Systems (MEMS 2008), 13–17 January 2008, pp. 810–813.
15 Svasek, P., Svasek, E., Lendl, B., and Vellekoop, M. (2004) Fabrication of miniaturized fluidic devices using SU-8 based lithography and low temperature wafer bonding. *Sens. Actuators A*, **115** (2–3), 591–599.
16 Bay, J., Hansen, O., and Bouwstra, S. (1996) Design of a silicon microphone with differential read-out of a sealed double parallel-plate capacitor. *Sens. Actuators A*, **53** (1–3), 232–236.
17 Brunschwiler, T., Michel, B., Rothuizen, H., Kloter, U., Wunderle, B., Oppermann, H., and Reichl, H. (2008) Interlayer cooling potential in vertically integrated packages. *Microsyst. Technol.*, **15**, 57–74.
18 Chen, J., Lu, Y., and Yang, J.P. (2005) Integrated fabrication of electrostatic microactuator for HDD R/W head positioning. *Microsyst. Technol.*, **11**, 8–10.
19 Bakke, T., Volker, B., Rudloff, D., Friedrichs, M., Schenk, H., and Lakner, H. (2006) Large scale, drift free monocrystalline silicon micromirror arrays made by wafer bonding. *Proc. SPIE*, **6114**, 611402.
20 Seo, S.W., Han, S., Seo, J.H., Kim, Y.M., Kang, M.S., Min, N.K., Choi, W.B., and Sung, M.Y. (2009) Microelectromechanical-system-based variable-focus liquid lens for capsule endoscopes. *Jpn. J. Appl. Phys.*, **48**, 052404.
21 Sparks, D., Queen, G., Weston, R., Woodward, G., Putty, M., Jordan, L., Zarabadi, S., and Jayakar, K. (2001) Wafer-to-wafer bonding of nonplanarized MEMS surfaces using solder. *J. Micromech. Microeng.*, **11**, 630–634.
22 Usenko, A.Y. and Carr, W.N. (1999) Silicon-on-insulator technology for microelectromechanical applications. *Semicond. Phys. Quantum Electron. Optoelectron.*, **2** (1), 93–97.
23 Traeger, R.K. (1976) Hermeticity of polymeric lid sealant. Proceedings of the 25th Electronics Components Conference, April 1976, pp. 361–367.
24 Reinert, W., Kahler, D., and Longoni, G. (2005) Assessment of vacuum lifetime in nL-packages. Proceedings of the 7th Electronic Packaging Technology Conference (EPTC 2005), 7–9 December 2005, pp. 225–230.
25 Vallin, Ö., Jonsson, K., and Lindberg, U. (2005) Adhesion quantification methods for wafer bonding. *Mater. Sci. Eng.*, **50** (4–5), 109–165.
26 Gösele, U., Tong, Q.-Y., Schumacher, A., Kräuter, G., Reiche, M., Plößl, A., Kopperschmidt, P., Lee, T.-H., and Kim, W.-J. (1999) Wafer bonding for microsystems technologies. *Sens. Actuators A*, **74**, 161–168.

27 Veenstra, T.T., Berenschot, J.W., Gardeniers, J.G.E., Sanders, R.G.P., Elwenspoek, M.C., and van den Berg, A. (1999) The use of selective anodic bonding to create near-zero dead volume membranes. *ECS Proc.*, **99-35**, 381–385.

28 Farrens, S. (2008) Precision wafer to wafer packaging using eutectic metal bonding. Proceedings of the SMTA Pan Pacific Microelectronics Symposium, Kauaii, HI, January 2008.

29 Katz, A., Lee, C.H., and Tai, K.L. (1994) Advanced metallization schemes for bonding of InP-based laser devices to CVD-diamond heatsinks. *Mater. Chem. Phys.*, **37**, 303–328.

30 Yin, L., Meschter, S.J., and Singler, T.J. (2004) Wetting in the Au–Sn system. *Acta Mater.*, **52** (10), 2873–2888.

31 Aasmundtveit, K.E., Wang, K., Hoivik, N., Graff, J.M., and Elfving, A. (2009) Au–Sn SLID bonding: fluxless bonding with high temperature stability to above 350 °C. European Microelectronics and Packaging Conference (EMPC 2009), 15–18 June 2009, pp. 1–6.

32 Brewer Science (2010) *Protective Coating Processes*, http://www.brewerscience.com/products/protek/ (accessed 26 August 2011).

33 Heschel, M. and Bouwstra, S. (1998) Conformal coating by photoresist of sharp corners of anisotropically etched throughholes in silicon. *Sens. Actuators A*, **70** (1–2), 75–80.

15
Three-Dimensional Integration

Philip Garrou, James Jian-Qiang Lu, and Peter Ramm

15.1
Definitions

Three-dimensional (3D) integration is commonly defined as bonding of stacked device substrates with vertical electrical interconnects between the device layers [1]. A device layer may consist of one particular integrated circuitry or one set of functional devices. Besides wafer thinning and through-silicon via (TSV) formation, wafer bonding is the key unit process of 3D integration technologies. The choice of wafer bonding technologies has to be made by considering the full process conditions such as temperature limitations and the sequence of the technology modules, which may vary for different applications. The large spectrum of corresponding application-specific 3D integration technologies can be reasonably classified in the following main categories:

- stacking of packaged devices (or substrates);
- stacking of embedded bare devices (without TSVs);
- 3D TSV technology.

Most of the applications discussed here are based on TSV technologies showing the capability for volume production of high-performance products.

The large variety of existing TSV processing flows is categorized, according to ITRS [2], explicitly corresponding to the order of TSV processing, wafer thinning, and wafer bonding. The order of TSV process with respect to device fabrication is one of the main criteria and today commonly characterized by fabrication of TSVs (see Figure 15.1):

- *prior* silicon front-end-of-line (FEOL) – "via first"
- *post* silicon back-end-of-line (BEOL) – "via last"
- *post* FEOL but *prior* BEOL – "via middle."

Handbook of Wafer Bonding, First Edition. Edited by Peter Ramm, James Jian-Qiang Lu, Maaike M.V. Taklo.
© 2012 Wiley-VCH Verlag GmbH & Co. KGaA. Published 2012 by Wiley-VCH Verlag GmbH & Co. KGaA.

Via First, Middle and Last Process Flows

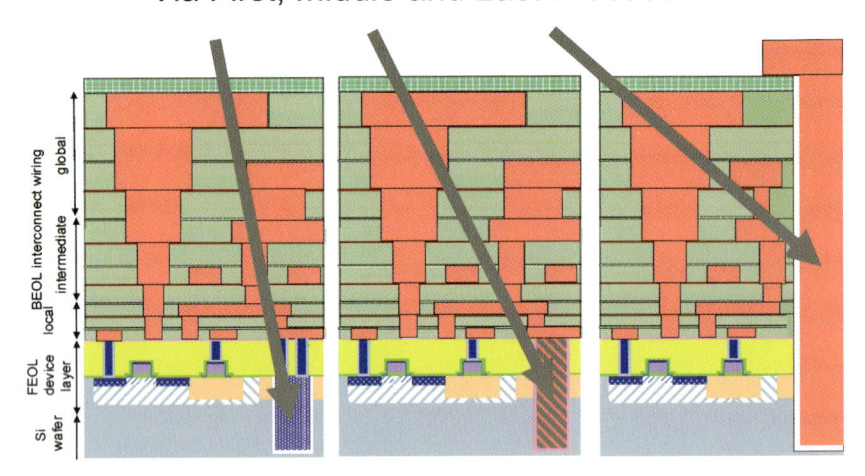

Figure 15.1 Schematic of TSV process flows [2].

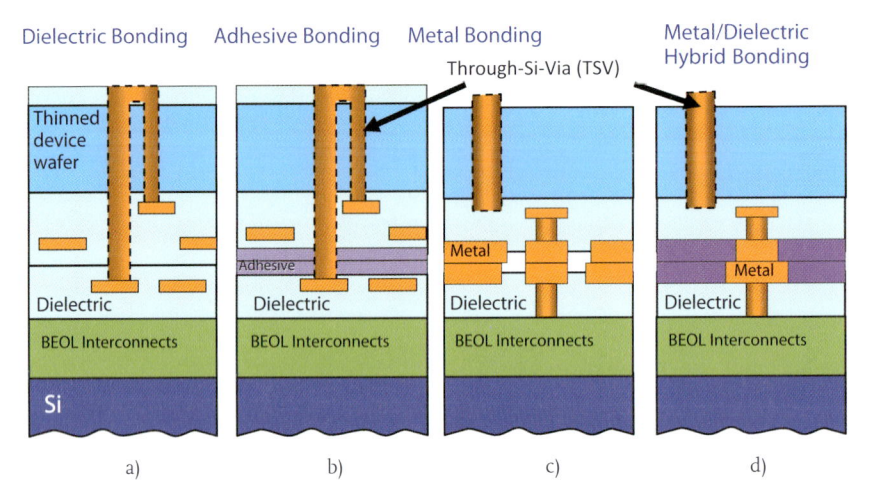

Figure 15.2 Schematic of main wafer bonding technologies for 3D integration.

15.2
Application of Wafer Bonding for 3D Integration Technology

Several wafer bonding technologies presented in this book are advantageously used in the fabrication of 3D integrated products. Figure 15.2 shows the most commonly applied techniques of oxide fusion, adhesive bonding, and metal bonding for 3D TSV integration. Here, the two bonding approaches on the left-hand side are the "via last" 3D integration technologies, where TSVs are formed after wafer bonding and thinning. The two bonding approaches on the right-hand side can be used to form the vertical electrical interconnects at the bonding interface during the bonding process, while TSVs can be made at any stage of the processing flow, depending upon the applications and processing constraints. In consequence the 3D integration technologies based on metal bonding, explicitly Cu–Cu and intermetallic compound bonding, are chosen for the majority of applications presented in the following sections.

Among many existing 3D integration approaches, two key 3D TSV technologies are discussed below concerning the interaction between bonding processes and 3D processing flows: TSV Cu–Cu bonding technology and TSV solid–liquid interdiffusion (SLID) technology.

TSV Cu–Cu bonding technology is one of the metal-bonding-based 3D approaches [3, 4]. Copper is the choice of BEOL interconnect material and the copper bonding process is compatible with BEOL processing. Currently two approaches to achieve a successful copper bond are the surface activated method and thermal compression method. The major difference between these two approaches is the bonding mechanism. The surface activated bonding method can enable two copper surfaces to be bonded at room temperature directly, while thermal compression copper bonding uses high temperature and force with simpler design and less cost consideration. However, surface activated bonding does require a high surface smoothness and post-bond annealing at high temperature to form a permanent bond, while thermal compression bonding does require a good surface preparation. Both methods require a temperature of around 350 °C to form a permanent copper bond. The details of the copper bonding processes are discussed in Chapter 9, as well as in Chapters 11–13 that deal with hybrid bonding. The key advantages of Cu–Cu bonding are: (i) electrical interconnects between the two stacked device substrate (i.e., "bonded interstrata via" or BISV, as called in Chapter 11) can be formed during the bonding process; (ii) the BISV density (not the TSVs) can be very high because these vias exist only at the bonding interface (hence they are very short, do not pass through the BEOL and silicon substrate, and hence do not consume BEOL or FEOL areas), and can be very well defined to submicrometer dimensions (this BISV density is largely limited by the alignment accuracy); and (iii) the copper bond once formed is robust and further thermal processes can only enhance the bond, not degrade it, because the copper melting temperature is over 1000 °C. Because of these advantages and the compatibility with copper TSV and copper BEOL interconnects, copper bonding has been extensively studied with either just copper pads (Figure 15.2c) or hybrid copper/

dielectric bonding interfaces (Figure 15.2d). However, it is difficult to form a seamless, low-resistive copper bond in atmosphere because of the copper oxidation. Therefore, SLID and eutectic metal bonding technologies are likewise applied in 3D manufacturing developments.

Based on SLID bonding, a modular 3D TSV technology has been developed by Fraunhofer Munich, targeting multiple 3D device stacks with high performance and small form factor requirements [5, 6]. Corresponding to this TSV SLID technology the TSVs are fabricated on partly or completely processed bare device wafers. TSVs are etched through dielectric layers and further into the silicon substrate of the devices, subsequently isolated and metallized (e.g., by deposition of copper or tungsten). It is a key advantage when the TSVs are pre-processed on the wafer level prior to wafer bonding: the process flow conditions such as the deposition temperatures for insulation and metallization are not limited by characteristics of the bonding layers. After complete pre-processing of the TSVs the device wafers are then temporarily bonded to a handling substrate as described in Chapters 16 and 19. Subsequently the temporarily bonded wafers are thinned until the metal-filled TSVs are exposed from the rear side. The thinned devices are then connected to a second device wafer by SLID as described in Chapter 10. It is crucial that both the applied temporary and permanent bonding procedures have to be chosen in order to withstand the process conditions of all applied TSV backside processes, that is, the process temperatures. For the case of the conventionally used Cu/Sn SLID system, the substrates to be bonded are deposited with Cu/Sn layer structures. At typical wafer bonding temperatures of 240–270 °C the deposited tin is completely transformed into Cu_3Sn intermetallic compound. This ε-phase is thermodynamically stable with a melting point above 600 °C. Using appropriate film thicknesses, the tin is consumed and the solidification is completed within a few minutes, leaving only unconsumed copper on both sides. The key advantage of the TSV SLID technology is that both the mechanical and the electrical interconnects are formed simultaneously – so the 3D integration process is completed. In principle, no post-processing of TSVs is required.

After removal of the handling substrate the TSVs are interconnected to the device metallization. With the TSV SLID technology, the 3D integration can be executed subsequently for a next device layer to be stacked. This key ability of modularity is a consequence of the fact that the melting temperature of the intermetallic compound is considerably higher than the wafer bonding temperature.

A schematic of the 3D stacking concept exemplifying the modular principle of the described TSV SLID technology is shown in Figure 15.3a, and the cross section of a 3D integrated test chip with tungsten-filled TSVs and Cu/Sn SLID bond is shown in Figure 15.3b.

The motivations and examples of applications of 3D integration technology are described in the Sections 15.3 and 15.4.

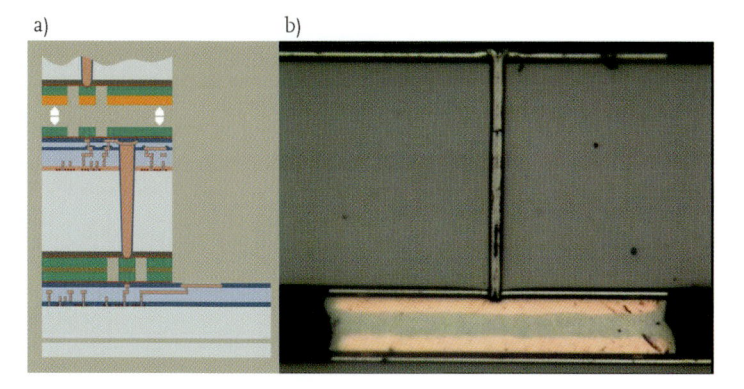

Figure 15.3 Application of wafer bonding for 3D integration: TSV SLID technology based on metal bonding by SLID [7]. (a) Schematic of the modular principle. (b) Cross section of a 3D integrated test chip with tungsten-filled TSVs and Cu/Sn SLID bond.

15.3
Motivations for Moving to 3D Integration

The performance and economic motivations for moving to 3D integration technology have been clear for several years. By stacking a thinned die, or circuit, on top of another we achieve the shortest interconnect distances which results in less power required to drive a signal, because the signal has less distance to travel. In addition entire layers of IP can be re-used, either through customization on a separate layer or by connection to other functions through an interposer. This is summarized as follows:

Performance
- shorter nets;
- lower capacitance interconnect;
- wide I/O bandwidth.

Volumetric efficiency (size/density)
- stacking separated functions.

Reduced power
- lower cap and shorter nets;
- fewer buffers.

Modularity
- IP reuse;
- lower design cost.

A comparison between an equivalent package-on-package (PoP) solution and a 3D TSV solution showed that the 3D TSV solution has significant advantages (Figure 15.4) [8].

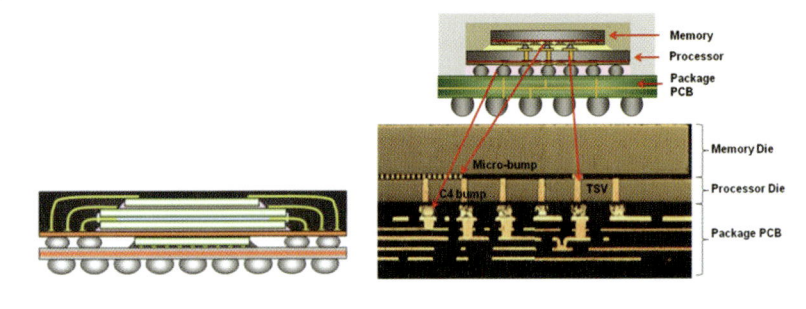

FC PoP   TSV-SiP with wide IO DRAM

	Conventional 3D Package (FC-PoP) with LPDDR2	TSV-SIP with Wide IO memory
Memory I/O Power Consumption	176 mW	44 mW

Figure 15.4 Power consumption comparison flip chip (FC) PoP versus 3D system-in-package (SiP) with TSV [8].

Changing the current chip fabrication infrastructure will surely not be a trivial or low-cost undertaking. Still, most feel it is worthwhile when compared to the projected more than $6 billion cost for new fabrication infrastructures at 22 nm and beyond.

Once the infrastructure is in place, it is hoped that 3D IC technology will reduce both risk and cost. The 3D IC approach should create economic benefits such as: reducing the time it takes to design and verify chips at the most advanced nodes; allowing the use older analog IP blocks rather than having to develop new IP blocks at the most advanced process nodes; and allowing the mixing of normally incompatible technologies (heterogeneous integration – more than Moore; Figure 15.5).

Three-dimensional integration is recognized as a key technology for heterogeneous products, demanding for smart system integration rather than extremely high interconnect densities. According to surveys on revenue forecast by Yole Développement such "more than Moore" products, for example heterogeneous microelectromechanical systems (MEMS)/IC systems, may even be the main driving markets for 3D TSV integration [9]. Many R&D activities worldwide are focusing on heterogeneous integration for novel functionalities [10]. Corresponding 3D integration technologies are still in evaluation at several companies and research institutions, such as the European "e-BRAINS" consortium, with a dedicated focus on reliability issues of such sophisticated heterogeneous products [11].

In the following section those 3D applications that are already or will soon be on the market are described.

Figure 15.5 Beyond Moore's law [2].

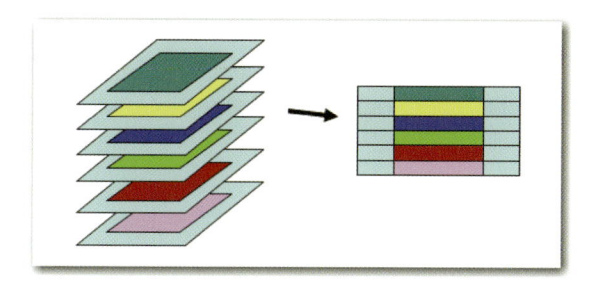

Figure 15.6 Full reconfiguration of an IC chip into stacked strata.

15.4
Applications of 3D Integration Technology

15.4.1
Three-Dimensional Applications by Evolution Not Revolution

In its ultimate manifestation, 3D IC integration will involve complete chip redesign where individual functions, or groups of functions that are process-compatible, would be fabricated in different strata and vertically bonded together to form the final functioning circuit as shown in Figure 15.6. A stratum may consist of one particular integrated circuitry or one set of functional devices. Bonding is one of the three key unit processes in 3D IC integration, the other two being interstrata interconnect formation and thinning. If the interstrata interconnects go through the thinned silicon substrate, they are called TSVs. In general, if the interstrata interconnects go through the layer of active silicon regardless of whether the device has a thinned silicon substrate (e.g., bulk CMOS circuit wafer) or does not have a thinned silicon substrate (e.g., SOI CMOS circuit wafer, where all bulk silicon substrate is removed), they can be still called TSVs.

However, this will more than likely take a decade or more to happen. Most technological advances in our industry are made in an evolutionary not a revolutionary manner since the microelectronics industry resists all materials and/or processing changes [12]. The following evolutionary steps are leading us to full 3D integration.

15.4.2
Microbump Bonding/No TSV

The first step towards full 3D IC integration has been chip-on-chip (CoC) technology where chips are thinned and face-to-face (F2F) bonded. For the first generation, when only two chips need to be bonded, this F2F solution requires no TSV. The F2F bonding is similar to flip chip bonding, but with microbumps. Cu/Sn or other eutectic alloy microbumps can be used for such a F2F bonding.

High-speed data transfer with high bandwidth is achieved when using CoC technology, because of the direct memory to logic interconnect. The microbumps provide more interconnects than wire bonding, and because they are only several dozen micrometers in diameter they offer low parasitic capacitance, resistance, and inductance making it easier to raise the operating frequency. Since individual memory chips are used, the memory capacity limitations of embedded DRAM can be minimized.

15.4.2.1 Infineon
Infineon's "SOLID" technology involves chip-to-wafer (C2W) bonding without TSV where wafers or chips are simply thinned and F2F bonded [13]. The term SOLID used for this specific process flow is based on the above described SLID. The SOLID technology uses Cu(Ag)Sn to bond C2W; this simple Cu(Ag)Sn metallurgy enables corrosion-resistant contacts with a high melting point and excellent reliability performance. The bottom chip must be larger (see Figure 15.7) and must have peripheral pads which allow subsequent I/O bonding of this mated structure by wire bonding or ball grid array-like structures. The technology was initially developed in close cooperation with Fraunhofer Munich for Infineon's chip card products division. Such cards would provide data storage for mobile phones, personal digital assistants, digital cameras, and music players. This C2W technology reportedly improved both smartcard capacity and security since die separation requires 600°C which destroys them and consequently the information stored in the chip [13, 14].

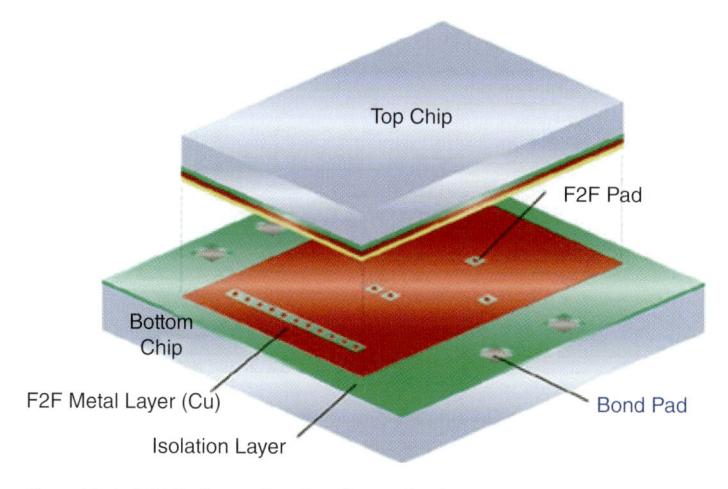

Figure 15.7 SOLID face-to-face bonding technology [13].

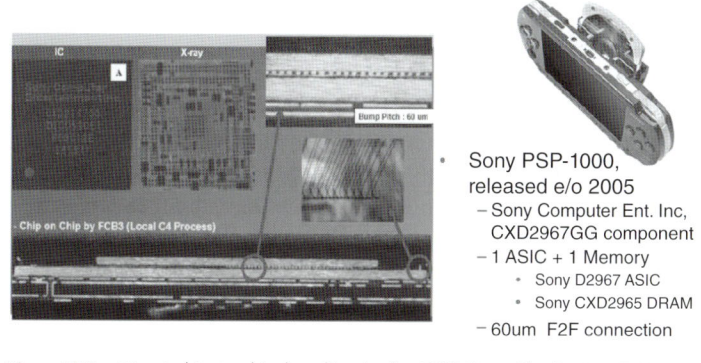

- Sony PSP-1000, released e/o 2005
  - Sony Computer Ent. Inc, CXD2967GG component
  - 1 ASIC + 1 Memory
    - Sony D2967 ASIC
    - Sony CXD2965 DRAM
  - 60um F2F connection

Figure 15.8 Direct chip-to-chip bonding in the 2005 Sony PlayStation [17].

15.4.2.2 Sony

Before 2005, the microcontroller for the Sony PlayStation was fabricated using a 90 nm embedded DRAM process technology. At the end of 2005 Sony switched to CoC, bonding the DRAM directly to the logic. Sony reportedly did not see lower cost coming from scaling from the 90 nm to a 65 nm merged DRAM process technology. They reported that "A massive capital investment would have been needed to drop the design rule to 65 nm ... and even if we had single-chipped with DRAM and the logic, each demanding a different manufacturing process, it would have taken a long time to get the yield up." [15, 16] CoC bonding in the 2005 Sony PlayStation is shown in Figure 15.8.

15.4.3
TSV Formation/No Stacking

Several applications such as power amplifiers and CMOS image sensors initially used TSV, but no stacking technology. While temporary adhesive bonding may be involved, no direct metal or metal alloy bonding is used. It is envisioned that further functional separation and subsequent bonding of such strata will be seen in image sensor devices in the future.

15.4.3.1 Power Amplifiers
For any power amplifier operating at high frequency, any ground lead inductance on the emitter reduces the gain of the amplifier due to transistor de-biasing. To overcome this, a TSV is formed from the backside of the substrate as the low-inductance ground to the package. This is a traditional technology for GaAs heterojunction bipolar transistor power amplifiers. In 2008 IBM announced commercialization of SiGe power amplifiers with similar TSVs to backside ground as shown in Figure 15.9 [18]; the tungsten-filled TSV reportedly delivers a more than 75% reduction in inductance compared to traditional wire bonding.

15.4.3.2 CMOS Image Sensors
TSVs also became quickly accepted and commercialized in CMOS image sensor (CIS) devices. CISs could not take the form factor advantages of wafer-level packaging (WLP) technology since WLP fabricates the interconnect structure (using solder balls) on the face of the silicon devices and mount face down, whereas CMOS sensing chips must mount face up. The obvious solution was to have the active area of the sensor face up and the interconnect on the backside through via-last, backside formation of TSVs. Shellcase (later purchased by Tessera) was the first to propose such technology with their opto-CSP technology [19].

Such a process sequence is shown in Figure 15.10.

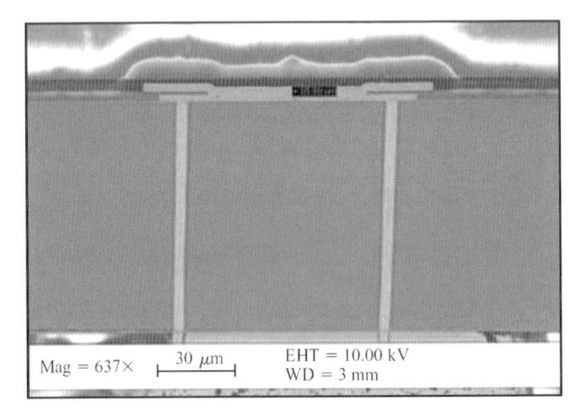

Figure 15.9 TSV in IBM SiGe power amplifier [18].

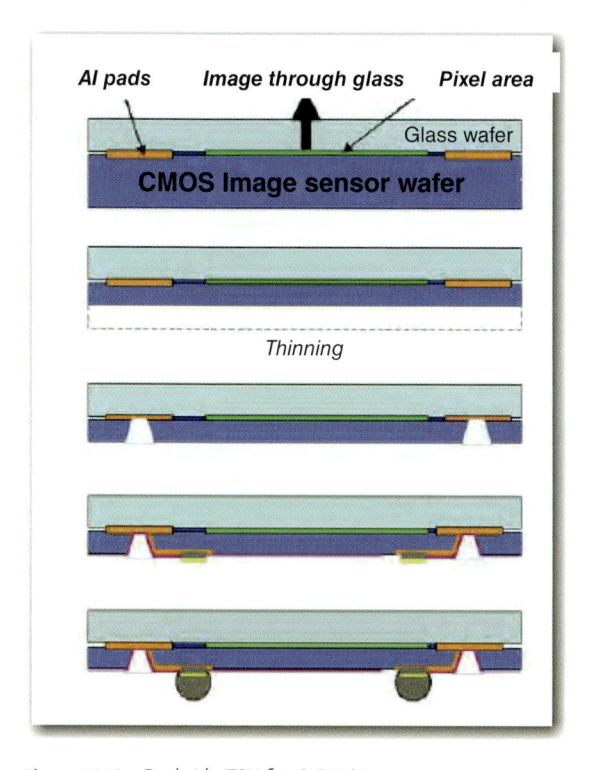

Figure 15.10 Backside TSV for CIS [20].

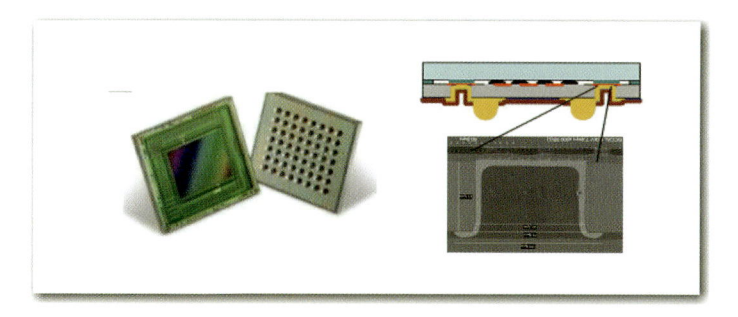

Figure 15.11 2 M pixel camera module of ST Microelectronics [20].

In 2008 ST Microelectronics announced a 2 M pixel mobile phone camera module VD6725 using TSVs, which was fabricated in its Crolle facility in both 200 and 300 mm formats. The wafers are thinned down to 70 µm and subsequently laminated to a 500 µm thick glass cover. Indications are that ST Microelectronics used its copper/benzocyclobutene redistribution technology to do backside wafer processing for these CIS devices. The device and cross section are shown in Figure 15.11.

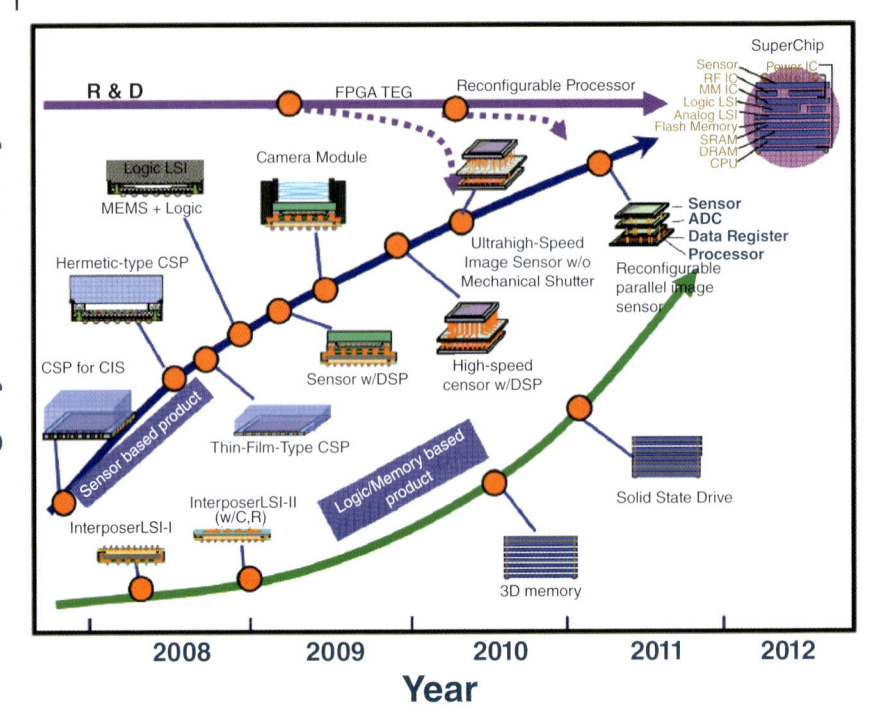

Figure 15.12 Zycube roadmap showing partitioned and stacked CISs [21].

So far no repartitioned, stacked CIS structures have been commercialized, but multilayer stacks are on published roadmaps of the image sensor vendors as shown in Figure 15.12.

Between 2007 and 2008 Toshiba, Zycube/Oki, ST Microelectronics, Aptina (Micron), and Samsung, all significant players in the CIS market, announced production using backside TSV architectures.

In 2008 the Toshiba TCM9000MD camera module was commercialized for mobile phones [22]. A micrograph and cross sections are shown in Figure 15.13. Figure 15.13a shows that the actual sensor area of the CIS chip is still significantly less than the total chip area. Thus, although the TSVs are reportedly allowing Toshiba to shrink the overall chip size by 73% as compared with non-TSV analogs, the "fill factor" is still very low and this chip begs for true 3D integration where the logic circuitry would be moved to a separate layer and bonded.

15.4.4
Memory

Memory has always been viewed as the key application for this 3D TSV market whether it is stacked memory for miniaturization or wide I/O memory stacked on logic for low latency.

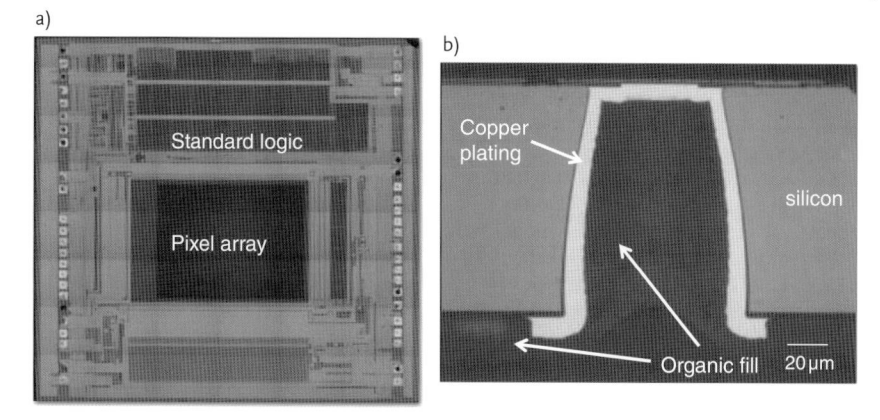

Figure 15.13 (a) Toshiba TCM9000MD CIS. (b) Close-up cross section of Toshiba TSV [22]. TSVs are about 75 μm wide and 90 μm deep.

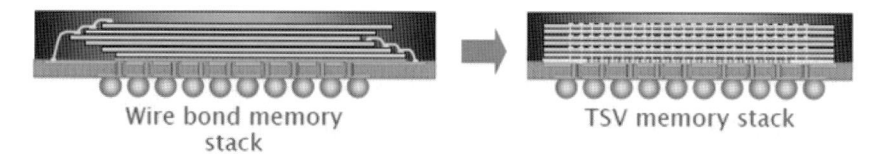

Figure 15.14 Migration from 3D packaging to 3D integration [9].

It is expected that in the next few years we will see the migration from wire-bonded stacked memory to 3D integrated stacks as shown in Figure 15.14.

15.4.4.1 Memory Categories

There are two basic memory categories: (i) system memory and (ii) storage devices.

System Memory System memory devices must quickly process large amounts of data to keep up with processor performance in PCs and servers. Low power consumption is very critical since the power required by memory is a relatively large portion of that used in server and mobile applications. As memory-intensive applications, such as cloud computing and virtualization, increase, energy efficiency will become more critical.

Power consumption in DRAM system memory has been driven by scaling. Smaller geometries require lower supply voltages, that is, 3.3 V for SDRAM, 2.5 V for DDR SDRAM, 1.8 V for DDR2 SDRAM, and 1.5 V for DDR3 SDRAM. As we approach 10 nm technologies supply voltages will decrease to 1.0 V or less. Past this point additional power reduction is unlikely because the manufacture of cell capacitors and cell-array transistors becomes more difficult as process technology enters into the sub-10 nm range. Thus new technologies such as wide IO TSV will be required to continue to achieve advancements.

Storage Devices Storage devices, such as hard-disk drives (HDDs), focus on very fast storage of large amounts of data in a small physical area. NAND flash memory has proliferated as the storage medium in devices such as music players. Recently NAND has also been applied for memory storage in PCs and servers. These solid-state drives (SSDs), consisting of a controller and NAND flash memory, are beginning to replace conventional HDDs.

Reduction of power consumption in storage devices is driven by the reduction of mechanical parts (platters, spindles, arms) because this is a significant portion of HDD total power consumption. Since SSDs do not have mechanical parts, power consumption is determined solely by the semiconductor components, that is, the controller and the NAND flash memory itself. Power consumption can reportedly be reduced 3.5 times by replacing a HDD with a SSD [8].

Looking more closely at the use of TSV technology as a means of reducing power consumption, Samsung has compared a conventional 3D package using FC PoP with low-power DDR2 (LPDDR2) memory to an equivalent system-in-package with wide IO memory [23]. One can see the decrease in memory consumption is significant (as shown in Figure 15.4).

Three-dimensional integrated DRAMs have evolved before the 3D integration of NANDs because NAND technology is so much more price-sensitive; the cost of any new processes added (such as 3D integration) into the NAND fabrication may lead to a manufacturer losing its NAND price advantage.

15.4.4.2 Repartitioning Memory

The current nonvolatile memory device architecture mainly consists of two functional blocks: the memory array composed of the array of memory cells; and "circuitry" composed of a group of other circuits which include charge pumps, I/O buffers, row and column decoders, sensing circuits, etc.

The memory block is designed and processed to maximize memory density and therefore requires the latest lithographic techniques, while the rest of the circuitry does not. For example, Losavio et al. have examined a 90 nm NOR-based flash memory as shown in Table 15.1 [23].

Since only memory cell fabrication requires leading-edge lithography, the opportunity exists to repartition the blocks onto two different strata, which would allow

Table 15.1 Requirements for 90 nm NOR-based flash [23].

	Area (%)	Litho Res.	No. of masks	Mask cost
Memory cells	50	High	25	High
Low-voltage circuits	22	Medium	20	Medium
High-voltage circuits	13	Medium	20	Medium
Charge pumps	10	Low	20	Low
I/O buffers	3	Low	20	Low
Others	5	Medium	25	Medium

15.4 Applications of 3D Integration Technology

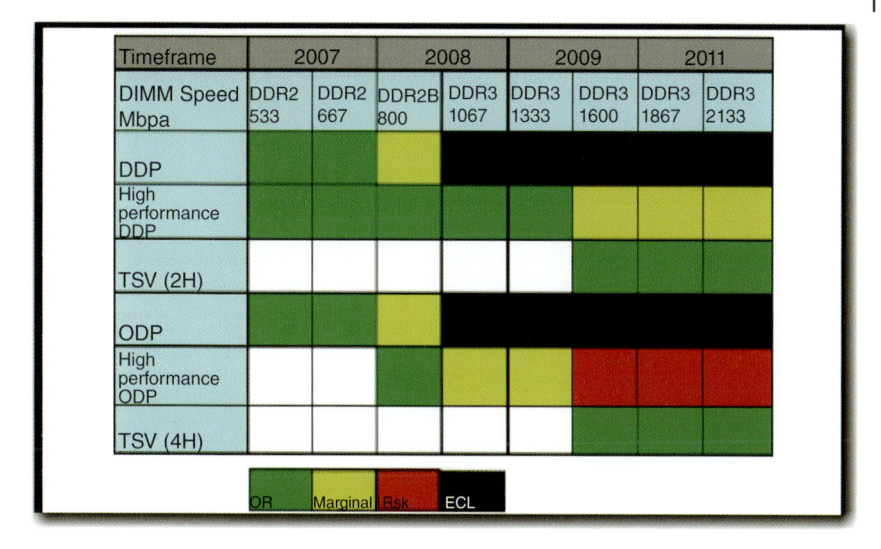

Timeframe	2007		2008		2009			2011
DIMM Speed Mbpa	DDR2 533	DDR2 667	DDR2B 800	DDR3 1067	DDR3 1333	DDR3 1600	DDR3 1867	DDR3 2133
DDP								
High performance DDP								
TSV (2H)								
ODP								
High performance ODP								
TSV (4H)								

OR | Marginal | Rsk | ECL

Figure 15.15 DDR memory fabrication/packaging roadmap [24].

one to optimize manufacturing, yield, and/or reliability of each function block on separate wafers. Wafers of the memory array blocks could be manufactured with leading-edge technology, while "circuitry" layers could be manufactured on an older, lower cost production line.

With current inter-die interconnection technology (i.e., wire bonding or bumping) this cannot be attempted because current technology does not make enough I/Os available for the connections to be made.

DDR Requirements The IBM server business has analyzed future DDR memory (Figure 15.15) and concluded that current standard packaging choices (dual-die or quad-die stack and wirebond packaging) are at risk, or will find it impossible to meet the speed requirements of 1333 and 1600 Mb s^{-1} for advanced DDR3 memory [24]. Its conclusion is that TSV-based memory stacks will be required to meet the performance criteria for these generations of DDR memory.

Samsung Samsung prototyped an 8 Gb DDR3 DRAM using 3D TSV technology [25]. The technology consists of an interface layer and multiple core layers configured with a single master and three slave chips connected by 300 TSVs as shown in Figure 15.16. The master is a 2 Gb × 4 DDR3 DRAM with additional multirank control circuits, and the slaves have 2 Gb memory cores and wafer-level test circuitry for a total device density of 8 Gb. The master acts as a buffer that isolates the channel and the slave chips. This allows the I/O data rate to increase to 1600 Mb s^{-1}, while it is limited to 1066 Mb s^{-1} in conventional quad-die package (QDP) structures. Redundant circuits, including delay-locked loop, input buffers, and clock circuitry, are eliminated in the slaves. Typical specifications are given in Table 15.2. An optical micrograph of the stacked chip is shown in Figure 15.17.

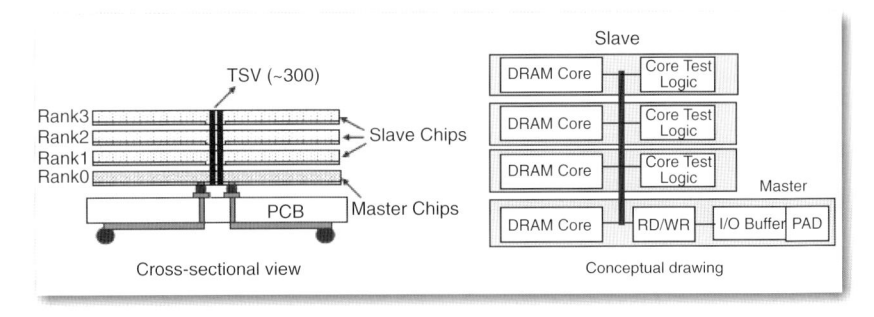

Figure 15.16 Samsung 1600 Mb s^{-1} DDR3 [25].

Table 15.2 Specifications for Samsung 1600 Mb s^{-1} DDR3 with TSVs [25].

Item	Specification
Type	DDR3
VDD	1.5 V
Density (mono)	2 Gb
Density (stack)	8 Gb (2 Gb × 4)
Date rate	1600 Mb s^{-1}
CL	10 cycles
Organization	×4

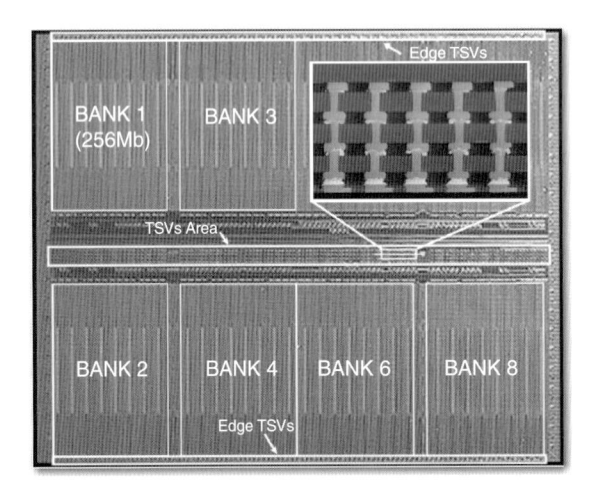

Figure 15.17 Optical micrograph of Samsung 1600 Mb s^{-1} DDR3 [25].

TSVs were found to add latency of one clock cycle at 1333 Mb s^{-1}. This latency increase, however, is small compared to the three or four cycles added in other module packaging solutions that were examined.

Samsung described a TSV repair scheme which uses redundant TSVs. If one of the TSVs fails, it is replaced by one of the redundant TSVs. Two additional TSVs per four-signal TSV group provide assembly yield of more than 98% for the 300 TSV devices. The chip area overhead due to the redundant TSVs is 0.3 and 0.6% for 40 and 60 μm pitch, respectively.

The chip bonding technology was not described in [26], but it seems that microbump bonding was used to connect the TSVs as shown in Figure 15.17.

In late 2010 Samsung announced that it would begin mass production of 8 GB DDR3 memory modules based on 4 Gb, 1.5 V, 40 nm DDR3 memory chips operating at 1333 MHz and 3D TSV chip stacking technology. The new technology reportedly offers a 53% power savings when compared to two 4 Gb DDR3 modules. Samsung announced plans to apply this technology to 30 nm-class and finer process nodes [26]. The modules would first see use in high-performance servers where lower power consumption and increasing memory capacity are important. Adoption in Dell's Precision M6500 mobile workstations is expected starting in 2012 [27].

In early 2011 Samsung researchers offered details of their wide I/O memory solutions for mobile products [28]. This addresses the "memory wall" problem where DRAM performance is constrained by the capacity of the data channel that sits between the memory and the processor. No matter how much faster the DRAM chip itself gets, the channel typically chokes on the capacity. Systems are not able to take advantage of new memory technologies because of this latency issue – they need more bandwidth. Previous generations of mobile DRAMs used a maximum of 32 pins for I/O. The new wide I/O solution, which has 512 I/O (up to 1200 total) pins, can transmit data at a rate of 12.8 Gb s^{-1}. It is expected to replace low-power DDR2 DRAM (LPDDR2) which runs at approximately 3.2 Gb s^{-1} according to Samsung. Following this wide I/O DRAM launch, Samsung aimed to provide 22 nm, 4 Gb wide I/O mobile DRAM sometime in 2013.

The dies shown in Figure 15.18 are stacked using $20 \times 17 \mu m^2$ microbumps on 50 μm pitch. The TSV diameter is 7.5 μm. TSV resistance and capacitance are in the range 0.22 to 0.24 Ω and 47.4 fF, respectively.

Elpida Elpida has prototyped similar 8 Gb DRAM technology using copper-filled TSVs [29]. The 8 Gb DDR3 SDRAM, operating at 1600 Mb s^{-1}, consists of eight 1 Gb DDR3 SDRAM chips and an interface chip. This 8 Gb DRAM uses a quarter of the standby power compared with current packaging solutions (Figure 15.19). As shown in Figure 15.19 and from previous published research papers, it seems that microbump bonding with SnAg/Cu (or other metals) as the bump material would be used.

Elpida reports that application of such 3D IC products can reduce power consumption of servers, data centers, and other high-capacity memory systems.

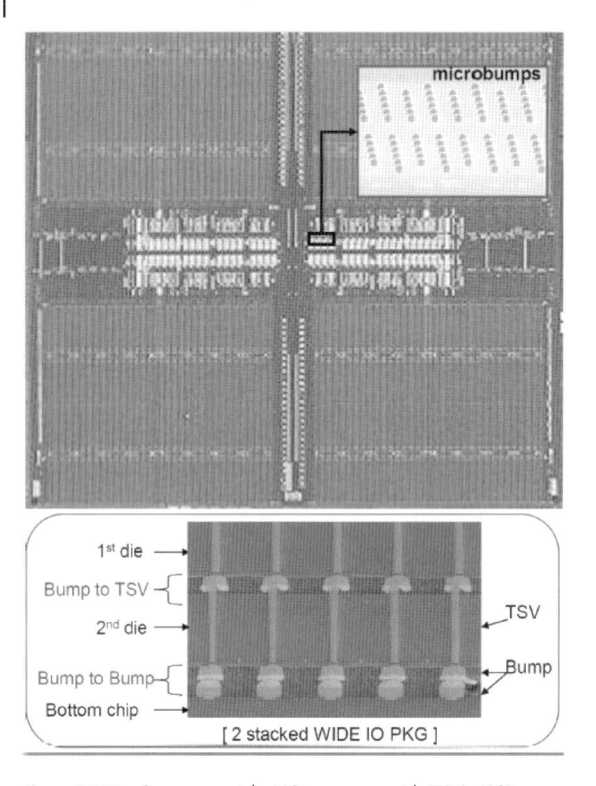

Figure 15.18 Samsung wide I/O memory with TSVs [28].

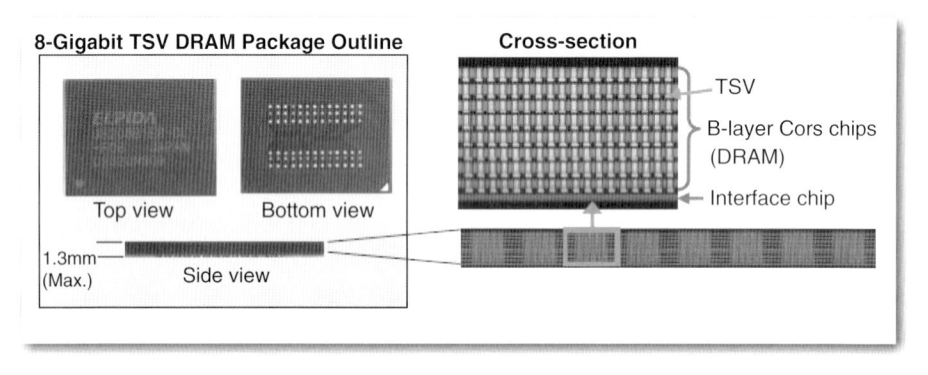

Figure 15.19 Elpida 8 Gb stacked DRAM prototype [29].

In the summer of 2010 Elpida announced a partnership with Powertech UMC [30] to develop memory on logic products. Elpida indicated that the partnership is needed to satisfy infrastructure requirements: "... the partnership can take responsibility for the final stacked device ... without the partnership, responsibility would be very difficult" [31].

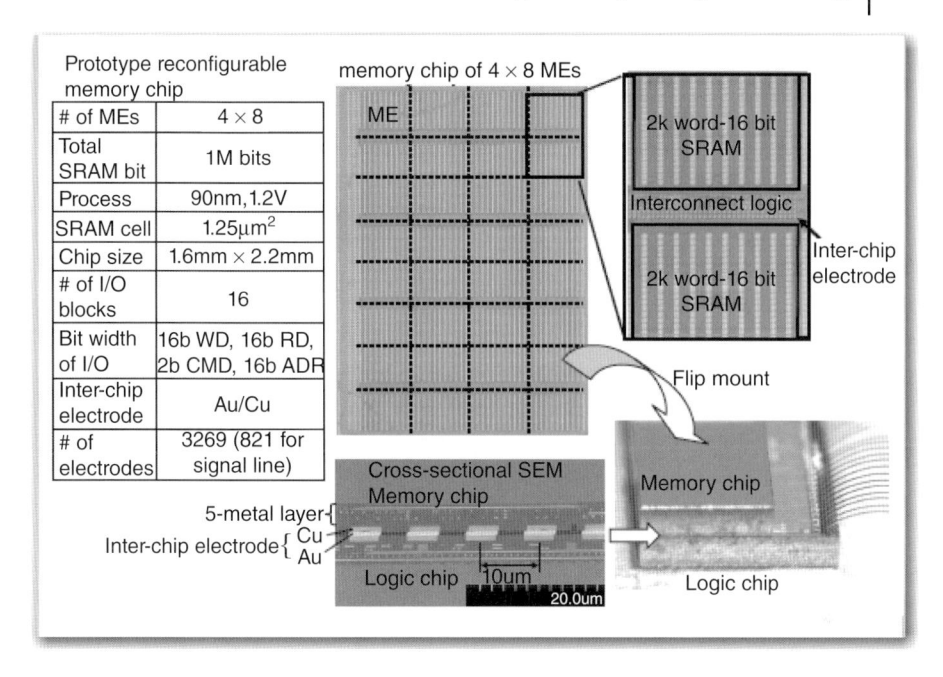

Figure 15.20 NEC CoC memory on logic solution [32].

NEC NEC memory on logic, CoC technology is called "chip stacked flexible memory" [32]. NEC reported that each functional IP core including 3D graphics and video codec can be connected to closely positioned local memory for fast access and wide bandwidth.

NEC developed high-density (10 μm pitch) interconnect technology. When such a memory chip configuration is used with a mobile phone system-on-chip, 50% of the on-chip SRAM of the system-on-chip can be eliminated. Figure 15.20 shows a micrograph of a memory chip prototype having 4 × 8 memory elements fabricated in a 90 nm process. On the chip surface, planarized copper metal structures with a thin gold bump (about 400 nm) are formed for interchip bonding pads. Gold electrodes of 5 μm^2 and 10 μm pitch on copper pads are located in logic areas. The Au–Au bump bonding enables the interchip connections. No indication of commercialization timing was given.

NEC has published extensively about its joint program with Elpida to develop vias-first memory technology with doped polysilicon-filled TSVs [33, 34]. NEC has recently announced that it has abandoned plans to use "via first" doped polysilicon as the conductor for DRAM TSVs, and would move to metal-filled structures "... because of the difficulty in making consistent ohmic contact to the backside of these poly-Si TSV" [35].

Tezzaron Tezzaron has been a pioneer in the 3D IC area, working on 3D memory solutions for more than a decade. Tezzaron's SuperContacts™ (Figure 15.21)

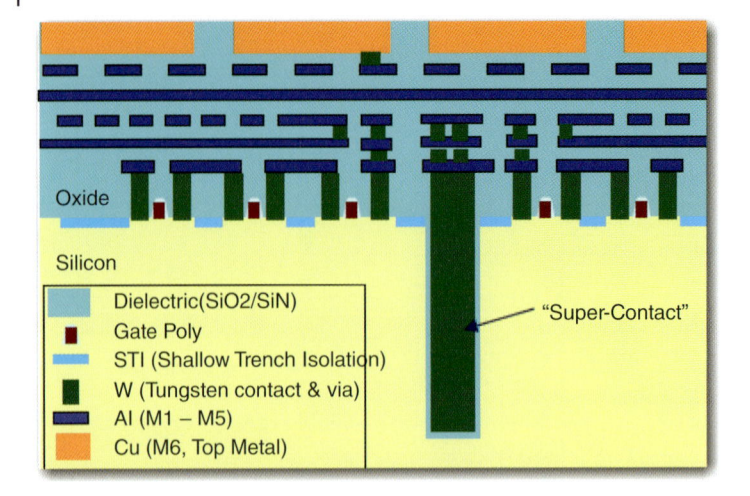

Figure 15.21 Tezzaron SuperContacts [36].

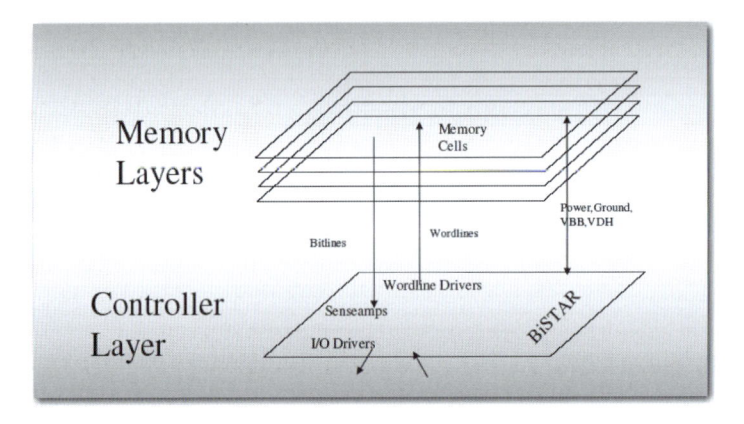

Figure 15.22 Tezzaron memory solution [37].

are 1.2 μm diameter tungsten-filled "via middle" TSVs [36]. The concept is to build the memory cells in one strata and the controller functions in another layer as shown in Figure 15.22. Chartered Semiconductor (now GlobalFoundries) and Tezzaron Semiconductor have been working together on the scale-up of these 3D IC memories.

Other Memory Suppliers In early 2011 Micron commented that it was currently "sampling products based on TSVs" and that "mass production for TSV-based 3-D chips is slated in the next 12–18 months" [38].

In 2011 Nanya Technology announced that it expected to start shipping its integrated 3D solutions in small volumes in 2011 [39].

Hynix has been working extensively on 3D integrated memory stacks and expects to bring such products to market within the next few years.

15.4.5
Memory on Logic

Memory on logic, especially wide I/O interface memory, appears on all foundry, outsourced assembly and test (OSAT), and identity management roadmaps as a key product driver for 3D IC technology.

Logic + memory stacking generically includes stacking cache or main memory onto a high-performance logic device. These applications typically do not require tight interstrata pitch. Examples have been published where 4 μm TSVs have been used for these connections [40], while traditional copper solder joints and Cu–Sn–Cu thermocompression interconnects were integrated as the bonding approach for this stacking process.

Simulations were run to compare the baseline performance of the Intel® Core™ 2 Duo to cases where a large SRAM or DRAM cache was stacked on top. Overall, this resulted in higher bandwidth and shorter latency access [40]. With stacked cache, the off-die bandwidth requirements are reduced three times, while simultaneously yielding a 66% average power reduction in average bus power due to reduced bus activity.

A simplifying aspect of stacking DRAM memory on logic is that thermal issues are minimal since DRAMs consume little power and the hotter die (processor) is in close proximity to the heat sink, as would normally be the case. An in-depth thermal calculation of a 92W microprocessor stacked with a DRAM showed the peak temperature only increased by 1.92 °C [40]

Black et al. have studied cache placement in multicore processors [41]. Options are shown in Figure 15.23. Option (a) consists of 4 MB of on-chip cache. Option (b) increases the L2 size to 12 MB SRAM and stacks the additional 8 MB L2 cache on top of the baseline processor die. The 8 MB stacked L2 is roughly the same size as the baseline die. Option (c) replaces the SRAM L2 with a denser DRAM L2. Typically DRAM is about eight times denser than an SRAM. Hence, they replaced the 4 MB L2 with a 32 MB stacked DRAM L2. In this option they remove the 4 MB L2 cache on the baseline processor die reducing the planar die dimensions by 50%.

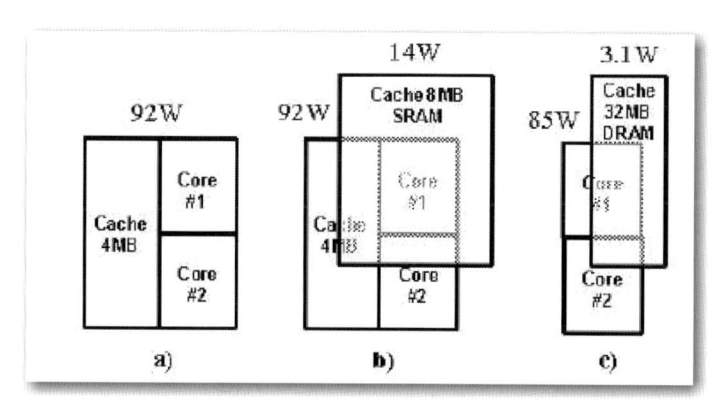

Figure 15.23 Memory stack options [41]

They found that increasing the last level cache capacity from 4 to 32 MB, on average, reduces bus bandwidth requirements by three times and cycles per memory access (CPMA) by 13% with peak CPMA reduction of 50%. There is also a 66% average power reduction in average bus power, due to reduced bus activity. Assuming a bus power consumption rate of $20\,\text{mW}\,\text{Gb}^{-1}\,\text{s}^{-1}$, 3D stacking of DRAM reduces bus power by 0.5 W.

Putswammy and Loh of Georgia Institute of Technology have reported a repartitioning of the Intel Core 2 processor to better understand the potential impact of 3D on latency [42]. The results are summarized in Table 15.3. The reduction of latency by repartitioning the cache and cores is significant.

It is thus expected that current methods of interconnecting memory and logic will migrate to 3D IC integration as shown in Figure 15.24, as manufacturing becomes available and costs come down.

15.4.6
Repartitioning Logic

Repartitioning of logic devices into separate strata will require significant advances in the current design and modeling tools available and a significant density of

Table 15.3 Impact of partitioning on latency for Intel Core 2 processor [42].

	Latency reduction (%)
Scheduler	32
ALU + bypass	36
Reorder buffer	52
Register file	53
L1 cache	31
L2 cache	51
Register alias table	36

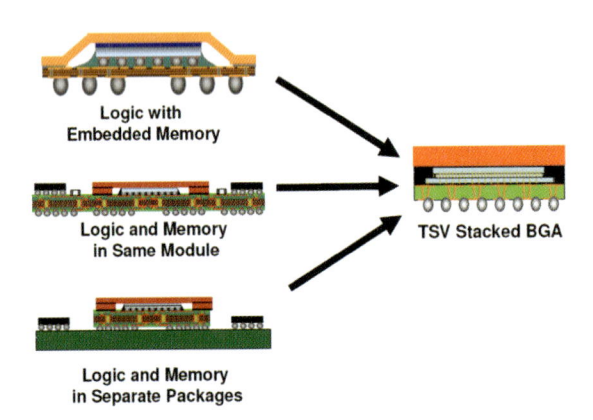

Figure 15.24 Migration to 3D IC memory on logic [43].

TSVs that today would be difficult to yield at reasonable cost. Thermal issues could be much more severe than for 3D systems containing just one logic stratum in close proximity to the heat sink. It is therefore likely that 3D logic repartitioning will be several years or more away.

15.4.7
Foundry and OSAT Activity

15.4.7.1 Foundry Activities
TSMC has announced that mass production of 3D chips with TSVs would start "with the 28 nm node" [44]. TSMC also announced the commercialization of silicon TSV interposers. For instance, Xilinx announced a single layer, multichip silicon interposer for its 28 nm 7 series FPGAs. Xilinx reported that chips would be fabricated by TSMC and assembled by Amkor [45, 46].

To support 3D IC customer designs, TSMC has released TSMC Reference Flow 11.0 which enables 3D IC integration to become part of the mainstream design flow. Both Cadence [47] and Synopsys [48] have contributed to these capabilities.

UMC has announced a three-way cooperation agreement with Elpida Memory and Powertech Technology (PTI) leveraging UMC's foundry logic technologies, Elpida's DRAM technology, and PTI's assembly skills to develop a total 3D IC logic + DRAM integrated solution. The total solution will include logic + DRAM interface design, TSV formation, wafer thinning, testing, and chip stacking assembly for customers. The resulting technology is expected to increase cost competitiveness, improve logic yield impact, and accelerate entry into the 3D IC market [30].

15.4.8
Other 3D Applications

Besides CMOS-based 3D integrations, there is a wide range of applications using wafer bonding technologies, particularly MEMS. In fact, the wafer bonding technology was initially developed for MEMS, and further improved for many other applications.

There have been significant advances for MEMS technology using wafer bonding, as described in Chapter 14. Yole Développement has a MEMS law: "One product, one process, one package," which still rules as in 2011 [49]. By 2020 Yole predicts: "it is likely that MEMS fabs will have developed internal standard process blocks but it will be fab-specific standard tools" [49]. There have been many publications describing the combination of CMOS with MEMS (e.g., [50] and many papers published in digests of International Electron Devices meetings, for example in 2008, 2009, and 2010). CMOS/MEMS is likely to be restricted to very specific applications where MEMS arrays will need very close electronic processing. For all other cases, it will depend on MEMS product cycle time, flexibility, cost, integration, market demand, and power consumption [49]. However, with the development of TSV technology, for a few applications ". . . users are willing to pay the higher cost to get the better performance and smaller size from the shorter connections. Yole projects demand for MEMS with 3D-TSV will reach several hundred thousand wafers a year by 2015" [51]. With the maturity of 3D technologies, more 3D integrated CMOS/MEMS products will be developed with low cost.

Moreover, as shown in Figure 15.5, heterogeneous integration technologies have been developed for functional diversification systems (more than Moore) [2], that is, integration of CMOS with other devices, such as analog/RF, solid-state lighting, high-voltage power, passives, sensors/actuators, biochips, and biomedical devices. This heterogeneous integration started with system-in-packaging technology, and is expected to evolutionarily move to 3D heterogeneous integration with TSVs and wafer bonding.

Infineon, Sensonor, SINTEF, and Fraunhofer Munich demonstrated 3D heterogeneous integration of an ultra-miniaturized tire pressure monitoring system (TPMS) within the European e-CUBES project [7, 52]. The key element of Infineon's TPMS, a 3D IC/MEMS stack, consisting of a microcontroller, an RF transceiver, a bulk acoustic wave resonator, and a pressure sensor from Sensonor, was processed at the wafer level in combination with Fraunhofer Munich's 3D TSV technology (see Section 15.2) and SINTEF's MEMS-specific 3D integration technology (Figure 15.25).

The requirements for TSVs are different between 3D ICs and MEMS/IC stacks. While 3D ICs typically require high TSV interconnect density and conductivity, the integration of MEMS with ICs, in most cases, does not need very high TSV performance. Rather, there is a strong requirement for very deep TSVs because the device substrates typically cannot be extremely thinned without violating the mechanical stability of the device or without breaking fragile mechanical structures.

Figure 15.25 Three-layer 3D IC/MEMS stacks for Infineon's TPMS wireless sensor nodes (e-CUBES®), composed of a microcontroller, an RF transceiver, a bulk acoustic wave resonator, and a pressure sensor [7].

15.5
Conclusions

The semiconductor industry has advanced exponentially for more than 40 years, predominantly by shrinking transistor dimensions. However, the semiconductor industry with the transistor scaling alone will not be able to overcome the performance and cost problems of future IC fabrication. Given technical and economic realities, most CMOS practitioners today have concluded that CMOS shrinkage following Moore's law will come to an end somewhere around the 22 nm node, depending on device and structure, and that 3D integration is one of the few avenues available to alleviate this problem until an alternative to CMOS technology is agreed upon. Three-dimensional ICs will most probably be based on TSV technology. Several corresponding full 3D processing flows have been demonstrated; however, to 2011, there were still no microelectronic products based on TSV technologies in the market – except CISs. Three-dimensional chip stacking of memory and logic devices without TSVs is already widely introduced in the market. Applying TSV technology for memory on logic will increase the performance of these advanced products and simultaneously improve the form factor. As such, we can expect 3D IC applications to continue to advance in the coming years in an evolutionary fashion. Three-dimensional processing technologies and equipment will be further developed for TSVs, alignment, thinning processes, and – in particular – wafer bonding. Several wafer bonding technologies presented in this book will be advantageously used in the fabrication of 3D integrated products. Three-dimensional integration technologies based on metal bonding, explicitly Cu–Cu bonding and SLID bonding may be the best choice for the majority of applications.

References

1 Garrou, P., Bower, C., and Ramm, P. (eds) (2008) *Handbook of 3D Integration*, Wiley-VCH, Weinheim.
2 Semiconductor Industry Association (2009) The international technology roadmap for semiconductors. SEMATECH, Austin, TX (2010 update).
3 Reif, R., Fan, A., Chen, K.-N., and Das, S. (2002) Fabrication technologies for three-dimensional integrated circuits. Proceedings of the International Symposium on Quality Electronic Design (ISQED).
4 Lu, J.-Q. (2009) 3D hyper-integration and packaging technologies for micro-nano-systems. *Proc. IEEE*, **97** (1), 18–30.
5 Ramm, P. and Klumpp, A. (2002) Method of vertically integrating electric components by means of back conditioning. US Patent 6,548,391, filed 18 March 2002 and issued 15 April 2003.
6 Ramm, P., Klumpp, A., Merkel, R., Weber, J., Wieland, R. (2004) Vertical system integration by using inter-chip vias and solid-liquid interdiffusion bonding. *Japn. J. Appl. Phys.*, **43** (7A), L829–L830.
7 Ramm, P., Klumpp, A., Weber, J., Lietaer, N., Taklo, M., W.D. Raedt, T. Fritzsch, and P. Couderc (2010) 3D integration technology: status and application development. Proceedings of ESSCIRC/ESSDERC 2010, Seville, Spain, pp. 9–16.
8 Kwon, O.H. (2011) Eco-friendly semiconductor technologies for healthy

living (plenary lecture). IEEE International Solid-State Circuits Conference (ISSCC), 20–24 February 2011, pp. 22–28.

9 Zinck, C. (2010) 3D integration infrastructure and market status. Proceedings of the IEEE International 3D System Integration Conference (3DIC 2010), Munich, edited by P. Ramm and E. Beyne (IEEE Xplore 978-1-4577-0526-7 ©2011 IEEE), http://ieeexplore.ieee.org/xpl/conhome.jsp?punumber=1002927.

10 Fernández-Bolaños, M. and Ionescu, A. (2010) Heterogeneous integration for novel functionality. IEEE International 3D System Integration Conference (3DIC 2010), 16–18 November 2010.

11 e-Brains Consortium (2011) European Large Scale Integrating Project Best-Reliable Ambient Intelligent Nanosensor Systems by Heterogeneous Integration (e-BRAINS), www.e-brains.org (accessed 26 August 2011).

12 Garrou, P. (2009) The evolution to 3-D IC integration. Proceedings of the IMAPS Device Packaging Conference, Scottsdale AZ.

13 Huebner, H., Eigner, M., Gruber, W., Klumpp, A., Merkel, R., Ramm, P., Roth, M., Weber, J., and Wieland, R. (2002) Face-to-face chip integration with full metal interface. Proceedings of the Advanced Metallization Conference (AMC 2002), San Diego, CA, p. 53.

14 Doe, P. (2005) 3D Interconnect Gets Real, Solid State Technology, January 2005.

15 Uno, M. (2007) Chip-on-Chip Offers Higher Memory Capacity, Speed. Nikkei Electronics Asia, February 2007, pp. 28–32

16 Wakiyama, S., Ozaki, H., Nabe, Y., Kume, T., Ezaki, T., and Ogawa, T. (2007) Novel low-temperature CoC interconnection technology for multichip LSI (MCL). Proceedings of the 57th Electronic Components and Technology Conference (ECTC 07), pp. 610–615.

17 Yoon, S.W., Yang, D.W., Koo, J.H., Padmanathan, M., and Carson, F. (2009) 3D TSV processes and its assembly/packaging technology. IEEE International Conference on 3D System Integration (3DIC 2009), San Francisco, CA.

18 Joseph, A. et al. (2008) TSV enable next generation SiGe power amps for wireless communication. IBM J. Res. Dev., **52**, 635–648.

19 Garrou, P. (2000) Wafer level chip scale packaging (WL-CSP): an overview. IEEE Trans. Adv. Packag., **23**, 198–205.

20 Guillou, Y., et al. (2009) 3D IC products using TSV for mobile phone applications: an industrial perspective. Proceedings of the IMAPS European Microelectronics Packaging Conference (EMPC).

21 Motoyoshia, M. and Koyanagi, M. (2008) 3D-LSI technology for image sensor. Proceedings of PIXEL 2008 International Workshop, Batavia, IL, September 2008.

22 Tomkins, G. (2009) Detailed reverse engineering analysis of TSV technology in CIS. Proceedings of the Nikkei Micro 3D Technology Conference.

23 Losavio, A. et al. (2007) Novel non-volatile memory architecture using through silicon via interconnections. Proceedings of the 2nd International Workshop on 3D System Integration, Munich.

24 Yoon, J. (2008) Advanced Si and DRAM Pkg technology: status and future directions. Proceedings of the SMTA Pan Pacific Symposium.

25 Kang, U. et al. (2009) 8 Gb DDR3 DRAM using through-silicon-vis technology. IEEE International Solid-State Circuits Conference (ISSCC 2009), pp. 130–131.

26 Micronews (2010) *New Samsung 8GB DDR3 Module Utilizes 3D TSV Technology*, http://www.i-micronews.com/news/Samsung-8GB-DDR3-module-utilizes-3D-TSV-technology,5940.html (accessed 26 August 2011).

27 ElectroIQ (2010) *Era of 3DIC Has Arrived with Samsung Commercial Announcement*, http://www.electroiq.com/index/packaging/packaging-blogs/ap-blog-display/blogs/ap-blog/post987_2675024352428682313.html.

28 Kim, J.-S. et al. (2011) A 1.2 V 12.8 GB/s 2 Gb mobile wide I/O DRAM with 4 × 128 I/Os using TSV-based stacking. IEEE International Solid-State Circuits Conference (ISSCC 2011), pp. 496–498.

29 Garrou, P. (2009) Elpida Develops 3-D Stacked 8 Gb DRAM, Semiconductor International, 4 September 2009.

30 Micronews (2010) *Elpida, PTI & UMC to Partner for 3DIC Commercialization of Logic + DRAM Stacks by 2011*, http://www.i-micronews.com/news/Elpida-PTI-UMC-partner-3DIC-commercialization-Logic+DRAM,5020.html (accessed 26 August 2011).

31 Micronews (2011) *Samsung, Elpida and Tohoku Univ Discuss 3D IC at RTI Conference*, http://www.i-micronews.com/lectureArticle.asp?id=6065 (accessed 26 August 2011).

32 Saito, H. *et al.* (2009) A chip-stacked memory for on-chip SRAM-rich SoCs and processors. IEEE International Solid-State Circuits Conference (ISSCC 2009), pp. 60–61.

33 Mitsuhashi, T. *et al.* (2006) Development of 3D packaging process technology for stacked memory chips. *MRS Symp. Proc.*, **970**, 0970-Y03-06.

34 Kawano, M., Takahasi, N., Kurita, Y., Soejima, K., Komuro, M., and Matsui, S. (2008) Three-dimensional packaging technology for stacked DRAM with 3-Gb/s data transfer. *IEEE Trans. Electron Devices*, **55**, 1614–1620.

35 Garrou, P. (2009) NEC Points to Ni for Memory 3D TSV, Perspectives from the Leading Edge, blog, Semiconductor International, 19 July 2009.

36 Patti, R. (2008) 3D integration at Tezzaron Semiconductor Corporation, in *Handbook of 3D Integration* (eds P. Garrou, C. Bower, and P. Ramm), Wiley-VCH Verlag GmbH, Weinheim, pp. 463–486.

37 Patti, R. (2008) Memory-logic 3D combination. Proceedings of the IMAPS Device Packaging Conference, Scottsdale, AZ.

38 LaPedus, M. (2011) *Micron COO Talks 450-mm, 3-D, EUV*, http://www.eetimes.com/electronics-news/4212072/Micron-COO-talks-450-mm–3-D–EUV (accessed 26 August 2011).

39 DigiTimes (2011) *Nanya, ITRI Team Up for 3D Stacked DRAM*, http://www.digitimes.com/news/a20110125PD207.html (accessed 26 August 2011).

40 Morrow, P. *et al.* (2006) Design and fabrication of 3D microprocessors. *MRS Symp. Proc.*, **970**, 0970-Y03-02.

41 Black, B. *et al.* (2006) Die stacking 3D microarchitecture. IEEE/ACM 39th International Symposium on Microarchitecture, December 2006, pp. 469–479.

42 Puttaswamy, K. and Loh, G. (2007) Thermal herding: microarchitecture techniques for controlling hotspots in high-performance 3D-integrated processors. 13th IEEE International Symposium on High Performance Computer Architecture, p. 193.

43 Huemoeller, R. (2009) 3D packaging – view from the SAT. RTI 3-D Integration for Semiconductor Integration and Packaging Conference, Burlingame, CA.

44 Oshita, J. (2010) *TSMC Announces 3D Chip Technologies Using TSVs*, http://techon.nikkeibp.co.jp/english/NEWS_EN/20101209/188050/ (accessed 26 August 2011).

45 Micronews (2010) *Xilinx Brings 3D TSV Interconnects to Commercialization Phase in Digital FPGA World*, www.i-micronews.com/lectureArticle.asp?id=5693 (accessed 26 August 2011).

46 Micronews (2011) *IBM and TSMC Discuss 3D IC vs Scaling*, http://www.i-micronews.com/news/IBM-TSMC-Discuss-3D-IC-vs-Scaling,6066.html (accessed 26 August 2011).

47 Cadence (2010) *Cadence Delivers TLM-Driven Design and Verification, 3D-IC Design and Integrated DFM Capabilities to TSMC Reference Flow 11.0*, http://www.cadence.com/cadence/newsroom/press_releases/pages/pr.aspx?xml=061110_tsmc (accessed 26 August 2011).

48 PR Newswire (2010) *Synopsys Delivers Comprehensive Design Enablement for TSMC 28-nm Process Technology with Reference Flow 11.0*, http://www.prnewswire.com/news-releases/synopsys-delivers-comprehensive-design-enablement-for-tsmc-28-nm-process-technology-with-reference-flow-110-95950634.html (accessed 26 August 2011).

49 Yole Développement (2011) MEMS: trends in MEMS manufacturing and packaging, technology report, January 2011.

50 Varadan, V.K. (ed.) (2004) *Smart Structures and Materials 2004: Smart Electronics, MEMS, BioMEMS, and Nanotechnology*, vol. 5389, SPIE, Bellingham, WA.

51 Mounier, E. (2010) *MEMS Technology Roadmap: Demand for Smaller, Lower Cost Devices Drives Major Technology Trends for Next Decade*, http://memsblog.wordpress.com/2010/12/02/mems-technology-roadmap-demand-for-smaller-lower-cost-devices-drives-major-technology-trends-for-next-decade/ (accessed 26 August 2011).

52 Ramm, P., Klumpp, A., Weber, J., and Taklo, M. (2010) 3D system-on-chip technologies for more than Moore systems. *J. Microsyst. Technol.*, **16**, 1051–1055.

16
Temporary Bonding for Enabling Three-Dimensional Integration and Packaging

Rama Puligadda

16.1
Introduction

The end of device scaling based on Moore's law or the possibility of prolonging this legendary law by the next generation of lithography technology has been the subject of debate for nearly a decade. Building three-dimensional (3D) devices has recently emerged as a plausible solution to the ever-increasing challenge to make smaller and faster devices with higher functionality. Three-dimensional integration and packaging require new ways to process and handle wafers. Some contention exists about the definitions of *3D integration* and *3D packaging*. Most commonly, making 3D structures by means of through-silicon vias (TSVs) is considered to be 3D integration, while making stacked packages and other stacked structures by means of external connections is termed 3D packaging [1]. Most of the discussion in this chapter pertains to 3D integration of devices.

Numerous ways of stacking devices exist, but they all have wafer-thinning processes in common and some approaches require thin-wafer handling. Wafers thinned to a thickness of 150 µm or more have been shown to survive handling without significant damage, but these wafers need highly sophisticated and often expensive equipment to avoid chipping and breakage. However, TSV formation and filling at thicknesses greater than 100 µm can be challenging and costly. Additionally, interconnect delay and heat dissipation issues can be better addressed with thinner substrates. To make TSV technology viable, wafers must be thinned to thicknesses of 100 µm or less. Thinned silicon or III–V wafers are extremely fragile and tend to warp and fold, the risk escalating with increasing wafer size. It is already highly desired for most 3D integrated circuit (IC) manufacturers to thin wafers to a thickness of 50 µm or less to realize the full benefit of 3D integration. Handling such thin wafers demands that each wafer be temporarily bonded to a carrier or handle wafer using a temporary bonding material. Alternatively, the device wafer can be bonded permanently or mounted directly onto another device wafer face-to-face, followed by thinning and processing. Examples of typical backside processing are described in the subsequent sections.

Handbook of Wafer Bonding, First Edition. Edited by Peter Ramm, James Jian-Qiang Lu, Maaike M.V. Taklo.
© 2012 Wiley-VCH Verlag GmbH & Co. KGaA. Published 2012 by Wiley-VCH Verlag GmbH & Co. KGaA.

Once a device wafer is attached to a carrier wafer, the wafer pair can be processed like a standard full-thickness wafer without additional modification of the processing equipment or the process flows [2–6]. When the backside processing is complete, the thinned wafer can be separated (released or debonded) from the carrier and attached to a dicing frame, where it is diced to create dies for the pick-and-place process. Alternatively in a wafer-to-wafer scheme, the entire device wafer can be bonded permanently to another device wafer before being released from the carrier wafer.

The following sections provide a description of different material and process options for temporary bonding. Several examples of integration processes completed using Brewer Science WaferBOND® HT technology are also covered.

16.2
Temporary Bonding Technology Options

Temporarily bonding a device wafer to a carrier and effectively releasing the device wafer are considered to be among the most critical and enabling technologies for achieving 3D integration of devices. The requirements of the temporary bonding materials heavily depend not only on the backside processes, but also on the condition of the device wafer and the processes it has been through prior to bonding. A "via-first" [7] device wafer presents different challenges from a device wafer in a "via-last" process. In a via-first process, the TSVs are fabricated in the device wafer prior to bonding. The higher coefficient of thermal expansion of copper, tungsten, or polysilicon used to fill the TSVs compared to that of silicon causes stress in the wafer, which becomes important when the wafer is thinned to 100 μm or less. In a via-last process, the TSVs are fabricated after the device wafer is thinned and bonded to a carrier wafer. The temporary bonding agent in this case must go through many processing steps including deep via etching, electroplating, etc. Schematics of the via-first and via-last processes are shown in Figure 16.1. In addition to the backside process, requirements of the adhesives depend on the condition of the device wafer. For instance, bumps on the device side impart greater challenges than a device wafer with low topography. Additionally, bumping on the backside of the thinned wafer requires special chucking for release and further handling.

Several different technology options are available today for bonding and debonding or demounting after processing, although the options described in this section are in different stages of development and testing. While tapes and waxes are commercially available and used industry-wide for a limited number of applications today, they typically do not survive high-temperature processes and harsh chemical environments. Additional limitations include high total thickness variation (TTV) and the presence of residues after release. The sections below focus mainly on temporary bonding agents that provide alternatives to tapes and waxes.

Via first

TSVs from front side — Temporary bonding — Thinning and backside processing — Debonding or release

Via last

Device wafer with CMOS — Temporary bonding — Thinning — TSV from backside and backside — Debonding or release

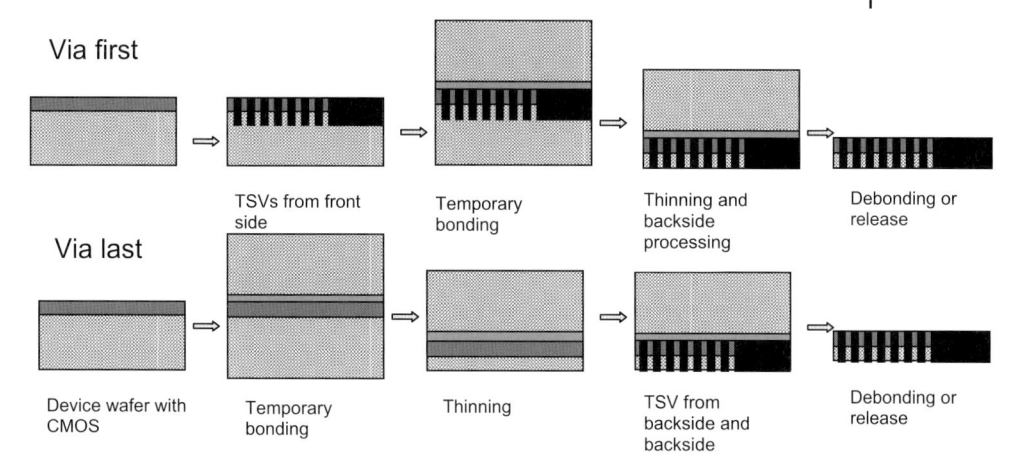

Figure 16.1 Schematics of via-first and via-last processes.

16.2.1
Key Requirements

The key requirements for temporary bonding agents or adhesives and the related bonding and debonding processes include the following.

- **Ease of application.** The process flow for application of the temporary bonding agent should enable high throughput and preferably use industry-known and accepted unit processes.

- **Low TTV.** The TTV of the adhesive is transferred to the thinned wafer during the thinning process. For a thermoplastic type of adhesive, reflow during the bonding process allows greater uniformity in the bond line.

- **Good adhesion to a wide variety of surfaces.** A typical device wafer has several different surfaces in contact with the adhesive. Poor wetting to any of these surfaces may lead to delamination or voiding in the bondline.

- **Bonding and debonding at appropriate temperatures.** Low bonding and debonding temperatures are advantageous for conserving thermal budgets of the device wafer. For obvious reasons, these processes should be performed below temperatures limited by device stability.

- **High thermal stability.** The adhesive layer(s) must remain stable through the high-temperature backside processes including debonding to avoid breakage failure and equipment contamination. Examples of high-temperature processes include dielectric deposition, annealing, metal deposition, permanent bonding, and solder reflow. Furthermore, outgassing from the adhesive layer(s) is an important consideration.

- **High chemical resistance to process chemicals.** The adhesive material must be resistant to process chemicals such as etchants and corrosive electroplating solutions yet remain soluble in the removal solutions after debonding.

- **Good mechanical strength to keep device wafer in place.** The mechanical stability of the adhesive becomes important during high-temperature processes where it must hold down the thinned, often highly stressed device wafer to prevent buckling or curling.

- **No damage to wafer during thinning, debonding, or handling.** Damage to device wafers during thinning can be mitigated to a large extent using one of the several edge protection schemes described in Section 16.3.2. In addition to damage that the grinding process can cause, significant damage can occur during handling after debonding if the wafer is not fully supported. Flattening or smearing of backside bumps is an additional concern and therefore requires special debonding and handling chucks.

- **No residues.** It is imperative that the bonding agents or adhesives must be completely removed from the surface of the wafer to avoid difficulties in downstream processing and to prevent negatively affecting the reliability of the device package.

16.2.2
Foremost Temporary Wafer Bonding Technologies

Several different temporary bonding approaches being developed today enable thin-wafer handling, some of which are commercially available for testing and development of 3D IC processes. The following sections describe the leading technologies in detail. While most of the systems use similar bonding technology, the debonding technology is different for each system.

16.2.2.1 Thermoplastic Adhesive, Slide-Off Debonding Approach

The materials used in this approach are primarily developed by Brewer Science Inc., and were originally designed to be compatible with equipment manufactured by EV Group (EVG) [8]. The adhesive solution is applied to either the carrier or the device wafer by spin-coating. An important consideration for the coating process is that the adhesive thickness must exceed the height of device features to ensure complete planarization. The remaining solvent in the film is then eliminated using a baking process. The device wafer and carrier are bonded in a vacuum chamber at a temperature above the softening point of the adhesive and at pressures ranging from 5 to 20 psi. This technology can be used with silicon, glass, or sapphire carriers of the same size as the device wafer. As a result, this approach requires no modification to the lithography, etch, or deposition tools.

After backside thinning and other processes on the device wafer, the wafer pair is heated to a temperature above the softening point of the adhesive, and the device wafer is separated from the carrier by sliding off the carrier wafer. The thin device

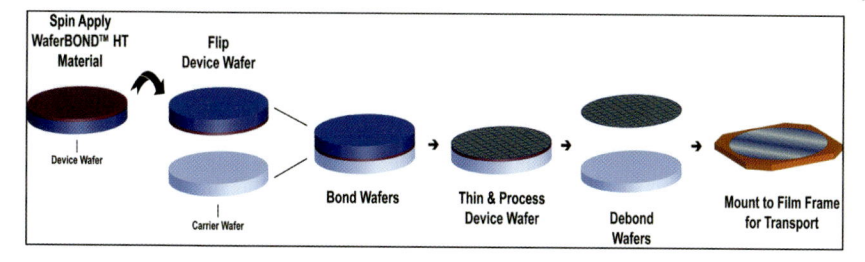

Figure 16.2 Process flow for slide-off debonding approach.

wafer is completely supported on a vacuum chuck in the debonding equipment. The device wafer and carrier are then cleaned on a specially designed chuck using an appropriate remover, which is typically supplied with the adhesive. The thin device wafer is then transferred to a dicing frame or a coin stack. Other output formats such as individual carriers have also been investigated.

The wafers are processed in a single-wafer-processing mode in each of these processing steps. The slide-off approach is shown in Figure 16.2. When automated, this approach requires an equipment set consisting of a bonder unit where the device wafer is coated and bonded and a debonder unit where the wafers are separated and cleaned.

16.2.2.2 Wafer Support System Using Ultraviolet-Curable Adhesive and Light-to-Heat Conversion Layer

This approach has been developed by 3M to be used in conjunction with specially designed equipment [9, 10]. A liquid adhesive is applied using a spin-coating process, and the adhesive planarizes the topography on the device wafer. The device wafer is then mounted onto a glass carrier. The glass carrier is coated with a material called a light-to-heat conversion (LTHC) layer before the carrier is laminated to the wafer. The adhesive is cured using ultraviolet (UV) light to create a bond. The LTHC layer can absorb certain wavelengths of light such that irradiating this layer with a focused laser beam can generate enough heat to decompose it. As a result, the wafer detaches from the glass carrier. The backside of the device wafer is first attached to supporting tape before rastering the laser through the glass side. The cured adhesive film is peeled from the device wafer after separation of the glass carrier. An overview of 3M's process is shown in Figure 16.3. This approach also requires an equipment set with two members, a "mounter" and a "demounter." Additional equipment may be required for preparing the glass carrier with the LTHC layer.

16.2.2.3 Single-Wafer Debonding by Adhesive Dissolution through Perforated Support Plate

The equipment and materials needed for this approach are being developed by Tokyo Ohka Kogyo, Japan. The adhesive is spin-coated onto the device wafer, and the adhesive's solvent is removed by baking. Following the coating process, the

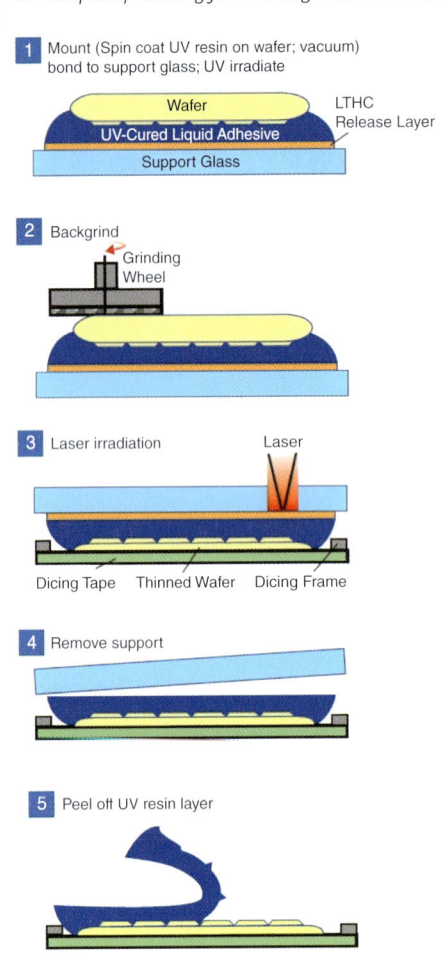

Figure 16.3 Process flow for 3M wafer support system.

device wafer is bonded to a perforated support plate. The supported wafer goes through subsequent backside processing such as grinding, etching, dielectric deposition, etc. Upon completion of backside processing, the wafer is separated by dissolution of the adhesive. Solvents or removal solutions penetrate through the perforations in the support plate and dissolve or etch the adhesive layer. The thin wafer is supported on a film frame through the demounting process. The thin wafer is further cleaned to remove any residues.

Dedicated equipment units for mounting and demounting of the support plate are also needed for this approach. Figure 16.4 shows an overall description of the process and equipment used in this approach.

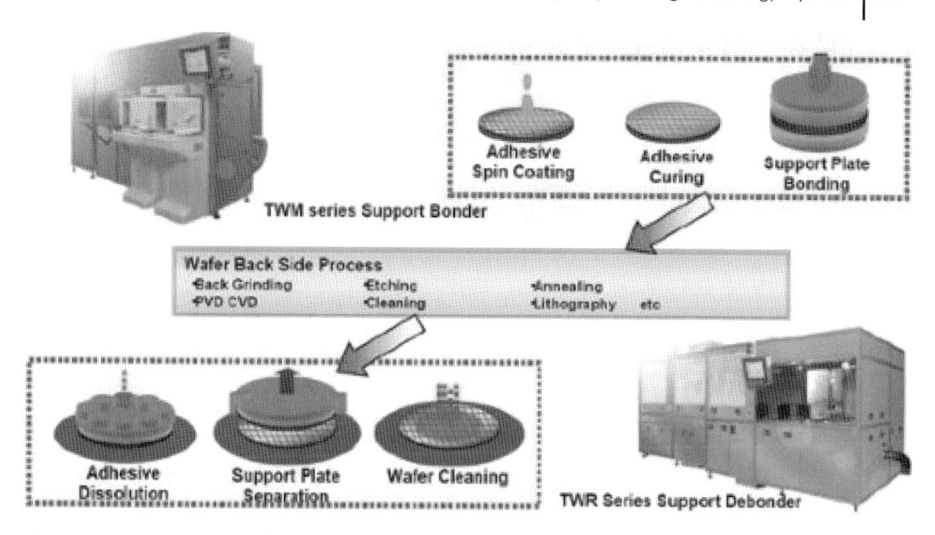

Figure 16.4 Overview [11] of single-wafer debonding process using dissolution through perforated wafer plate.

16.2.2.4 Debonding Using a Release Layer and Liftoff

This technology, developed by Thin Materials AG, Germany, involves separating the thinned wafer from the carrier by a liftoff process. The device wafer in this approach [11] is coated with a very thin (a few hundred nanometers thick) release layer. The release layer is deposited in a two-step process. First, a precursor material is spin- or spray-coated onto the wafer and is then subjected to a plasma treatment to give it nonstick characteristics. This step is followed by spin-coating a silicone-based elastomeric material and curing at temperatures near 200 °C. Following thinning and backside processing, the backside of the device wafer is attached to a film frame, and the carrier wafer is lifted off the wafer. Figure 16.5 shows a schematic of the overall process.

16.2.2.5 Room Temperature, Low-Stress Debonding

Developed recently by Brewer Science, this approach utilizes two zones to bond the device wafer to the carrier wafer. It uses a material in the center (one of the zones) to fill in the gap while not adhesively bonding them together. The outer edge is adhesively bonded to ensure the wafers remain together during subsequent processing and to prevent corrosive process chemicals from penetrating the bond line between the wafer stack. The two zones are achieved by treating the carrier wafer to give it a nonstick quality except at the periphery. This technique allows the adhesive coated on the device wafer to adhere only to the edge zone of the carrier. Figure 16.6 shows a schematic of the process used to achieve selective adhesion at the edges. Unlike the approach described previously, the carrier wafer is peeled away after thinning and backside processing while the device wafer is supported on a vacuum chuck.

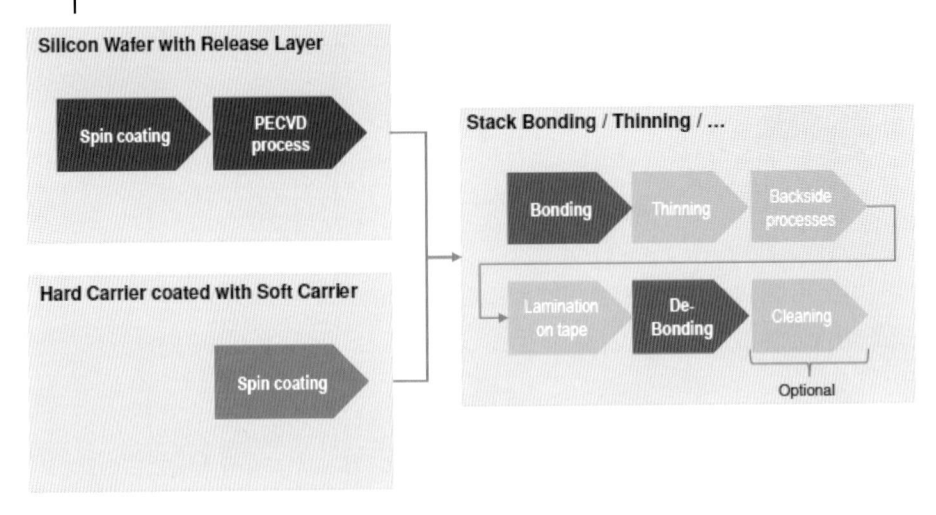

Figure 16.5 Overall [12] description of release using liftoff process.

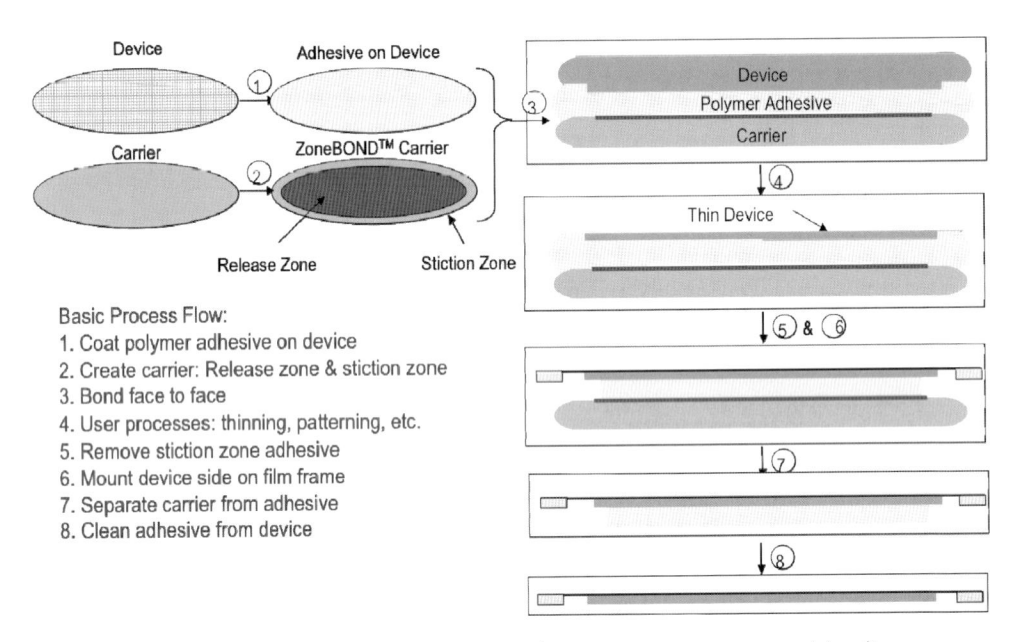

Basic Process Flow:
1. Coat polymer adhesive on device
2. Create carrier: Release zone & stiction zone
3. Bond face to face
4. User processes: thinning, patterning, etc.
5. Remove stiction zone adhesive
6. Mount device side on film frame
7. Separate carrier from adhesive
8. Clean adhesive from device

Figure 16.6 Overview of Brewer Science's low-stress, room temperature debonding process.

16.3
Boundary Conditions for Successful Processing

This section deals with boundary conditions established empirically by engineers for ensuring successful processing of device wafers on the backside.

16.3.1
Uniform and Void-Free Bonding

Uniformity of the bond is one of the factors that affect the TTV of the thinned wafer. In a via-first process, a nonuniform bonding layer may cause metal filling in some vias to be exposed before the rest during backside thinning, resulting in smearing of metal across the surface. On the other hand, in a via-last situation, nonuniform thickness of the wafer might result in loss of contact with metallization on the front side. Contribution of temporary bonding layer uniformity to TTV can be minimized by (i) utilizing a spin-coating process to deposit the temporary bonding layer and (ii) ensuring that chucks on the bonder are flat and leveled. A drop micrometer is frequently used to measure total stack thickness at several points on the wafer to ensure low variation.

A void in the bond line is a more severe problem. Scanning acoustic microscopy is an effective method to detect voids in the bonding layer. A large void trapped between the two wafers can expand in a subsequent high-temperature, low-vacuum process and cause delamination of the device wafer from the carrier. In extreme cases, it can cause breakage of the thinned wafer. Additionally, small voids can agglomerate during the high-temperature processes to make larger voids. Voids caused by air trapped during the bond process can be avoided by drawing a vacuum in the bond chamber before bringing the two wafers into contact. Another common source of voids in the adhesive is the outgassing from the device wafer surface. This can be alleviated by treatment of the device wafers at high temperature for a limited time before coating with adhesive.

16.3.2
Protection of Wafer Edges during Thinning and Subsequent Processing

Edge chipping and breakage is one of most common failure modes for thinning processes. Grinding a device wafer from the backside creates a very sharp edge that is prone to chipping and cracking. A study conducted by Brewer Science [13] compared the efficacy of the most common edge protection schemes. Control wafer pairs were simply bonded and thinned. Figure 16.7 shows each of the four schemes.

All four of the protection schemes showed improvement with regard to edge chipping. The size and number of chips were markedly less than those of the control pairs.

Statistical analysis of the data depicted in Figure 16.8 showed that the edge pre-trimming process developed by DISCO® is the most effective followed by pre-thinned carriers.

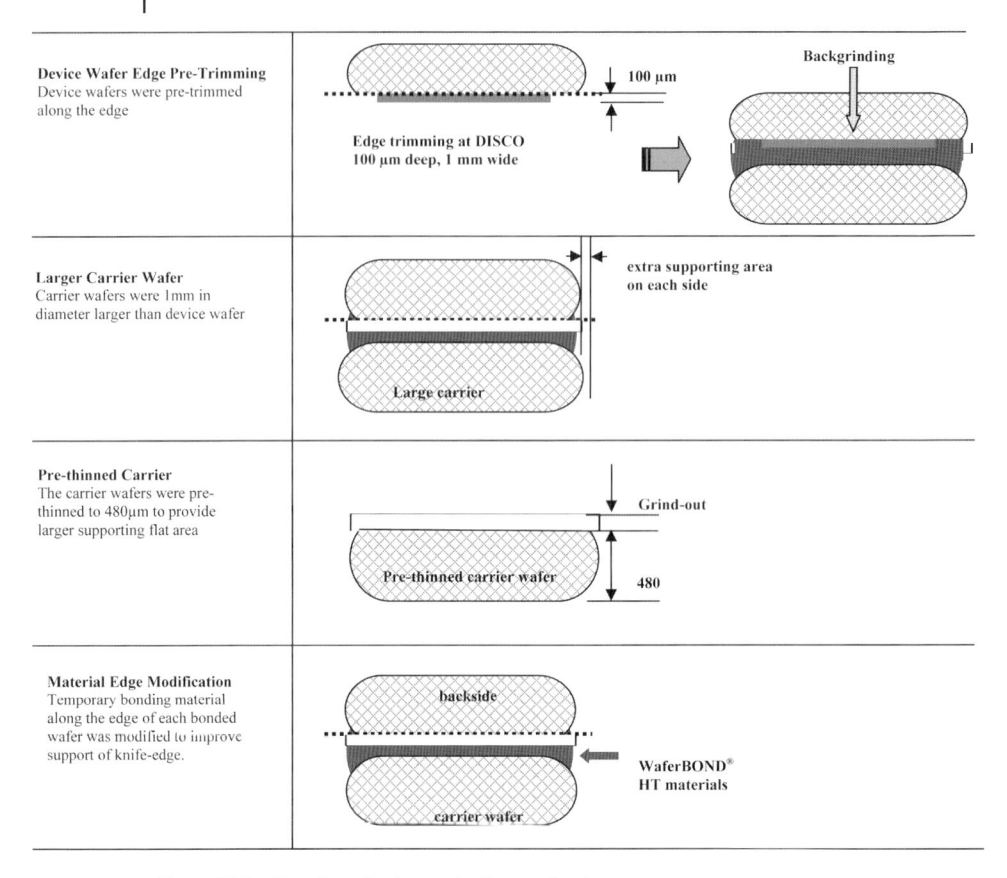

Device Wafer Edge Pre-Trimming Device wafers were pre-trimmed along the edge	Edge trimming at DISCO 100 μm deep, 1 mm wide → Backgrinding, 100 μm
Larger Carrier Wafer Carrier wafers were 1mm in diameter larger than device wafer	extra supporting area on each side — Large carrier
Pre-thinned Carrier The carrier wafers were pre-thinned to 480μm to provide larger supporting flat area	Grind-out — Pre-thinned carrier wafer, 480
Material Edge Modification Temporary bonding material along the edge of each bonded wafer was modified to improve support of knife-edge.	backside / carrier wafer ← WaferBOND® HT materials

Figure 16.7 Overview of edge protection methods.

16.4
Three-Dimensional Integration Processes Demonstrated with Thermomechanical Debonding Approach

This section presents examples where temporary bonding is successfully integrated into a 3D TSV process flow.

16.4.1
Via-Last Process on CMOS Image Sensor Device Wafers

A complete TSV process flow [14–16] (shown in Figure 16.9) was realized on active CMOS image sensor device wafers from ST Microelectronics using temporary bonding material from Brewer Science Inc. WaferBOND® HT 10.10 temporary bonding material was integrated into the process flows to achieve an aspect ratio of 1:1, the TSV diameter being 65 μm. Bonding to borosilicate carriers was carried out using EVG® 520.

16.4 Three-Dimensional Integration Processes Demonstrated | 339

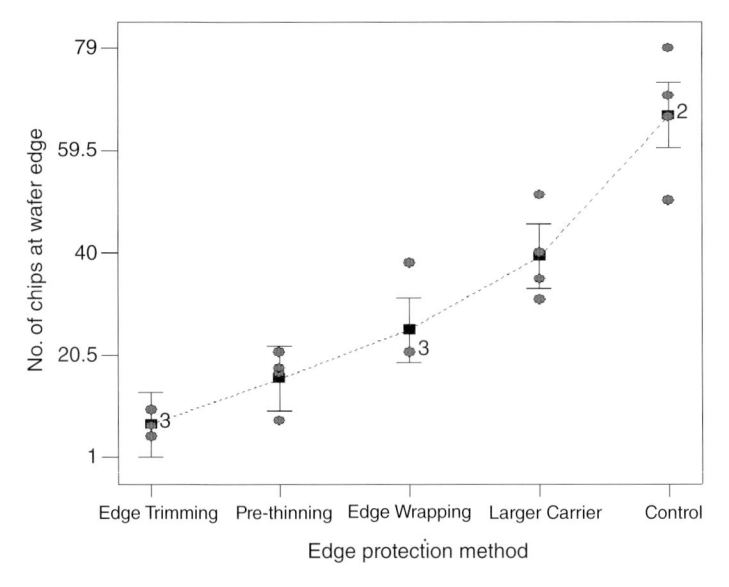

Figure 16.8 Results of a comparison study of known edge protection methods.

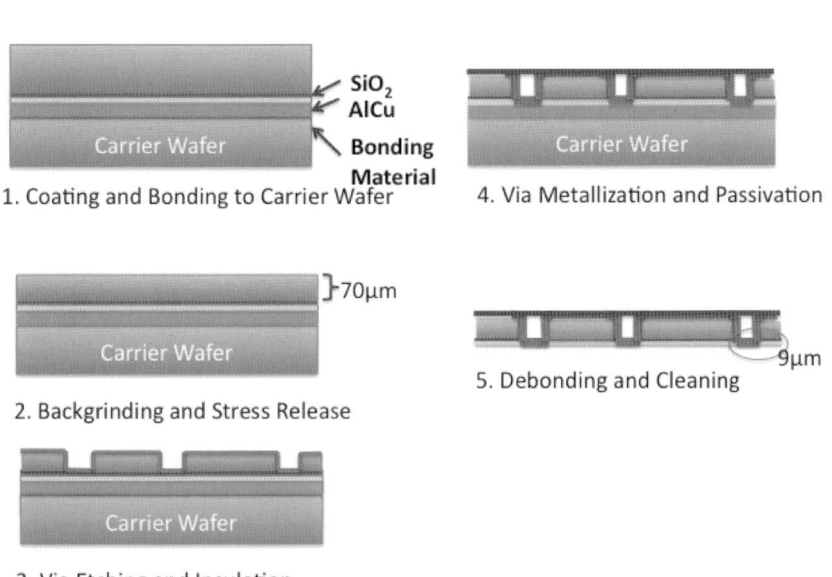

1. Coating and Bonding to Carrier Wafer
2. Backgrinding and Stress Release
3. Via Etching and Insulation
4. Via Metallization and Passivation
5. Debonding and Cleaning

Figure 16.9 Overview of via-last process flow.

Post-bonding processes included the following steps.

- **Backgrinding.** The wafers were thinned to 70 μm. Edge extraction of 1.5 μm was used to prevent chipping of the thinned wafer.

- **Via etching and insulation.** Via etching was performed using the BOSCH process to achieve vias of 65 μm in diameter through the wafer. TSVs with an aspect ratio of 1 : 1 were insulated using a silane-based plasma-enhanced chemical vapor deposition (PECVD) process at 150 °C.

- **Via metallization by electroplating and passivation.** Vias were partially filled by copper electroplating. The vias were then protected using a passivation layer of photosensitive benzocyclobutene cured at 250 °C for an hour under nitrogen atmosphere.

- **Debonding.** The processed device wafer was debonded using slide-off debonding at 180 °C on an EVG® debonder. The debonded wafers maintained their integrity after the debonding process. The slide-off debonding process did not affect the edges of the wafer to cause any other physical damage to the wafer. No residue was observed on the front side of the CMOS image sensor wafer after the cleaning process. A scanning electron microscopy image of the debonded and cleaned wafer is shown in Figure 16.10.

- **Electrical testing.** A simple "daisy chain" was designed to evaluate the impact of the TSV process on the electrical yield. The daisy chain included two aluminum front-side pads and two backside TSVs with an aspect ratio of 1 : 1. The resistance measurements were obtained with a four-point probe. Results showed that device performance did not degrade due to the debonding and cleaning processes. Figure 16.11 shows a comparison of electrical performance of a device wafer before and after debonding.

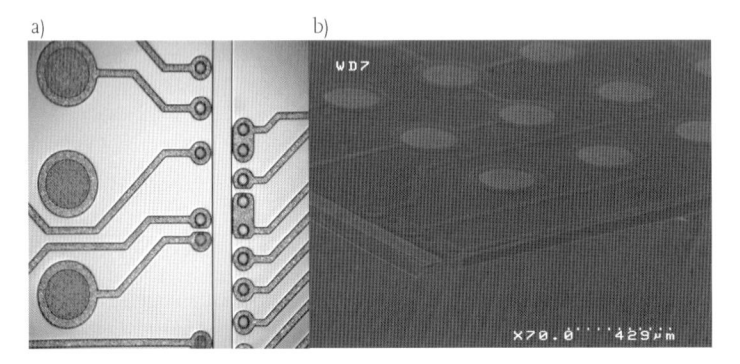

Figure 16.10 (a) Optical microscopy and (b) scanning electron microscopy images of copper pads and lines after debonding and cleaning.

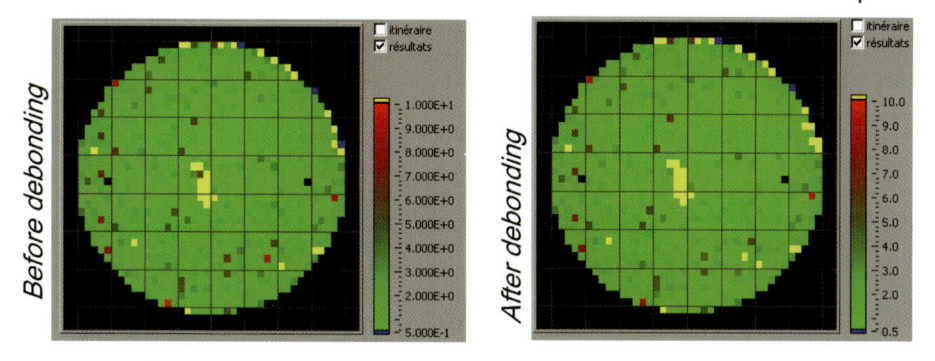

Figure 16.11 Electrical performance of a device wafer before and after debonding.

16.4.2
Via-Last Process with Aspect Ratio of 2:1

TSVs with an aspect ratio of 2:1 were also realized on active wafers at Leti [17]. Wafers bonded with WaferBOND® HT 10.10 temporary bonding material were exposed to temperatures higher than 200 °C. These wafers had copper pillars 10 μm high on the front side. The wafers were thinned by grinding and polishing to a thickness of 120 μm to achieve the 2:1 aspect ratio while via diameter was maintained at 65 μm. Via insulation for the high-aspect-ratio vias in this process scheme was achieved with a tetraethyl orthosilicate (TEOS) PECVD process at 255 °C for 3 min. The active wafers in this case were not CMOS image sensor wafers and could therefore withstand temperatures higher than 200 °C.

Noteworthy factors included the following.

- **Bumping.** To avoid damage to the thinned wafers, bumping was performed before debonding. Two different types of bumping processes were used on 120 μm thick wafers to make (i) 100 μm tin–silver–copper solder balls and (ii) 30 μm copper pillars capped with tin–silver–copper solder.

- **Debonding.** To minimize stress on the backside bumps, the debonding temperature was increased to 200 °C. Additionally, special chucks were used on an EVG® 850DB debonder to allow debonding without any damage to the bumps or device wafer.

Figure 16.12 shows that backside solder balls maintained shape and integrity after debonding and cleaning of the thin wafer.

16.4.3
Via-Last Process with 50 μm Depth Using High-Temperature TEOS Process

The previous example illustrated the robustness of temporary adhesives through a high-temperature process, but the TSV process was carried out with wafers

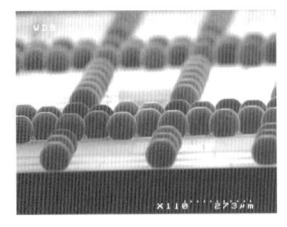

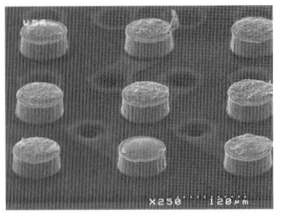

Figure 16.12 Backside solder balls maintain shape and integrity after debonding and cleaning of a thin wafer.

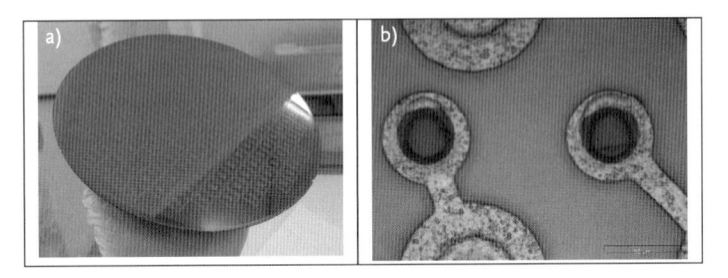

Figure 16.13 (a) Entire device wafer; (b) microscopic image of copper pads.

120 μm thick. True TSV processes are in fact performed on wafers less than 50 μm thick, and via insulation and other processes require adhesives that are robust and capable at higher temperatures. This example [18] describes a TSV process demonstrated at CEA-Leti using 50 μm thinned active wafers [6] using a high-temperature via insulation step. The new temporary adhesive used here was designed to have higher temperature capability than WaferBOND® HT 10.10 material.

The main steps in the via-last TSV process are shown in Figure 16.9. Via insulation in this case was performed at 255 °C for 3 min. The entire process flow was successfully achieved on all the wafers. The temporary bonding agent experienced no degradation throughout the process. Additionally, no delamination of the device wafer from the carrier was observed through backside processing. Figure 16.13a shows an image of a device wafer after completion of the backside process. Figure 16.13b shows a microscope image of the TSV and rerouting pads before debonding.

16.4.4
Die-to-Wafer Stacking Using Interconnect Via Solid–Liquid Interdiffusion Process

Interconnect via solid–liquid interdiffusion (ICV-SLID) soldering technology [19] has been developed by Fraunhofer EMFT (formerly IZM-Munich) for metal

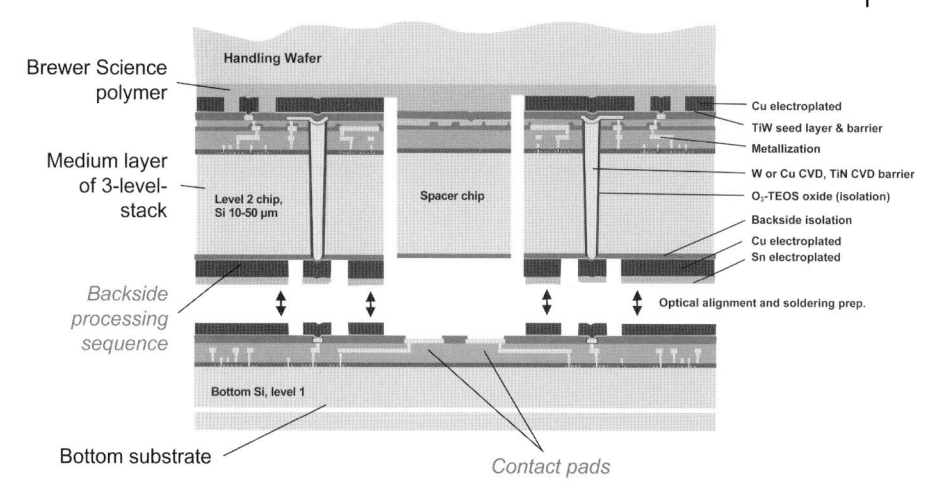

Figure 16.14 Schematic of die-to-wafer bonding using ICV-SLID process.

bonding in post-front-end or via-first 3D integration processes. A chip-to-wafer stacking process [20] with wafer-level transfer based on this technology was demonstrated using WaferBOND® HT 10.10 material. Figure 16.14 shows a schematic of the die-to-wafer stacking process. The device wafer with TSVs was temporarily bonded to a carrier wafer and processed on the backside. The dies were separated by a sawing process while still supported on the temporary carrier. The dies were then permanently bonded to the bottom wafer using the SLID process. The tungsten-filled vias in the top die were connected to the metallization on the bottom wafer by a solder–metal system. A solvent that penetrated the bonding layer along the dicing streets removed the temporary bonding material. The result is shown in the detailed image of the two-level stack in Figure 16.15. The measurement pads on the topside of the medium layer were thus accessible for testing the two-level daisy chains. Subsequent to debonding and testing, a third level was added to the medium layer.

16.5
Concluding Remarks

In this chapter, temporary bonding technologies for thinning and subsequent processing of thinned wafers for 3D integration have been described. Section 16.2 introduced the requirements of a temporary bonding system and the various technology options that are currently being developed. Section 16.3 elaborated best-known methods for successful bonding, thinning, and thin-wafer processing.

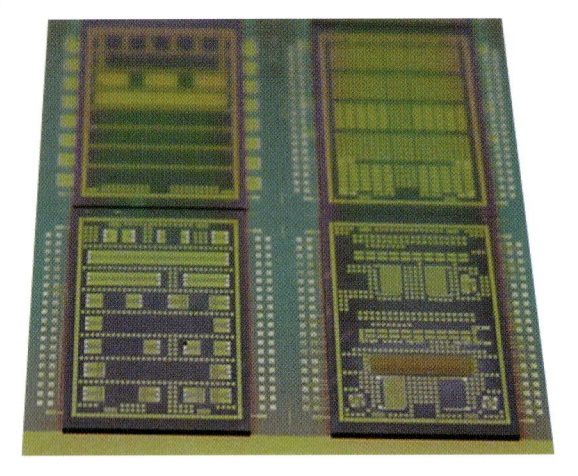

Figure 16.15 Top view of medium layer bonded on bottom substrate.

Section 16.4 presented examples of TSV processing that were successfully demonstrated using thermoplastic adhesives and slide-off debonding.

Acknowledgments

The author would like to acknowledge the work done by our partners at Leti and, in particular, the contributions made by Amandine Jouve. The author would also like to acknowledge the work done by 3DP project team members at Brewer Science Inc. Cooperation and input from Mark Privett and Catherine Frank is gratefully acknowledged. The author would also like to express appreciation for the support extended by Armin Klumpp from Fraunhofer EMFT.

References

1 Garrou, P., Bower, C., and Ramm, P. (eds) (2008) *Handbook of 3D Integration: Technology and Applications of 3D Integrated Circuits*, Wiley-VCH Verlag GmbH, Weinheim, p. 1.
2 Puligadda, R. *et al.* (2007) High-performance temporary adhesives for wafer bonding applications. *MRS Symp. Proc.*, **970**, 0970-Y04-09.
3 Hosali, S. *et al.* (2008) Through silicon via fabrication, backgrind and handle wafer technologies, in *Wafer Level 3-D ICs Process Technology* (eds C.S. Tan, R.J. Gutmann, and L.R. Reif), Springer, New York, pp. 85–115.
4 Hermanowski, J. (2009) Thin wafer handling – study of temporary wafer bonding materials and processes. Electronic Components and Technology Conference.
5 Pargfrieder, S. *et al.* (2009) 3D Integration by Through-Silicon-Via (TSV) Processing Enabled by Temporary Bonding and Debonding Technology. Advanced Packaging, April 2009.

6 Wolf, J.M. *et al.* (2008) 3D process integration – requirements and challenges. *MRS Symp. Proc.*, **112**, 3–15.
7 Garrou, P., Bower, C., and Ramm, P. (eds) (2008) *Handbook of 3D Integration: Technology and Applications of 3D Integrated Circuits*, Wiley-VCH Verlag GmbH, Weinheim, pp. 27–28.
8 Mathias, T. (2007) Ultrathin-wafer processing utilizing temporary bonding and debonding technology. Proceedings of the International Wafer-Level Packaging Conference.
9 Kessel, C.R. *et al.* (2007) Wafer thinning with the 3M wafer support system. Proceedings of the International Wafer-Level Packaging Conference.
10 Sukengwu, M.H. *et al.* (2008) Temporary bonding of wafer to carrier for 3D-wafer level packaging. 10th Electronics Packaging Technology Conference.
11 Garrou, P. (2009) Temporary Bonding for 3DIC Thinning and Backside Processing. Perspectives from the Leading Edge, May and June 2009.
12 Boudaden, J. and Piekass, M. (2008) Carrier technology for wafer thinning. Be Flexible forum, Munich.
13 Bai, D. *et al.* (2009) Edge protection of temporarily bonded wafers during backgrinding. *ECS Trans.*, **18** (1), 757–762.
14 Jouve, A. *et al.* (2008) Facilitating ultrathin wafer handling for TSV processing. 10th Electronics Packaging Technology Conference.
15 Sillon, N. (2009) Enabling technologies for 3D integration: from packaging miniaturization to advanced stacked ICs. Electronic Components and Technology Conference.
16 Charbonnier, J. *et al.* (2009) Integration of a temporary carrier in a TSV process flow. Electronic Components and Technology Conference.
17 Jouve, A. *et al.* (2009) Material improvement for ultra-thin wafer handling in TSV creation and PECVD process. 3DIC Conference, San Francisco.
18 Cheramy, S. *et al.* (2009) 3D integration process flow for set-top box applications: description of technology and electrical results. EMPC Conference, Rimini, Italy.
19 Ramm, P. (2008) Through silicon via technology – processes and reliability for wafer-level 3D system integration. Electronic Components and Technology Conference, pp 841–846.
20 Jouve, A. (2008) Temporary bonding for SLID-solder process. Presented at Be Flexible forum, Fraunhofer IZM.

17
Temporary Adhesive Bonding with Reconfiguration of Known Good Dies for Three-Dimensional Integrated Systems

Armin Klumpp and Peter Ramm

There is a variety of reasons for the application of three-dimensional (3D) integration, such as improved system performance, lower power consumption, and smaller form factor (see Chapter 15). Three-dimensional integration concepts have been proposed for more than three decades [1–4] and numerous 3D process flows which were demonstrated are feasible [5]. However, there is a strong demand for considering the manufacturing costs of corresponding 3D integrated products. Certainly, production processes based on wafer-level fabrication have a comparatively favorable cost structure. However, for many applications wafer yield and chip area issues will be problematic for full wafer stacking approaches to 3D integration before the infrastructures for full wafer 3D approaches become mature. In consequence, die-to-wafer stacking concepts utilizing known good dies (KGDs) only are advantageous for 3D integration of devices with low wafer yield and/or nonidentical chip sizes. For cost-effective production of 3D integrated systems it is essential to combine KGDs in order to enable high yield in stack assembly. Moreover, for a number of applications there is a need for 3D stacking of devices from different production wafers showing different die and/or wafer sizes (Figure 17.1). This chapter describes some suitable die-to-wafer stacking methods developed by Fraunhofer Munich [6] and corresponding handling processes.

KGDs are selected and arranged on a temporary handling substrate as a kind of reconfiguration. The application of this handling approach for 3D integration is discussed below concerning the interaction between the temporary bonding process and an exemplary 3D processing flow of a through-silicon via (TSV) technology described in Chapter 15. The TSV technology is based on solid–liquid-interdiffusion (SLID) bonding as described in Chapter 10.

17.1
Die Assembly with SLID Bonding

SLID bonding and its application for 3D integration are described in detail in Chapters 10 and 15, respectively. In this section the constraints for die assembly with SLID bonding are discussed. One of the constraints is valid for any types of

Handbook of Wafer Bonding, First Edition. Edited by Peter Ramm, James Jian-Qiang Lu, Maaike M.V. Taklo.
© 2012 Wiley-VCH Verlag GmbH & Co. KGaA. Published 2012 by Wiley-VCH Verlag GmbH & Co. KGaA.

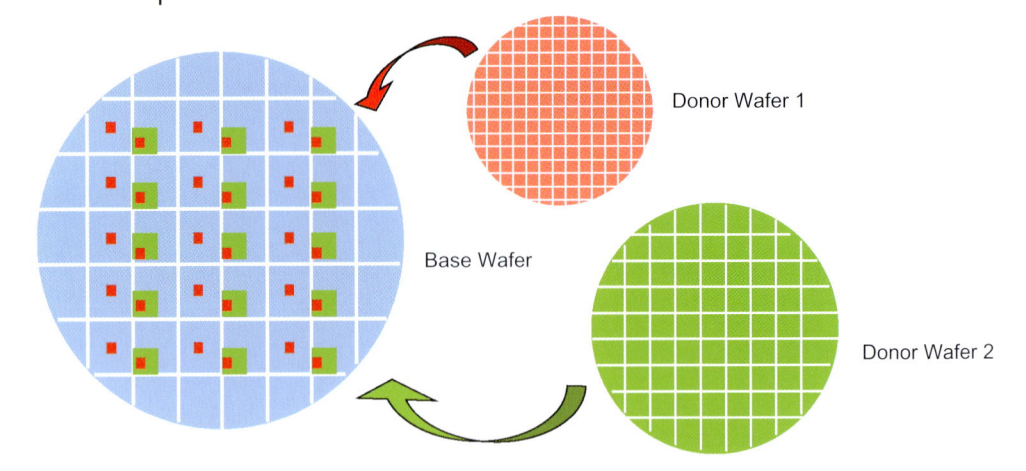

Figure 17.1 Assembly of 3D integrated systems on a base wafer by using KGDs of different production wafers possibly showing nonidentical die and wafer sizes.

dies that need to be assembled using the SLID metal combination where the metals tend to form oxides. A metal combination of copper and tin is well suited and often used (see Chapter 10). Here especially, the copper interface needs to be protected from oxidation during handling and processing. It is essential for proper assembly to provide good electrical contact and mechanically sound interfaces. Tin melts above approximately 240 °C; placing of KGDs on a base wafer with a flip-chip or die bonder at elevated temperatures would lead to the covering of the copper structure interfaces on the base wafer with copper oxide (because today no commercially available bonder operates under chemically inert atmosphere).

Another constraint is due to the diffusion rate of intermetallic compound formation. A relatively low diffusion rate may reduce the usually very high placement rates of flip-chip or die bonder equipments. Therefore, the SLID assembly and the reconfiguration of KGDs are separated into two different process steps. The first step is the rearrangement (so-called reconfiguration) of KGDs at low temperature and at a high placement rate. The second step is the SLID bonding for a comparatively long processing time, but – and this is the essential advantage – using wafer-to-wafer assembly leading to simultaneity for all transferred chips.

17.2
Reconfiguration

Reconfiguration populates the KGDs at the pitch of the target chips on the base wafer onto the temporary handling substrate or carrier [6]. Using a silicon wafer as carrier substrate is preferable since the silicon wafer has the same expansion coefficient as the base wafer and can be purchased with a very small thickness

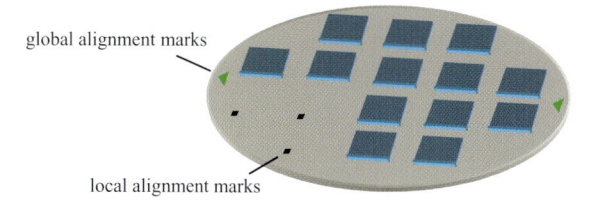

Figure 17.2 Partially populated carrier. Chips are aligned along the local alignment marks; global alignment marks are used for wafer-to-wafer transfer of the chips on the carrier.

variation. The silicon carrier substrate is polished on one side and if necessary additional structures can be added (e.g., within a CMOS technology line). Due to the two-step process flow, the carrier needs chip alignment marks and global wafer-to-wafer alignment marks (Figure 17.2).

At Fraunhofer EMFT two main principles are pursued in terms of temporary handling substrates: electrostatic clamping (see Chapter 19) and gluing. For gluing, the temporary handling substrate is spin-coated with a poly(methyl methacrylate) (PMMA) type of polymer and cured (i.e., complete removal of solvent as well as densification due to desorption of low-molecular-weight polymer parts). The applied PMMA polymer softens at reasonably low temperatures and is therefore used as thermoplastic glue for the chips. The die bonder keeps the carrier at the softening temperature level of the polymer, while all chips are transferred from either a dicing frame or a waffle pack. After transfer, the carrier is cooled to fix the chips. The carrier can then be handled by the wafer-to-wafer bond equipment that is used for the SLID assembly step. This principle can be used for small chips as well for larger chips, but for the latter rotation errors may be important for placement accuracy.

17.3
Wafer-to-Wafer Assembly by SLID Bonding

SLID assembly is performed as a wafer-to-wafer step using the global alignment marks. In this way the interface conditioning of the copper patterns is equal, that is, with identical and short delay time between interface pretreatments and SLID bonding. Bonding is executed in an inert atmosphere. Typical interface conditioning methods are, for example, wet chemistry to remove copper oxides (Figures 17.3a and 17.4) or chemical reduction by *in situ* hydrogen treatment at elevated temperature (below the softening temperature of the polymer). With tin as the low-melting metal of the SLID system the bond temperature is above the softening point of the gluing polymer and the chips are released from the carrier. This is advantageous for those chips that are slightly inclined relative to the carrier surface. Releasing by the polymer gives the chance to level these chips again (of course at the expense of alignment accuracy). The thickness of the polymer is in the same range as the tin layer thickness, so even in the unlikely case of dipping the chip

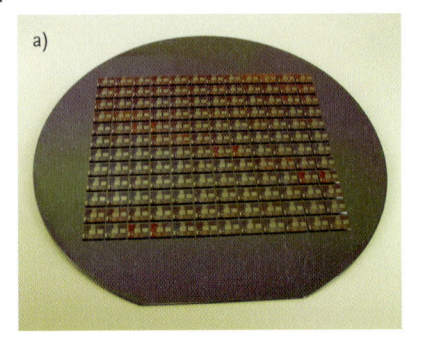

Figure 17.3 Completely populated carriers: standard silicon wafers of (a) 150 mm and (b) 200 mm diameter. The 150 mm carrier gives an example of intense oxidation of copper structures due to the elevated placement temperature (blue appearance of chip surface). The blue color on the carrier itself is due to spin-coated PMMA type of polymer.

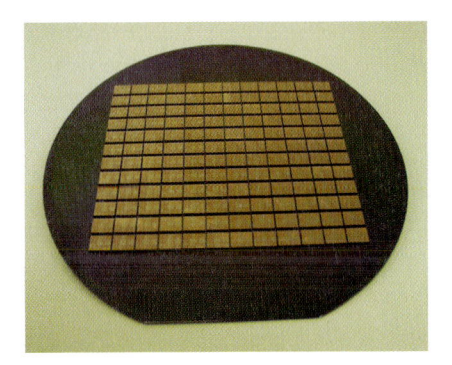

Figure 17.4 Example for pretreatment of chips on carrier; wet chemical removal of copper oxide results in clean interfaces enabling excellent SLID bonding results.

in very low-viscosity polymer there is still a wetting of the opposite copper surface by tin. Once all these precautions have met the resulting SLID bond is rigid enough to process this new wafer, for example by thinning and further plasma processing. Figure 17.5 shows an example of a pattern of chips transferred from carrier to base wafer. Figure 17.6 shows an example with 700 μm thick chips soldered face down on the base wafer, followed by thinning down to 50 μm for opening TSVs from the backside. Additional layer deposition in combination with lithography and etching prepares this new surface for adding the next layer of a three-level 3D integrated test device.

The two levels shown in Figure 17.6 are part of a test device where the electrical characterization of the 3D integrated systems is enabled by use of conventional measurement pads (visible in the areas between the stacked top chips).

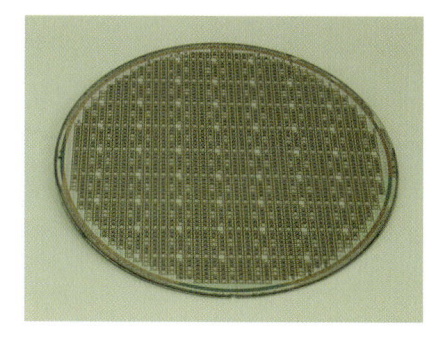

Figure 17.5 Example of a sophisticated pattern of chips transferred from carrier to base wafer; chips are thinned down to 300 μm before reconfiguration on the carrier and subsequently soldered on the base wafer by SLID bonding.

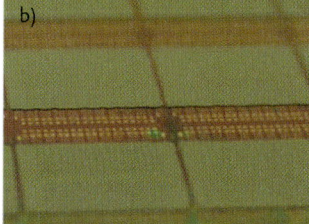

Figure 17.6 Die thinning after transfer to the base wafer: (a) detailed view of 700 μm thick top chips, stacked face down on the base wafer by SLID bonding; (b) thinning down to 50 μm to open TSVs from the backside; (c) comparison between original thickness and final thickness after thinning.

17.4
Reconfiguration with Ultrathin Chips

The reconfiguration method described above is well suited for sufficiently rigid chips that are not deformed when placed into the soft polymer. The related thickness depends on the chip size because the die bonder has to apply a greater force for larger chips. Especially in the case of chip thickness around and below 50 μm this can be critical. For 3D integration using TSVs, wafers (correspondingly devices) are typically thinned down to a thickness below 50 μm (in extreme cases, down to 10 μm). Therefore, appropriate handling of thin silicon wafers or chips is essential. In this region of silicon thickness no sufficiently robust and reliable handling concepts of free-standing silicon have been reported in the literature. Thinned wafers and thinned chips also undergo warpage as a result of stress in the dielectric layer structures of a device. The amount of warpage depends on the remaining silicon thickness, and usually thinner chips show higher warpage. In consequence, handling of stand-alone thin silicon is complicated or even

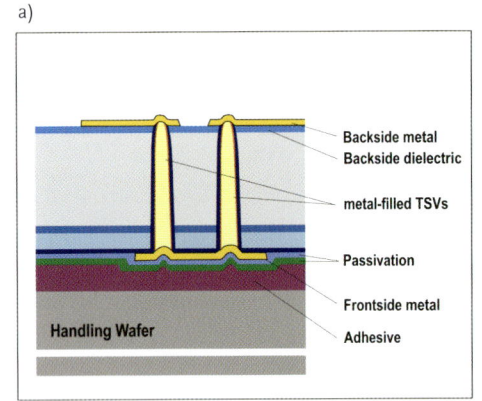

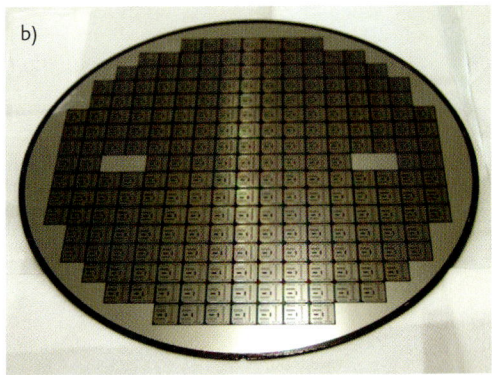

Figure 17.7 (a) Schematic of permanent carrier support for extremely thinned silicon wafers, including TSVs and redistribution metal layer on backside of the thinned wafer. (b) Total view of a 200 mm carrier with 17 μm silicon wafer on top; the last layer is the AlSiCu metal as reconfiguration layer.

impossible. For this reason the handling procedures applied at Fraunhofer EMFT are all pursued with an additional support by a silicon wafer or chip carrier which is released only after safe transfer to the base wafer. These methods (Figure 17.7a) are robust and reliable, but a silicon wafer is completely consumed: the device wafer to be thinned is glued face down to the carrier, again with the PMMA type of polymer. Now the thinning procedure including the backside redistribution and SLID soldering metal layers are applied. If the temperature restriction from the polymeric glue is fulfilled, this sandwich of carrier, glue, and thin device wafer cannot be distinguished from an ordinary wafer (Figure 17.7b). Instead of releasing at that process level, the stack of carrier and thin device wafer is diced and treated as a conventional chip with standard thickness. After redistribution on a carrier (carrier-2) – as already described in Section 17.2 – the chips are transferred to the base wafer. Finally chip, chip carrier, and carrier-2 are released at the same time by dissolving the polymer in a solvent bath (acetone is sufficient). For illustration, Figure 17.8 shows thin dies stacked on devices of the base wafer, where some die carriers are still in place on top of transferred thin chips.

17.5
Conclusion

Cost-effective die-to-wafer stacking concepts utilizing KGDs only are needed for many applications of 3D integration, that is, if it is necessary to combine devices from different production wafers showing different die or/and wafer sizes. Corresponding temporary adhesive bonding processes have been developed by Fraun-

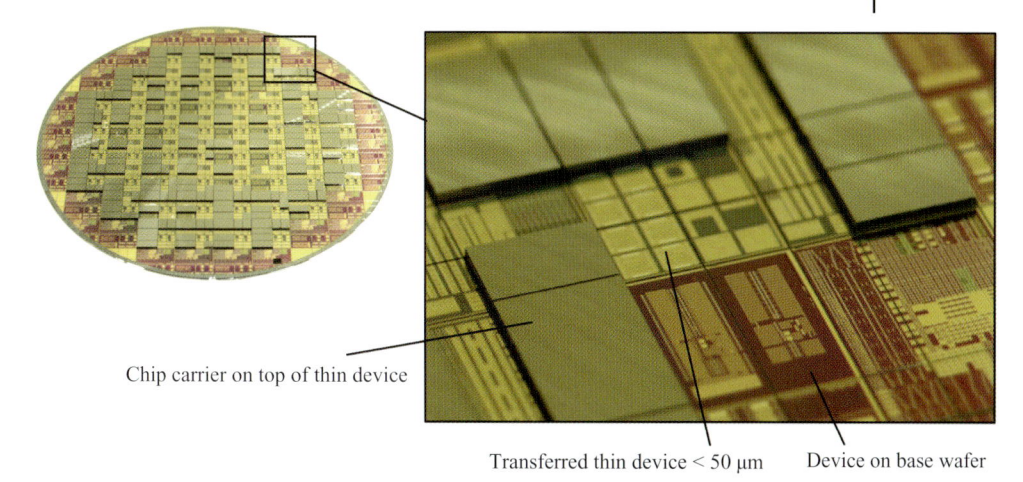

Figure 17.8 Handling of thin chips with carrier chip. Intermediate phase during solvent removal of handling carrier shows target device, transferred thin chip device, and still remaining carrier chips on top of thin device chips. Removal rate of carrier chip depends on chip size and device topography [5].

hofer Munich, enabling handling and reconfiguration of ultrathin KGDs for cost-efficient fabrication of 3D integrated products. With the use of temporary carriers processing can be carried out with standard equipment, because the complexity of the task is advantageously reduced to standard chip and wafer dimensions. The application of the developed temporary bonding technologies for 3D integration was shown exemplarily for a 3D processing flow of a TSV technology based on SLID bonding. While this metal bonding processes is best suited for multiple die-to-wafer stacking technologies, it shows a constraint, however, due to a relatively low diffusion rate of intermetallic compound formation, which may reduce the placement rates of die bonder equipment. Advantageously, the reconfiguration of KGDs by temporary adhesive bonding and the permanent bonding process are separated into two different process steps, where the assembly of the dies for the SLID process is done by wafer-to-wafer technology.

Acknowledgments

The authors would like to acknowledge Christof Landesberger, Karl-Reinhard Merkel, Lars Nebrich, Josef Weber, and Robert Wieland from Fraunhofer EMFT in Munich, and colleagues at Fraunhofer IZM, especially Kai Zoschke and Thomas Fritzsch. The results discussed in this chapter are partly based on the "KASS" project, funded by the German Government under support no. 01M3163, and the "e-CUBES" project, funded by the European Commission under support no. IST-026461.

References

1 Anthony, T.R. (1981) Forming electrical interconnections through semiconductor wafers. *J. Appl. Phys.*, **52**, 5340–5349.
2 Koyanagi, M. (1989) Roadblocks in achieving three-dimensional LSI. Proceedings of the 8th Symposium on Future Electron Devices, pp. 50–60.
3 Hayashi, Y. (1990) Method of stacking semiconductor substrates for fabrication of three-dimensional integrated circuit. US Patent 5,087,585, filed 11 July 1990 and issued 11 February 1992.
4 Ramm, P. and Buchner, R. (1995) Method of making a vertically integrated circuit. US Patent 5,766,984, filed 22 September 1995, issued 16 June 1998.
5 Garrou, P., Bower, C., and Ramm, P. (eds) (2008) *Handbook of 3D Integration*, Wiley-VCH Verlag GmbH, Weinheim.
6 Ramm, P. and Buchner, R. (1995) Method of making a three-dimensional integrated circuit. US Patent 5,563,084, filed 22 September 1995, issued 8 October 1996; Ramm, P. and Buchner, R. (1995) Method of making a three-dimensional integrated circuit. US Patent 5,877,034, filed 22 September 1995, issued 2 March 1999.

18
Thin Wafer Support System for above 250 °C Processing and Cold De-bonding

Werner Pamler and Franz Richter

18.1
Introduction

Stacking of chips will be used for three-dimensional chip integration to increase performance and to reduce footprint at the same time. Especially with through silicon via (TSV) technology, interconnect lines will become shorter, and hence the communication speed will increase accordingly. Not only individual chip integration, but also system integration of memory and logic chips in a three-dimensional way will provide new opportunities to develop next generations of compact, high-performance electronic devices. Wafer thickness will be below 50 µm to accommodate TSVs, thus providing new market opportunities for equipment and materials providers [1].

Nonstacked devices like logic circuits or CMOS image sensors also require wafer thinning for performance improvement. Highly integrated packages of logic, sensor, and memory chips are utilizing TSV connect technology to reduce package size significantly. Further integration towards fully three-dimensional integrated, multifunction chips will require further reduction in wafer thickness below 50 µm.

All ultrathin wafers of less than 100 µm will have to be handled throughout the process on a temporary carrier. Those carriers have to be fully compatible with standard semiconductor manufacturing processes used during backside patterning. Not only interconnects, but also integrated passive components may be produced on the wafer backside while mounted to the carrier.

It is most critical for temporary wafer bonding that the bond can be released after thinning of the device wafer. Several approaches have been used so far. In one technique, a thermoplastic layer is used for bonding. The adhesion force of this layer is weakened upon heating to elevated temperatures which allows separation of the wafers. In another method, the carrier is released by exposing the adhesive to intense laser irradiation. In a third method, the adhesion layer can be dissolved by chemicals applied through holes in a special, perforated carrier wafer. Unfortunately, the temperature range for subsequent processing may be too low for the first method, and the other approaches require expensive carriers.

Handbook of Wafer Bonding, First Edition. Edited by Peter Ramm, James Jian-Qiang Lu, Maaike M.V. Taklo.
© 2012 Wiley-VCH Verlag GmbH & Co. KGaA. Published 2012 by Wiley-VCH Verlag GmbH & Co. KGaA.

In the recent years, an alternative method for temporary wafer bonding has been developed which avoids these drawbacks. It is based on a release layer and an elastomer adhesive, and is the subject of this chapter.

18.2
Process Flow

An overview of this new bonding/de-bonding process is illustrated schematically in Figure 18.1. As with all temporary bonding approaches, two wafers are involved. One is the so-called "device wafer," which contains active structures and needs to be thinned. The other is the "carrier" needed for stabilization of the thinned device wafer. Usually, low-cost, blanket silicon wafers are used for the carrier in the elastomer technique, but glass wafers are suitable as well.

In the first step, the device wafer is coated with a release layer. This film is tailored such as to provide a sufficiently weak adhesion to the device wafer surface. In the second step, the carrier wafer is coated by the elastomer adhesive which adheres strongly to the carrier. Then both wafers are aligned to each other and bonded together at the release layer/elastomer interface. Thinning of the device wafer, for example by back-grinding, and backside processing of the thinned device wafer surface then follow. For de-bonding the stack is fixed by a vacuum chuck, and the carrier can be pulled off from the device wafer at room temperature because of the weak adhesion of the release layer to the device wafer. The individual process steps are described in more detail in the following subsections.

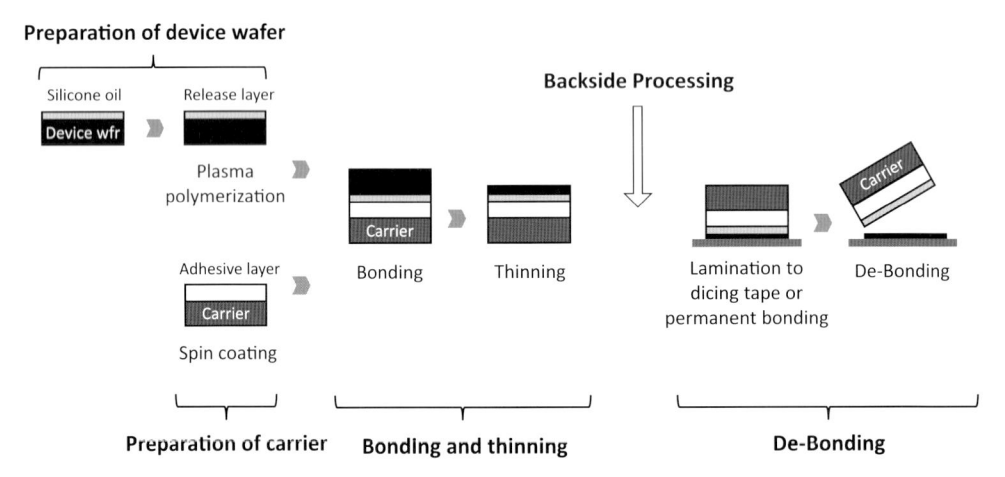

Figure 18.1 Schematic process flow, involving preparation of device wafer (black) and carrier (dark gray), bonding and thinning, as well as de-bonding.

18.2.1
Release Layer Processing

The release layer is of fundamental importance for the de-bonding process. It modifies the surface of the device wafer in such a way that the elastomer-coated carrier wafer can be removed from the device wafer easily enough. On the other hand, the release layer mixes with the elastomer glue applied to the carrier to form a combined film. The resulting glue layer holds the device and carrier wafer stack together, but has a weakest link at the device wafer interface [2].

Analogies exist for the release layer in everyday life. In the kitchen, an oil layer is used to prevent meat from sticking to a pan; or the fabrication of work pieces from mold iron requires coating of the mold with a special release layer.

In the current implementation of the process, the release layer is created using two steps. In the first step, the device wafer is spin-coated with silicone oil of less than 100 nm in thickness. This oil consists of relatively short chains of silicone molecules and provides a weak adhesion to the substrate. A plasma treatment follows in order to solidify the oil by crosslinking. Ideally, the final structure shows a gradient from an oily structure at the device wafer surface to an SiO_2-like, strongly crosslinked silicone near the top [3, 4]. The former is responsible for the weak adhesion to the substrate while the latter bonds easily to the elastomer coating of the carrier wafer.

18.2.2
Carrier Wafer Processing

In order to glue device and carrier wafers together, the latter is spin-coated by a liquid silicone-based layer ("elastomer"). The thickness of the elastomer, typically 100 μm, is chosen such that elevated features of the device wafer are completely immersed in the elastomer after bonding. In this way surface topographies of the device wafer can be leveled effectively, and even surfaces with bumps higher than 100 μm have been processed successfully using sufficiently thick elastomer layers.

Due to the finite viscosity of the liquid elastomer during spin-coating, a region of greater thickness forms around the wafer edge, called an edge bead (Figure 18.2). While detrimental for usual photolithographic resist coating processes, the edge bead is essential for the force-free bonding step described below. The height of the edge bead is roughly 50% of the overall elastomer thickness.

18.2.3
Bonding Process

The coated device and carrier wafers are aligned face-to-face and center-to-center in a bonding chamber. Usually, the device wafer is mounted face down at a close distance above the elastomer-coated carrier facing upwards. After evacuating the process chamber to less than 1 mbar, the two wafers are brought into contact (Figure 18.3). It is important that the upper device wafer first touches the

Figure 18.2 Edge bead after spin-coating of a carrier wafer by a liquid elastomer.

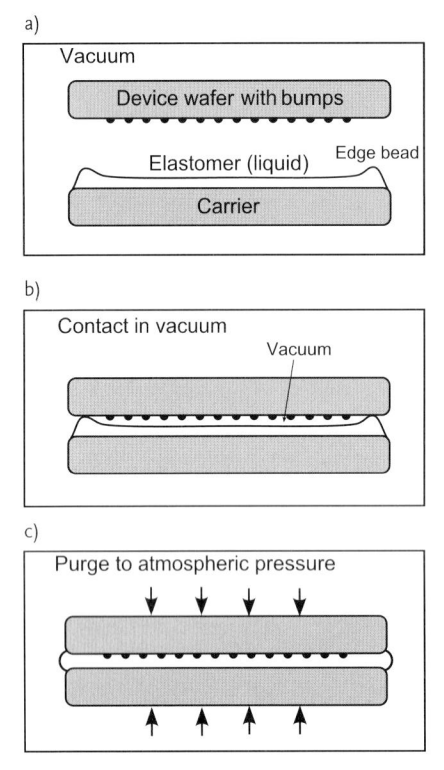

Figure 18.3 Bonding process. (a) Device and carrier wafers are facing each other and are carefully aligned. (b) When brought in contact in vacuum, the pocket between both wafers is sealed off by the edge bead. (c) After venting, both wafers are squeezed together by means of atmospheric pressure. The edge bead prevents air from entering the pocket seen in (b).

elastomer coating of the carrier wafer at the top of the edge bead. In this way, the pocket between both wafers is sealed off from the environment by the edge bead. Afterwards the chamber is vented, and the wafers are squeezed together just by atmospheric pressure. This closes the space between the wafers. Simultaneously, the sealing effect of the edge bead prevents the purge gas from entering the seam. It should be emphasized that no additional mechanical force is needed. Thus, the bonding equipment can be much simpler than that providing an external force.

Then the wafer stack is heated above 150 °C. At elevated temperatures solidification of the liquid elastomer occurs by crosslinking of the molecular chains of the silicone. This is achieved by special catalysts added to the silicone. Since the upper region of the release layer resembles this polymeric structure, a good chemical bond is formed between the release layer coating of the device wafer and the elastomer coating of the carrier wafer.

The mechanical properties of the polymerized elastomer layer are critical for achieving defect-free thinning, easy de-bonding, and low total thickness variation (TTV) of the thinned wafer.

As the elastomer wets the surface of the carrier as well as the surface of the device wafer covered by the release layer, the rounded edges of device wafer and carrier facing each other are covered with elastomer. In order to protect the very thin device wafer edge after back-grinding a sufficient centering accuracy of device wafer to carrier is required. Taking the SEMI standards for wafer dimensions (www.semi.org) into account, the off-center tolerance should be less than ±50 µm.

18.2.4
Thinning

Usually, mechanical back-grinding is employed to thin the device wafer down to the desired final thickness [5, 6]. Other thinning methods can be used as well. If required, backside processing of the thinned device wafer surface follows to provide backside contacts, additional interconnects, TSVs, bumps, etc.

When using mechanical back-grinding the elastomer has to be modified in a way so as to allow the grinding wheel to cut into it without applying significant forces to the outside sharp edge of the thinned device wafer. With the elastomer covering the rounded outside rim of the device wafer the knife-sharp edge created during thinning of the device wafer is embedded into the elastomer, thus protecting the edge against mechanical damages (Figure 18.4).

Thinning the wafer down to a thickness smaller than the radius of the device wafer rim on the bottom side leads to a reduction of the device wafer diameter, hence reducing the risk of an unsupported, overhanging area of the thinned device wafer due to misalignment. Nevertheless, in order to enlarge the process window, it is highly recommended to use an oversized carrier in the case of glass carriers or to apply an edge-trimming step in the case of silicon carriers.

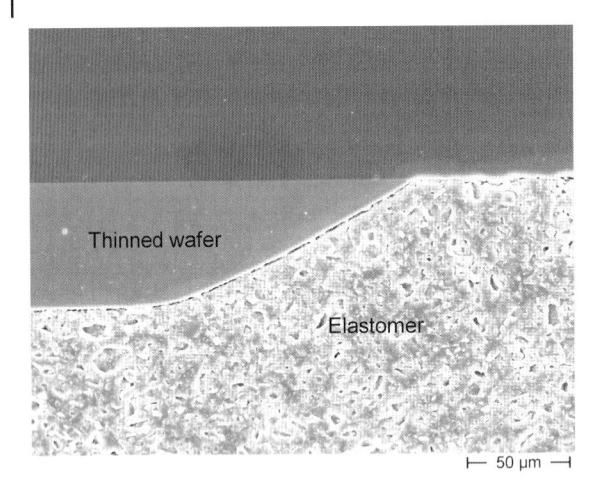

Figure 18.4 Scanning electron microscopy image of a wafer edge protected by elastomer.

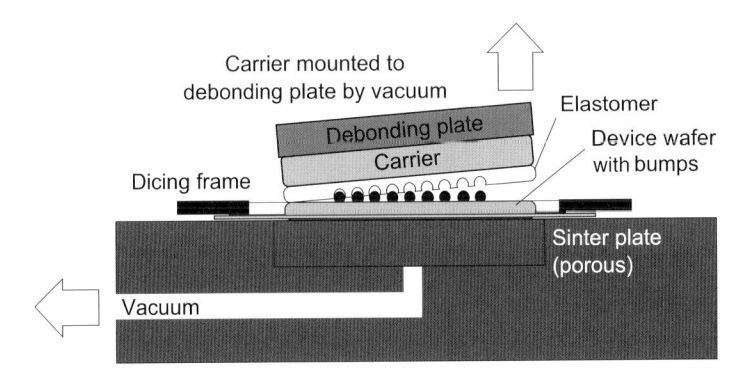

Figure 18.5 De-bonding process. The bonded stack is held in a fixed position by means of vacuum applied through a porous plate. A slightly bendable vacuum chuck ("de-bonding plate") is used to pull off the carrier from the thinned device wafer.

18.2.5
De-bonding Process

De-bonding is performed in a purely mechanical way (Figure 18.5) without heating the stack ("cold de-bonding") or using chemicals. This can be achieved because of the oily interface layer on the device wafer which resembles a mechanical weakness in the bonded stack. Usually, the thinned wafer side of the bonded stack is laminated onto a standard dicing tape. In order to keep the sensitive thinned wafer in a fixed, flat, and very stable position, the tape is sucked down by vacuum onto a flat, porous plate. Using a slightly flexible, soft, bendable vacuum chuck (de-bonding plate), the carrier wafer can be pulled off by lifting it from one side. The same basic procedure can be applied if the thinned wafer has been bonded to a second carrier in order to allow backside processing of the device wafer (see below). Since the thinned wafer is fixed during de-bonding, the risk of breakage is practically eliminated.

Due to usage of silicones, the surface is in a hydrophobic state after de-bonding. If this is not desired a short plasma clean can be used to make the surface hydrophilic again.

18.2.6
Equipment

The following equipment is needed for the elastomer bonding/de-bonding process:

- spin coater
- plasma chamber
- vacuum chamber for low-cost, force-free bonding
- hot plate
- de-bonder.

Except for the de-bonder, all components are standard tools.

18.3
Properties

18.3.1
Device Wafer Thickness

At present, wafer thicknesses of 50 µm can be handled successfully on a routine basis. But thicknesses as low as 30 µm have already been demonstrated.

18.3.2
Thickness Uniformity

The thickness uniformity of the thinned wafer is characterized by the TTV which is defined as the difference between the largest and the smallest wafer thickness. Mechanical parameters of the elastomer layer, such as elastic modulus and Shore hardness, are critical in obtaining low TTV values. The TTV increases, for example, if the elastomer is too soft, that is, the composition of the elastomer is not correct. In many cases, TTV values of the stack have been observed to be less than 10 µm over the entire wafer. Using optical interferometric measurements the TTV of thinned wafers has been shown to be only a few micrometers.

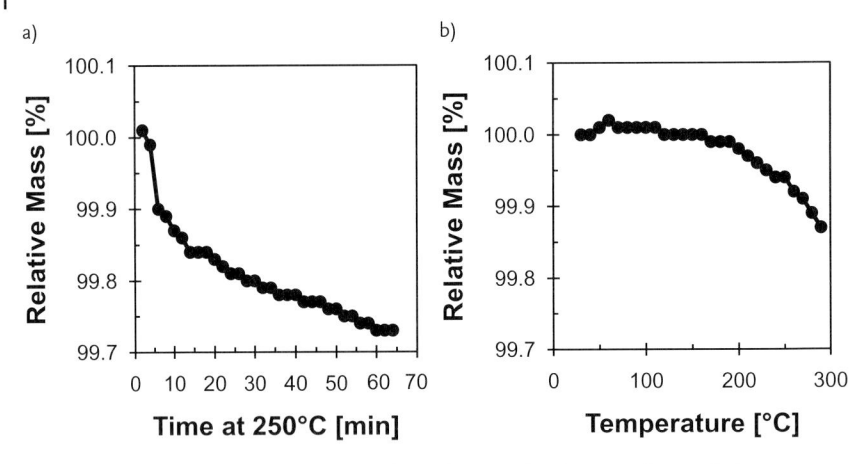

Figure 18.6 Thermogravimetric analysis of a wafer coated with a partially cured elastomer. The elastomer mass is measured as a function of (a) heating time and (b) temperature (ramp rate of 10 °C min^{-1}). Note the strongly expanded scale of the vertical axes.

18.3.3
Stability

The fully cured elastomer adhesive is stable up to temperatures of 400 °C. Since, however, the temperature of bonded stack processing is usually limited to lower temperatures, it is necessary to use an elastomer that is only partially cured. At temperatures that are too high, this might lead to outgassing caused by decomposition of the silicone chains. This effect can be reduced by tailoring the composition of the elastomer. Additional issues can result from delamination which originates in the mismatch of thermal expansion coefficients between the silicon wafer and the glue layer by more than two orders of magnitude, as is observed for many types of adhesives. Currently, stability has been demonstrated for 2 h at 250 °C on a routine basis. Development is on-going to raise the maximum allowable temperature even further.

Figure 18.6 shows thermogravimetric analysis [7] data for a typical elastomer sample. In Figure 18.6a, the temperature is held constant at 250 °C, while in Figure 18.6b a temperature ramp from 25 to 300 °C is applied. In both cases, only a very small mass loss of less than 0.3% is observed (note the strongly expanded scale of the vertical axes in these figures). This low outgassing rate is due to the fact that the material is solidified as silicone rubber and does not contain any solvents.

18.4
Applications

In the following subsections, a series of applications of the elastomer bonding technique are described.

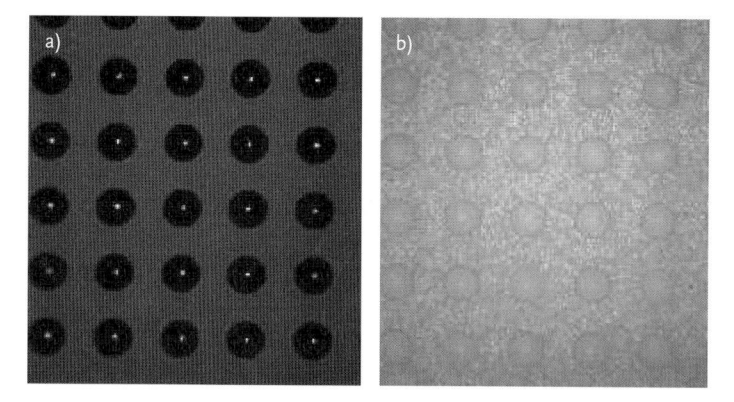

Figure 18.7 (a) De-bonded bumped device wafer and (b) the corresponding elastomer-coated carrier wafer.

18.4.1
Bonding of Bumped Wafers

Since the elastomer layer is thick enough to level large surface topographies, it is possible to bond bumped wafers by means of the elastomer technique. Only the thickness of the elastomer has to be adapted to the bump height. Successful de-bonding has been demonstrated for bumps as high as 130 μm. Figure 18.7 shows a de-bonded bumped wafer and the corresponding carrier wafer. The elastomer on the carrier wafer exhibits a negative imprint of the bumped wafer surface.

18.4.2
Packaging of Ultrathin Dies

Temporary wafer bonding can be employed to fabricate ultrathin wafers for dicing. For this purpose, the bonded wafer stack is laminated onto a dicing tape. After de-bonding, the device wafer is cut by standard techniques, and the individual dies can be picked from the tape and mounted to the package. Extremely thin dies can be fabricated in this way.

During the dicing process, however, chipping of the ultrathin wafers along the sawing lines cannot be avoided. This problem can be overcome by a technique called dicing before grinding (DBG) [8]. In this case, the device wafer is cut before the bonding process when it still has its original thickness. It is essential not to cut it fully through, but only slightly deeper than the desired thickness after grinding. This precut wafer is bonded to the carrier as usual, filling the dicing grooves with the elastomer. When the wafer is ground to its final thickness the dies are separated automatically without the need for cutting the extremely sensitive thinned wafer. This technique is made possible because the de-bonding step does not require a lateral force and, therefore, leaves the individual dies in place.

18.4.3
TSV Processing

With the upcoming three-dimensional integration techniques it becomes necessary to process the backside of the thinned device wafers. Again the device wafer is bonded to the carrier and thinned conventionally. Backside processes are performed on the thinned wafer surface, for example formation of bumps or contact pads. This is possible because the thinned wafer is supported by the carrier.

Further processing can be done in two ways. If the newly processed backside will be mounted permanently to an already-formed three-dimensional stack the dicing tape technique can be used as described above. If the original front side of the wafer is used for permanent bonding, however, a second temporary bonding step is required. For this purpose, the stack is bonded to a second carrier wafer by using the elastomer technique. The adhesion strength of the second release layer needs to be chosen so as to be stronger than that of the first release layer. By means of this technique, de-bonding opens the interface of the device wafer to the first carrier and exposes the original wafer surface.

18.4.4
Re-using the Carrier

Another application of release layers with different adhesion strengths enables re-use of the carrier wafer. For this purpose, the carrier is coated with a release layer before elastomer spin-coating. This release layer is adjusted such that it provides stronger adhesion than usual on the device wafer. After de-bonding, the elastomer can be peeled off easily from the carrier, allowing re-use of the carrier for other stacks.

18.5
Conclusions

The release layer/elastomer technique is a new technique for temporary wafer bonding. It offers a series of unique features:

- force-free bonding;
- cold de-bonding without chemicals;
- standard equipment, except for the de-bonder;
- low-cost silicon carriers, no special glass carriers needed, re-usable;
- bumped wafer processing;
- stability up to 250 °C, probably even higher;
- low outgassing;
- small TTV;
- DBG or etching before grinding.

Acknowledgments

This chapter would not have been possible without the ideas and contributions of Jamila Boudaden, Werner Brennenstuhl, Andreas Jakob, Wolfgang Keller, Andreas Luible, Madeleine Piekaß, and Klaus Vissing.

References

1 Garrou, P.H., Bower, C.H., and Ramm, P. (eds) (2008) *Handbook of 3D Integration*, Wiley-VCH Verlag GmbH, Weinheim.
2 Jakob, A., Vissing, K.-D., and Stenzel, V. (2006) US Patent 2006/00166464.
3 Wochnowski, H., Klyszcz-Nasko, H., Baalmann, A., and Vissing, K. (2001) Method for producing a permanent demoulding layer by plasma polymerization on the surface of a moulded-part tool, a moulded-part tool produced by said method and the use thereof. US Patent 6,949,272, filed 5 June 2001, issued 27 September 2005.
4 Dölle, C., Papmeyer, M., Ott, M., and Vissing, K.(2009) *Langmuir*, **25**, 7129.
5 Pei, Z.J., Fisher, G.R., and Liu, J. (2008) *Int. J. Machine Tools Manuf.*, **48**, 1297.
6 Thompson, T.E. (2008) *Chip Scale Rev.*, **12**, 28.
7 Heal, G.R. (2002) Thermogravimetry and derivative thermogravimetry, in *Principles of Thermal Analysis and Calorimetry* (ed. P.J. Haines), Royal Society of Chemistry, Cambridge, pp. 10–54.
8 Lieberenz, T.H. and Martin, D. (2006) *Chip Scale Rev.*, **10**, 51.

19
Temporary Bonding: Electrostatic

Christof Landesberger, Armin Klumpp, and Karlheinz Bock

Oppositely charged materials attract each other by means of electrostatic forces. This physical principle has been used in the semiconductor industry for many years. So-called electrostatic wafer chucks (ESCs) are widely used in plasma etching equipments to reversibly fix a wafer substrate onto a pedestal. Interesting benefits of electrostatic attraction techniques are their applicability to high-temperature processes, capability for handling wafers under vacuum conditions, and of course the lack of any adhesive bonding material, which would need to be applied before and removed and dissolved after such temporary bonding procedures. Electrostatic mechanisms allow one to switch on and off bonding forces just by charging and discharging electrostatic electrodes.

In order to reach the goal of a mobile electrostatic carrier system for thin wafer handling and processing the development task is to prepare a carrier plate that has the dimensions of a standard semiconductor wafer and which also keeps its electrostatic status over a long period of time after disconnection of an external power supply.

This chapter shows that electrostatic carrier support systems open the door to a new and powerful manufacturing concept for backside processing of thin and ultrathin semiconductor substrates.

19.1
Basic Principles: Electrostatic Forces between Parallel Plates

This section summarizes the basic theoretical formulas to calculate electrostatic forces in a plate capacitor. Regarding applications of electrostatic wafer bonding, such a configuration will be called "unipolar chucking." This means that an electrical potential is applied between the wafer and a carrier substrate, the wafer itself remaining in a charged state.

Section 19.1.2 explains the "bipolar configuration" of electrostatic wafer clamping. According to this concept the carrier substrate has electrodes of both polarities and the wafer itself is not connected to an external power supply. Electrostatic attraction is provoked just by separation of the influence charges at the surface of an electrical conductive or semiconductive substrate.

19.1.1
Electric Fields and Electrostatic Forces in a Plate Capacitor

Standard textbooks on physics explain the calculation of capacitance, electric fields, and resulting electrostatic forces between two charged plates separated by a distance d and filled by a material of dielectric constant ε_r. Although this seems to be a simple and well-known task different formulas for the resulting electrostatic force are derived and used by different authors. Davey and Klimpke [1] explain that the formulas commonly used in textbooks are only valid for liquid dielectrics which can fill the void created by a differential displacement of the plates. Following those considerations and calculation we may write for the closure force F on parallel plates

$$F = \frac{Q^2}{2\varepsilon_0 A} \tag{19.1}$$

where A is the area of one plate, Q is the accumulated charges on each plate, and ε_0 is the dielectric constant of vacuum ($\varepsilon_0 = 8.85 \times 10^{-12}\,\mathrm{A\,s\,V^{-1}\,m^{-1}}$).

The capacitance of a plate capacitor is known to be

$$C = \frac{A\varepsilon_0\varepsilon_r}{d} \tag{19.2}$$

Capacitance C, charge Q, and voltage V are related by $Q = CV$. Then the clamping force F can be written as

$$F = \frac{A\varepsilon_0\varepsilon_r^2 V^2}{2d^2} \tag{19.3}$$

In the case of a fluid dielectric material Eq. (19.3) would simply show a linear factor ε_r instead of ε_r^2 [1]. Equation (19.3) shows that high bonding forces require large voltages at minimum thickness d of the insulating layer. Materials of high dielectric strength ε_r would be favorable. However, in a practical application of electrostatic wafer bonding a small air gap between both plates (e.g., wafer substrates) will be inevitable. As the dielectric constant of air is close to 1 (like vacuum) the principal benefit of high-ε materials could be lost. In order to estimate possible consequences we calculate the change in the capacitance on introducing a double-layer material with dielectric constants ε_1 and ε_2 (Figure 19.1). The capacitance of the double-layer capacitor is given by

$$C = \frac{A\varepsilon_0\varepsilon_1\varepsilon_2}{d_1\varepsilon_2 + d_2\varepsilon_2} \tag{19.4}$$

which is equivalent to a series connection of two capacitors of different dielectric materials and layer thicknesses. In the case of a small air gap where $d_2 \ll d_1$ and $\varepsilon_2 = 1$, Eq. (19.4) gives again the capacitance of a capacitor having just one dielec-

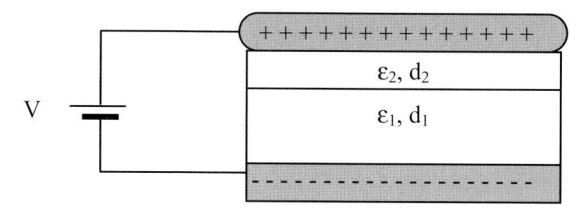

Figure 19.1 Standard unipolar capacitor configuration with double layer dielectrics ε_1 and ε_2 with thicknesses d_1 and d_2.

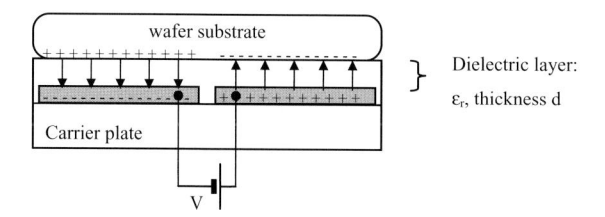

Figure 19.2 Bipolar capacitor configuration: charges are influenced at the backside of a wafer substrate which is placed onto the bipolar carrier plate.

tric (Eq. (19.2)). This shows that an air gap is tolerable as long as the thickness of such a gap is small with respect to the second dielectric layer.

The electrical field E in a dielectric layer is given by $E = Q/\varepsilon_0\varepsilon_r A$. Due to the reciprocal dependence of ε_r the field is stronger in layers of lower ε_r. Consequently, a small air gap actually will increase the attractive force between the plates. This was also shown by Yoo *et al.* by means of finite element analysis [2].

From these theoretical considerations we conclude the following aspects for electrostatic wafer bonding in a "unipolar" configuration:

- Applying an electrical field of $50\,V\,\mu m^{-1}$ between two wafer plates of 200 mm in diameter and using a dielectric material of $\varepsilon_r = 3$ will result in an attractive force of 3200 N. A field of $100\,V\,\mu m^{-1}$ will result in a force four times higher.
- Electrical insulation properties of dielectric layers are of crucial relevance for high bonding forces; the ratio V/d needs to be maximized.
- High-ε_r materials result in high charge storage capabilities (=high capacitance).
- Low-ε_r materials allow for high-strength electric fields; this is equivalent to high bonding forces.
- Thin air gaps between the substrates are not a critical issue.

19.1.2
Electrostatic Attraction in a Bipolar Configuration

Figure 19.2 shows the electrode configuration of a bipolar electrostatic chucking unit. We consider the area of the bottom plate to be divided into two large-area electrodes which will be charged oppositely by a voltage V. We estimate each electrode area to be one-half of the wafer area A. A semiconductor wafer substrate placed on top of the dielectric layer of the carrier plate represents the upper plate. Figure 19.2 is not drawn to scale. Actually, the thickness of the dielectric layer will be much smaller than the distance between the electrodes. Therefore, lateral electrical fields between the electrodes are neglected.

The amount of charge Q_{bp} for the bipolar configuration depends on its capacitance. In order to estimate the capacitance C_{bp} of the bipolar configuration we use Eq. (19.2) and consider the electrode configuration as a series of two capacitors, each one having size $A/2$ and dielectric thickness d, which is again equivalent to one plate capacitor of thickness $2d$ and electrode area $A/2$. Thereby, we get for the capacitance of the bipolar electrode configuration

$$C_{bp} = \frac{\varepsilon_0 \varepsilon_r A}{4d} \tag{19.5}$$

Subscript "bp" stands for "bipolar" and should indicate that the amount of charge at each electrode will be different from that of the unipolar configuration described above.

The resulting electrostatic force can be derived by using Eqs. (19.1) and (19.5) and $Q_{bp} = C_{bp} V$:

$$F_{bp} = \frac{\varepsilon_0 \varepsilon_r^2 A V^2}{8d^2} \tag{19.6}$$

Values of C_{bp} and F_{bp} are just one-fourth of the corresponding values for the unipolar plate configuration. Nevertheless, the electrostatic attractive force is still strong enough to securely attract wafer substrates. The bipolar configuration offers the main advantage that direct charging of the device wafer is avoided.

Equations (19.5) and (19.6) should be regarded as estimations. This is for two reasons. First, we neglected the presence of nonuniform electric fields which will actually be present in the lateral region between the electrodes. Second, we assumed that we are able to influence the same amount of electric charge (electrons and holes) at the backside of the wafer substrate as were charged to the electrode areas. It should be kept in mind that for some semi-insulating wafer materials the latter assumption might be too optimistic and the actually achievable electrostatic bonding forces might be slightly lower than of Eq. (19.6).

19.1.3
Johnsen–Rahbek Effect

In 1923 Johnsen and Rahbek investigated the attractive force between a semiconductor plate (as a dielectric of low resistivity) and a metal plate in the presence of an electric field [3]. They found that the actual attractive force is much higher than what would be calculated from Eq. (19.3). The experimental effect was explained by the assumption that permanent charges are located in the dielectric layer directly at the contact interface of the electrodes. Thereby, the distance between the charged plates is no longer given by the thickness of the dielectric, but by a certain gap which might be in the range of a few nanometers resulting in a much higher attractive force. Due to the low-resistivity dielectric material small electric currents across the interface of the electrode plates were detected by Johnsen and Rahbek. Furthermore, the force was strongly dependent on the surface smoothness of both plates. A short review of these effects can be found in [4].

Currently, conventional ESCs are often divided into Coulomb-type ESCs and Johnsen–Rahbek-type ESCs. In Johnsen–Rahbek-type ESCs the dielectric layer above the electrodes has a volume resistivity in the range 10^9–$10^{13}\,\Omega\,cm$ [5, 6]. In contrast, Coulomb-type ESCs use highly insulating dielectrics (e.g., SiO_2) which have a resistivity in the range 10^{13}–$10^{17}\,\Omega\,cm$.

It should be kept in mind that also in the case of strong electrical insulators the resistivity will decrease with increasing temperature. Therefore, both types of ESCs might show Johnsen–Rahbek effects when used at temperatures above 300 °C.

19.2
Technological Concept for Manufacture of Mobile Electrostatic Carriers

One technical aim is a mobile electrostatic carrier (e-carrier) for thin wafer handling and processing. In order to maintain all standard wafer handling robot and cassette systems, the carrier plate should have size and shape of a standard wafer substrate. A handling sequence will look as follows. A thin wafer is placed onto an e-carrier, a power supply is connected to the contact pads of the e-carrier, and, thereby, electrostatic fields and forces are generated. After this charging step power supply and cables are removed, whereas the polarization, the corresponding electrostatic field, and the attractive bonding forces remain. The stacked wafer pair is now ready for further processing.

According to this patented concept [7] a long duration time of clamping force requires a high capacitance of the plate capacitor configuration and a perfect electrical insulation of the electrode areas. Elimination of leakage currents, especially at elevated temperatures, is crucial for the applicability of e-carrier technology. So, we are looking for Coulomb-type behavior instead of Johnsen–Rahbek-type behavior (see Section 19.1.3).

In order to achieve a strong clamping force the ratio V/d must be maximized (Eqs. (19.3) and (19.6)). Thin-film technology on silicon wafer substrates represents an optimum choice for the manufacture of high-performance e-carriers in terms of insulation properties, strong electrostatic forces, high-temperature stability, thermal properties, and compatibility with standard wafer fabrication environments.

19.2.1
Selection of Substrate Material

In principle electrode areas for electrostatic attraction can be prepared on practically any rigid or flexible substrate material. However, surface flatness and surface planarity play a crucial role in enabling strong electrostatic forces and minimizing the risk for electric breakthrough.

Furthermore, the intended use of mobile e-carriers in a semiconductor manufacturing environment leads to a variety of additional requirements in terms of compatibility with thin-film semiconductor technology, such as wafer-shaped substrate form and the prohibition of substances which are generally not allowed in wafer fabrication, for example noble metals, iron, copper, alkali ions, and others.

Further constraints arise from specific processes that need to be carried out at a thin wafer that is electrostatically clamped onto the carrier substrate. For instance, plasma processes require sufficient backside cooling. Therefore, high thermal conductivity may be a key. Regarding processes above 200 °C the coefficient of thermal expansion (CTE) of substrate material and thin wafer material must be very close to each other in order to avoid bending and delamination of the stacked wafer pair. This is the reason why supporting silicon on SiO_2 glass carriers (e.g., quartz or alkali-free glasses) is limited to process temperatures below some 150 °C. Boron silicate glass substrates would offer a CTE close to that of silicon. However, due to the content of alkali ions, it is generally not compatible with standard CMOS fabrication technology.

Electrically insulating substrates facilitate the preparation of through-substrate vias and also do not require an insulation layer between electrodes and substrate. On the other hand, conductive substrates (e.g., silicon) result in a higher electrical capacitance of the wafer–carrier stack and therefore improve the long-term polarization status of such types of e-carriers.

Applicability of ceramic materials like aluminum nitride (AlN) or silicon carbide (SiC) is limited by the fact that these substrates generally exhibit a large bow. Also, they are quite expensive, at least at large substrate diameters.

Handling of thin and fragile gallium arsenide (GaAs) wafers might be accomplished by e-carriers made of single-crystalline sapphire (Al_2O_3) substrates in order to join wafer and substrate materials of nearly the same thermal expansion behavior.

The use of polymer-based substrates is restricted to low-temperature processing. Plastic materials also show poor form stability and generally no plane-parallel surfaces. More expensive polymer materials such as poly(ether ether ketone) or polyimide might be of interest for applications in wet-etch processes.

Taking into account all these aspects of material properties it can be concluded that carriers based on silicon wafers actually combine optimum form properties with high thermal conductivity, reasonable cost, and of course optimum compatibility with silicon wafer fabrication environments.

19.2.2
Selection of Thin-Film Dielectric Layers

According to the basic relation between electrostatic force and the dielectric layers on top of the electrodes of an e-carrier, the main requirements for the dielectric materials are high electrical insulation at minimum layer thickness and preferably

Table 19.1 Dielectric constants of various electrically insulating materials.

Type of material	Dielectric material	Dielectric constant
Glass (SiO$_2$)	Thermally grown SiO$_2$	3.8 to 4.5
	PECVD[a)] SiO$_2$	5
	Pyrex	5.6
	BPSG glass	4 to 5
Ceramic	PECVD[a)] silicon nitride (Si$_3$N$_4$)	6 to 8
	Titanium dioxide (TiO$_2$)	80 to 170
	Al$_2$O$_3$	8 to 10
Polymers	Polyimide	3.4
	Epoxy	2 to 3
High-ε materials (ferroelectrics)	SrTiO$_3$	200 to 300
	BaTiO$_3$ ($T_c = 120\,°C$)	100 to 1250

a) Plasma-enhanced chemical vapor deposition.

a high dielectric constant in order to increase the capacitance of the wafer stack configuration and, thereby, improve the duration of electrostatic attraction.

Table 19.1 gives a selection of dielectrics that could be of interest for this application. It is well known that SiO$_2$ layers offer very good electrical insulation properties. Thermally grown SiO$_2$ layers could be used on silicon wafer substrates. Chemical vapor deposited or plasma-enhanced chemical vapor deposited layers of SiO$_2$ can be used for surface insulation. The dielectric constants of different glass layers depend on their specific manufacture process. In any case, ε_r of SiO$_2$ glass will be below 6.

Higher ε_r values would be offered by silicon nitride (Si$_3$N$_4$), alumina (Al$_2$O$_3$), or titanium dioxide (TiO$_2$). Ferroelectric materials, such as various types of titanates, would allow for values of dielectric constant far above 100. However, these very high ε_r values only occur below the specific Curie temperature T_c. High-ε_r materials are really interesting for the preparation of mobile e-carriers as they would enable a very long duration of polarization as well as high bonding forces even at moderate voltages. However, their actual physical properties critically depend on their specific deposition process. Dielectric constant, electrical insulation, and surface roughness need to be optimized as a whole when selecting the best material for e-carriers. Furthermore, the temperature dependence of these parameters must be taken into account. Unfortunately, there is a lack of experimental data for the electrical resistance of many dielectric materials for the temperature range 300–500 °C.

Finally, there is another very important technological constraint on the selection of dielectric materials: layer deposition must not exhibit defects, for example pinholes, scratches, or spikes. A single spot of low resistivity anywhere at the surface of an e-carrier substrate will lead to an electrical breakthrough later on. It will be an interesting task for the near future to find out which dielectric material and which deposition process will result in optimum performance in terms of both charge storage capability and electrical insulation.

19.2.3
Electrode Patterns: Materials and Geometry

When selecting a conductive material for electrodes and contacts several requirements need to be considered: metal ions must not diffuse or migrate into dielectric insulation layers; large-area metal layers must not induce a strong bow to the substrate; and the material should be stable against oxidation in air. Furthermore, if e-carriers are designed for high-temperature applications then all the properties listed before should be preserved when an e-carrier substrate is heated to 400 °C.

At Fraunhofer EMFT we tested layers of titanium metal and the alloy titanium tungsten (TiW) and thin layers of polycrystalline silicon. The last mentioned enables a completely metal-free system that may be important for handling of silicon wafers in the so-called front end of line. Even small traces of titanium at the top surface of an e-carrier, as could be possible when working with titanium or TiW, could degrade an electrical device. Therefore, these materials are only used for dedicated backend handling purpose. The advantage of using titanium or TiW is the very thin thickness required of the electrode layers; when working with polysilicon the electrode layer needs to be thicker to have sufficient conductivity. As a consequence, at the (relatively sharp) edges of structured polysilicon, electrical breakthrough has a higher probability of occurring compared to the safer very shallow structures of titanium or TiW.

There are two principal geometries for the bipolar electrode layout: first, at least two large areas covering more or less the complete wafer surface; second, fine patterned finger-like structures of many conductive lines. In the first case the region of insulating areas on the wafer surface is minimized. Electrostatic fields run perpendicular between the e-carrier electrode and the backside of a semiconductor wafer placed thereon. In contrast, finger-like electrode patterns result in a large fraction of inhomogeneous electrostatic fields between neighboring and oppositely charged electrode lines. The resulting electrostatic force is lower than in the case of large-area electrodes.

In order to activate electrostatic fields, e-carrier substrates need to be connected to a power supply at specifically designed contact pads. In principle these contact pads can be located anywhere at the front or rear side of an e-carrier. However, they must be accessible to contact needles after clamping a device substrate on top of the e-carrier. Also, material deposition or etch removal at the contact pads during device wafer processing must not occur. Therefore, the preferred location for contact pads generally will be the backside of e-carriers. There are two main possibilities to achieve such contact geometry: preparation of through-substrate vias or preparation of conducting lines around the edge of the carrier substrate.

Through-substrate vias may be prepared by laser treatment, wet etching, plasma dry etching, or ultrasonic drilling. In the case of silicon substrates optimum smoothness of sidewalls of via holes can be achieved by anisotropic etching.

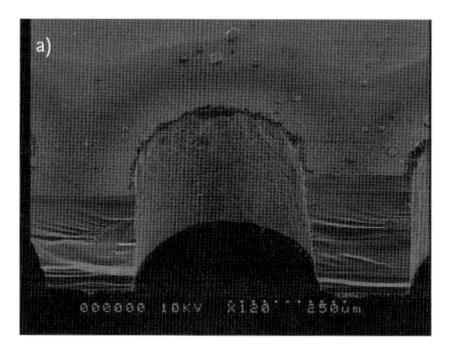

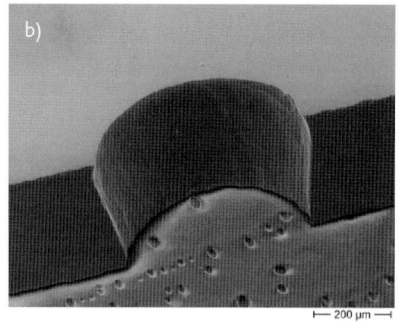

Figure 19.3 Scanning electron microscopy images of laser-machined through-substrate via (a) after drilling and (b) after subsequent cleaning.

However, such a process requires an additional lithography step to define the position and diameter of via holes. As an e-carrier needs just two contact pads at the backside, the number of vias will be two or maybe four. Therefore, for reasons of cost and simplicity it was decided to prepare vias by laser drilling. In order to minimize deterioration of the surface a thermal oxide was grown on the silicon wafer substrate before the formation of holes. Figure 19.3 shows scanning electron microscopy cross-sectional images of through-substrate holes in a silicon wafer after laser drilling and after chemical cleaning.

Finally, the protective oxide layer was removed. Then a new 1000 nm thick thermal oxide was grown resulting in insulated surfaces as well as insulated via side walls. Metallization of vias was done by sputtering TiW from both sides of the wafer substrate. Thereby, metal deposition of the electrode areas and electrical interconnects through via holes are realized in one step.

An alternative way to realize backside contacts was proposed by Wieland [8]. According to his concept electrically conductive wires are prepared over the wafer edge. Thereby, front-side electrodes can be connected to backside contact pads without the necessity of preparing through-substrate vias. Metal layers or polysilicon can be used for interconnects over the wafer edge. In the case of polysilicon the wafer edges are already covered due to the process characteristics. When using sputtered layers of titanium or TiW it is necessary to select process parameters that provide a large internal divergence of the material transfer. In this case the wafer edge is covered as well. Subsequent double-sided lithography and wet chemical etching define the position of the "over edge" contacts.

19.2.4
Examples of Mobile Electrostatic Carriers

Mobile e-carriers have been prepared at Fraunhofer EMFT on ceramic substrates [9], glass and silicon wafers, sapphire [10], and also on polymer substrates. Figure 19.4 shows an example of mobile e-carriers based on silicon wafers of 150 mm in

Figure 19.4 Mobile electrostatic carriers based on silicon wafer substrates prepared at Fraunhofer EMFT.

diameter with backside contact pads and through-silicon via (TSV) interconnects [11]. The e-carrier in the background of Figure 19.4 also shows the mirror image of the rewiring between TSV and contact pads originating from the backside of the e-carrier in the front.

It should also be mentioned that a variety of names have been proposed and are in use to emphasize the specific features of these new types of e-carriers: "smart carrier," "transfer-ESC" (T-ESC®; "Transfer-ESC" is a registered trademark of Protec Carrier Systems GmbH, see www.protec-carrier.com), "MEC" (mobile electrostatic carrier, [12]), and also "e-carrier."

19.3
Characterization of Electrostatic Carriers

19.3.1
Electrical and Thermal Properties, Leakage Currents

The experiments described in the following were carried out with bipolar mobile e-carriers with TSVs and backside contact pads. The substrate material was a silicon wafer with a diameter of 150 mm. In order to determine the duration of electrostatic fields and their dependence on temperature a thin silicon wafer was attached onto an e-carrier and the wafer stack was placed onto the heated chuck of a wafer probing equipment. After charging of e-carriers at a voltage of 100 V the external power supply was disconnected and the wafer stack was heated to the final test temperatures of 150 and 200 °C. The remaining voltage at the backside contact pads was measured by an electrometer (Keithley 617) after certain periods of time. Results are shown in Figure 19.5. After 1 h at room temperature the

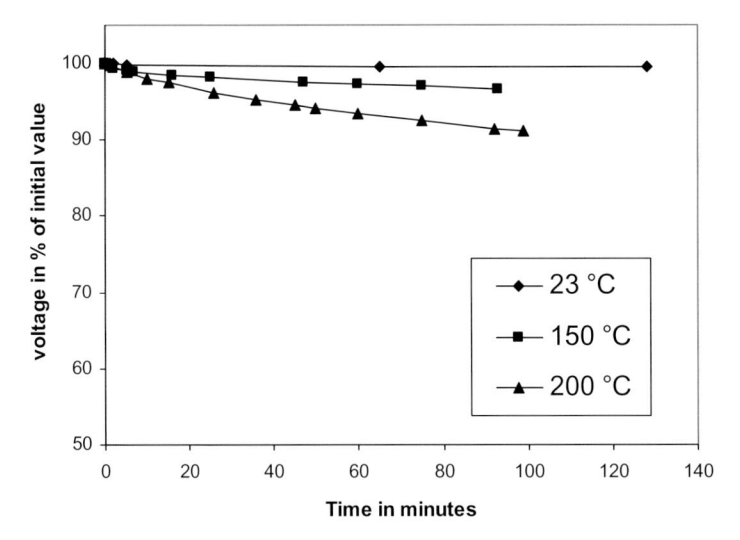

Figure 19.5 Time-dependent decay of clamping voltage of e-carriers at three temperatures as indicated.

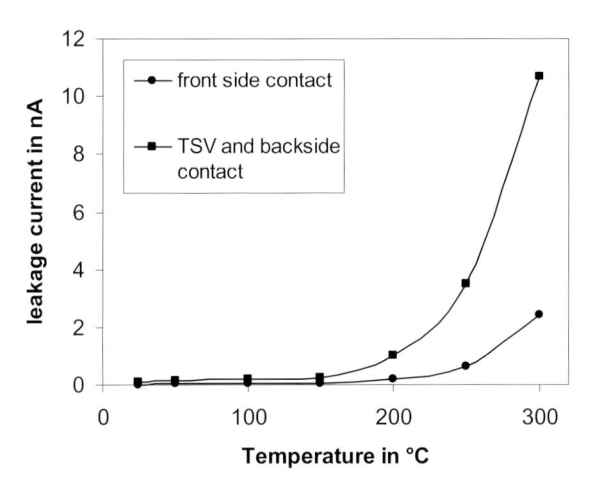

Figure 19.6 Temperature dependence of leakage currents measured at electrostatically clamped wafer stacks at 100 V DC. The graph shows the strong relevance of perfect insulation of TSV interconnects.

voltage is reduced by less than 1%; after 1 h at 200 °C the voltage is reduced by 6.5%. Such low decays in voltage practically do not influence the safe clamping status of the stacked wafer pairs.

As a second test method we measured the leakage current at 100 V DC while heating the wafer stack to 300 °C. The measuring unit was a precision semiconductor parameter analyzer (Agilent 4156 A). Figure 19.6 shows the measured increase of leakage currents with increasing temperature. In the case of e-carriers with TSVs the leakage current is increased by a factor of five compared to e-carriers with just front-side contact pads. This clearly shows that further improvements of the high-temperature behavior of e-carriers can be achieved by improving the electrical insulation of TSV contacts.

In principle, the measurement of leakage currents at 300 °C can be used to estimate the duration of charge storage. The electric capacitance of the e-carrier was measured to be 140 nF. A leakage current of 11 nA would discharge this capacitance within approximately 20 min. However, the holding capability of e-carriers also depends on the polarization of the dielectric layers. Actually, even after discharge of the electrodes strong holding forces can be detected. Possible explanations of this behavior would be the existence of a permanent polarization of dielectrics or the presence of trapped charge carriers (Johnsen–Rahbek effect) which will result in oppositely charged surfaces which are separated by the thin air gap of a few nanometers between the surface of the e-carrier and the thin wafer attached onto it. This effect would explain why strong attractive forces may be present although the remaining voltage at the electrodes is in the range of just some tens of volts. So, the observed increase of leakage current at 300 °C does not necessarily prevent the holding capabilities even at much higher temperatures. Further research work will be carried out to experimentally verify and extend the high-temperature behavior of such e-carriers.

19.3.2
Possible Influence of Electrostatic Fields on CMOS Devices

A frequently asked question concerns the possible influence of electrostatic fields on the electrical properties of CMOS devices. To clarify this point we measured 200 μm thin CMOS devices before and after chucking on mobile e-carriers. The threshold voltage V_t of the CMOS transistors was used for the detection of possible influences. Repeatability of the measurement setup was 0.08%. First, we measured the threshold voltage V_t when placing the active side of the wafer face up onto the e-carrier. Then we placed a CMOS wafer face down onto an e-carrier. In this configuration the distance between IC devices and the electrodes of e-carrier substrate is just approximately 5 μm. In order to simulate a long-term attachment the wafer was electrostatically clamped for more than 16 h. Changes in V_t were found to be 0.1% [11]. Such small deviations may also be induced by small changes in the temperature of the laboratory environment. So it can be concluded that applied electrostatic fields generally do not affect the characteristics of CMOS transistors. Similar observations were found for thinned GaAs devices by Stieglauer et al. [12].

It should be mentioned that the electric field induced by the e-carrier is in the range 50–100 V μm^{-1}. This is lower by a factor of 10–20 than the typical breakdown voltage of a SiO$_2$ gate dielectric of standard CMOS transistors. This might be the reason for the observed insensitivity of transistor properties against the electric fields of the e-carrier plate.

19.4
Electrostatic Carriers for Processing of Thin and Flexible Substrates

Manufacturing technologies for thin wafers of thicknesses of 50 to 150 µm have become a basic need for a wide variety of new microelectronic products. Among them, power devices, discrete semiconductors, optoelectronic components, and integrated circuits for radio-frequency identification systems represent some of the most important applications. Further technological developments are targeting stacked-die assemblies, vertical system integration, and many new ideas in the field of microelectromechanical system devices. These aims require new handling techniques for processes that have to be performed at the backside of thin or fragile substrates.

Electrostatic attraction offers a simple and powerful technique for temporary bonding of thin or flexible substrates in semiconductor manufacturing. Mobile e-carriers provide mechanical support during manufacture of fragile substrates and increase production yield due to elimination of wafer breakage. A unique advantage of the electrostatic handling technique is its capability of reversibly bonding a fragile substrate just by simple charging and discharging procedures.

The following subsections describe examples of processing for which mobile e-carriers have already been used successfully. The selected examples also illustrate the specific requirements and advantages when mobile e-carriers are used in the temporary bonding technique for thin wafer processing.

19.4.1
Handling and Transfer of Thin Semiconductor Wafers

The photographs in Figure 19.7 illustrate a typical handling sequence for temporary bonding of thin semiconductor wafers onto an e-carrier. First, an e-carrier is placed onto a charging unit (laboratory type). Backside contact pads of the e-carrier are connected from the rear side. Then a thin wafer substrate (irregular bow is due to the small thickness of some 70 µm) is attached onto the e-carrier. After applying a voltage in the range 100–200 V at the electrodes the thin wafer is attracted onto the e-carrier. Finally, the clamped wafer stack is handled by a vacuum tip just like in the case of a standard wafer substrate. In this configuration

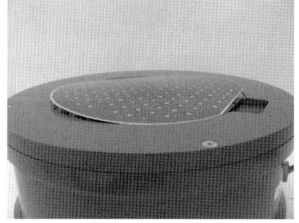

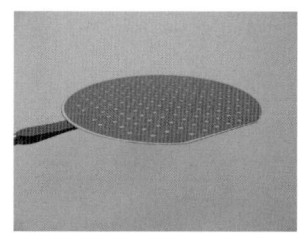

Figure 19.7 Sequence of photographs showing thin wafer handling by means of a mobile electrostatic carrier.

the thin and fragile wafer can easily enter fabrication equipments and undergo further processing steps.

Mobile e-carriers also offer new possibilities for transferring wafers between different kinds of support substrates. For instance, one can use reversibly adhesive tapes for a first process sequence (e.g., wafer grinding and etching) and then directly transfer the thinned wafer onto an e-carrier [13]. Thereby, thin wafers will always be connected to a rigid support and transfer times may be in the range of a minute.

19.4.2
Wafer Thinning and Backside Metallization

Wafer thinning is most often done by grinding and a subsequent stress relief process which may comprise either wet-chemical spin-etching, chemomechanical polishing, or plasma dry etching. The applicability of e-carriers during grinding and spin-etching has already been reported [14].

Some of the most prominent examples in thin wafer technology are related to power devices. These products require the deposition and sintering of a metal layer at the rear side of a very thin semiconductor wafer (thickness range of 50–150 μm). Sputter deposition of metal usually is followed by a sinter process in the temperature range 350–430 °C. Temporary electrostatic bonding of device and carrier wafer is able to withstand these process conditions. Here, e-carriers really take advantage of the electrostatic bonding principle. As no adhesives are involved critical high-temperature effects such as outgassing or polymer degradation cannot occur.

19.4.3
Electrostatic Carriers in Plasma Processing

Many semiconductor processes take place in a plasma process chamber, for instance layer deposition (dielectrics, metal) or etching and structuring of layers or silicon substrates. The applicability of mobile e-carriers to such process environments has already been proven. However, it has to be taken into account that a plasma is an electrically conductive atmosphere. Therefore, discharge of the electrodes of e-carriers must be prevented. In the case of backside contact pads the wafer table must not allow for short circuiting of these pads. This can be done either by preparing specifically designed insulating materials (e.g., ceramics) at the front side of the chuck table where the pads of e-carriers might be in mechanical contact or by placing these pads above the region of helium backside cooling channels.

19.4.4
Electrostatic Carriers Enable Bumping of Thin Wafers

Solder bumps are widely used for flip chip assemblies and chip-scale package technologies. However, thinning of wafers with solder balls on top of them having a diameter of 100–250 µm is a challenging task. Actually, wafer thickness below 250 µm cannot be reached by backside grinding due to the high risk for wafer breakage. Mobile e-carriers offer the possibility to change the process sequence: first perform wafer thinning; then place the thin wafer onto an e-carrier; and finally run the bumping sequence (stencil printing of solder paste, reflow) by means of mobile e-carrier. According to this patented concept [15], manufacture of 55 µm thin test wafers having solder balls of a diameter of 150 µm on top was demonstrated [16].

19.4.5
Electrostatic Carriers in Wet-Chemical Environments

Of course, electrostatic attraction is also active in fluids. However, there is a physical principle that must be taken into account: the tendency of dielectric materials to move towards the regions of highest strength of the electrical field. In the case of e-carrier wafer stacks the strongest fields exist between thin wafer and e-carrier. Therefore, fluids will try to penetrate along the interface between wafer and carrier substrate. This is especially true for polar molecules; for instance water shows a dielectric constant $\varepsilon_r = 80$. Generally, interface wetting will not be tolerable due to unwanted etching effects or contamination problems. One possibility would be to develop a polymer sealing surface along the edge of an e-carrier which will reduce the penetration of liquids. Another idea would be the application of a temporary sealing ring after clamping of thin wafer and e-carrier along the wafer edge [11]. The latter concept would be preferred if the e-carrier were to be allowed to have a slightly larger diameter than the thin wafer attached onto it.

19.4.6
Electrostatic Handling of Single Dies

Electrostatic die handling is very similar to full wafer handling except that there is a finer meandered electrical gap between the two electrodes. This defines the minimum size of a chip, needing to bridge the gap between the two electrodes to safely remain fixed on the carrier. Additionally the implemented stress in the thinned chips determines the maximum electrically induced force that is needed to keep the chips flat [17]. So here we are in the regime of strongly application-determined geometry of the electrodes.

The electrostatic bonding principle may also be used to arrange a large number of single chip components onto a temporary handle substrate. If this is performed by an aligned die-placing process then the whole configuration can be used for "wafer-level chip-to-wafer" die bonding.

19.4.7
Processing of Foils and Insulating Substrates

Electrostatic attraction is very well suited for temporary bonding of foil substrates with a conductive layer on front or rear side. The flexibility of web substrates such

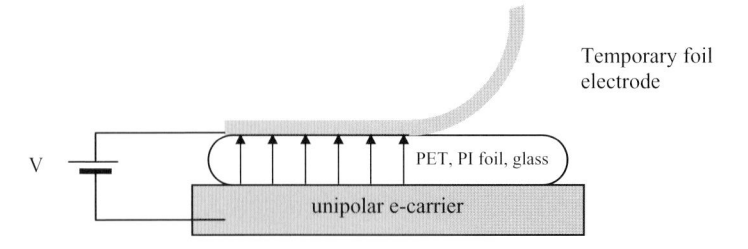

Figure 19.8 Method for temporary bonding of insulating materials by electrostatic forces. The upper electrode foil is removed after charging the unipolar electrostatic carrier. (PET, poly(ethylene terephthalate); PI, polyimide.)

as poly(ethylene terephthalate) or polyimide ensures close contact between foil and e-carrier and therefore allows for strong holding forces. Both unipolar and bipolar electrode configurations can be used.

Furthermore, it has also been found that even nonconductive materials, for example thin plates of borosilicate glass or bare polyimide sheets, can be clamped by electrostatic forces [18]. The principle is shown in Figure 19.8. An insulating substrate (e.g., polymer tape or thin glass sheet) is placed onto a unipolar e-carrier plate and then a flexible foil electrode is laminated on top. By applying an electrical voltage between foil electrode and bottom plate the electric field will attract the electrodes against each other and, thereby, press the insulating substrate onto the carrier plate. The electric field also induces a polarization into the dielectric substrate. After disconnection of the power supply the upper foil electrode is delaminated. Due to the remaining polarization in the dielectric sheet the insulating substrate is kept clamped for a certain period of time. It has been found that this polarization-based bonding technique is strong enough for subsequent spin-coating and annealing processes performed at the insulated substrate at temperatures up to 150 °C.

19.5
Summary and Outlook

Electrostatic forces enable an easy and powerful technique for temporary bonding of thin and flexible substrates. Due to the lack of any polymeric adhesive, mobile e-carrier technology can be applied to process temperatures up to 400 °C at present and probably even above. Next development steps need to encounter automatic equipment for both transfer of thin wafers as well as for controlled charging and discharging of electrostatic support systems. Further investigations will target the introduction of materials with high dielectric constants and superior properties in terms of electrical insulation.

One of the first applications in semiconductor manufacturing might be handling, electrical testing, and shipment of very thin and brittle device wafers. Different applications are under evaluation at present. Another application scenario will be the deposition and sintering of metal layers at the rear side of power devices and discrete semiconductors.

Looking slightly further into the future, the mobile e-carrier technique is expected to open the door to new process schemes in the field of three-dimensional system integration, handling and processing of large-area flexible foil substrates, and new production technologies for generative and rapid prototyping as well as for the realization of very thin high-performance solar cell modules.

References

1 Davey, K. and Klimpke, B. (2002) Computing forces on conductors in the presence of dielectric materials. *IEEE Trans. Educ.*, **45** (1), 95–97.
2 Yoo, J., Choi, J.-S., Hong, S.-J., Kim, T.-H., and Lee, S.J. (2007) Finite element analysis of the attractive force on a coulomb type electrostatic chuck. Proceedings of the International Conference on Electrical Mechanical Machines and Systems, Seoul, Korea.
3 Atkinson, R. (1969) A simple theory of the Johnsen–Rahbek effect. *Br. J. Appl. Phys., Ser. 2*, **2**, 325–332.
4 Balakrishnan, C. (1950) Johnsen–Rahbek effect with an electronic semi-conductor. *Br. J. Appl. Phys.*, **1** (8), 211–213.
5 Qin, S. and McTeer, A. (2007) Wafer dependence of Johnsen–Rahbek type electrostatic chuck for semiconductor processes. *J. Appl. Phys.*, **102**, 064901.
6 Kanno, S. and Usui, T. (2003) Generation mechanism of residual clamping force in a bipolar electrostatic chuck. *J. Vac. Sci. Technol. B*, **21** (6), 2371–2377.
7 Landesberger, C., Bleier, M., and Klumpp, A. (2001) Mobile holder for a wafer. US Patent 7,027,283, filed 30 July 2001, issued 11 April 2006; Landesberger, C., Bleier, M., and Klumpp, A. (2001) Mobile holder for a wafer. European Patent EP 1305821 B1, filed 30 July 2001, issued 8 October 2008.
8 Wieland, R. and Bollmann, D. (2005) Bipolarer Trägerwafer und mobile, bipolare, elektrostatische Waferanordnung. German Patent DE 102005056364 B3.
9 Bock, K., Landesberger, C., Bleier, M., Bollmann, D., and Hemmetzberger, D. (2005) Characterization of electrostatic carrier substrates to be used as a support for thin semiconductor wafers. International Conference on Compound Semiconductor Manufacturing Technology (GaAs Mantech), New Orleans, LA, April 2005.
10 Landesberger, C., Bock, K., Stieglauer, H., and Häupl, K. (2009) Electrostatic carriers towards industrial application. Workshop on Thin Semiconductor Devices, Munich, Germany, November 2009.
11 Landesberger, C., Wieland, R., Klumpp, A., Ramm, P., Drost, A., Schaber, U., Bonfert, D., and Bock, K. (2009) Electrostatic wafer handling for thin wafer processing. European Microelectronics and Packaging Conference & Exhibition, Rimini, Italy, June 2009.
12 Stieglauer, H. et al. (2010) Mobile electrostatic carrier (MEC) for a GaAs wafer backside manufacturing process. CS Mantech Conference, Portland, OR, 17–20 May 2010.
13 Landesberger, C., Bonfert, D., Bock, K., and Häupl, K. (2007) Advances in mobile electrostatic carrier technique. Workshop on Thin Semiconductor Devices, Munich, Germany.
14 Raschke, R. (2009) New transferable electrostatic carriers for applications in vacuum up to 400 °C. Workshop on Thin Semiconductor Devices, Munich, Germany.

15 Landesberger, C., and Ostmann, A. (2004) Verfahren zum Bearbeiten eines Halbleitersubstrats. German Patent DE 102004021259 B4.
16 Landesberger, C., Scherbaum, S., and Bock, K. (2007) Carrier techniques for thin wafer processing. Conference on Compound Semiconductors Manufacturing Technology, Austin, TX, May 2007.
17 Wieland, R., Hacker, E., Landesberger, C., Ramm, P., and Bock, K. (2008) Thin substrate handling by electrostatic force. Conference on Smart Systems Integration, Barcelona, Spain.
18 Landesberger, C. (2007) Method and apparatus for electrostatic fixing of substrates with molecules which can be polarized. European Patent EP 1905071 A1, filed 22 March 2007, issued 2 April 2008.

Index

a

accelerometer 286
adhesion energy 82, 223
adhesion promoter 36, 44
adhesive bonding 216
adhesive injection method 141 f
adhesive wafer bond, permanent 54
adhesive wafer bonding *see also* polymer adhesive wafer bonding 32 ff
– bond chamber for 22
– features of 19
– for producing three-dimensional LSI 140
– for three-dimensional intergration 301
– SOG-based 19 ff
– with thermoplastic polymer (HD-3007) 56
– with thermosetting polymers for permanent wafer bonds (BCB) 54 f
– with thermosetting polymers for temporary wafer bonds (mr-I 9000) 54 f
adhesive, liquid organic 141
adsorption theory 36
advanced MEMS device 284 ff
aligned bonding 244 f
alkali-containing glass 64, 68
aluminum nitride substrate 205, 208
aluminum/silicon oxide DBI™ hybrid bonding 270
annealing temperature 101
– for direct bonded wafer pairs 86, 88
– in plasma-activated bonding 110 f
anodic bonding 63 ff
– bonding current 67
– characterization of the bond quality 69
– effect on flexible micromechanical structures 71
– electrical degradation of devices during 71
– electrical effects of 74
– features of 63
– formation of oxide layer 66
– glasses for 68
– in high vacuum 70
– layer contacting 64 ff
– mechanism of 64 ff
– of patterned wafers for MEMS 290 ff
– prevention of electrical degradation during 73
– reactions in the interface regions during 66
– with thin films 75
application
– of adhesive wafer bonding with SOG layers 29 f
– of anodic bonding 63
– of copper bonded wafers 178
– of copper/tin SLID bonding 201
– of direct wafer bonding 93 f
– of direct wafer bonding in ultrahigh vacuum 90
– of elastomer bonding/de-bonding technique 363
– of electrostatic temporary wafer bonding 379 f
– of eutectic gold-tin bonding 120, 133, 150 f, 157
– of glass frit bonding 14 ff
– of gold/tin SLID bonding 204
– of hybrid copper/BCB bonding platform 220
– of hybrid copper/BCB redistribution layer bonding 218 f
– of plasma-activated bonding 111
– of polymer adhesive wafer bonding 34
– of three-dimensional integration technology 306 ff
– of metal/silicon oxide DBI™ hybrid bonding 273 f
atmospheric-pressure plasma 107, 112

Index

b

back-end-of-the-line (BEOL) process *see also* TSV processing   216, 238, 303
backgrinding   339 f, 359
backside contact formation   359, 375 ff, 379
backside metallization   380
backside processing   329 ff, 334 f, 342, 356
backside solder bump   341
backside TSV   311
BCB *see* benzocyclobutene
benzocyclobutene (BCB)   40, 54, 217, 283
BEOL interconnect   303
BEOL *see* back-end-of-the-line process
binder burn-out   10
binder polymerization   10
bipolar chucking   369
BISV *see* bonded interstrata via
blanket copper-to-copper direct bonding   161 ff, 239 f, 245
blister test   109
blistering   93
bond chamber
– for adhesive wafer bonding   22
bond strength
– in direct copper/silicon dioxide bonding   241, 246
– of anodic wafer bonds   69
– of copper bonded wafers   168, 246
– of glass frit bonds   14
– of MEMS   287
– of plasma-treated wafers   103
– of SLID bonds   207 f
– of SOG layer bonds   23 ff
– techniques for testing the   108, 240, 245, 248
bond wave propagation   82
bonded interstrata via (BISV)   215, 218, 303
bonding
– of bumped wafers   363
– of carrier and device wafer   358 f
– of patterned surfaces   243
bonding current   67
bonding interface   14
– analysis of, by electron dispersive X-ray (EDX)   250
– analysis of, by *in situ* thermal evolution X-ray reflectivity (XRR)   248 f
– dislocations in the   92
– electrical characterization of copper/copper direct   252
– in DBI™   266
– in direct copper/silicon dioxide bonding   240 ff

– in the hybrid copper/BCB system   224 ff
– properties of   90, 251 ff
bonding pressure   56, 58
– in copper/tin SLID bonding   193 f
– in localized wafer bonding   51
– in thermocompression copper bonding   167
bonding strength *see* bond strength
bonding temperature   48, 56, 58
– in copper/tin SLID bonding   193 f
– in thermocompression copper bonding   167
– of anodic wafer bonding   63
– of anodic wafer bonding with glass thin films   75 f
– of polymer adhesive wafer bonding   33
boundary layer   35
bridge dicing   283
B-stage polymer   42, 54
bubble *see* interface defect   93
bump reflow   121 ff
bumping   341

c

C2W bonding *see* chip-to-wafer bonding
capillary force   83
capped inertial sensor   15
carrier wafer   330, 338, 356, 359
– material for   372
– re-use of   364
– processing of   357
cavity in micromechanical devices, vacuum-sealed   70
characterization of wafer bond quality
– criteria for   108
– of copper bonded wafers   169 ff
– of plasma-activated wafer bonds   110 f
chemical bonding   36
chemical-mechanical polishing (CMP)   24, 31, 114, 167, 218, 222, 239
chip alignment   349, 381
chip design based on three-dimensional integration   307, 312
chip stacked flexible memory   319
chip-on-chip (CoC) technology   308
chip-to-wafer (C2W) bonding   308
chip-to-wafer stacking using ICV-SLID soldering   342
chip-to-wafer thermode bonding   135
chuck   112
– electrostatic wafer chuck (ESC)   367, 371
– for glass frit bonding   13
– for polymer adhesive wafer bonding   44
– vacuum   356, 360

Index | 387

CIS *see* CMOS image sensor
clamping force   368, 371
CMOS
– compatibility to bonding methods   114, 204, 262
– low-k three dimensional integration   276 f
– possilble influence of electrostatic fields on   378
– shrinkage   307
CMOS image sensor (CIS)   114, 310 ff
– three-dimensional backside illuminated   273 ff
– via-last process on   338 f
CMOS technology   178
CMOS wafer   29, 30, 94, 271, 273
CMP *see* chemical-mechanical polishing
CoC memory on logic solution   319 f
CoC technonolgy *see* chip-on-chip technology   308 f
coefficient of thermal expansion (CTE)   39, 269
contact resistance of copper/copper direct bonding interfaces   251 ff
contamination removal   103, 166 f
copolyester   41
copper bonded layer
– morphology of   163, 165
– oxide distribution in   164
copper DBI™   270
copper dishing   239 f
copper microbump   202
copper nanorod   177
copper/copper bonding   161 f
– alignment accopperracy of   171 f
– application in three-dimensional integration   217 f
– direct   177, 238 ff
– for three-dimensional TSV intergration   303 f
– low-temperature   176 f
– nonblanket   174 f
copper/dielectric hybrid bonding   176
copper/silicon dioxide DBI™ hybrid bonding   270
copper/silicon dioxide direct bonding   240 ff
– advantages of   237
copper/tin phase diagram   187
copper/tin SLID process   184 f, 346
– application of   201 ff
– bond formation   189, 191 f
– bonding parameters used in   193
– copper/tin-to-copper/tin symmtetric bonding   198

– fluxless oxidation-free   196 ff
– required material properties for   190
copper-to-copper bonding *see also* copper/copper direct bonding   216, 242, 245 ff, 257
covalent bond   34
crack-opening method   30, 108 f
creeping *see* viscoelastic effect
CTE mismatch   46 ff, 53, 102, 155 f, 204, 208, 267
CTE *see* coefficient of thermal expansion
curing   37

d

3D BSI-CIS *see* CMOS image sensor, three-dimensional backside illuminated
daisy chain structure   148 f, 154 ff, 210, 256 f, 270 f, 340
damascene CMP process   239
damascene metal/silicon dioxide DBI® process flow   263
damascene-patterned copper   176, 218
DBC substrate *see* direct bonded copper substrate
DBD *see* dielectric barrier discharge
DBI™ technology (direct bond interconnect)   177, 261 f
DCB technique *see* double cantilever technique
DDR memory   315 ff
debonding   331
– room temperature, low stress   335 f
– single-wafer   333
– slide-off   332
– using a release layer and liftoff   335 f
– 355
– process of cold   360 f
Debye force   83
deglazing   9
degradation of MOS
– by high electric fields   73 f
– by sodium contamination   72
deposition
– of polymer adhesives   44
– of thin glass layers   75
device encapsulation   282 ff
device layer   94, 135, 301, 304
device wafer   356, 359
– thickness of   361
– via first   330
– via last   330
dicing   282
dicing before grinding (DBG) approach   363

die assembly with SLID bonding 347 f
die bonding 120
– die-to-wafer bonding 148, 238, 255
– die-to-wafer stacking usind ICV-SLID soldering 342 ff
dielectric barrier discharge (DBD) 89, 107, 11
dielectric bonding 303
dielectric layer material 373 f
diffusion bonding 250
diffusion copper bonding *see also* thermocompression copper bonding 162 f
diffusion theory 36
dipole-dipole interaction 34
direct bond interconnect *see* DBI™ technology
direct bonded copper (DBC) substrate 205, 208
direct wafer bonding 81 ff
– in advanced substrates for microelectronics 93, 290
– in ultrahigh vacuum 89, 92
– low-temperature 88
– of patterned wafers for MEMS 287 ff
– of silicon wafers 102
– physical characterization 82 ff, 91
– principle of 101
– properties of wafers for 82
– surface chemistry of 82 ff, 91
– techniques for 84 f
– using hydrophilc surfaces 84 f, 90, 102
– using hydrophobic surfaces 86, 102
dislocation in bonded hydrophobic wafers 92
double cantilever beam (DCB) technique 241, 245
DRAM system
– by Elpida 317 f
– by NED 317 f
– by Samsung 315 f

e

e-carrier *see* mobile electrostatic carrier
edge chipping 337
edge pretrimming 337
edge protection 337, 360
elastomer adhesive 36, 38, 362
elastomer coating layer 357 ff
electric field 368
electrical characterization of bonded interconnects 251 ff
electrical degradation 71 f
electrical testing using daisy chains 339

electrode material for electrostatic temporary wafer bonding 374
electron probe microanalysis (EPMA) of microbumps 146 ff
electroplating of gold/tin solder 121, 127, 130 f, 294 ff
electrostatic bonding *see* anodic bonding
electrostatic clamping 349, 367, 371
– of foils 381
– of insulating substrates 381
– of thin wafers for bumping 380 f
electrostatic die handling 381
electrostatic force 65, 71, 83
– in a plate capacitor 368 f
electrostatic interaction 36, 65
electrostatic wafer chuck (ESC) 367, 371
encapsulation
– hermetic 120
– of bolometer pixels 205
– of MEMS 282 f, 293
– of surface micromachine sensors 15
– on wafer level 283
– vacuum, on wafer-level 70 f
epoxy adhesive 40, 357 ff, 361
– injection of 146, 151
eutectic bonding 119 ff
– advantages of 119 f
– preconditioning in 125 f, 132
eutectic gold/indium bonding 139 ff, 146
– fabrication process for LSI test chips applying 149 f
eutectic gold/tin bonding 120 ff
– of patterned wafers for MEMS 293 ff

f

F2F bonding *see* face-to-face bonding
face-to-face (F2F) bonding 308
feedthrough 283
FEOL process *see* front-end-of-line process
Ferro FX-11-036 5
field-assisted sealing *see* anodic bonding
filler for glass frit materials 4, 8
flexible interlayer 26 f, 31
flexible substrate 371, 379
flip-chip bonding 294, 347
flip-chip reflow soldering 121, 127
flip-chip/underfill hybrid bond 261
fluoropolymer 41
fluxless bonding 196 ff
foundry acitivty 323
four-point bending technique 222 f, 245, 340
fracture toughness 109, 112

frictional nonreflowable surface structure 53
front-end-of line (FEOL) process *see also* TSV processing 301
fully processed wafer 3
fusion bonding *see* direct wafer bonding 81

g
GaAs wafer, SOG-coated 28
GaAs/InP wafer pair bonded with SOG layer 29
GaAs/Si wafer pair bonded with SOG layer 25 f
germanium-coated wafer 95
germanium-on-insulator wafer (GOI) *see* germanium-coated wafer
glass crystallization 9
glass frit
– deposition of the 4
– material 4 f
– paste 5
– wetting temperature 11
glass frit bond characterization 14
glass frit wafer bonding 3 ff
– advantages of 14, 17
– application of 14 ff
– basic principles of 3
– fusing effect of surface layers 13
– influence of structure with on printed glass frit thickness 7
– multiple-temperatur glass-conditioning process 9
– parameter field evaluation of bonding process 12
– steps of wafer bond process 11 f
– thermal transformation of printed paste into glass for bonding 8 ff
– using screen printing technology 5 f
glass material for anodic bonding 68
glass polarization 64
glass pre-melting 10
glass transition 37
glass transition temperature 37
glass-frit-bonded gyroscope 4
glueing 349
GOI *see* germanium-on-insulator wafer
gold/silicon dioxide DBI™ hybrid bonding 270 f
gold/tin SLID process *see also* solid-liquid interdiffusion bonding 185
– application of 204
– bond formation 199
– properties of bonding material 199 f
– thermomechanical properties of gold-tin phases 200
– wafer-level bonding using 206
gold/tin solder bump 121 ff, 127
gold/tin solder deposition 293 f
gold/tin phase diagram 120, 188
grain boundary 250
gyroscope 286

h
hardening 37
hermetic sealing at wafer level 203
hermeticity 110
– of anodic wafer bonds 69
hermeticity test *see* leak test
heterogeneous MEMS 306
high-density indium/gold microbump 152
hot-melt *see also* thermoplastic polymer 37
hybrid bonding 237 ff, 261 ff
– comparison of heterogeneous and homogeneous bonds 261
hybrid copper/BCB bonding 217 f
– bonding experiments using structured silicon wafers 228
– copper/BCB surface profile 223, 229
– electrical characterization of 231
– evaluation of processing issues 222
– topography accomodation in 227 ff
– topography of bonding interfaces in 225 ff
– using partially copperred BCB 222, 228
– wafer bonding steps 220
hybrid copper/BCB bonding platform *see also* three-dimensional integration platform 220
hybrid metal/dielectric bonding 215 ff
hybrid metal/polymer bonding 216, 218, 220
hybrid polymer 36, 38
hybrid surface 267 f
hydrogen bonding 83
hydrophilic surface 66, 84, 86, 90, 108, 288
hydrophobic surface 66, 84, 86 ff, 288

i
ICP-RIE plasma *see* inductively coupled plasma RIE
IMC *see* intermetallic compound
indium/gold microbump 139, 143
– bonding principle of 144
– current-voltage characteristic of 148 f, 155

– EPMA mapping of   146 ff
– formation by planarized liftoff method   144 f, 153
– high-density technology   152
– resistance of   154 f
indium/gold phase diagram   143
inductively coupled plasma (ICP) RIE   104 f, 115
integration of heterogeneous materials   102 ff
interchip via (ICV)-SLID technology   201 f, 342
interconnect   201 f, 210, 237, 266, 268 f
interconnect via solid-liquid interdiffusion (ICV-SLID) soldering technology   342
interdiffusion   181
interface defect   23, 91, 225 ff
interface sealing mechanism   250
interlocking structure   53
intermediate bonding layer
– effect of layer thickness   47, 67
– for low-temperature wafer bonding   88
– glass as   3 ff
– glass film as   63
– polymers as   32 ff
– SOG-based   21 ff
intermetallic compound (IMC)   181
– formation of, in copper/tin SLID bonding   185, 189 ff
– formation of, in gold/tin SLID bonding   187, 199
interstrata interconnection   215, 219, 306
– approaches to   216
interstrata via-chain structure using hybrid copper-BCB bonding   226 f
ionic bond   34
ion-implanted wafer   93

j
Johnsen-Rahbek effect   370 f, 378

k
Keesom force   83
Kelvin structure   251 f
keyed alignment structure   53
known good die (KGD)   345 f

l
laminate   285, 292
large-scale integration (LSI)   139
laser Doppler velocimeter sensor   112
layer fusing *see also* glass frit bonding   13
layer-transfer by hydrogen-induced splitting   93

leak test   14, 110, 113, 203
leakage current   13, 72, 74, 371, 376 ff
liquid crystal polymer   41
liquid lens   286
London force   83
low-melting point glass   4
low-pressure plasma   105 f, 112, 114
low-temperature processing of wafer bonds   19, 30
low-temperature wafer bonding   101, 112, 168, 178, 237
LSI *see* large-scale integration

m
MC test *see* micro-chevron test
mechanical interlocking   36
memory device   313
memory on logic   321, 322, 323
memory stacking   321
MEMS capping   4
MEMS *see* microelectromechanical systems
metal bonding   119 ff, 237, 303
metal microbump   140 ff
metal oxide patterned surface bonding   238 f
metal/adhesive hybrid bonding *see* hybrid copper/BCB bonding   262
metal/direct hybrid bonding   261 f, 303
metal/non-adhesive hybrid bonding *see* metal/direct hybrid bonding   262
metal/silicon nitride DBI^{TM} hybrid bonding   261, 271 f
metal/silicon oxide DBI^{TM}
– alignment   265 f
– application of   273 f
– bond components   266 f
– influence of metal polishing rate   269
– influence of the CTE   269
– influence of the stabilitiy of the native metal oxide   269
– influence of the yield strength of the metal   269 f
– metals for   270 f
– surface activation   265
– surface contact   267
– surface fabrication for   263
– surface patterning   264
– surface roughness   264
– surface termination   265
– surface topography   264
metal/silicon oxide hybrid bonding *see* metal/silicon oxide DBI®
metallic lead-through   15
metal-oxide-semiconductor (MOS)   72

metal-oxide-semiconductor field-effect-
   transistor (MOSFET)   93, 151, 178
microactuator   286
microbump bonding   308
microbump daisy chain   148 f
microbump pitch   153
micro-chevron test (MC test)   109
microelectromechanical systems (MEMS)
   3, 34, 63, 81, 95, 112, 203, 220, 281 ff
– electrical degradation of, during anodic
   bonding   71 ff
– encapsulation of   282
– fabrication of, using anodic bonding
   290 ff
– fabrication of, using eutectic gold/tin
   bonding   293 ff
– fabrication of, using fusion bonding
   287 ff
– post-processing of bonded wafers   285
– protection during wafer dicing   282
– requirements of, for the bonding process
   286
– routing of electrical signal lines   282
– short circuiting   291
– stacking of several wafers   284
– wafer bonding techniques for the
   fabrication of   281
microelectronics, surface planarization in
   20
microfluidic chip   286
microfluidics packaging   113 f
micromechanical device   70, 112 f, 286
micromirror   286
microphone   286, 292
microvoid   49
mobile electrostatic carrier (e-carrier)   371,
   375
– electrical and thermal properties of   376
– for processing of thin and flexible
   substrates   379
– in plasma processing   380
– in wet-chemical environments   381
– leakage currents of   376
Moore's law   307
MOS see metal-oxide-semiconductor
MOSFET see metal-oxide-semiconductor
   field-effect-transistor
multicore processor   321
multilayer stacking based on copper bonding
   172 ff

n

nanoelectromechanical system (NEMS)
   203

nano-imprint lithographic process   49
nano-imprint resist   40, 54
negative-bias temperature instability (NBTI)
   75
NEMS see nanoelectromechanical system
nickel DBI™   271
nickel/silicon oxide DBI™ hybrid bonding
   271 ff
nonplanar metal/silicon oxide DBI™ process
   flow   263

o

optical aligment   120
optical microsystem   112 f
organic/indium-gold hybrid bonding   142
organic/metal hybrid bonding   140 f
OSAT acticitvy   323
outsourced assembly and test (OSAT)
   321
oxide bonding   216, 238 f, 263, 303
– direct   140
– post-oxide bonding anneal   177
oxide layer   290
oxide layer formation   66, 90
oxide-to-oxide bonding   216

p

PAB see plasma-activated bonding
packaged device stacking see also
   three-dimensional integration   301 ff
package-on-package (PoP) technology   305
parylene   41
passivation layer   197 f
patterned bonding   161, 231, 239
PEEK see polyetheretherketone
phase diagram of a binary metal system
   182
phase transformation of gold/tin solder
   121 ff
photoresist   40
pixelated three-dimensional integrated
   circuit   273 ff
planarization   144, 239
planarized liftoff method   144 f, 153
planarized stud metal/silicon oxide DBI®
   process flow   263
plasma   104
plasma activation   104, 106 f
plasma printing   107
plasma processing   380
plasma reactor for low-pressure activation
   106
plasma torch   107
plasma treatment of silicon surfaces   103

plasma-activated bonding (PAB)   101 ff
– application of   111 ff
– atmospheric pressure   106
– classification of   105
– compatibility to CMOS process   115
– low-pressure   105
– of a pressure sensor   112
– of patterned wafers for MEMS   290
– process flow   108
– requirements on surface quality for   101
plate capacitor
– bipolar configuration   369
– electric fields in a   368
– unipolar configuration   369
polyetheretherketone (PEEK)   41
polyimide   41, 56, 144, 153
polymer adhesion mechanism   34 f
polymer adhesive   34
– for wafer bonding   38 f
– patterned   51
– physical drying   36
– properties of   36 ff, 46
– UV-curable   50
polymer adhesive wafer bonding, see also adhesive bonding   33 ff
– advantages of   33, 58, 217
– application of   34, 217
– based on BCB for three-dimensional integration   217 f
– equipment   43 f
– influencing parameters on   46 ff
– localized   50 ff
– process steps of   45, 55, 57
– program for   56, 58
– wafer-to-wafer alignment   52 f
PoP technology, see package-on-package technology
post-bonding annealing   165, 168, 177, 216, 241 f, 246 ff, 255, 303
post-bonding process   340
post-CMP cleaning   221, 227, 231 f
post-processing of bonded wafers   285, 292
power amplifier   310
pressure in anodically bonded cavities   70
pressure sensor   70, 112, 284, 286
pre-thinned carrier wafer   338
production of SOI wafers   93

r
reactive ion etching (RIE) plasma   104 f
reconfiguration of known good dies   347 ff
reconfiguration with ultrathin chips   351
redistribution layer   219
reflow soldering   127 ff

release layer   356
release layer processing   357
remote plasma   105
repartitioning logic   322 f
repartitioning memory   314
RIE plasma, see reactive ion etching plasma
routing of electrical signal lines   282 f

s
screen printing
– for glass frit bonding   5 f
– influence of structure with on printed glass frit thickness   7
– self-aligned   6 f
sealing capacity   287
sealing pressure   14
self-alignment   128 f
semiconductor wafer direct bonding (SWDB)   81, 93 f
sequential plasma mode   105, 113
shielding of microelectronic systems   73
side-pin, contact by   291
silde-off debonding   332
silent discharge, see dielectric barrier discharge
silicate SOG layer   23 ff
silicon lid for MEMS application   134 f
silicon surface   83
– hydrophilic   85
– hydrophobic   86
silicon wafer   348
– SOG-coated   22
silicon/Pyrex glass wafer pair   63
silicon-glass anodic bonding   64, 75 f
silicon-on-insulator (SOI) substrate   81, 90, 285
siloxane SOG film   20
silver/tin fluxless SLID system   185, 187
SLID, see solid-liquid interdiffusion bonding
slip line   289
sodium contamination caused by anodic bonding   72
SOG, see spin-on glass
SOI substrate, see silicon-on-insulator substrate
solder alloy   119 f
solder reaction   121 ff, 128
sol-gel process   19
solid bridging caused by impurities   83
SOLID technology   308
solid-liquid interdiffusion bonding (SLID)   177, 181 f, 347
– advantages of   181, 211
– bonding process   189, 191 f

– comparison with soldering process 182 ff, 211
– critical thickness of bonding material 191, 199
– electrical reliability of 207
– electromigration in 207
– fluxless bonding 196 ff
– for three-dimensional TSV integration 303 f
– interface conditioning method 349
– mechanical bond strength 207
– metal systems for 183, 186
– surface pretreatment 195 f
– temperature cycling test 208 f, 211
– void formation 209
– wafer-to-wafer assembly 349
spin coating 19, 21 ff, 44 ff, 51, 55, 144, 221, 296, 332 ff, 349, 357 ff, 364
spin-on glass (SOG) material
– deposition of 19 ff
– for surface planarization 20, 28 f
– for wafer bonding 21 ff
spray coating 19, 21 ff, 30, 44, 51 f, 55, 296, 335
SSOI, see strained silicon-on-insulator wafer
stacked cache 321
storage device 313
strained silicon-on-insulator (SSOI) wafer 94
stratum 307
stress compensation 30
structured surface 11
stud pull test 14 f
substrate 81, 90, 93 f
surface activation of wafers 88 f, 265
surface cleaning 108, 167, 196, 288
surface contamination 19
surface defects of patterned wafers 19
surface energy, measurement of 110
surface etching 288
surface interaction during wafer bonding 82 f
surface layer
– fusing effect of 13
– of bonded wafers 285
– profile of copper/BCB CMP surface 223
– requirements on, for fusion bonding 288
– topography accomodation capability of partially cured BCB 227 ff
surface passivation 177, 196
surface patterning 50, 52, 220, 223, 264
surface planarization 239
– global 20
– local 20

surface termination 265
surface wetting 36
surface-activated copper bonding 162, 177, 303
SWDB, see semiconductor wafer direct bonding
system memory device 313
system-on-a-chip 95

t
tape 330
temporary bonding
– adhesive 54 f, 347 ff
– key requirements for bonding material 331 f
– technologies 330
temporary handle wafer bonding 33, 356
temporary handling substrate 347, 348, 371 f
temporary wafer bonding 33, 141, 329 ff
– boundary conditions 336
– debonding using a release layer and liftoff 335 f
– elastomer bonding/cold de-bonding approach 357 ff
– electrostatic 367 ff
– geometry of electrodes for electrostatic 374
– protection of wafer edges 337 ff, 360
– release layer/elastomer technique 357 ff, 364
– room temperature, low stress debonding 335 f
– single-wafer debonding by adhesive dissolution through perforated support plate 333 ff
– slide-off debonding approach using thermoplastic adhesives 332
– thermomechanical debonding approach 338 f
– wafer support system using UV-curable adhesive and light-to-heat conversion layer 333 f
tensile test 109
Tezzaron SuperContacts™ 319 f
thermal conditioning of glass frit paste 8 ff
thermal mismatch 13, 21, 25 f, 29, 63, 68 f, 163
thermal reliability 171
thermocompression copper bonding 162 f, 183
– bond chamber conditions 167
– bond duration 167
– bond pressure 167

– bond profile   168
– bonding temperature   167
– chip seal design   168
– fabrication of copper bond pads   166 f
– image analysis of the copper bonded interface   170
– microstructure evolution during   164 f
– nitrogen annealing   168
– orientation evolution during   165
– pattern density   169
– size of copper bond pad   169
– structure of copper bond pad   168
– surface preparation of copper bond pads   166 f
thermode soldering   130 ff
thermoplastic adhesive for temporary wafer bonding   332
thermoplastic glue   351
thermoplastic polymer   36, 38, 42, 56
thermosetting polymer   36, 38, 42, 54
thin capping by wafer bonding   15
thin film of glass for anodic bonding   75
thin wafer handling   33, 329, 353 f, 379 f
– support system   355 ff, 367 f
– use of e-carriers for bumping thin wafers   380 f
three-dimensional bonding   139 ff
three-dimensional chip stacking   101
three-dimensional heterogeneous integration   275 f
three-dimensional hyper-integration, see also three-dimensional integration   215 ff
three-dimensional IC integration based on copper bonding   174 ff
three-dimensional IC/MEMS stack   324
three-dimensional integrated system assembly   348
three-dimensional integration   132 f, 161, 172, 261
– advantages of   305
– classification of   301
– comparison with PoP technology   305
– heterogeneous   324
– key unit processes   301
– of CMOS with other devices   322 f
– processing steps with hybrid copper/BCB wafer bonding   221
– requirements on bonding quality   216
– schematic of wafer bonding technologies for   302
– strategies for   215 ff
– using temorary wafer bonding   329 ff, 347 ff
– utilizing a thermomechanical debonding approach   338 f
three-dimensional integration platform   216
– using hybrid copper/BCB redistribution layer bonding   217 f
three-dimensional interconnects   220, 237, 264, 266 f, 273 f, 277
three-dimensional LSI   139 f
– test chip fabrication by eutectic indium-gold bonding   149
– wafer-to-wafer bonding techniques for   139
three-dimensional microprocessor chip   151
three-dimensional packaging   329
three-dimensional shared memory chip   151
three-dimensional stacked image sensor   152
through-silicon via (TSV)   140, 145, 151, 201, 215, 238, 282
through-strata via (TSV)   215
through-substrate via (TSV)   33, 301 f, 330
– fabrication of, in electrostatic temporary wafer bonding   374 f
tire pressure monitoring system (TPMS)   324
tool plate for glass frit bonding   13
topography accomodation   227 ff
total thickness variation (TTV)   330, 358, 361
– effect of uniformity on   337
TPMS, see tire pressure monitoring system
transient liquid-phase bonding   133, 135 f
TSV formation without stacking   310 f
TSV memory stack   312
TSV processing   301, 364
TSV stacked BGA   321

u

ultrafine leak test   70
ultrahigh vacuum (UHV)   89, 92
ultrahtin buried oxide (UT-BOX) layer   94
ultrathin chip handling   352 f
ultrathin dye packaging   363
ultrathin wafer   3555
under-bump metallization (UBM)   130 ff, 182
unipolar chucking   367
UV-curable adhesive   333

v

vacuum chuck   356, 360
van der Waals force   83
van der Waals interaction   34

via etching 339 f
via insulation 339 f
via metallization 339 f
via-first process 301, 319, 330 f, 337, 342
via-last process 301, 310, 330 f
– flow of a 338 ff
– using high-temperature TEOS process 341 f
– with aspect ration of 2:1 341
viscoelastic effect 37
viscous flow of surface layer 103
void 243
– bonding void 225 ff
– crosslink-percentage-controlled 229
– formation of 209, 225 ff, 289
– in bonded wafer pairs 285
– in temporary wafer bonding processes 337
– Kirkendall 124, 128, 132, 209 f
– nanovoid 225 ff
– progression of voiding types 230

w

wafer alignment, *see also* wafer-to-wafer alignment 141, 146, 265
wafer bond characterization 108
wafer bonding process integration 287 ff
wafer bonding programm 56
wafer bonding, *see also* wafer-to-wafer bonding
– bonding of SiC chips to substrates 205
– copper/copper direct bonding 140
– copper-tin/copper-tin direct bonding 140
– direct metal bonding for three-dimensional integration 140
– mechanical stress caused by microbumps 156
– of CMOS wafers to compound semiconductor wafers 31
– of thermally mismatched wafers 21, 25 f

– organic/metal hybrid bonding 140
– oxide/metal hybrid bonding for three-dimensional integration 140
– requirements on materials for 38 f
– surface planarization in 20
– with planarization SOG 28 f
– with polymers 39
– with silicate SOG layers, *see also* adhesive wafer bonding 21 ff
wafer support system 333 f
wafer thinning 215, 242, 301, 329, 359, 380
– protection of wafer edges during 337 ff
wafer-level assembly 133
wafer-level bonding 181 ff, 282
wafer-level integration 161
– using eutectic gold/tin soldering 132 f
wafer-level packaging (WLP) 310
wafer-level solid-liquid interdiffusion bonding 181 ff
wafer-to-wafer alignment
– in copper/silicon dioxide hybrid bonding 244 f
– in glass frit bonding 11, 13
– in polymer adhesive wafer bonding 52 f
– in temporary bonding processes 357 f
wafer-to-wafer assembly using SLID bonding 350 ff
wafer-to-wafer bonding 135, 140, 148
wafer-to-wafer shifting
– prevention of 13, 52
wax 41, 330
weak boundary layer theory 36
wetting temperature 11, 13
wide I/O memory 315, 318
wire bond memory stack 312
WLP, *see* wafer-level packaging

y

yield strength 269 f